A Comprehensive Approach to Cell Biology

A Comprehensive Approach to Cell Biology

Editor: Larry Madison

STATES
ACADEMIC PRESS
www.statesacademicpress.com

States Academic Press,
109 South 5th Street,
Brooklyn, NY 11249, USA

Visit us on the World Wide Web at:
www.statesacademicpress.com

ISBN: 978-1-63989-002-6 (Hardback)

Cataloging-in-Publication Data

A comprehensive approach to cell biology / edited by Larry Madison.
p. cm.
Includes bibliographical references and index.
ISBN 978-1-63989-002-6
1. Cytology. 2. Cells. 3. Biology. I. Madison, Larry.
QH581.2 .C66 2022
571.6--dc23

Table of Contents

Preface

Cell biology is a field of biology that is concerned with the study of cellular structure and function using techniques, such as cell culture, microscopy, cytometry, and cell fractionation. The field encompasses the study of both eukaryotic and prokaryotic cells, including the mechanisms and processes of cell cycle, biochemistry, cellular communication, metabolism, and composition. Research in cell biology takes an interdisciplinary approach by integrating other sciences such as genetics, molecular genetics, immunology, medical microbiology, cytochemistry, and molecular biology. This book contains some path-breaking studies in the field of cell biology. From theories to research to practical applications, case studies related to all contemporary topics of relevance to cell biology have been included herein. This book includes contributions of experts and scientists which will provide innovative insights into this field.

The researches compiled throughout the book are authentic and of high quality, combining several disciplines and from very diverse regions from around the world. Drawing on the contributions of many researchers from diverse countries, the book's objective is to provide the readers with the latest achievements in the area of research. This book will surely be a source of knowledge to all interested and researching the field.

In the end, I would like to express my deep sense of gratitude to all the authors for meeting the set deadlines in completing and submitting their research chapters. I would also like to thank the publisher for the support offered to us throughout the course of the book. Finally, I extend my sincere thanks to my family for being a constant source of inspiration and encouragement.

Editor

1

IL-2 and IL-15 dependent thymic development of Foxp3-expressing regulatory T lymphocytes

Cécile Apert, Paola Romagnoli✉, Joost P. M. van Meerwijk✉

CPTP, Université de Toulouse, CNRS, Inserm, UPS, Toulouse, France
✉ Correspondence: Paola.Romagnoli@inserm.fr (P. Romagnoli), Joost.van-Meerwijk@inserm.fr (J. P. M. van Meerwijk)

ABSTRACT

Immunosuppressive regulatory T lymphocytes (Treg) expressing the transcription factor Foxp3 play a vital role in the maintenance of tolerance of the immune-system to self and innocuous non-self. Most Treg that are critical for the maintenance of tolerance to self, develop as an independent T-cell lineage from common T cell precursors in the thymus. In this organ, their differentiation requires signals from the T cell receptor for antigen, from co-stimulatory molecules, as well as from cytokine-receptors. Here we focus on the cytokines implicated in thymic development of Treg, with a particular emphasis on the roles of interleukin-2 (IL-2) and IL-15. The more recently appreciated involvement of TGF-β in thymic Treg development is also briefly discussed. Finally, we discuss how cytokine-dependence of Treg development allows for temporal, quantitative, and potentially qualitative modulation of this process.

KEYWORDS regulatory T cells, thymus, differentiation, IL-2, IL-15

INTRODUCTION

Several populations of T lymphocytes with immunosuppressive activity have been identified to date (Josefowicz et al., 2012a; Kim and Cantor, 2011; Sakaguchi et al., 2010; Shevach, 2011; Vuddamalay and van Meerwijk, 2017; Xu et al., 2016). The most critical population of Treg expresses Foxp3: Humans and mice carrying mutations in the gene encoding this forkhead/winged helix transcription factor develop a rapidly lethal autoimmune and inflammatory pathology (Bennett et al., 2001; Brunkow et al., 2001; Fontenot et al., 2003). Treg therefore prevent the development of autoimmune pathology and chronic inflammation in interfaces with the environment such as the intestines, lungs, and skin.

As other T lymphocytes, activation of Treg requires that they recognize antigenic peptides presented by cell-surface expressed molecules encoded in the major histocompatibility complex (MHC). Autoimmune pathology develops upon activation of autospecific conventional T cells (Tconv) and suppression of these immune-responses involves autospecific Treg (Kieback et al., 2016; Yang et al., 2015). The antigen-specificity of Treg involved in the prevention of chronic inflammatory diseases remains less well defined, but antigens from commensals as well as from infectious agents are most certainly recognized. It therefore appears appealing to postulate that Treg preventing autoimmune pathology develop mostly in the thymus where autospecific T cell precursors can recognize self-antigens presented by thymic stromal cells (Klein and Jovanovic, 2011). It needs to be emphasized, however, that autospecific Treg can also differentiate from Tconv precursors in peripheral lymphoid organs (Sun et al., 2015). The Treg preventing chronic inflammation likely mostly differentiate in peripheral (lymphoid) organs where Tconv involved in immune-responses, and therefore specific for the agents that provoked the inflammation, differentiate into Treg under tolerogenic conditions (Bilate and Lafaille, 2012; Josefowicz et al., 2012b). Nevertheless, colonic Treg express T cell receptors for antigen (TCR) that were also found on thymic Treg, suggesting that Treg involved in the prevention of chronic inflammation may develop in the thymus (Cebula et al., 2013). Recent data showing that peripheral Treg recirculate back to the thymus and that they represent a very substantial proportion of thymic Treg in adult mice (Cowan et al., 2016; Thiault et al., 2015), however, urge for a reinvestigation of this issue.

In addition to signals via the TCR, cytokine signaling plays a central role in Treg differentiation in the thymus and the periphery. In this review, we focus on Treg development in the thymus and will discuss TCR- and cytokine receptor-derived signals. We will discuss how and where these distinct signals intersect to drive development of a Treg repertoire which efficiently protects the organism from autoimmune pathology.

TCR-SIGNALING AND THYMIC TREG-DIFFERENTIATION

Within the thymus two anatomically and functionally distinct compartments can be distinguished; the cortex and the medulla. Hematopoietic precursors enter the thymus at the cortico-medullary junction. During their differentiation, thymocytes will first migrate into the cortex, then back to the medulla before leaving the thymus at the cortico-medullary junction (Love and Bhandoola, 2011). Thymocytes acquire the expression of the heterodimeric TCRαβ. The genes encoding the TCR undergo random somatic rearrangements that allow for the generation of a wide repertoire of TCR capable of recognizing a vast variety of rapidly evolving infectious agents. Distinct TCR-based selection processes ensure that only functional and harmless T cells leave the thymus (Kurd and Robey, 2016; Lucas et al., 2016; Ohigashi et al., 2016).

In the thymic cortex, only thymocytes bearing a TCR with a certain affinity for peptide/MHC complexes presented by cortical thymic epithelial cells (cTEC) survive. This process, called positive selection, ensures the selection of developing T cells expressing a "functional" TCR and contributes to the generation of a repertoire of T cells expressing TCR that can recognize antigens in the context of self MHC-molecules. Positively selected thymocytes rapidly upregulate the chemokine receptor CCR4 which attracts these cells to CCL17 and CCL22-producing, Sirpα$^+$ dendritic cells (DC) and to the medulla (Hu et al., 2015). Too high affinity interactions of the TCR with peptide/MHC complexes, expressed by presumably DC, lead to a first wave of negative selection of autospecific and therefore dangerous T cell-precursors (McCaughtry et al., 2008; Stritesky et al., 2013). Thus deleted T cell-precursors are likely specific for ubiquitously expressed and potentially for epithelial antigens. T lymphocytes that developed in a thymus in which MHC molecules were exclusively expressed by cTEC are strongly autoreactive, suggesting that cTEC contribute little, if at all, to negative selection (Capone et al., 2001; Laufer et al., 1996). Interestingly, high affinity interactions in the cortex may also lead to the differentiation of Treg (Liston et al., 2008; Ribot et al., 2007), but Treg-differentiation in the cortex remains a little-studied process. Thymocytes surviving cortical selection upregulate the chemokine-receptor CCR7 and migrate to the medulla in response to CCL19 and CCL21 produced by medullary thymic epithelial cells (mTEC) (Ueno et al., 2004).

In the thymic medulla, thymocytes interact via their TCR with peptide/MHC complexes presented by mTEC, DC, and B lymphocytes. High avidity interactions of thymocytes with medullary stromal cells can lead to their negative selection through induction of apoptosis ("deletion") or clonal anergy (Kyewski and Klein, 2006). High avidity interactions of developing T cells with all cited types of thymic stromal cells also lead to the differentiation of Treg (Klein et al., 2009) and the majority of developing Treg is localized in the medulla rather than in the cortex (Thiault et al., 2015). Combined, the processes of negative selection and Treg-differentiation in the thymus play a major role in the establishment of immune-tolerance to self-antigens.

It has therefore been clearly established that high avidity interactions of the developing T cells with thymic stromal cells can lead to two quite distinct outcomes: negative selection and differentiation of Treg. It remains less clear what tips the balance between these two fates. The mere affinity of the TCR with peptide/MHC complexes is not responsible. In mice expressing a single TCR specific for a non-self-antigen, no Treg develop. When the recognized antigen is introduced as a transgene, deletion as well as Treg-differentiation take place (Klein and Jovanovic, 2011). Other factors must therefore be involved. The observation that the distinct medullary stromal cell-types induce Treg-differentiation at distinct doses of antigenic peptide *in vitro*, suggests that the stromal cell with which thymocytes interact *in vivo* may in part determine the choice between deletion and Treg-differentiation (Wirnsberger et al., 2009). This may be due to the distinct surface expression levels of ligands for e.g. CD28, CD27 or other members of the TNF-receptor superfamily, or other molecules involved in deletion and Treg-differentiation (Coquet et al., 2013; Mahmud et al., 2014; Tai et al., 2005; Tang et al., 2003). However, even in experimental systems in which agonist peptide/MHC ligand was presumably exclusively presented by a single stromal cell-type, i.e., mTEC, deletion as well as Treg-differentiation were observed (Aschenbrenner et al., 2007). At least two explanations can be proposed. First, Treg-lineage commitment may take place independently of the thymocyte's TCR (Pennington et al., 2006) and distinct selection criteria for Tconv and Treg precursors determine development of these two populations. Second, heterogeneity among mTEC (and potentially the other stromal cell-types) (Brennecke et al., 2015; Meredith et al., 2015) may be involved. These issues would merit further investigation.

EPIGENETIC MODIFICATIONS AND THYMIC DEVELOPMENT OF TREG

Epigenetic gene regulation, such as DNA methylation and histone modifications, is implicated in lineage specification and maintenance. Several groups have demonstrated that DNA demethylation at conserved non-coding sequence within the *Foxp3* locus ensures the stability of its expression in thymic derived Treg (Floess et al., 2007; Kim and Leonard, 2007; Zheng et al., 2010). It was shown that DNA methylation is lost during the last (i.e., Foxp3-expressing) stages of thymic Treg-development through oxidation of 5-methylcytosine and other intermediates in the demethylation process. It was suggested that two enzymes, TET2 and TET3, initiate this reaction (Toker et al., 2013). Indeed, in *Tet2/Tet3* double deficient mice, in which regulatory regions remain methylated, Foxp3 expression is unstable and Treg lose their suppressive functions (Yue et al., 2016). Interestingly, Treg-specific demethylated regions (TSDRs) are also found in other genes encoding for factors essential for Treg function, such as CD25, CTLA-4, Eos, and GITR (Ohkura et al., 2012). While TCR signaling is required for demethylation of TSDRs, *Foxp3* gene expression is dispensable. These data indicate that to establish Treg lineage two independent but complementary molecular mechanisms are in play: *Foxp3* gene expression and epigenetic changes (Ohkura et al., 2012). Establishment of a Treg epigenetic landscape may therefore precede and promote *Foxp3* gene expression. CpG demethylation (or initiation of this process) in the TSDR or CNS2 of the *Foxp3* gene strictly correlated with expression of this gene, yielding little insight into this question (Toker et al., 2013; Yue et al., 2016). However, the recently described binding of a global chromatin organizer, Satb1, to another regulatory region of the *Foxp3* locus (CNS0) in immature CD4/CD8 double positive thymocytes and its requirement for Treg development suggest that early epigenetic modifications control the expression of Foxp3 and Treg signature genes (Kitagawa et al., 2017). How the expression and activity of Satb1 are regulated remains to be determined.

INVOLVEMENT OF IL-2 AND IL-15 IN TREG DIFFERENTIATION IN THE THYMUS

Early studies with mice genetically deficient in production of the "T cell growth factor" IL-2 or expression of its receptor surprisingly showed that these animals developed severe autoimmune pathology instead of immunodeficiency (Sadlack et al., 1995; Suzuki et al., 1995; Willerford et al., 1995). Initially, defects in IL-2 dependent activation induced cell-death (AICD) of autoreactive T cells were suspected. However, complementation of mice deficient in IL-2 or its receptor with WT Treg prevented pathology (Suzuki et al., 1999; Wolf et al., 2001). The latter data indicated that a lack of Treg or Treg-functional capacity was responsible for the lymphoproliferation and lethal autoimmune pathology in mutant mice. It was later appreciated that IL-2 plays a major role in Treg homeostasis. The role of IL-2 in the differentiation of Treg from Tconv precursors in peripheral lymphoid organs (and potentially in tissues) and in survival and function of mature Treg has recently been discussed (Chinen et al., 2016) and is beyond the scope of this review.

One of the earliest indications that IL-2 may play a role in the thymic development of Treg came from studies by Malek and colleagues showing that mice in which the IL-2Rβ was exclusively expressed by developing thymocytes, survived substantially longer than IL-2Rβ-deficient animals (Malek et al., 2000). Later studies showed that substantially reduced proportions of mature $CD4^{+}CD25^{+}$ regulatory thymocytes developed in IL-2Rβ-deficient mice and that differentiation of Treg with *in vitro* immunosuppressive activity was fully restored in mice in which IL-2Rβ expression was restricted to thymocytes (Burchill et al., 2007; Fontenot et al., 2005; Malek et al., 2002; Soper et al., 2007; Vang et al., 2008).

Several other studies further addressed the role for IL-2 in Treg development in the thymus using distinct mutant mice. Rudensky and colleagues found approximately half the number of Treg in the thymus of IL-2-deficient as compared to that of WT mice (Fontenot et al., 2005). However, other studies found practically normal numbers of Treg in IL-2 deficient mice (Burchill et al., 2007; D'Cruz and Klein, 2005). Similarly discordant results were found in studies with mice deficient in the IL-2Rα chain (D'Cruz and Klein, 2005; Fontenot et al., 2005; Soper et al., 2007). It therefore appears that IL-2 may play a role in Treg development, but that it is not strictly required.

How then to explain the substantially reduced numbers of Treg in the thymus of IL-2Rβ deficient mice, a very consistent observation? The IL-2Rβ chain is shared between the receptors for IL-2 and for IL-15. The latter cytokine may therefore play an important role in Treg development. Farrar and colleagues found approximately half the number of Treg in the thymus of IL-15 deficient mice as compared to WT animals (Burchill et al., 2007). In IL-2 and IL-15 doubly deficient thymuses, they found as few Treg as in the thymus of IL2Rβ-deficient mice. However, Ziegler and colleagues found normal proportions of Treg in the thymus of mice lacking expression of the IL-15Rα chain, which is only involved in signalling via the receptor for IL-15 (Soper et al., 2007). It therefore appears likely that IL-15 is involved in Treg development in the thymus but, again, that it is not absolutely required.

Therefore, whereas in mice lacking expression of both IL-2 and IL-15 or both of their receptors substantially reduced numbers of Treg were consistently found in the thymus, in mice deficient for one or the other cytokine or receptor discordant results were obtained. These data therefore clearly demonstrate that the cytokines IL-2 and IL-15 play important roles in the thymic development of Treg. What then explains the discordant results when only one of the two cytokines or receptors is missing? In transgenic mice in which the green-fluorescent protein is expressed under control of the *Rag2* promoter ("Rag2-GFP mice"), newly developing T cells are GFP^{+} and cells that are "older" do no longer express GFP (Yu et al., 1999). Using Rag2-GFP mice, it was observed that only a small proportion of thymic Treg are actually newly developing cells (Cuss and Green, 2012; McCaughtry et al., 2007; Thiault et al., 2015). It was therefore important to assess the roles for IL-2 and IL-15 in *de novo* development of Treg. Seddon and colleagues used a particularly original approach to study *de novo* T cell development in which the developmental blockade of T cells in ZAP-70 deficient mice can be lifted by inducing ZAP70 expression using a TetOn transgenic system (Marshall et al., 2014). It was thus unambiguously shown that in absence of signaling through the receptors for IL-2 or of IL-15 Treg development was severely hampered. These results strongly indicate important and non-redundant roles for IL-2 and IL-15 in Treg development in the murine thymus. Importantly, using *in vitro* culture systems, Sousa and colleagues demonstrated quantitatively important and non-redundant roles for IL-2 and IL-15 in the development of Foxp3-expressing human Treg from immature thymic precursors (Caramalho et al., 2015).

THE MECHANISMS BY WHICH IL-2 AND IL-15 CONTROL TREG-DEVELOPMENT IN THE THYMUS

To investigate how IL-2 and IL-15 impinge on Treg development in the thymus, it was important to determine at what stage these cytokines act. Treg and Tconv develop from common precursors. The $CD4^{+}CD8^{-}TCR^{high}$ ("CD4SP") stage of development is apparently the last one shared between Treg and Tconv. The CD4SP stage can be further dissected into four subpopulations using expression of CD25 and of Foxp3 as criterion. In cultures of electronically sorted CD4SP $CD25^{-}Foxp3^{-}$ thymocytes, no $Foxp3^{+}$ cells develop and the addition of IL-2 or IL-15 does not change this result (Lio and Hsieh, 2008). By contrast, some $CD25^{+}Foxp3^{-}$ cells expressed Foxp3 upon *in vitro* culture and this Treg differentiation was greatly increased by addition of IL-2 or IL-15 (Lio and Hsieh, 2008; Tai et al., 2013; Vang et al., 2008). Similar results were obtained using human CD4SP $CD25^{+}$ $CD127^{+}$ (in large majority $Foxp3^{-}$) thymocytes (Caramalho et al., 2015). These results suggested that CD4SP $CD25^{+}Foxp3^{-}$ thymocytes, which develop independently of IL-2 and IL-15 (Marshall et al., 2014), are the Treg-precursors that require IL-2 or IL-15 for their final steps of differentiation. However, in absence of signaling through the receptor for IL-15 (but not IL-2) the development of another CD4SP subpopulation, $CD25^{-}Foxp3^{+}$ cells, was found to be reduced (Marshall et al., 2014). Upon *in vitro* culture of WT CD4SP $CD25^{-}Foxp3^{+}$ thymocytes in presence of IL-2, these cells become $CD25^{+}Foxp3^{+}$ Treg (Marshall et al., 2014; Tai et al., 2013).

These and other data indicate that CD4SP $CD25^{+}Foxp3^{+}$ Treg can develop via two distinct pathways. In the first pathway, TCR-signaling allows CD4SP $CD25^{-}Foxp3^{-}$ thymocytes to become $CD25^{+}Foxp3^{-}$ cells which, when stimulated through their receptors for IL-2 or IL-15, will become $CD25^{+}Foxp3^{+}$ Treg. In the second pathway, signaling through the TCR and the IL-15 receptor allows CD4SP $CD25^{-}Foxp3^{-}$ thymocytes to become $CD25^{-}Foxp3^{+}$ cells which, when stimulated through their receptor for IL-2, will become $CD25^{+}Foxp3^{+}$ Treg. Interestingly, experiments in which a Nur77-GFP reporter indicated the affinity of TCR-peptide/MHC interactions suggested that the development of CD4SP $CD25^{+}Foxp3^{-}$ precursors involves higher affinity interactions with stromal cells than differentiation of $CD25^{-}Foxp3^{+}$ precursors (Marshall et al., 2014).

In both proposed developmental pathways, the cytokines IL-2 and IL-15 therefore appear to induce expression of Foxp3. Signaling downstream of the receptors for IL-2 and IL-15 involves the JAK3-STAT5 pathway. Consistent with the roles for these cytokines in Treg differentiation in the thymus, in mice with a T cell-specific deficiency in STAT5 expression a severe defect in Treg development was observed (Burchill et al., 2007). Moreover, transgenic expression of constitutively active STAT5 rescued Treg development in IL-2Rβ-deficient mice (Burchill et al., 2003; Burchill et al., 2007). The *Foxp3* gene contains several conserved potential STAT5 binding sites, some of which demonstrated binding of STAT5 in Treg but not Tconv (Burchill et al., 2007; Yao et al., 2007). It therefore appears that signaling through the receptors for IL-2 and IL-15 induces, via STAT5 activation, expression of Foxp3.

An alternative but not mutually exclusive explanation for the action of IL-2 (and potentially IL-15) was reported by Singer and colleagues. They showed that Foxp3 expression promotes cell-death by inducing expression of pro-apoptotic (e.g., PUMA and BIM) and repressing pro-survival proteins (e.g., Bcl-2) (Tai et al., 2013). *In vitro* experiments showed that IL-2 protected $Foxp3^{+}$ but not $Foxp3^{-}$ CD4SP cells from apoptosis by increasing the expression of Bcl-2, which was mediated by STAT5. Though intriguing, these results will need to be reconciled with the observation that transgenic expression of Foxp3 restored Treg development and prevented autoimmune pathology in IL-2Rβ deficient mice (Soper et al., 2007).

IL-2 and IL-15 appear not implicated in the induction of the demethylation of the regulatory region of the *Foxp3* locus of Treg precursors. Compared to the strong TSDR demethylation observed in $CD25^{+}Foxp3^{-}$ cells stimulated via their TCR in the presence of IL-2, precursors cultured *in vitro* with IL-2 or IL-15 alone display only a minor TSDR demethylation (8%–14%) (Toker et al., 2013).

THE CELLULAR ORIGINS OF THE IL-2 AND IL-15 INVOLVED IN TREG DEVELOPMENT

Identification of the cellular sources of the IL-2 and IL-15 involved in Treg differentiation may reveal important regulatory mechanisms of this process (*cf. infra*). In hematopoietic chimeras in which only radioresistant stromal cells carried an invalidating mutation in the *Il2* gene, normal numbers of Treg developed, strongly suggesting that the IL-2 involved in Treg development is produced by (radiosensitive) hematopoietic cells (Marshall et al., 2014). Probably the most obvious candidates for the cells producing IL-2 in the thymus are (developing) T cells. IL-2 would then potentially act in an autocrine manner to stimulate Treg differentiation. In mixed hematopoietic chimeras, *Il2*-mutant thymocytes can efficiently differentiate into Treg when other hematopoietic cells can produce IL-2 (Tai et al., 2005), suggesting that an autocrine loop does not play a major role. Robey and colleagues showed that IL-2 derived from DC added to *in vitro* thymic slice cultures could drive Treg development (Weist et al., 2015). In the human thymus, IL-2 protein was found in the cortex and the medulla, and the *Il2* gene appeared expressed mostly by mature T cells (Caramalho et al., 2015). No evidence was found for IL-2-production by human thymic dendritic cells. It will now be important to identify the precise thymic cell-type(s) that produce(s) the IL-2 involved in Treg differentiation using conditional *Il2* KO mice (Popmihajlov et al., 2012) bred to mice expressing the Cre-recombinase under control of appropriate promoters, e.g., those controlling the genes encoding CD11c, CD4, or $p56^{lck}$.

The identity of the thymic population producing the IL-15 involved in Treg development is probably clearer, at least in the mouse. In hematopoietic chimeras in which radioresistant cells could not present, via their IL-15 receptor α-chain, IL-15 to developing thymocytes, strongly reduced proportions of Treg developed (Marshall et al., 2014). The thymic development of iNKT cells also requires IL-15, which is presented by radioresistant stromal (but not dendritic) cells (Castillo et al., 2010). Using reporter mice, Ikuta and coworkers showed that, in the thymus, IL-15 is produced mainly by mTEC and, at lower levels, by some pericytes and blood endothelial cells (Cui et al., 2014). Flow-cytometry analysis identified mTEC expressing high levels of MHC class II ("$mTEC^{high}$") as the main IL-15 producing stromal cell-type in the thymus. By contrast, dendritic cells and macrophages did not produce IL-15. Lefrançois and coworkers demonstrated, using reporter mice, that very immature (and therefore cortical) thymocytes can produce high levels of IL-15 and that these cells rapidly lose this capacity during T cell development (Colpitts et al., 2013). However, given that Treg mainly develop in the thymic medulla suggests that this source of IL-15 is irrelevant to Treg differentiation. Combined, the published data strongly indicate that the IL-15 involved in Treg development in the murine thymus is produced and presented by $mTEC^{high}$. Intriguingly, these stromal cells ectopically express tissue-antigens known to be involved in selection of the Treg's TCR-repertoire (Aschenbrenner et al., 2007; Malchow et al., 2013; Takaba et al., 2015; Yang et al., 2015). Also in the human thymus, IL-15 was mainly found in the medulla. As in mice, mTEC expressed the *Il15* gene. However, also B cells, monocytes/macrophages, and NK cells produced this cytokine (Caramalho et al., 2015). Given that B lymphocytes can induce T cell-tolerance in the mouse thymus, the latter observation may prove of importance (Yamano et al., 2015).

The distinct signals involved in T cell differentiation in peripheral tissue are not necessarily derived from the same cell. Thus, differentiation of Th2 cells, specialized in the defense against large extracellular parasites such as helminthes, requires TCR triggering through interactions with peptide/MHC complexes, presented by dendritic cells, and the Th2-polarizing cytokine IL-4, which can be derived from type 2 innate lymphoid cells (Pelly et al., 2016). Signals involved in T cell development and activation do not need to be delivered all at once neither and can be accumulated through multiple sequential interactions with different antigen-presenting cells in a multistep process (Khailaie et al., 2014; Kisielow and Miazek, 1995; Valitutti et al., 1995; Wilkinson et al., 1995). Similar mechanisms may act within the thymus to drive Treg differentiation. In the "two-step model" for Treg differentiation in the thymus proposed by Hsieh and colleagues, the first step involves TCR-triggering which leads to upregulation of the high-affinity receptor for IL-2. The second step involves engagement of the latter receptor by IL-2 which, in the experimental setup used by these authors, was not derived from the same stromal cell (Lio and Hsieh, 2008). Using biphoton microscope analysis of explanted thymic tissue, Robey and colleagues demonstrated that thymocytes migrate very rapidly during their development, which is consistent with the notion that involved stimuli may be derived from distinct stromal cells (Le Borgne et al., 2009). On the other hand, in order to drive Treg differentiation in explanted WT thymic tissue, the dendritic cells that presented the peptide/MHC complexes involved in TCR triggering, that were dropped onto the cultures, also needed to produce IL-2 (Weist et al., 2015). It remains therefore unclear if TCR and IL-2-receptor engagements take place simultaneously or sequentially, and if the involved ligands are expressed by the same or by distinct stromal cells.

HOW THE REQUIREMENT FOR IL-2 AND IL-15 MAY ALLOW MODULATION OF THE GENERATION OF THE TREG REPERTOIRE

The involvement of IL-2 and IL-15 may allow for control of Treg development in the thymus in quantitative and qualitative terms. Transgenic mice expressing constitutively active STAT5 had three- to seven-fold increased numbers of Treg, suggesting that STAT5 activation (through the receptors for IL-2 or IL-15) is a limiting factor in Treg differentiation in the thymus (Burchill et al., 2008; Burchill et al., 2007). In mutant mice lacking the pro-apoptotic proteins Puma and Bim or expressing anti-apoptotic Bcl-2, substantially increased numbers of Treg were found in the thymus, showing that the large majority of developing Treg normally die (Tai et al., 2013). In mixed hematopoietic chimeras in which growing numbers of Treg precursors expressed the common cytokine receptor γ-chain ("γ_c", a constituent of, among others, the receptors for IL-2 and IL-15), the numbers of thymic Treg saturated at approximately 10^6 cells, suggesting that availability of IL-2 and IL-15 quantitatively controls Treg development (Tai et al., 2013). Using cultures of explanted thymic tissue, Robey and colleagues showed that thymic Treg limited the differentiation of new Treg by competing for IL-2 (Weist et al., 2015). In *in vitro* cultures of human thymocytes, addition of IL-2 or IL-15 substantially increased the development of Treg (Caramalho et al., 2015). Interestingly, also in TCR-ligand doubly transgenic mice, the absolute numbers of Treg found in the thymus were well below the numbers expected if all autospecific T cell-precursors had differentiated into Treg (reviewed in ref. Romagnoli et al., 2008), but the potential involvement of limiting availability of cytokines has, to our knowledge, not been demonstrated.

It therefore appears that the availability of IL-2 (and potentially IL-15) may be a limiting factor in Treg development. IL-2 thus appears to quantitatively regulate the numbers of Treg developing in the thymus. Importantly, the availability of IL-2 appears to diminish with age. T cells activated in the periphery during immune responses can migrate back to the thymus (Hale and Fink, 2009) and Treg are not an exception to this phenomenon. Probably because of their repertoire enriched in autospecific cells and the resulting massive activation of peripheral Treg (Fisson et al., 2003; Hsieh et al., 2004; Romagnoli et al., 2002), a very high proportion of thymic Treg are cells that had re-entered the thymus from the periphery (Cowan et al., 2016; Thiault et al., 2015). We showed that recirculating Treg inhibit the differentiation of their precursors and that they do so by limiting the availability of IL-2. Importantly, the proportion of recirculating Treg among thymocytes grows substantially with age, which results in a strongly fading *de novo* development of Treg (Thiault et al., 2015). Treg recirculation may also be modulated by e.g., infection, stress of other factors, which thus would potentially lead to increased or decreased Treg differentiation in the thymus.

Intriguingly, recirculating Treg appeared to control IL-2-dependent but not IL-2-independent Treg differentiation in the thymus (Thiault et al., 2015). Given that the proportion of recirculating Treg among thymocytes substantially increases with age, IL-2- (but not IL-15-) dependent Treg differentiation will strongly decline. It is tempting to speculate that, with age, Treg differentiation will decreasingly depend on IL-2-producing dendritic cells and progressively on interactions with IL-15-producing mTEC. The potential consequences of this shift on e.g., selection of the TCR-repertoire expressed by developing Treg would merit thorough examination.

INVOLVEMENT OF OTHER γ_C AND RELATED CYTOKINES IN TREG DEVELOPMENT

In mice carrying mutations in the gene encoding the common cytokine receptor γ-chain γ_c, very low proportions of Treg develop in the thymus (Bayer et al., 2008; Burchill et al., 2008; Burchill et al., 2007; Tai et al., 2013). Since γ_c is a constituent of the receptors to the cytokines IL-2, IL-4, IL-7, IL-9, IL-15, and IL-21, "γ_c cytokines" other than IL-2 and IL-15 may be involved in Treg development in the thymus. IL-4Rα-deficient mice appear to have normal proportions of thymic Foxp3$^+$ Treg (Vang et al., 2008). Similar to IL-2 and IL-15, IL-7 induced the conversion of murine CD4SP CD25$^+$Foxp3$^-$ precursors into Foxp3$^+$ Treg in *in vitro* cultures (Vang et al., 2008). IL-7 also drives Treg differentiation of human non-regulatory thymocytes stimulated in presence of TGF-β, but the relevance of this observation for physiological Treg development in the thymus requires clarification (Caramalho et al., 2015). IL-7, presumably produced by thymic epithelial cells (Shitara et al., 2013), plays a major role in early stages of T-lymphocyte development. Therefore, mice deficient in IL-7 or the IL-7Rα chain have severely reduced thymocyte-counts and, consequently, absolute numbers of thymic Treg (Bayer et al., 2008; von Freeden-Jeffry et al., 1995). However, the proportion of Foxp3$^+$ cells among CD4SP thymocytes appeared unaltered (Bayer et al., 2008; Vang et al., 2008). To circumvent the effects of IL-7 on early T cell-development, Ikuta and colleagues generated mice in which the gene encoding the IL-7Rα cahin is deleted at the CD4$^+$CD8$^+$ stage of thymocyte development. The proportion of Foxp3$^+$ Treg among CD4SP thymocytes appeared unaltered in these mice (Tani-ichi et al., 2013). It therefore appears unlikely that IL-7 plays a role in the final stages of Treg differentiation in the thymus. This conclusion is coherent with the observation that mice deficient in IL-7 or its receptor do not develop immunopathology (von Freeden-Jeffry et al., 1995).

To the best of our knowledge, the involvement of the γ_c cytokines IL-9 and IL-21 in Treg development in the mouse thymus has not been reported. In humans, IL-9 appears to play a role in T cell-development, but involvement in Treg differentiation was not investigated (De Smedt et al., 2000). *In vitro* stimulation of non-regulatory thymocytes in presence of TGF-β leads to Foxp3$^+$ Treg differentiation in presence of IL-2 or IL-15, but not of IL-4 nor of IL-21 (Caramalho et al., 2015). Whereas very convincing evidence has been reported for a key-role for IL-2 and IL-15 in intrathymic Treg differentiation, evidence that other γ_c cytokines are involved is therefore currently lacking.

Thymic stromal lymphopoietin (TSLP) is an IL-7-related cytokine, the receptor of which shares the IL-7Rα chain with the receptor for IL-7. It was reported that human Hassall's corpuscles produce TSLP which activates thymic dendritic cells to upregulate expression of CD80 and CD86. *In vitro*, these dendritic cells then appeared to induce the differentiation of Treg from thymocyte precursors (Hanabuchi et al., 2010; Watanabe et al., 2005). Whereas IL-7 induces the differentiation of murine Treg from CD4SP CD25$^+$Foxp3$^-$ precursors, TSLP does not (Vang et al., 2008). Deficiency of the IL-7Rα chain, and therefore in signaling through the TSLP-R, leads to strongly reduced thymocyte numbers but not to a specific Treg defect (Bayer et al., 2008). Durum and colleagues argued that Treg development requires IL-7 or TSLP mediated stimulation through IL-7Rα (Mazzucchelli et al., 2008). However, it remains unclear why these authors found normal absolute numbers of Treg in the thymus from mice in which radioresistant thymic stromal cells did not produce IL-7 (cf. ref. Shitara et al., 2013). Whereas in humans TSLP may have an indirect role in the thymic generation of Treg, it appears therefore unlikely that TSLP plays a role in this process in mice.

INVOLVEMENT OF TGF-β IN TREG DEVELOPMENT IN THE THYMUS

Transforming growth factor-β (TGF-β) plays a central role in the differentiation of conventional T cells into Treg in the periphery ("pTreg") (Bilate and Lafaille, 2012). Treg that differentiated as an independent lineage in the thymus ("tTreg") appear different from pTreg, e.g., in the methylation status of their *Foxp3* locus (Floess et al., 2007). Moreover, the Foxp3 locus contains at least one TGF-β response element (i.e., CNS1) (Zheng et al., 2010). Even if CNS1 appears dispensable for Treg development in the thymus, TGF-β may play a role in the thymic differentiation of Treg. This hypothesis was investigated in mice in which TGF-β mediated signaling was perturbed. In TGF-β deficient mice, Treg differentiation in the thymus appeared normal (Fahlen et al., 2005; Marie et al., 2005). Thymocytes deficient in expression of the TGF-βRII also developed normally into Foxp3$^+$ Treg (Li et al., 2006; Marie et al., 2006). Nevertheless, TGF-β appears to play an important role in thymic Treg differentiation: In the first week of life of mice in which developing thymocytes do not express the TGF-βRI, hardly any Treg differentiate (Liu et al., 2008). However, these few Treg then appear to be activated to proliferate in an IL-2 dependent manner, which appears to explain the finding by several groups of normal proportions of Treg in adult TGF-β (signaling)-deficient animals. It was subsequently suggested that, rather than favoring Treg-differentiation, TGF-β has an anti-apoptotic effect on Treg (Ouyang et al., 2010). However, mice in which the *Tgfbr1* gene was inactivated in differentiated Foxp3$^+$ Treg have normal numbers of these cells in the thymus. The latter finding supports a role for TGF-β in differentiation rather than in survival of Treg (Konkel et al., 2014). Interestingly, the involved TGF-β was apparently produced by thymic macrophages, dendritic cells, and epithelial cells in response to apoptosis of developing thymocytes, which potentially explains why Treg develop later than Tconv during ontogeny (Konkel et al., 2014). Other mechanisms may modulate TGF-β production in the thymus and thus quantitatively influence tTreg differentiation.

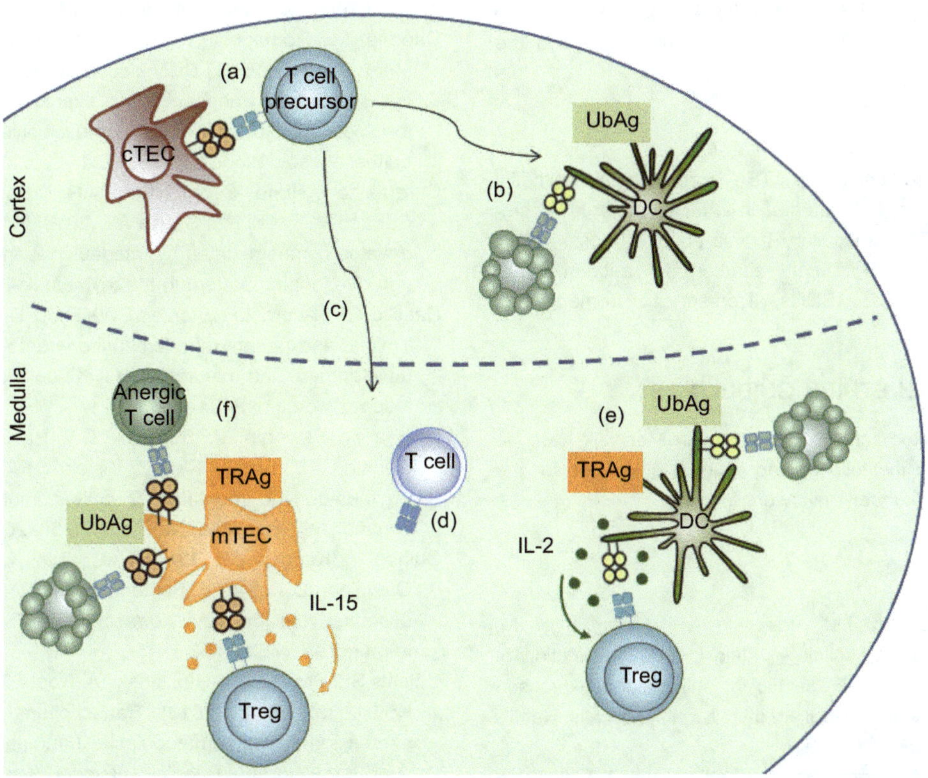

Figure 1. A model for thymic development and selection of Treg. Rescue from programmed cell-death of probably common Tconv/Treg precursors requires interaction of their TCR with MHC/peptide complexes expressed on the surface of cortical thymic epithelial cells (cTEC) (a). The thus positively selected precursor-population undergoes a first wave of negative selection (i.e., induction of apoptosis) in the cortex, depleting it of cells specific for the "ubiquitous" antigens (UbAg) presented by cortical dendritic cells (DC) (b). Surviving thymocytes upregulate CCR7 and CCR4 and migrate to the medulla (c) where precursors that recognize MHC/self-peptide complexes with low affinity develop into fully mature Tconv (d). T cell-precursors that recognize MHC/self-peptide complexes with higher affinity have three distinct destinies; become anergic, die by apoptosis, or differentiate into Treg. The latter processes appear to especially concern cells specific for the peripheral-tissue restricted antigens (TRA) expressed by mTEC and presented by mTEC or, upon transfer, by thymic DC. The signals, that determine which of the three distinct outcomes a given autospecific T cell-precursor will adopt, remain incompletely identified, but the IL-2 produced by DC (e) and mTEC-derived IL-15 (f) drive precursors into the Treg lineage.

In conclusion, substantial progress has been made in deciphering the roles of IL-2 and IL-15 (and other cytokines) in thymic differentiation of Treg (Fig. 1). The regulation of IL-2 bioavailability by peripheral Treg that recirculate back to the thymus reveals a connection between peripheral immune responses and thymocyte differentiation and offers a new perspective on Treg function. Activated peripheral Treg, therefore, not only control the homeostasis of the immune system in the periphery, but also the development of new Treg in the thymus. Studying the involvement of cytokines in the development of Treg and other T cell-subsets will shed light on modulation, by peripheral immune-responses, of T cell differentiation and TCR-repertoire selection in the thymus.

ACKNOWLEDGEMENTS

We thank the members of the TMIT team for critical discussions on the topic of the involvement of cytokines in thymic Treg development. This work was financially supported by the "Fondation pour la Recherche Médicale" (FRM-DEQ20160334920), the "Agence Nationale pour la Recherche" (ANR-16-CE15-0015-01), and the IdEx Toulouse.

ABBREVIATIONS

AICD, activation induced cell-death; cTEC, cortical thymic epithelial cells; DC, dendritic cells; IL-2, interleukin-2; MHC, major histocompatibility complex; mTEC, medullary thymic epithelial cells; Tconv, conventional T cells; TCR, T cell receptors for antigen; Treg, regulatory T lymphocytes; TSDRs, Treg-specific demethylated regions

COMPLIANCE WITH ETHICS GUIDELINES

Cecile Apert, Paola Romagnoli, and Joost van Meerwijk have no conflict of interest. All institutional and national guidelines for the care and use of laboratory animals were followed.

REFERENCES

Aschenbrenner K, D'Cruz LM, Vollmann EH, Hinterberger M, Emmerich J, Swee LK, Rolink A, Klein L (2007) Selection of Foxp3(+) regulatory T cells specific for self antigen expressed and presented by Aire(+) medullary thymic epithelial cells. Nature Immunol 8:351–358

Bayer AL, Lee JY, de la Barrera A, Surh CD, Malek TR (2008) A function for IL-7R for CD4+CD25+Foxp3+ T regulatory cells. J Immunol 181:225–234

Bennett CL, Christie J, Ramsdell F, Brunkow ME, Ferguson PJ, Whitesell L, Kelly TE, Saulsbury FT, Chance PF, Ochs HD (2001) The immune dysregulation, polyendocrinopathy, enteropathy, X-linked syndrome (IPEX) is caused by mutations of FOXP3. Nature Genet 27:20–21

Bilate AM, Lafaille JJ (2012) Induced CD4+Foxp3+ regulatory T cells in immune tolerance. Annu Rev Immunol 30:733–758

Brennecke P, Reyes A, Pinto S, Rattay K, Nguyen M, Kuchler R, Huber W, Kyewski B, Steinmetz LM (2015) Single-cell transcriptome analysis reveals coordinated ectopic gene-expression patterns in medullary thymic epithelial cells. Nature Immunol 16:933–941

Brunkow ME, Jeffery EW, Hjerrild KA, Paeper B, Clark LB, Yasayko SA, Wilkinson JE, Galas D, Ziegler SF, Ramsdell F (2001) Disruption of a new forkhead/winged-helix protein, scurfin, results in the fatal lymphoproliferative disorder of the scurfy mouse. Nature Genetics 27:68–73

Burchill MA, Goetz CA, Prlic M, O'Neil JJ, Harmon IR, Bensinger SJ, Turka LA, Brennan P, Jameson SC, Farrar MA (2003) Distinct Effects of STAT5 Activation on CD4+ and CD8+ T Cell Homeostasis: Development of CD4+CD25+ Regulatory T Cells versus CD8+ Memory T Cells. J Immunol 171:5853–5864

Burchill MA, Yang J, Vogtenhuber C, Blazar BR, Farrar MA (2007) IL-2 receptor beta-dependent STAT5 activation is required for the development of Foxp3+ regulatory T cells. J Immunol 178:280–290

Burchill MA, Yang J, Vang KB, Moon JJ, Chu HH, Lio CW, Vegoe AL, Hsieh CS, Jenkins MK, Farrar MA (2008) Linked T cell receptor and cytokine signaling govern the development of the regulatory T cell repertoire. Immunity 28:112–121

Capone M, Romagnoli P, Beermann F, MacDonald HR, van Meerwijk JPM (2001) Dissociation of thymic positive and negative selection in transgenic mice expressing major histocompatibility complex class I molecules exclusively on thymic cortical epithelial cells. Blood 97:1336–1342

Caramalho I, Nunes-Silva V, Pires AR, Mota C, Pinto AI, Nunes-Cabaco H, Foxall RB, Sousa AE (2015) Human regulatory T-cell development is dictated by Interleukin-2 and -15 expressed in a non-overlapping pattern in the thymus. J Autoimmun 56:98–110

Castillo EF, Acero LF, Stonier SW, Zhou D, Schluns KS (2010) Thymic and peripheral microenvironments differentially mediate development and maturation of iNKT cells by IL-15 transpresentation. Blood 116:2494–2503

Cebula A, Seweryn M, Rempala GA, Pabla SS, McIndoe RA, Denning TL, Bry L, Kraj P, Kisielow P, Ignatowicz L (2013) Thymus-derived regulatory T cells contribute to tolerance to commensal microbiota. Nature 497:258–262

Chinen T, Kannan AK, Levine AG, Fan X, Klein U, Zheng Y, Gasteiger G, Feng Y, Fontenot JD, Rudensky AY (2016) An essential role for the IL-2 receptor in Treg cell function. Nat Immunol 17:1322–1333

Colpitts SL, Stonier SW, Stoklasek TA, Root SH, Aguila HL, Schluns KS, Lefrancois L (2013) Transcriptional regulation of IL-15 expression during hematopoiesis. J Immunol 191:3017–3024

Coquet JM, Ribot JC, Babala N, Middendorp S, van der Horst G, Xiao Y, Neves JF, Fonseca-Pereira D, Jacobs H, Pennington DJ et al (2013) Epithelial and dendritic cells in the thymic medulla promote CD4+Foxp3+ regulatory T cell development via the CD27-CD70 pathway. J Exp Med 210:715–728

Cowan JE, McCarthy NI, Anderson G (2016) CCR7 controls thymus recirculation, but not production and emigration, of Foxp3(+) T Cells. Cell reports 14:1041–1048

Cui G, Hara T, Simmons S, Wagatsuma K, Abe A, Miyachi H, Kitano S, Ishii M, Tani-ichi S, Ikuta K (2014) Characterization of the IL-15 niche in primary and secondary lymphoid organs in vivo. Proc Natl Acad Sci USA 111:1915–1920

Cuss SM, Green EA (2012) Abrogation of CD40-CD154 signaling impedes the homeostasis of thymic resident regulatory T cells by altering the levels of IL-2, but does not affect regulatory T cell development. J Immunol 189:1717–1725

D'Cruz LM, Klein L (2005) Development and function of agonist-induced CD25+Foxp3+ regulatory T cells in the absence of interleukin 2 signaling. Nat Immunol 6:1152–1159

De Smedt M, Verhasselt B, Kerre T, Vanhecke D, Naessens E, Leclercq G, Renauld JC, Van Snick J, Plum J (2000) Signals from the IL-9 receptor are critical for the early stages of human intrathymic T cell development. J Immunol 164:1761–1767

Fahlen L, Read S, Gorelik L, Hurst SD, Coffman RL, Flavell RA, Powrie F (2005) T cells that cannot respond to TGF-beta escape control by CD4(+)CD25(+) regulatory T cells. J Exp Med 201:737–746

Fisson S, Darrasse-Jeze G, Litvinova E, Septier F, Klatzmann D, Liblau R, Salomon BL (2003) Continuous activation of autoreactive CD4+ CD25+ regulatory T Cells in the steady state. J Exp Med 198:737–746

Floess S, Freyer J, Siewert C, Baron U, Olek S, Polansky J, Schlawe K, Chang H-D, Bopp T, Schmitt E et al (2007) Epigenetic control of the foxp3 locus in regulatory T cells. PLoS Biol 5:e38

Fontenot JD, Gavin MA, Rudensky AY (2003) Foxp3 programs the development and function of CD4(+)CD25(+) regulatory T cells. Nat Immunol 3:3

Fontenot JD, Rasmussen JP, Gavin MA, Rudensky AY (2005) A function for interleukin 2 in Foxp3-expressing regulatory T cells. Nat Immunol. 6:1142–1151

Hale JS, Fink PJ (2009) Back to the thymus: peripheral T cells come home. Immunol Cell Biol 87:58–64

Hanabuchi S, Ito T, Park WR, Watanabe N, Shaw JL, Roman E, Arima K, Wang YH, Voo KS, Cao W et al (2010) Thymic stromal lymphopoietin-activated plasmacytoid dendritic cells induce the generation of FOXP3+ regulatory T cells in human thymus. J Immunol 184:2999–3007

Hsieh CS, Liang Y, Tyznik AJ, Self SG, Liggitt D, Rudensky AY (2004) Recognition of the peripheral self by naturally arising CD25+ CD4+ T cell receptors. Immunity 21:267–277

Hu Z, Lancaster JN, Sasiponganan C, Ehrlich LI (2015) CCR4 promotes medullary entry and thymocyte-dendritic cell interactions required for central tolerance. J Exp Med 212:1947–1965

Josefowicz SZ, Lu LF, Rudensky AY (2012a) Regulatory T cells: mechanisms of differentiation and function. Annu Rev Immunol 30:531–564

Josefowicz SZ, Niec RE, Kim HY, Treuting P, Chinen T, Zheng Y, Umetsu DT, Rudensky AY (2012b) Extrathymically generated regulatory T cells control mucosal TH2 inflammation. Nature 482:395–399

Khailaie S, Robert PA, Toker A, Huehn J, Meyer-Hermann M (2014) A signal integration model of thymic selection and natural regulatory T cell commitment. J Immunol 193:5983–5996

Kieback E, Hilgenberg E, Stervbo U, Lampropoulou V, Shen P, Bunse M, Jaimes Y, Boudinot P, Radbruch A, Klemm U et al (2016) Thymus-derived regulatory T cells are positively selected on natural self-antigen through cognate interactions of high functional avidity. Immunity 44:1114–1126

Kim HJ, Cantor H (2011) Regulation of self-tolerance by Qa-1-restricted CD8(+) regulatory T cells. Sem Immunol 23:446–452

Kim HP, Leonard WJ (2007) CREB/ATF-dependent T cell receptor-induced FoxP3 gene expression: a role for DNA methylation. J Exp Med 204:1543–1551

Kisielow P, Miazek A (1995) Positive selection of T cells: rescue from programmed cell death and differentiation require continual engagement of the T cell receptor. J Exp Med 181:1975–1984

Kitagawa Y, Ohkura N, Kidani Y, Vandenbon A, Hirota K, Kawakami R, Yasuda K, Motooka D, Nakamura S, Kondo M et al (2017) Guidance of regulatory T cell development by Satb1-dependent super-enhancer establishment. Nature Immunol 18:173–183

Klein L, Jovanovic K (2011) Regulatory T cell lineage commitment in the thymus. Sem Immunol 23:401–409

Klein L, Hinterberger M, Wirnsberger G, Kyewski B (2009) Antigen presentation in the thymus for positive selection and central tolerance induction. Nat Rev Immunol 9:833–844

Konkel JE, Jin W, Abbatiello B, Grainger JR, Chen W (2014) Thymocyte apoptosis drives the intrathymic generation of regulatory T cells. Proc Natl Acad Sci USA 111:E465–E473

Kurd N, Robey EA (2016) T-cell selection in the thymus: a spatial and temporal perspective. Immunol Rev 271:114–126

Kyewski B, Klein L (2006) A central role for central tolerance. Annu Rev Immunol 24:571–606

Laufer TM, DeKoning J, Markowitz JS, Lo D, Glimcher LH (1996) Unopposed positive selection and autoreactivity in mice expressing class II MHC only on thymic cortex. Nature 383:81–85

Le Borgne M, Ladi E, Dzhagalov I, Herzmark P, Liao YF, Chakraborty AK, Robey EA (2009) The impact of negative selection on thymocyte migration in the medulla. Nat Immunol 10:823–830

Li MO, Sanjabi S, Flavell RA (2006) Transforming growth factor-beta controls development, homeostasis, and tolerance of T cells by regulatory T cell-dependent and -independent mechanisms. Immunity 25:455–471

Lio CW, Hsieh CS (2008) A two-step process for thymic regulatory T cell development. Immunity 28:100–111

Liston A, Nutsch KM, Farr AG, Lund JM, Rasmussen JP, Koni PA, Rudensky AY (2008) Differentiation of regulatory Foxp3+ T cells in the thymic cortex. Proc Natl Acad Sci USA 105:11903–11908

Liu Y, Zhang P, Li J, Kulkarni AB, Perruche S, Chen W (2008) A critical function for TGF-beta signaling in the development of natural CD4+CD25+Foxp3+ regulatory T cells. Nat Immunol 9:632–640

Love PE, Bhandoola A (2011) Signal integration and crosstalk during thymocyte migration and emigration. Nat Rev Immunol 11:469–477

Lucas B, McCarthy NI, Baik S, Cosway E, James KD, Parnell SM, White AJ, Jenkinson WE, Anderson G (2016) Control of the thymic medulla and its influence on alphabetaT-cell development. Immunol Rev 271:23–37

Mahmud SA, Manlove LS, Schmitz HM, Xing Y, Wang Y, Owen DL, Schenkel JM, Boomer JS, Green JM, Yagita H et al (2014) Costimulation via the tumor-necrosis factor receptor superfamily couples TCR signal strength to the thymic differentiation of regulatory T cells. Nat Immunol 15:473–481

Malchow S, Leventhal DS, Nishi S, Fischer BI, Shen L, Paner GP, Amit AS, Kang C, Geddes JE, Allison JP (2013) Aire-dependent thymic development of tumor-associated regulatory T cells. Science 339:1219–1224

Malek TR, Porter BO, Codias EK, Scibelli P, Yu A (2000) Normal lymphoid homeostasis and lack of lethal autoimmunity in mice containing mature T cells with severely impaired IL-2 receptors. J Immunol 164:2905–2914

Malek TR, Yu A, Vincek V, Scibelli P, Kong L (2002) CD4 regulatory T cells prevent lethal autoimmunity in IL-2Rbeta-deficient mice. Implications for the nonredundant function of IL-2. Immunity 17:167–178

Marie JC, Letterio JJ, Gavin M, Rudensky AY (2005) TGF-beta1 maintains suppressor function and Foxp3 expression in CD4 +CD25+ regulatory T cells. J Exp Med 201:1061–1067

Marie JC, Liggitt D, Rudensky AY (2006) Cellular mechanisms of fatal early-onset autoimmunity in mice with the T cell-specific targeting of transforming growth factor-beta receptor. Immunity 25:441–454

Marshall D, Sinclair C, Tung S, Seddon B (2014) Differential requirement for IL-2 and IL-15 during bifurcated development of thymic regulatory T cells. J Immunol 193:5525–5533

Mazzucchelli R, Hixon JA, Spolski R, Chen X, Li WQ, Hall VL, Willette-Brown J, Hurwitz AA, Leonard WJ, Durum SK (2008) Development of regulatory T cells requires IL-7Ralpha stimulation by IL-7 or TSLP. Blood 112:3283–3292

McCaughtry TM, Wilken MS, Hogquist KA (2007) Thymic emigration revisited. J Exp Med 204:2513–2520

McCaughtry TM, Baldwin TA, Wilken MS, Hogquist KA (2008) Clonal deletion of thymocytes can occur in the cortex with no involvement of the medulla. J Exp Med 205:2575–2584

Meredith M, Zemmour D, Mathis D, Benoist C (2015) Aire controls gene expression in the thymic epithelium with ordered stochasticity. Nat Immunol 16:942–949

Ohigashi I, Kozai M, Takahama Y (2016) Development and developmental potential of cortical thymic epithelial cells. Immunol Rev 271:10–22

Ohkura N, Hamaguchi M, Morikawa H, Sugimura K, Tanaka A, Ito Y, Osaki M, Tanaka Y, Yamashita R, Nakano N et al (2012) T cell receptor stimulation-induced epigenetic changes and Foxp3 expression are independent and complementary events required for Treg cell development. Immunity 37:785–799

Ouyang W, Beckett O, Ma Q, Li MO (2010) Transforming growth factor-beta signaling curbs thymic negative selection promoting regulatory T cell development. Immunity 32:642–653

Pelly VS, Kannan Y, Coomes SM, Entwistle LJ, Ruckerl D, Seddon B, MacDonald AS, McKenzie A, Wilson MS (2016) IL-4-producing ILC2s are required for the differentiation of TH2 cells following Heligmosomoides polygyrus infection. Mucosal Immunol 9:1407–1417

Pennington DJ, Silva-Santos B, Silberzahn T, Escorcio-Correia M, Woodward MJ, Roberts SJ, Smith AL, Dyson PJ, Hayday AC (2006) Early events in the thymus affect the balance of effector and regulatory T cells. Nature 444:1073–1077

Popmihajlov Z, Xu D, Morgan H, Milligan Z, Smith KA (2012) Conditional IL-2 gene deletion: consequences for T cell proliferation. Front Immunol 3:102

Ribot J, Enault G, Pilipenko S, Huchenq A, Calise M, Hudrisier D, Romagnoli P, van Meerwijk JPM (2007) Shaping of the autoreactive regulatory T cell repertoire by thymic cortical positive selection. J Immunol 179:6741–6748

Romagnoli P, Hudrisier D, van Meerwijk JPM (2002) Preferential recognition of self-antigens despite normal thymic deletion of CD4+CD25+ regulatory T cells. J Immunol 168:1644–1648

Romagnoli P, Ribot J, Tellier J, van Meerwijk JPM (2008) Thymic and peripheral generation of CD4+Foxp3+ regulatory T cells. In: Jiang S (ed) Regulatory T cells and clinical application. Springer Science+Business Media, New York, pp 29–55

Sadlack B, Lohler J, Schorle H, Klebb G, Haber H, Sickel E, Noelle RJ, Horak I (1995) Generalized autoimmune disease in interleukin-2-deficient mice is triggered by an uncontrolled activation and proliferation of CD4+ T cells. Eur J Immunol 25:3053–3059

Sakaguchi S, Miyara M, Costantino CM, Hafler DA (2010) FOXP3+ regulatory T cells in the human immune system. Nat Rev Immunol 10:490–500

Shevach EM (2011) Biological functions of regulatory T cells. Adv Immunol 112:137–176

Shitara S, Hara T, Liang B, Wagatsuma K, Zuklys S, Hollander GA, Nakase H, Chiba T, Tani-ichi S, Ikuta K (2013) IL-7 produced by thymic epithelial cells plays a major role in the development of thymocytes and TCRgammadelta+ intraepithelial lymphocytes. J Immunol 190:6173–6179

Soper DM, Kasprowicz DJ, Ziegler SF (2007) IL-2Rbeta links IL-2R signaling with Foxp3 expression. Eur J Immunol 37:1817–1826

Stritesky GL, Xing Y, Erickson JR, Kalekar LA, Wang X, Mueller DL, Jameson SC, Hogquist KA (2013) Murine thymic selection quantified using a unique method to capture deleted T cells. Proc Natl Acad Sci USA 110:4679–4684

Sun J, Furio L, Mecheri R, van der Does AM, Lundeberg E, Saveanu L, Chen Y, van Endert P, Agerberth B, Diana J (2015) Pancreatic beta-cells limit autoimmune diabetes via an immunoregulatory antimicrobial peptide expressed under the influence of the gut microbiota. Immunity 43:304–317

Suzuki H, Kundig T, Furlonger C, Wakeham A, Timms E, Matsuyama T, Schmits R, Simard J, Ohashi P, Griesser H et al (1995) Deregulated T cell activation and autoimmunity in mice lacking interleukin-2 receptor beta. Science 268(5216):1472–1476

Suzuki H, Zhou YW, Kato M, Mak TW, Nakashima I (1999) Normal regulatory alpha/beta T cells effectively eliminate abnormally activated T cells lacking the interleukin 2 receptor beta in vivo. J Exp Med 190:1561–1572

Tai X, Cowan M, Feigenbaum L, Singer A (2005) CD28 costimulation of developing thymocytes induces Foxp3 expression and regulatory T cell differentiation independently of interleukin 2. Nat Immunol 6:152–162

Tai X, Erman B, Alag A, Mu J, Kimura M, Katz G, Guinter T, McCaughtry T, Etzensperger R, Feigenbaum L et al (2013) Foxp3 transcription factor is proapoptotic and lethal to developing regulatory T cells unless counterbalanced by cytokine survival signals. Immunity 38:1116–1128

Takaba H, Morishita Y, Tomofuji Y, Danks L, Nitta T, Komatsu N, Kodama T, Takayanagi H (2015) Fezf2 orchestrates a thymic program of self-antigen expression for immune tolerance. Cell 163:975–987

Tang Q, Henriksen KJ, Boden EK, Tooley AJ, Ye J, Subudhi SK, Zheng XX, Strom TB, Bluestone JA (2003) Cutting edge: CD28 controls peripheral homeostasis of CD4+CD25+ regulatory T cells. J Immunol 171:3348–3352

Tani-ichi S, Shimba A, Wagatsuma K, Miyachi H, Kitano S, Imai K, Hara T, Ikuta K (2013) Interleukin-7 receptor controls development and maturation of late stages of thymocyte subpopulations. Proc Natl Acad Sci USA 110:612–617

Thiault N, Darrigues J, Adoue V, Gros M, Binet B, Perals C, Leobon B, Fazilleau N, Joffre OP, Robey EA et al (2015) Peripheral regulatory T lymphocytes recirculating to the thymus suppress the development of their precursors. Nat Immunol 16:628–634

Toker A, Engelbert D, Garg G, Polansky JK, Floess S, Miyao T, Baron U, Duber S, Geffers R, Giehr P et al (2013) Active demethylation of the Foxp3 locus leads to the generation of stable regulatory T cells within the thymus. J Immunol 190:3180–3188

Ueno T, Saito F, Gray DH, Kuse S, Hieshima K, Nakano H, Kakiuchi T, Lipp M, Boyd RL, Takahama Y (2004) CCR7 signals are essential for cortex-medulla migration of developing thymocytes. J Exp Med 200:493–505

Valitutti S, Muller S, Cella M, Padovan E, Lanzavecchia A (1995) Serial triggering of many T-cell receptors by a few peptide-MHC complexes. Nature 375:148–151

Vang KB, Yang J, Mahmud SA, Burchill MA, Vegoe AL, Farrar MA (2008) IL-2, -7, and -15, but not thymic stromal lymphopoeitin, redundantly govern CD4+Foxp3+ regulatory T cell development. J Immunol 181:3285–3290

von Freeden-Jeffry U, Vieira P, Lucian LA, McNeil T, Burdach SE, Murray R (1995) Lymphopenia in interleukin (IL)-7 gene-deleted mice identifies IL-7 as a nonredundant cytokine. J Exp Med 181:1519–1526

Vuddamalay, Y., and van Meerwijk, J. (2017). CD28neg and CD28low CD8+ regulatory T cells: Of Mice and Men. Front Immunol 8.

Watanabe N, Wang YH, Lee HK, Ito T, Wang YH, Cao W, Liu YJ (2005) Hassall's corpuscles instruct dendritic cells to induce CD4+CD25+ regulatory T cells in human thymus. Nature 436:1181–1185

Weist BM, Kurd N, Boussier J, Chan SW, Robey EA (2015) Thymic regulatory T cell niche size is dictated by limiting IL-2 from antigen-bearing dendritic cells and feedback competition. Nat Immunol 16:635–641

Wilkinson RW, Anderson G, Owen JJ, Jenkinson EJ (1995) Positive selection of thymocytes involves sustained interactions with the thymic microenvironment. J Immunol 155:5234–5240

Willerford DM, Chen J, Ferry JA, Davidson L, Ma A, Alt FW (1995) Interleukin-2 receptor α chain regulates the size and content of the peripheral lymphoid compartment. Immunity 3(4):521–530

Wirnsberger G, Mair F, Klein L (2009) Regulatory T cells differentiation of thymocytes does not require a dedicated antigen-presenting cell but is under T cell-intrinsic developmental control. Proc Natl Acad Sci USA 106:10278–10283

Wolf M, Schimpl A, Hunig T (2001) Control of T cell hyperactivation in IL-2-deficient mice by CD4(+)CD25(-) and CD4(+)CD25(+) T cells: evidence for two distinct regulatory mechanisms. Eur J Immunol 31:1637–1645

Xu Z, Ho S, Chang CC, Zhang QY, Vasilescu ER, Vlad G, Suciu-Foca N (2016) Molecular and cellular characterization of human CD8 T suppressor cells. Front Immunol 7:549

Yamano T, Nedjic J, Hinterberger M, Steinert M, Koser S, Pinto S, Gerdes N, Lutgens E, Ishimaru N, Busslinger M et al (2015) Thymic B cells are licensed to present self antigens for central T cell tolerance induction. Immunity 42:1048–1061

Yang S, Fujikado N, Kolodin D, Benoist C, Mathis D (2015) Regulatory T cells generated early in life play a distinct role in maintaining self-tolerance. Science 348:589–594

Yao Z, Kanno Y, Kerenyi M, Stephens G, Durant L, Watford WT, Laurence A, Robinson GW, Shevach EM, Moriggl R et al (2007) Nonredundant roles for Stat5a/b in directly regulating Foxp3. Blood 109:4368–4375

Yu W, Nagaoka H, Jankovic M, Misulovin Z, Suh H, Rolink A, Melchers F, Meffre E, Nussenzweig MC (1999) Continued RAG expression in late stages of B cell development and no apparent re-induction after immunization. Nature 400:682–687

Yue X, Trifari S, Aijo T, Tsagaratou A, Pastor WA, Zepeda-Martinez JA, Lio CW, Li X, Huang Y, Vijayanand P et al (2016) Control of Foxp3 stability through modulation of TET activity. J Exp Med 213:377–397

Zheng Y, Josefowicz S, Chaudhry A, Peng XP, Forbush K, Rudensky AY (2010) Role of conserved non-coding DNA elements in the Foxp3 gene in regulatory T-cell fate. Nature 463:808–812

2

Novel matrine derivative MD-1 attenuates hepatic fibrosis by inhibiting EGFR activation of hepatic stellate cells

Yi Feng✉**, Hai-yan Ying, Ying Qu, Xiao-bo Cai, Ming-yi Xu, Lun-gen Lu**✉

Department of Gastroenterology, Shanghai General Hospital, Nanjing Medical University, Shanghai 200080, China
✉ Correspondence: orchidfy@hotmail.com (Y. Feng), Lungenlu1965@163.com (L. Lu)

ABSTRACT

Matrine (MT), the effective component of *Sophora flavescens Ait*, has been shown to have anti-inflammation, immune-suppressive, anti-tumor, and anti-hepatic fibrosis activities. However, the pharmacological effects of MT still need to be strengthened due to its relatively low efficacy and short half-life. In the present study, we report a more effective thio derivative of MT, MD-1, and its inhibitory effects on the activation of hepatic stellate cells (HSCs) in both cell culture and animal models. Cytological experiments showed that MD-1 can inhibit the proliferation of HSC-T6 cells with a half-maximal inhibitory concentration (IC_{50}) of 62 μmol/L. In addition, MD-1 more strongly inhibits the migration of HSC-T6 cells compared to MT and can more effectively induce G_0/G_1 arrest and apoptosis. Investigating the biological mechanisms underlying anti-hepatic fibrosis in the presence of MD-1, we found that MD-1 can bind the epidermal growth factor receptor (EGFR) on the surface of HSC-T6 cells, which can further inhibit the phosphorylation of EGFR and its downstream protein kinase B (Akt), resulting in decreased expression of cyclin D1 and eventual inhibition of the activation of HSC-T6 cells. Furthermore, in rats with dimethylnitrosamine (DMN)-induced hepatic fibrosis, MD-1 slowed the development and progression of hepatic fibrosis, protecting hepatic parenchymal cells and improving hepatic functions. Therefore, MD-1 is a potential drug for anti-hepatic fibrosis.

KEYWORDS matrine derivative, hepatic stellate cell, hepatic fibrosis, epidermal growth factor receptor, signal transduction pathway

INTRODUCTION

Hepatic fibrosis is a pathological process comprising atypical hyperplasia (dysplasia) of intrahepatic connective tissue and excessive accumulation of extracellular matrix (ECM) that can ultimately develop into hepatic cirrhosis. The main pathogenic factors that induce hepatic fibrosis include chronic hepatitis virus (i.e., HBV and HCV) and parasite infection (e.g., schistosomes), alcoholic liver disease, fatty liver disease, cholestasis, drug-induced liver injury, and autoimmune hepatitis. The persistent presence of these factors causes the production and activation of hepatic fibrosis-initiating cells, especially the activation of hepatic stellate cells (HSCs). Other cells that can be activated are fibroblasts, liver sinusoidal endothelial cells, and Kupffer cells surrounding the portal vein; mesenchymal stem cells derived from bone marrow; blood fibrocytes; and hepatocytes or biliary epithelial cells that have undergone epithelial-mesenchymal transition (EMT) (Fausther et al., 2013; Iwaisako et al., 2014b; Pinzani, 2015; Koo et al., 2016). These cells can proliferate and transdifferentiate into myofibroblasts (MFs) and then synthesize and secret a large amount of ECM, causing hepatic fibrosis and cirrhosis (Fagone et al., 2015).

Studies of the mechanisms underlying the development of hepatic fibrosis indicate that many factors associated with liver injuries can induce HSC activation (Seki and Brenner, 2015). Hepatic parenchymal cell injury can be caused by a variety of pathological factors and induce the synthesis and release of a variety of cytokines, including transforming growth factor β (TGF-β), epidermal growth factor (EGF), platelet-derived growth factor (PDGF), endothelin (ET), fibroblast growth factor (FGF), connective tissue growth factor (CTGF), and leptin, which can further activate the TGF-β/Smad, epidermal growth factor receptor (EGFR), and Ras/ERK signaling pathways to promote HSC proliferation and activity (Cuevas et al., 2011; Handy et al., 2011; Shimada et al., 2011; Liu et al., 2013). After activation, HSCs can transdifferentiate into MFs, which express α-smooth muscle actin (α-SMA), secrete ECM components (including type I and type II collagens and proteoglycan) that cause ECM accumulation, and promote the development of hepatic fibrosis. Therefore, inhibition of hepatic fibrosis through HSC inactivation is a major direction in the prevention and treatment of hepatic cirrhosis.

Two major treatments are used against hepatic fibrosis: those that eliminate pathogenic factors that cause hepatic fibrosis (e.g., antiviral therapy and glucocorticoid treatment of autoimmune or alcoholic liver diseases) and those that target ECM accumulation by decreasing its synthesis and facilitating its degradation. Clinical treatment measures for anti-hepatic fibrosis are currently limited. Therefore, it is crucial to explore new types of anti-hepatic fibrosis drugs. Matrine (MT), the effective component of *Sophora flavescens Ait*, has been shown to have anti-inflammation, immune-suppressive, anti-tumor, and anti-hepatic fibrosis activities. MT also has protective effects against acute liver injury in mice and rats (Zhang et al., 2011; Liu et al., 2014). However, the clinical pharmacological effects of this drug still need to be strengthened due to its relatively low efficacy and short half-life. In the present study, we structurally modified MT and obtained a novel thio derivative called MD-1 ($C_{30}N_4H_{40}SO_5F$) through thiosulfate and side chain Michael addition to prepare maleate for anti-hepatic fibrosis experiments. We examined the effects of MD-1 on HSC activation and inhibition of hepatic fibrosis in rats. In addition, the possible mechanism of action of this novel MT derivative was investigated to determine if it is a potential clinical treatment option for hepatic fibrosis.

RESULTS

Inhibition of HSC-T6 cell proliferation by novel thio derivatives of MT

A series of novel thio derivatives of MT, including MD-1, MD-2, and MD-3, were obtained from structural modifications to MT (Fig. 1A, also see Supplementary Information for details). First, we determined the effects of MT and its derivatives on the proliferation of HSC-T6 cells at a concentration of 100 μmol/L. MT treatment decreased the survival rate to 62.9% ± 10.0% (n = 3) compared to control (114.2% ± 8.5%, n = 3; $P < 0.001$, Fig. 1B). Among the three derivatives of MT, MD-1 had the strongest inhibitory effect on HSC-T6 cells, decreasing the cell survival rate to 24.2% ± 5.0%, which is significantly lower than that of MT (n = 3; $P = 0.004$). For MD-2 the cell survival rate was 27.4% ± 1.6% (n = 3; $P < 0.01$ compared to MT). The inhibitory effect of MD-3 on HSC-T6 cells was equivalent to that of MT (47.2% ± 8.1%, n = 3; $P = 0.10$). We also examined the effects of different concentrations of MD-1 on the proliferation of HSC-T6 cells. The survival rates of HSC-T6 cells decreased as the MD-1 concentration increased. The IC_{50} values of MD-1 and MT on HSC-T6 cells were 62 μmol/L and 128 μmol/L, respectively (Fig. 1C).

Inhibitory effects of MD-1 on HSC-T6 cell motility

The inhibitory effects of MD-1 and MT on the migration of HSC-T6 cells were observed using the Transwell assay. To exclude the influence of inhibited cell proliferation, the experimental concentrations of MD-1 and MT were set at their corresponding IC_{50} values to compare their effects on cell migration at the same cell survival rate (50%). After 24 h of treatment, the difference in HSC-T6 cell migration between 62 μmol/L MD-1 and 128 μmol/L MT was not significant, but both significantly reduced motility compared to control (Fig. 2B; $P < 0.05$). However, after 48 h, the inhibitory effect of MD-1 on cell migration was significantly stronger than that of MT ($P < 0.05$, n = 3, Fig. 2A and 2B). Comparing the difference between 24 h and 48 h groups, the results showed that there was a significant different increase in MT-treated cells at 48 h compared with that at 24 h ($P < 0.01$), but not in MD-1-treated cells, suggesting a stronger effect of MD-1 on inhibiting HSC-T6 cell migration.

Effects of MD-1 on cell cycle and apoptosis in HSC-T6 cells

The effects of MD-1 on cell cycle and apoptosis in HSC-T6 cells were detected by flow cytometry. Compared to the control group, MD-1- and MT-treated HSC-T6 cells had an increased percentage of cells in G_0/G_1 phase and a decreased percentage of cells in the S phase (Fig. 3A). The change in the cell cycle induced by MD-1 was even more significant than the change induced by MT (G_0/G_1: 47.2% ± 3.2% in MD-1 group, 38.6% ± 2.5% in MT group, and 25.2% ± 3.0% in control; S: 33.6% ± 4.5% in MD-1 group, 36.6% ± 2.6% in MT group, and 58.9% ± 2.2% in control; G_2/M: 18.5% ± 6.0% in MD-1 group, 24.9% ± 4.5% in MT group, and 17.4% ± 1.4% in control; n = 3 in each group). These results suggest that MD-1 could induce G_0/G_1 arrest in HSC-T6 cells and decrease the number of cells entering mitosis (Fig. 3A, right). The apoptosis rate in control HSC-T6 cells was 2.5% ± 1.3% (n = 3). After MD-1 or MT treatment for 48 h, the apoptosis rate increased to 25.6% ± 4.8% (n = 3; $P = 0.001$) or 7.6% ± 2.5% (n = 3; $P < 0.05$), respectively. The effect of MD-1 on the induction of apoptosis in HSC-T6 cells was significantly stronger than that of MT ($P = 0.005$; Fig. 3B, right).

Effects of MD-1 on the EGFR signaling pathway

After biotin-labeled MD-1 was prepared, the localization of MD-1 in HSC-T6 cells and its relationship with EGFR were

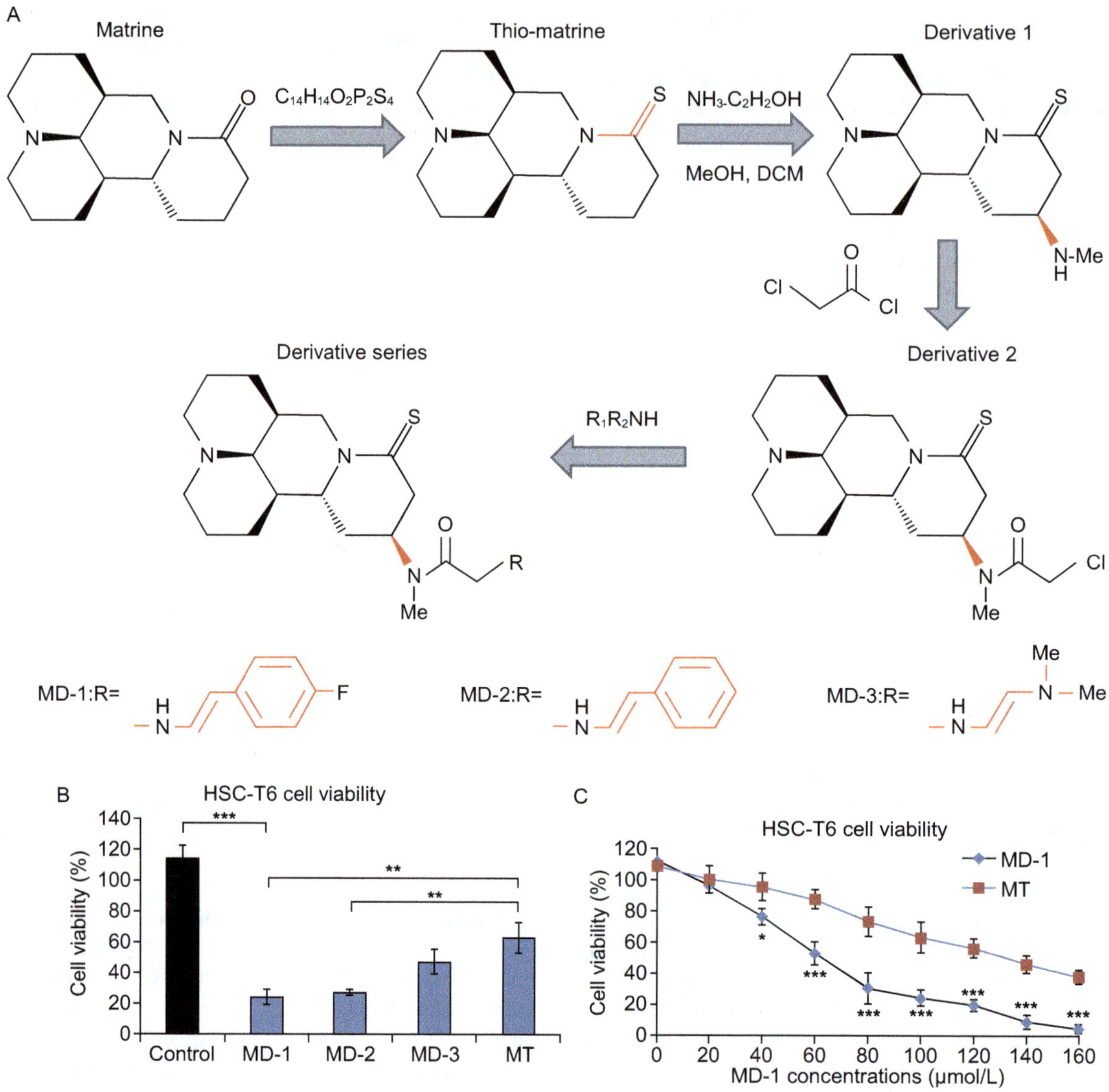

Figure 1. Novel thio derivatives of MT inhibit the proliferation of HSC-T6 cells. (A) The chemical structure of MD-1, MD-2, and MD-3. (B) The inhibitory effects of MD-1, MD-2, and MD-3 on the proliferation of HSC-T6 cells (MD-1 group: 24.2% ± 5.0%; MD-2 group: 27.4% ± 1.6%; MD-3 group: 47.2% ± 8.1%; MT group: 62.9% ± 10.0%; control group: 114.2% ± 8.5%; n = 3 in each group). (C) A concentration-dependent plot of the effect of MD-1 and MT on the proliferation of HSC-T6 cells. IC_{50} of MD-1 and MT was 62 μmol/L and 128 μmol/L, respectively.

examined by immunofluorescence labeling and co-immunoprecipitation assay. The localization of MD-1 was consistent with EGFR on the surface of the cell membrane (Fig. 4A). After knocking down the expression of EGFR in HSC-T6 cells using an EGFR-shRNA expression plasmid, immunofluorescence labeling showed that MD-1 binding to cell membranes decreased accordingly (Fig. 4B). Co-immunoprecipitation blots also showed a positive band of MD-1-Biotin was precipitated by EGFR in the MD-1-treated cells, but not in the parental cells (Fig. 4C). These results suggest that the target molecule of MD-1 in HSC-T6 cells was EGFR and that the drug exerted its function by binding to EGFR.

We also detected the expression of EGFR downstream signaling proteins. After HSC-T6 cells were treated with MD-1, the phosphorylation levels of cell proliferation- and motility-associated EGFR and Akt significantly decreased (p-EGFR: $P < 0.001$; t-EGFR: $P < 0.05$; p-AKT: $P < 0.01$; $n = 3$ in each group, Fig. 4D), and the expression levels of cell cycle regulatory protein cyclin D1 and apoptosis-associated

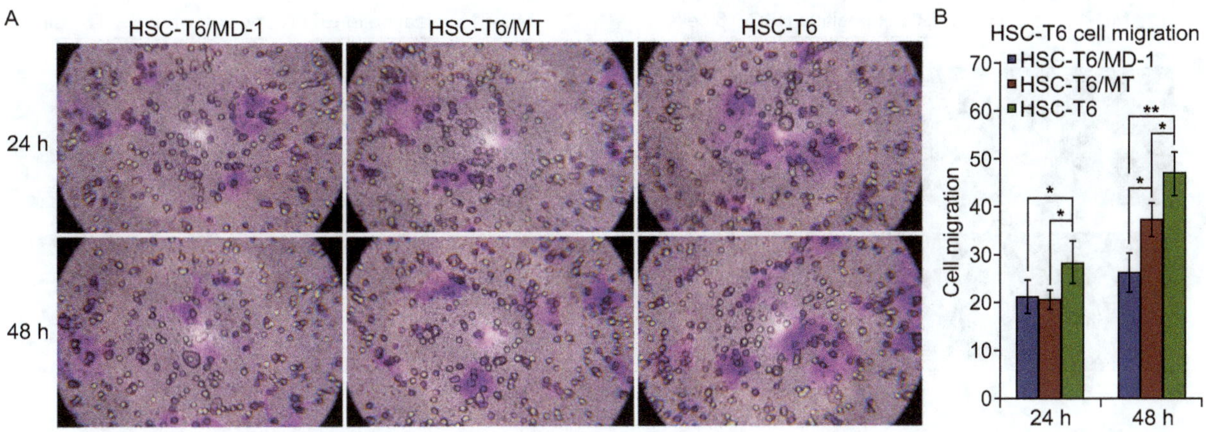

Figure 2. Inhibitory effects of MD-1 on the motility of HSC-T6 cells. (A) Top: Representative picture to show the migration ability of HSC-T6 cells at the same cell viability (50%) after 24 h of drug action (MD-1, MT, and control). Bottom: After 48 h of drug action. (B) Statistics for the migration ability of HSC-T6 cells. *$P < 0.05$; **$P < 0.01$.

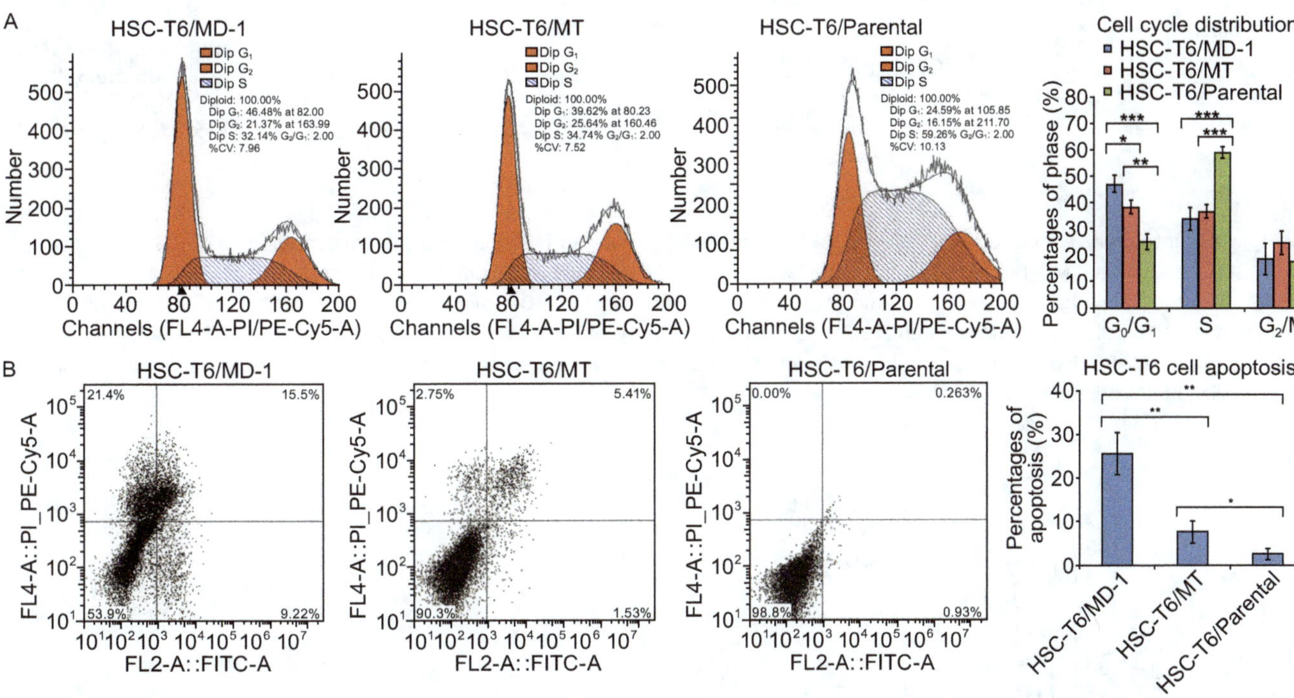

Figure 3. Inhibition of cell cycle and induction of apoptosis by MD-1 in HSC-T6 cells. (A) Left: The cell cycle distribution of HSC-T6 cells in the presence of MD-1 or MT at 62 μmol/L was examined by flow cytometry. Right: Statistics for cell cycle distribution. *$P < 0.05$; **$P < 0.01$; ***$P < 0.001$. (B) Left: The cell apoptosis rate was detected with flow cytometer in MD-1, MT and control group. *$P < 0.05$; **$P < 0.01$; ***$P < 0.001$.

protein p-Smad also decreased ($P < 0.01$ and $P < 0.05$, respectively, Fig. 4D). These results suggest that the EGFR and Akt signaling pathways are key factors in regulating the molecular mechanisms by which MD-1 inhibited HSC-T6 cell proliferation, migration, and cell cycle arrest and apoptosis, and that cyclin D1 and p-Smad were also involved.

Effects of MD-1 on ECM synthesis and secretion in HSC-T6 cells

After activation, HSCs transdifferentiated into MFs, which expressed α-SMA and secreted ECM. qRT-PCR demonstrated that MD-1-treated HSC-T6 cells expressed significantly decreased levels of α-SMA mRNA ($P < 0.01$, $n = 3$,

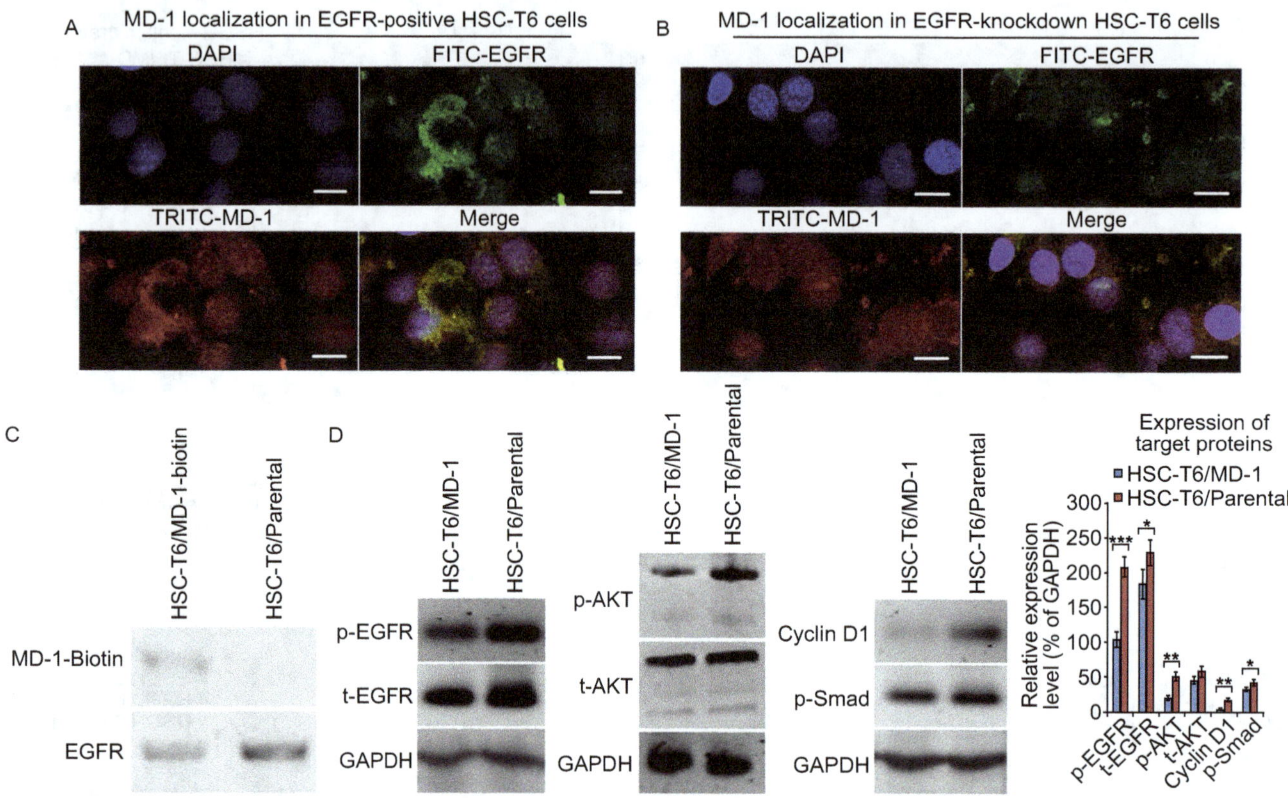

Figure 4. Inhibitory effects of MD-1 on the EGFR signaling pathway in HSC-T6 cells. (A) Immunostaining of EGFR-positive HSC-T6 cells showed MD-1 localization. Nuclei are stained with DAPI. Scale bar: 10 μm. (B) Similar to (A), but in EGFR-knockdown HSC-T6 cells. Scale bar: 10 μm. (C) Co-immunoprecipitation blots showed MD-1-Biotin was precipitated by EGFR in the MD-1-treated HSC-T6 cells, but not in the parental cells. (D) Representative Western blots to show the expression of EGFR, AKT, cyclin D1, and p-Smad. GAPDH was used as the loading control. Densitometry was performed to determine the relative expression levels normalized to GAPDH. **P* < 0.05; ***P* < 0.01; ****P* < 0.001.

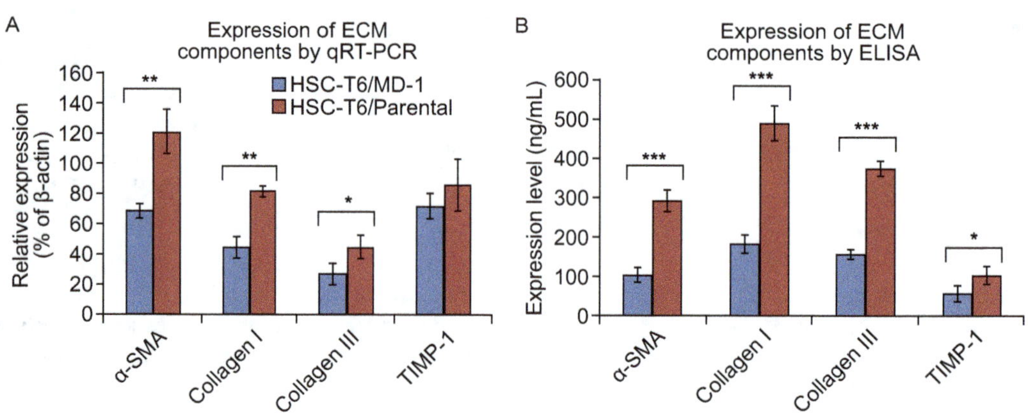

Figure 5. Inhibitory effects of MD-1 on ECM synthesis and secretion in HSC-T6 cells. (A) Relative expression levels of ECM components, including α-SMA, collagen I and III, and TIMP-1, as detected by real-time quantitative PCR (qRT-PCR). **P* < 0.05; ***P* < 0.01. (B) Expression of ECM components in the collected cell supernatant as determined by ELISA. **P* < 0.05; ****P* < 0.001.

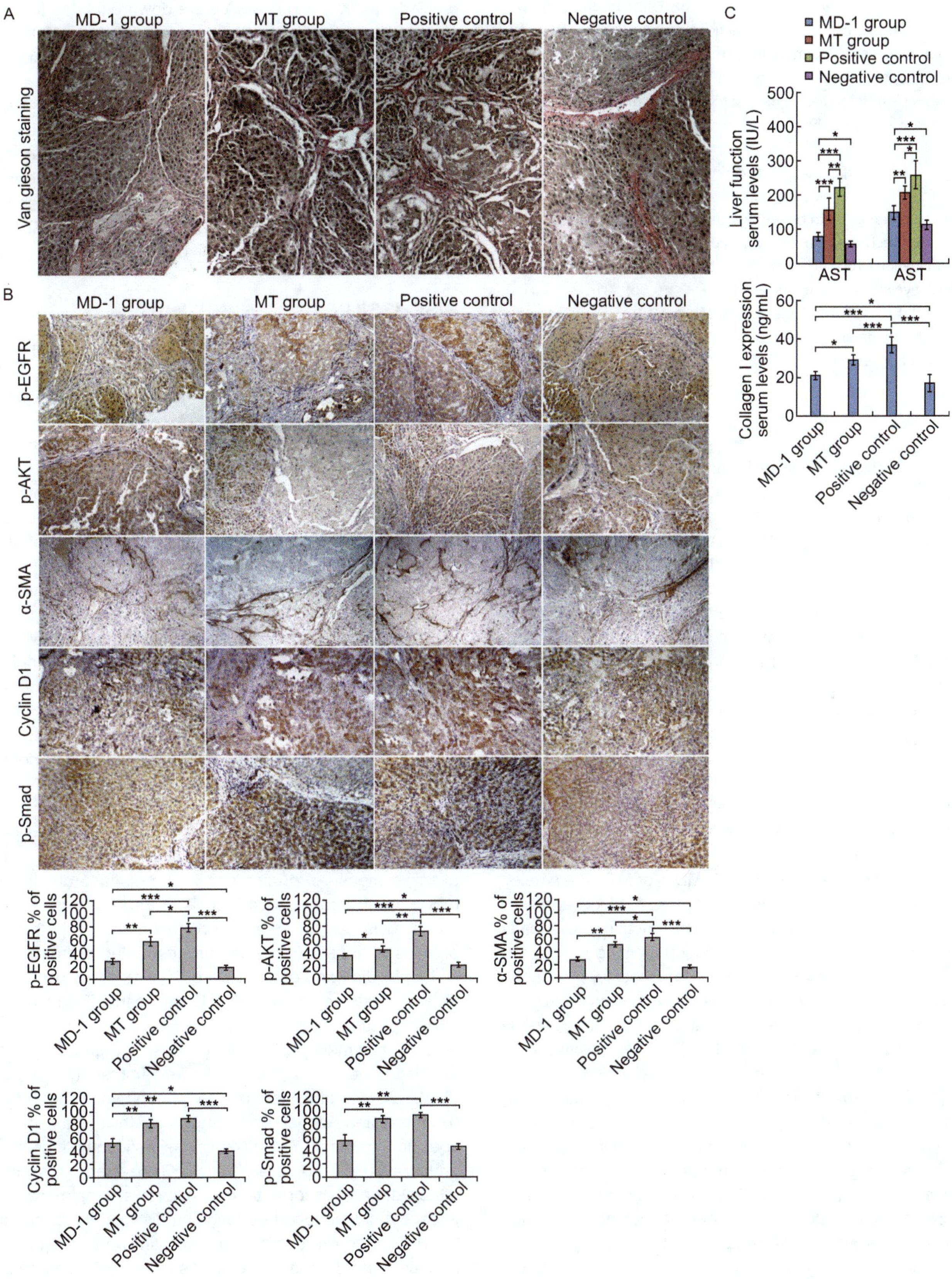
A
MD-1 group
MT group
Positive control
Negative control
Van gieson staining
B
MD-1 group
MT group
Positive control
Negative control
p-EGFR
p-AKT
α-SMA
Cyclin D1
p-Smad
p-EGFR % of positive cells
p-AKT % of positive cells
α-SMA % of positive cells
Cyclin D1 % of positive cells
p-Smad % of positive cells
C
MD-1 group
MT group
Positive control
Negative control
Liver function serum levels (IU/L)
AST
AST
Collagen I expression serum levels (ng/mL)

◄ **Figure 6. Inhibitory effects of MD-1 on DMN-induced hepatic fibrosis in rats.** (A) The sectioned rat liver tissues were stained with Van Gieson staining reagent to show the proliferated collagen fibers (stained with red) segmented and surrounded pseudolobules in the positive control group, and MD-1 treatment significantly reduced the collagen fibers. Original magnification: 200×. (B) Top: Expression of p-EGFR, p-AKT, α-SMA, cyclin D1, and p-Smad in sectioned rat liver was examined by immunohistochemistry. The percentages of positive cells (stained with brown) were counted within five fields of view for each section using a 20× objective lens. Original magnification: 100×. Bottom: Statistics for p-EGFR, p-AKT, α-SMA, cyclin D1, and p-Smad. *$P < 0.05$; **$P < 0.01$; ***$P < 0.001$. (C) Top: Detection of serum ALT and AST levels on fully automated biochemical analyzer. Bottom: Detection of collagen I by ELISA. *$P < 0.05$; **$P < 0.01$; *** $P < 0.001$.

Fig. 5A). The expression of intracellular type I and type III collagens was also downregulated ($P < 0.01$ and $P < 0.05$, respectively, Fig. 5A), and the expression of tissue inhibitor of metalloproteinase 1 (TIMP-1) was slightly decreased ($P = 0.20$, $n = 3$, Fig. 5A). ELISA showed that the levels of α-SMA, type I and III collagen, TIMP-1 in the MD-1-treated cells were significantly decreased (Fig. 5B). These results indicated that HSC activation could be significantly suppressed by MD-1, resulting in decreased ECM synthesis and secretion.

Effects of MD-1 on DMN-induced hepatic fibrosis in rats

A hepatic fibrosis model was established in rats by repeated intraperitoneal injection of DMN to induce hepatic fibrosis. During the initiation and progression of hepatic fibrosis, the rats were continuously treated with MD-1 or MT for 4 weeks. Two months after completing the treatment, blood and liver tissues were collected to examine liver function and hepatic fibrosis indicators. Van Gieson collagen staining of the liver tissue sections showed that DMN successfully induced hepatic fibrosis, proliferated collagen fibers segmented and surrounded pseudolobules, and hepatocytes had spotty necrosis and lymphocyte infiltration (Fig. 6A, positive control). Livers from MD-1-treated rats had significantly lower collagen fiber proliferation than controls, as well as narrow septae and incomplete segmentation of pseudolobules. However, the hepatocyte necrosis was not significant (Fig. 6A, MD-1 group). Based on pathological changes, the curative effect of MT was significantly less than that of MD-1 (Fig. 6A). Immunohistochemistry showed that the expression of p-EGFR, p-AKT, α-SMA, and cyclin D1 was significantly decreased in livers from the MD-1 group compared to the positive control group, but the expression was still higher than that of the normal control group (negative control group, Fig. 6B and 6C). Compared to the positive control group, p-EGFR, p-AKT, and α-SMA were downregulated in the MT group, whereas the reductions in cyclin D1 and p-Smad were not significant (Fig. 6B). ELISA showed that serum ALT, AST, and type I collagen levels were increased in the positive control group but significantly decreased in the MD-1 group while still being higher than the levels in the negative control group (Fig. 6C). In the MT group, ALT and AST were significantly decreased, but the reduction in type I collagen was not significant (Fig. 6C). Our results suggest a better inhibition of hepatic fibrosis in the MD-1 group than the MT group, both of which were more effective than the positive control.

DISCUSSION

Hepatic cirrhosis resulting from chronic liver disease is difficult to reverse and effective clinical treatments are lacking (Tsochatzis et al., 2014; Crossan et al., 2015). Because early-stage hepatic fibrosis can be reversed, active screening of new treatment methods for early interventions for hepatic fibrosis can effectively reduce the incidence of hepatic cirrhosis, or even hepatic carcinoma, improving the quality of life for many people. Among cells involved in hepatic fibrosis, HSCs (also known as Ito cells or fat-storing cells) in the perisinusoidal space play a critical role in the initiation and development of hepatic fibrosis (Page et al., 2015; Koo et al., 2016). HSCs are activated by a variety of cytokines to proliferate and transdifferentiate into MFs and then secrete a large amount of ECM, such as collagen fibers. The interaction among cytokines, cells, and ECM promotes the initiation and progression of hepatic fibrosis, finally resulting in hepatic cirrhosis and cancer (Iwaisako et al., 2014a; Iwaisako et al., 2014b; Wallace and Friedman, 2014; Tang et al., 2015). Because the causes of the majority of liver diseases, such as chronic hepatitis B and autoimmune hepatitis, are difficult to remove, targeting the pathogenic mechanism underlying hepatic fibrosis may be an effective measure to intervene in the disease development process and treat hepatic fibrosis. If the occurrence of hepatic fibrosis can be delayed or prevented, the pathological process of hepatic injury can be attenuated or treated.

Lamivudine and entecavir are antiviral drugs used for the clinical treatment of hepatitis B that can effectively inhibit HBV replication, with stronger binding and inhibitory effects on DNA polymerase. Therefore, these drugs can inhibit the development of hepatitis and hepatic fibrosis (Papachrysos et al., 2015). Inhibition of the phosphoinositide 3-kinase (PI3K)-related signal transduction pathway using PI3K-specific inhibitor LY294002 can significantly inhibit HSC proliferation (Wang et al., 2013). Gene therapy is also an effective measure for anti-hepatic fibrosis. Treatment of a rat thioacetamide-induced hepatic fibrosis model with a human MMP-1 adenovirus plasmid was shown to significantly attenuate the degree of hepatic tissue fibrosis, decrease the hydroxyproline level and the number of active HSCs, promote hepatocyte proliferation, and improve liver tissue structure (Iimuro et al., 2003). As a novel potential method, bone marrow mesenchymal stem cell (BMSC) transplantation has become a focus of attention (Jang et al., 2015). Studies in a murine hepatic fibrosis model showed that BMSCs injected into the tail vein can reach the liver and reduce the degree of hepatic fibrosis (Sakaida et al., 2004). However, the specific mechanism underlying the differentiation of BMSCs into hepatocytes and the reduction in hepatic fibrosis is still not clear, and an influence of the local microenvironment could not be fully excluded. Because intrahepatic inflammation is persistent in the majority of patients, BMSCs circulate to the liver, where they are activated by a variety of cytokines and induced to differentiate into MFs, which become a factor in the promotion of hepatic fibrosis. Therefore, more systemic in-depth studies on gene regulation and the microenvironment in BMSC differentiation are needed to facilitate the effective control of BMSC proliferation and accurately directed differentiation, allowing BMSCs to be effectively applied in clinical treatment.

MT is a quinolizidine compound with the chemical formula is $C_{15}H_{24}N_{20}$. Its molecular backbone is a heterodimeric quinolizidine ring. MT has anti-inflammatory, antiviral, antifibrotic, anti-arrhythmic, and immune-suppressive functions. The drug is mainly used for the treatment of viral hepatitis, hepatic fibrosis, and arrhythmia, but it also has certain efficacy in the prevention and treatment of tumors (Zhang et al., 2001; Zhang et al., 2011; Liu et al., 2014). MT can relieve pathological injury to liver parenchymal cells and non-parenchymal cells and has significant inhibitory effects on the proliferation of HSCs and fibroblasts. The extensive pharmacological activities of MT suggest that its mechanism of action is complex. A thio derivative of MT, MASM, can act directly on ribosomal protein S5 (RPS5) *in vitro* and in rats, reducing the phosphorylation levels of Ser473 and Thr308 in Akt and inhibiting HSC activation (Xu et al., 2014). However, the pharmacological activity of MT is not high, making it necessary to modify its structure to screen for derivatives that have high activities and low toxicity. The aforementioned MASM study confirmed that the pharmacological activity of a thio derivative of MT is greatly increased to that of MT and the toxicities equivalent (Hu et al., 2010; Xu et al., 2014). We performed modifications on the basis of thio derivatives of MT. The methylamino group at position 18 was acetylated to improve its stability, and an amino side chain was introduced to increase it activity. This produced the novel MT derivatives MD-1, MD-2, and MD-3, which had good stability and strong activity.

MD-1, MD-2, and MD-3 had large differences in their activities based on the different side chain groups. At 100 μmol/L, MD-1 reduced the survival rate of HSC-T6 cells to around 20%, MD-2 reduced the cell survival rate to 27.4%, and the inhibitory effect of MD-3 (47.2% cell survival rate) was equivalent to that of MT (62.9% cell survival rate, Fig. 1). These results suggest that MD-1 and MD-2 had inhibitory effects on HSC-T6 cells, whereas MD-3 did not significantly improve upon the inhibitory effect of MT on HSC-T6 cells. MD-1 significantly inhibited the proliferation and migration of HSC-T6 cells and induced G_0/G_1 arrest and apoptosis. Although the mechanism underlying hepatic fibrosis is complex, the multi-functional transmembrane glycoprotein EGFR specifically interacts with EGF and TGF-β1, causing its dimerization and regulating cell growth, proliferation, and differentiation (Voon et al., 2013). The EGFR-related signal transduction pathways are activated in HSCs in liver injury and chronic liver disease to promote the development and progression of hepatic fibrosis. Therefore, we focused on studying the effect of MD-1 on the EGFR-related signal transduction pathways. Immunofluorescence showed that the target molecule of MD-1 in HSC-T6 cells was EGFR. MD-1 interacted with EGFR on the surface of cell membranes, inhibiting EGFR phosphorylation. Inhibition of the phosphorylation of downstream protein kinases, such as Akt, affected the expression and activity of target proteins that regulate cell proliferation, migration, cell cycle, and apoptosis, such as cyclin D1 and p-Smad, finally changing the biological behaviors of cells. MD-1 reduced the synthesis and secretion of ECM components, such as type I collagen and type III collagen, in HSC-T6 cells, thereby exerting its anti-hepatic fibrosis activity. In the DMN-induced hepatic fibrosis model, MD-1 treatment delayed the development and progression of hepatic fibrosis, protected liver parenchymal cells, and improved liver function. Although the present study focused on the effect of MD-1 by inhibiting EGFR activation, other signaling pathways, such as the Ras/ERK pathway, may also be involved in hepatic fibrosis. Therefore, there are further studies needed to be carried out on the mechanisms of MT derivatives.

In summary, the present study reports a novel synthesized MT derivative, MD-1, that can significantly inhibit HSC activity, induce HSC apoptosis, and decrease the secretion of ECM components by HSCs. The drug has a protective effect on liver parenchymal cells in a rat DMN-induced hepatic fibrosis model. The possible mechanism by which MD-1 exerts its biological functions may be via EGFR binding on the cell surface, inhibiting its function and blocking the EGFR-related downstream signaling pathways. Thus, MD-1 is a potential clinical drug for anti-hepatic fibrosis.

MATERIALS AND METHODS

Cell culture

The rat HSC-T6 cell line was a gift from the Molecular Cancer Research Laboratory in the Eastern Hepatobiliary Surgery Hospital of Second Military Medical University (Xu et al., 2015). Cells were cultured in DMEM (GIBCO, New York, USA) containing 10% fetal bovine serum (FBS) in a 5% CO_2 atmosphere at 37°C. MT and its derivatives were synthesized by the School of Pharmacy, Second Military Medical University. The powder form of each compound (2 mg) was added to 200 μL DMSO until completely dissolved, then 1800 μL ddH_2O added to obtain a working solution of 1 mg/mL for future use.

Cell proliferation

HSC-T6 cells were cultured to the logarithmic phase and then inoculated onto 96-well plates (10^4 cells/well) for 24 h. Different gradient concentrations of MT and its derivatives MD-1, MD-2, and MD-3 were added. Each concentration group had eight replicate wells. After cells were cultured for another 24 h, cell proliferation was detected using the Cell Counting Kit-8 (CCK-8) reagent kit (Dojindo Molecular Technologies, Inc., Shanghai, China). Based on the IC_{50} values of the three drugs, MD-1, the one with the strongest activity, was selected for other experiments.

Cell motility

Transwells were placed in 24-well plates and HSC-T6 cells placed in the top chamber (2×10^5 cells/200 μL). The bottom chamber contained 500 μL culture medium containing 10% FBS. Cells in the top chambers were treated with MD1 or MT at the corresponding IC_{50} (62 μmol/L or 128 μmol/L, respectively). A control group without drug was used for comparison. After 48 h of culture, cells in the top layer of the Transwell were wiped and stained with 0.1% crystal violet for 15 min. Three fields were randomly selected under a light microscope (200× magnifications) for cell counting and photography. Experiments were performed in three biologically independent replicates.

Detection of cell cycle and cell apoptosis

HSC-T6 cells were inoculated onto a 6-well plate at 5×10^5 cells/well. MD-1 or MT was added at 62 μmol/L. After 48 h, the cells were collected and washed twice with pre-cooled PBS. Some cells were fixed in pre-cooled 75% ethanol in a 4°C refrigerator overnight, washed with PBS twice, and stained with propidium iodide (PI) containing RNase in the dark for 30 min. Cell cycle phase was detected using a flow cytometer (FACS420, BD Biosciences, San Jose, CA). The other cells were used for annexin V/PI staining. The apoptosis rate was also determined by flow cytometry.

Binding and inhibiting EGFR

The shRNA plasmid pGenensil-shEGFR, targeting the rat EGFR gene, and a negative control vector, pGenesil-shMC, were constructed by Wuhan Genesil Biotechnology Co., Ltd. (Wuhan, China). The 21-nt sequence of EGFR-shRNA targeted base pairs 2268–2290 (5′-gga tat taa agg aaa cag aat-3′) of the EGFR gene (GenBank: M37394.2). A mock control shRNA vector (shMC: 5′-gac ttc ata agg cgc atg cat-3′) was concomitantly constructed. HSC-T6 cells were inoculated on a 6-well plate at 5×10^5 cells/well. The shRNA plasmids were transfected into HSC-T6 cells using Lipofectamine 2000 (Invitrogen Corporation Shanghai Representative Office, Shanghai, China). After 48 h of continuous culture, cells were screened using G418 (400 μg/mL). The obtained cells were named HSC-T6-shEGFR and HSC-T6-shMC cells.

HSC-T6, HSC-T6-shEGFR, and HSC-T6-shMC cells were inoculated onto 6-well plates at 5×10^5 cells/well. Biotin-labeled MD-1 (MD-1-Biotin) was added at a concentration of 62 μmol/L. After culturing for another 48 h, the cells were collected. A part of cells were lysed in RIPA protein lysis buffer and the harvested proteins subjected to Western blot for detection of EGFR expression. A part of cells were smeared onto slides for immunofluorescence double labeling to localize MD-1-Biotin and EGFR. The working concentration of FITC-conjugated goat anti-rat EGFR and TRITC-conjugated anti-biotin antibodies was 1:1000 (Cell Signaling Technology, Inc., Beverly, MA). The rest of cells were prepared for the co-immunoprecipitation assay using the corresponding specific antibodies for MD-1-Biotin and EGFR.

Detection of EGFR signal transduction pathway-associated proteins

HSC-T6 cells were inoculated onto 6-well plates at 5×10^5 cells/well. MD-1 was added at a concentration of 62 μmol/L. After 48 h, the cells were collected and lysed in RIPA protein lysis buffer (Life Technologies Corporation, New York, USA). Total protein was extracted for Western blot detection of downstream proteins in the EGFR signal transduction pathway. The primary antibodies included EGFR, p-EGFR, AKT, p-AKT (Cell Signaling Technology, Danvers, MA), Survivin, cyclin D1 (Abcam Inc., Cambridge, MA), and p-Smad (Santa Cruz Biotech Inc., CA).

Detection of ECM proteins

HSC-T6 cells were inoculated onto 6-well plates at 5×10^5 cells/well. MD-1 was added at a concentration of 62 μmol/L. After 48 h, the cells and culture supernatant were collected. Cells were treated with TRIzol (Life Technologies Corporation, New York, USA) to extract total RNA for real-time quantitative RT-PCR (qRT-PCR) of ECM indicators using the PrimeScript RT Reagent Kit (TaKaRa Inc., Dalian, China). The primer sequences were: β-actin (537 bp), upstream 5′-ACC CAC ACT GTG CCC ATC TAT G-3′ and downstream 5′-AGA GTA CTT GCG CTC AGG A-3′; α-SMA (120 bp), upstream 5′-CCG AGA TCT CAC CGA CTA CC-3′ and downstream 5′-TCC AGA GCG ACA TAG CAC AG-3′; type III collagen (438 bp), upstream 5′-AGG CCA ATG GCA ATG TAA AG-3′ and downstream 5′-TAT TGG GTG AA A CAG CA-3′; TIMP-1 (250 bp), 5′-TCC CCA GAA ATC GAG AC-3′ and downstream 5′-TCA GAT TAT GCC AGG GAA CC-3′. The cell culture supernatant was used to detect ECM indicators by ELISA (Shanghai Westang Bio-Tech Co., Ltd, Shanghai, China).

Construction of the rat hepatic fibrosis model

A total of 20 male SD rats (Shanghai SLAC Laboratory Animal Co., Ltd, Chinese Academy of Sciences, Shanghai, China) at 4 weeks of age and a body weight of 80–100 g were randomly divided into four groups of five animals each: normal control group, model control group, MT group, and MD-1 group. Except for the normal control group, the rats were intraperitoneally injected with 1% DMN solution (10 μg/kg). Injections were performed three times each week (Monday, Wednesday, and Friday) for four consecutive weeks. The normal control group received an equal volume of normal saline. Starting from week 5, rats in the MT and MD-1 groups were treated with the drug of interest via intragastric administration (62 μmol/L/kg) three times each week (Monday, Wednesday, and Friday) for four consecutive weeks. The blank control group and model control group received of an equal volume of normal saline via intragastric administration. After the drug treatment was finished, animals were continuously fed for 2 months.

Rats were sacrificed after anesthesia. Blood and liver tissues were collected for evaluation of liver function and liver fibrosis indicators. The liver was fixed in 10% formalin for 6 h, embedded in paraffin, and sectioned. Staining was performed using the Van Gieson staining reagent kit (Maxim Biotechnology Development Co. Ltd, Fuzhou, China) according to the manufacturer's instructions. Fibrosis was observed under a microscope. p-EGFR, p-AKT, α-SMA, cyclin D1, and p-Smad were detected by immunohistochemistry. Serum ALT, AST, and type I collagen were detected by ELISA. All animal experiments were approved by the Animal Ethics Committee of Second Military Medical University (Shanghai, China).

Statistical analysis

Experimental data are presented as mean ± SD. One-way ANOVA was performed using SPSS (version 18.0). The least significant differences (LSD) test was performed if variances were homogeneous, and Dunnett's T3 method was used if variances were heterogeneous. A P value < 0.05 was considered significant.

ACKNOWLEDGEMENTS

This work was sponsored by the National Natural Science Foundation of China (Grant Nos. 81270518, 81500463, and 81470858).

ABBREVIATIONS

α-SMA, α-smooth muscle actin; BMSC, bone marrow mesenchymal stem cell; CTGF, connective tissue growth factor; DMN, dimethylnitrosamine; ECM, extracellular matrix; EGF, epidermal growth factor; EGFR, epidermal growth factor receptor; EMT, epithelial-mesenchymal transition; ET, endothelin; FBS, fetal bovine serum; FGF, fibroblast growth factor; HSCs, hepatic stellate cells; IC_{50}, half-maximal inhibitory concentration; MFs, myofibroblasts; MT, matrine; PDGF, platelet-derived growth factor; PI, propidium iodide; PI3K, phosphoinositide 3-kinase; TGF-β, transforming growth factor β.

COMPLIANCE WITH ETHICAL STANDARDS

Yi Feng, Hai-yan Ying, Ying Qu, Xiao-bo Cai, Ming-yi Xu, and Lun-gen Lu declare that they have no conflict of interest. All institutional and national guidelines for the care and use of laboratory animals were followed.

AUTHOR CONTRIBUTION

Y.F., M.Y.X., and L.G.L. designed the research; Y.Q. helped with cellular experiments; H.Y.Y. and X.B.C. helped with animal experiments; Y.F. and L.G.L. wrote the paper; L.G.L. supervised the project.

REFERENCES

Crossan C, Tsochatzis EA, Longworth L, Gurusamy K, Davidson B, Rodriguez-Peralvarez M, Mantzoukis K, O'Brien J, Thalassinos E, Papastergiou V et al (2015) Cost-effectiveness of non-invasive methods for assessment and monitoring of liver fibrosis and cirrhosis in patients with chronic liver disease: systematic review and economic evaluation. Health Technol Assess 19:1–409, v–vi

Cuevas MJ, Tieppo J, Marroni NP, Tunon MJ, Gonzalez-Gallego J (2011) Suppression of amphiregulin/epidermal growth factor receptor signals contributes to the protective effects of quercetin in cirrhotic rats. J Nutr 141:1299–1305

Fagone P, Mangano K, Mammana S, Pesce A, Pesce A, Caltabiano R, Giorlandino A, Portale TR, Cavalli E, Lombardo GA et al (2015) Identification of novel targets for the diagnosis and treatment of liver fibrosis. Int J Mol Med 36:747–752

Fausther M, Lavoie EG, Dranoff JA (2013) Contribution of myofibroblasts of different origins to liver fibrosis. Curr Pathobiol Rep 1:225–230

Handy JA, Fu PP, Kumar P, Mells JE, Sharma S, Saxena NK, Anania FA (2011) Adiponectin inhibits leptin signalling via multiple mechanisms to exert protective effects against hepatic fibrosis. Biochem J 440:385–395

Hu H, Wang S, Zhang C, Wang L, Ding L, Zhang J, Wu Q (2010) Synthesis and in vitro inhibitory activity of matrine derivatives towards pro-inflammatory cytokines. Bioorg Med Chem Lett 20:7537–7539

Iimuro Y, Nishio T, Morimoto T, Nitta T, Stefanovic B, Choi SK, Brenner DA, Yamaoka Y (2003) Delivery of matrix metalloproteinase-1 attenuates established liver fibrosis in the rat. Gastroenterology 124:445–458

Iwaisako K, Jiang C, Zhang M, Cong M, Moore-Morris TJ, Park TJ, Liu X, Xu J, Wang P, Paik YH et al (2014a) Origin of myofibroblasts in the fibrotic liver in mice. Proc Natl Acad Sci USA 111:E3297–E3305

Iwaisako K, Taura K, Koyama Y, Takemoto K, Asagiri M (2014b) Strategies to detect hepatic myofibroblasts in liver cirrhosis of different etiologies. Curr Pathobiol Rep 2:209–215

Jang YO, Jun BG, Baik SK, Kim MY, Kwon SO (2015) Inhibition of hepatic stellate cells by bone marrow-derived mesenchymal stem cells in hepatic fibrosis. Clin Mol Hepatol 21:141–149

Koo JH, Lee HJ, Kim W, Kim SG (2016) Endoplasmic reticulum stress in hepatic stellate cells promotes liver fibrosis via PERK-mediated degradation of HNRNPA1 and up-regulation of SMAD2. Gastroenterology 150(181–193):e188

Liu Y, Liu H, Meyer C, Li J, Nadalin S, Konigsrainer A, Weng H, Dooley S, ten Dijke P (2013) Transforming growth factor-beta (TGF-beta)-mediated connective tissue growth factor (CTGF) expression in hepatic stellate cells requires Stat3 signaling activation. J Biol Chem 288:30708–30719

Liu Y, Xu Y, Ji W, Li X, Sun B, Gao Q, Su C (2014) Anti-tumor activities of matrine and oxymatrine: literature review. Tumour Biol 35:5111–5119

Page A, Paoli PP, Hill SJ, Howarth R, Wu R, Kweon SM, French J, White S, Tsukamoto H, Mann DA et al (2015) Alcohol directly stimulates epigenetic modifications in hepatic stellate cells. J Hepatol 62:388–397

Papachrysos N, Hytiroglou P, Papalavrentios L, Sinakos E, Kouvelis I, Akriviadis E (2015) Antiviral therapy leads to histological improvement of HBeAg-negative chronic hepatitis B patients. Ann Gastroenterol 28:374–378

Pinzani M (2015) Pathophysiology of liver fibrosis. Dig Dis 33:492–497

Sakaida I, Terai S, Yamamoto N, Aoyama K, Ishikawa T, Nishina H, Okita K (2004) Transplantation of bone marrow cells reduces CCl_4-induced liver fibrosis in mice. Hepatology 40:1304–1311

Seki E, Brenner DA (2015) Recent advancement of molecular mechanisms of liver fibrosis. J Hepatobiliary Pancreat Sci 22:512–518

Shimada H, Staten NR, Rajagopalan LE (2011) TGF-beta1 mediated activation of Rho kinase induces TGF-beta2 and endothelin-1 expression in human hepatic stellate cells. J Hepatol 54:521–528

Tang WP, Akahoshi T, Piao JS, Narahara S, Murata M, Kawano T, Hamano N, Ikeda T, Hashizume M (2015) Basic fibroblast growth factor-treated adipose tissue-derived mesenchymal stem cell infusion to ameliorate liver cirrhosis via paracrine hepatocyte growth factor. J Gastroenterol Hepatol 30:1065–1074

Tsochatzis EA, Bosch J, Burroughs AK (2014) Liver cirrhosis. Lancet 383:1749–1761

Voon DC, Wang H, Koo JK, Chai JH, Hor YT, Tan TZ, Chu YS, Mori S, Ito Y (2013) EMT-induced stemness and tumorigenicity are fueled by the EGFR/Ras pathway. PLoS One 8:e70427

Wallace MC, Friedman SL (2014) Hepatic fibrosis and the microenvironment: fertile soil for hepatocellular carcinoma development. Gene Expr 16:77–84

Wang FP, Li L, Li J, Wang JY, Wang LY, Jiang W (2013) High mobility group box-1 promotes the proliferation and migration of hepatic stellate cells via TLR4-dependent signal pathways of PI3K/Akt and JNK. PLoS One 8:e64373

Xu WH, Hu HG, Tian Y, Wang SZ, Li J, Li JZ, Deng X, Qian H, Qiu L, Hu ZL et al (2014) Bioactive compound reveals a novel function for ribosomal protein S5 in hepatic stellate cell activation and hepatic fibrosis. Hepatology 60:648–660

Xu Y, Peng Z, Ji W, Li X, Lin X, Qian L, Li X, Chai X, Wu Q, Gao Q et al (2015) A novel matrine derivative WM130 inhibits activation of hepatic stellate cells and attenuates dimethylnitrosamine-induced liver fibrosis in rats. Biomed Res Int 2015:203978

Zhang B, Liu ZY, Li YY, Luo Y, Liu ML, Dong HY, Wang YX, Liu Y, Zhao PT, Jin FG et al (2011) Antiinflammatory effects of matrine in LPS-induced acute lung injury in mice. Eur J Pharm Sci 44:573–579

Zhang JP, Zhang M, Zhou JP, Liu FT, Zhou B, Xie WF, Guo C (2001) Antifibrotic effects of matrine on in vitro and in vivo models of liver fibrosis in rats. Acta Pharmacol Sin 22:183–186

3

Apaf1 inhibition promotes cell recovery from apoptosis

Anna Gortat[1], Mónica Sancho[1], Laura Mondragón[1], Àngel Messeguer[2], Enrique Pérez-Payá[1,3], Mar Orzáez[1]✉

[1] Laboratory of Peptide and Protein Chemistry, Centro de Investigación Príncipe Felipe, 46012 Valencia, Spain
[2] Department of Chemical and Biomolecular Nanotechnology, Instituto Química Avanzada de Cataluña (CSIC), 08034 Barcelona, Spain
[3] Instituto de Biomedicina de Valencia, IBV-CSIC, 46010 Valencia, Spain
✉ Correspondence: morzaez@cipf.es (M. Orzáez)

ABSTRACT

The protein apoptotic protease activating factor 1 (Apaf1) is the central component of the apoptosome, a multiprotein complex that activates procaspase-9 after cytochrome *c* release from the mitochondria in the intrinsic pathway of apoptosis. We have developed a vital method that allows fluorescence-activated cell sorting of cells at different stages of the apoptotic pathway and demonstrated that upon pharmacological inhibition of Apaf1, cells recover from doxorubicin- or hypoxia-induced early apoptosis to normal healthy cell. Inhibiting Apaf1 not only prevents procaspase-9 activation but delays massive mitochondrial damage allowing cell recovery.

Anna Gortat and Mónica Sancho have contributed equally to this work.

KEYWORDS Apaf1, Apaf1 inhibitors, apoptosis, apoptosome, autophagy, cell recovery

INTRODUCTION

Apoptosis is executed by highly regulated cellular pathways and is crucial for the removal of cells with damaged DNA or organelles, viral or bacterial infections as well as superfluous or ectopic cells. Two major signaling pathways have been described for apoptosis induction. The extrinsic pathway is initiated through external stress signals that induce the binding of plasma membrane proteins of the death receptor family (Peter and Krammer, 2003). The intrinsic pathway can be initiated by a number of factors including growth factor withdrawal and genotoxic stress which, with the participation of members of the Bcl-2 family of proteins, lead to mitochondrial outer membrane permeabilization (MOMP) (Brenner and Mak, 2009). This is followed by cytochrome *c* (Cyt *c*) release from mitochondria and formation of the cytosolic multiprotein complex termed the apoptosome. Assembly of the apoptosome triggers signaling cascades that activate the caspase family of cysteine aspartyl proteases. These caspases are essential to the initiation and execution of the apoptotic process. Apoptosis dysregulation is at the root of a variety of diseases. Apoptosis resistance has been causally linked to cancer and autoimmune diseases, and conversely, excessive apoptosis promotes pathological conditions related to stroke, ischemia-reperfusion damage and degenerative diseases (Green and Kroemer, 2005). While the development of new anti-cancer therapies relies on inducing apoptosis (Debatin, 2006), progress has been much slower in the discovery of inhibitors of unwanted apoptosis. Initial drug discovery efforts were directed towards the identification of executioner caspase-3 inhibitors (Linton, 2005). The caspase-3 inhibitors showed promising potential in several animal models of alcoholic liver diseases, hepatitis B and C virus infection (Kunstle et al., 1997), myocardial infarction and infarct size in stroke (Cheng et al., 1998; Yaoita et al., 2000), sepsis (Hotchkiss and Karl, 2003), neuronal death and in spinal cord injury (Springer et al., 1999). However, caspase inhibitor-based drugs are experimenting slow pharmacological advance due to the reported side effects especially in terms of hepatotoxicity (Schotte et al., 1999; Van Noorden, 2001; Hoglen et al., 2004). In addition, in the development of apoptosis inhibitors it should be considered, the necessity of maintaining cell functionality in order to provide long term survival rather than simply inhibiting death. Some lines of evidence showed that merely executioner caspase inhibition would not reach this objective. In a model of Parkinson's disease it was reported successful inhibition of neuronal cells death upon treatment with Z-Val-Ala-Asp (OMe)-fluoromethylketone (zVAD-fmk), an irreversible caspase inhibitor; however, the loss of neurite or reduction of dopamine uptake was not improved with this treatment (von Coelln et al., 2001).

When reviewing the apoptosis signaling pathway there are several points susceptible of intervention to develop unwanted cell death inhibitors. In this sense, the formation of the apoptosome offered evidences to be considered as an interesting target as it is downstream from the mitochondrial events that characterize the pathway but upstream of effector caspases (e.g., caspase-3/7) that are the executors of cell death. The main constituent of the apoptosome is Apaf1 a multidomain protein that activates upon binding to Cyt *c* released from mitochondria after MOMP (Srinivasula et al., 1998). Mochizuki et al. (2001) reported that an adeno-associate virus vector-based delivery of an Apaf1 dominant negative inhibitor was able to inhibit apoptosis-mediated degeneration of nigrostriatal neurons in a MPTP model of Parkinson's disease. Gao et al. (2009), by means of the overexpression of the specific Apaf1 inhibitory protein (AIP), have reported that inhibition of Apaf1 in an animal model of neonatal hypoxic-ischemic brain injury resulted in an attenuated brain tissue loss. In Apaf1 deficient cells, Ferraro et al. have demonstrated that such cells can turn-on readjustments of metabolic pathways to survive apoptotic stimulus while the depolarized state of mitochondria is reverted (Ferraro et al., 2008). Small molecules that inhibit Apaf1 are another promising approach for developing unwanted apoptosis inhibitors. We have reported on a family of small molecules that inhibits apoptosis by interfering with the apoptosome activity (Malet et al., 2006; Mondragon et al., 2008; Mondragon et al., 2009; Santamaria et al., 2009; Orzaez et al., 2014; Sancho et al., 2014b). In particular, SVT016426 was as efficient as the caspase inhibitor zVAD-fmk inhibiting the intrinsic apoptotic pathway. Here we show that the apoptosis inhibition provided by the Apaf1 inhibitor SVT016426 at the level of apoptosome contributes to maintain functional cells, thus raising hope for the development of future treatments of unwanted pathological apoptosis. Understanding the physiology of cell death has allowed the development of mechanistic approaches for the development of apoptosis-related drugs. However to properly face death prevention and most importantly cell recovery from early apoptosis stages, we have to understand not only how cells die but also how cells recover. We report here on a method to distinguish and to classify living cells at different stages of apoptosis. The possibility of isolating cells at an early apoptotic phase allowed us to identify autophagy as the molecular mechanism that facilitates SVT016426-dependent cell recovery.

RESULTS

Apaf1 inhibition provides survival to cells induced to execute apoptosis

Direct damage to cells causes individual cell death that depending on the number of cell loss can result on tissue or organ failure; e.g. cardiac damage that occurs late after chemotherapy (months or even a year or more) is one of the major side effects of doxorubicin (Doxo) treatment, a drug that is one of the most widely used anticancer drugs for solid tumors (Takemura and Fujiwara, 2007). In other cases, as stroke or tissue infarction, a hypoperfused, hypoxic, metastable region, named the penumbra, is formed around the core of necrotic cell death. The penumbra region retains structural integrity but has a compromised functionality and its long term recovery defines the basis for stroke and/or tissue infarction therapy (Yuan, 2009). We asked whether Apaf1 inhibition by SVT016426 could have application in hypoxia and Doxo-induced cell death. Chemical inhibitors of Apaf1, as SVT016426, inhibit the apoptosome-dependent induction phase in different cells induced to execute apoptosis (Malet et al., 2006; Mondragon et al., 2008; Mondragon et al., 2009; Orzaez et al., 2014). Then, we initially analyzed the ability of SVT016426 to inhibit apoptosome activity in HeLa cell extracts. Incubation of the cytosolic S100 cell extract with dATP and Cyt *c* restored the apoptotic pathway through induction of the apoptosome formation (Fearnhead, 2001); this restoration was followed using a fluorogenic substrate for caspases (Ac-DEVD-afc). SVT016426 treatment inhibited Apaf1-induced activation of caspase activity (Fig. 1A). We also analyzed target-specificity of SVT016426 in a model of Doxo-induced apoptosis in HeLa cells. For this purpose, we considered the use of small interfering RNA (siRNA)-based silencing of Apaf1 (Fig. 1B) and analyzed the activity of SVT016426 in Doxo-induced cell death in the presence or absence of Apaf1 in the cells. When HeLa cells transfected with a control random siRNA were treated with Doxo we obtained close to 60% of Doxo-induced cell death. However in the presence of SVT016426 death decreased to a 40% of the cell population (Fig. 1C). In contrast, Doxo-induced cell death was not inhibited by SVT016426 in Apaf1 siRNA-based knockdown cells (Fig. 1C). It should be mentioned here that in the absence of Apaf1, Doxo induced a caspase-independent cell death in these cells as it was described previously (Miyazaki et al., 2001; Andreu-Fernandez et al., 2013; Sancho et al., 2014a). These cell viability results were well correlated with caspase-9 processing (Fig. 1B) and measurements of caspase-3 activity (Fig. 1D) suggesting that SVT016426 inhibitory capacity was dependent on the levels of Apaf1 in the cell. These observations imply that SVT016426-mediated inhibition of Apaf1 results in pathway responses and cellular phenotypic effects compatible with an Apaf1-selective inhibition of apoptosis. Then SVT016426 not only inhibited caspase activity but also inhibited cell death. The SVT016426-induced cell death

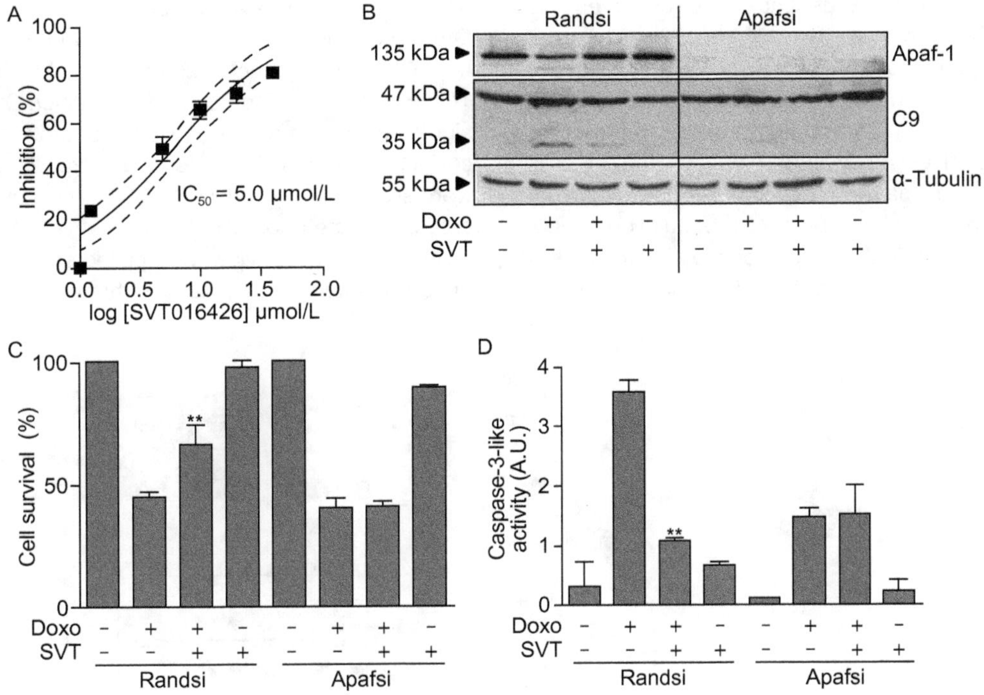

Figure 1. Chemical inhibition of Apaf1 provides cell survival. (A) The IC_{50} of SVT016426 was analyzed using HeLa S100 cell extract-based assay. Caspase-3-like activity on HeLa S100 extracts was measured in the presence or absence of increasing concentrations of SVT016426. Data are presented as mean inhibition percentage of control ± SD, $n = 3$. Curve fitting was performed using GraphPad Prism 3.0 software. (B) Apaf1 was silenced in HeLa cells by siRNA (100 nmol/L) methodology. Immunoblotting for Apaf1, caspase-9 and α-tubulin (loading control) was provided in cell extracts silenced using a random siRNA (Randsi) or Apaf1 siRNA (Apafsi). Cell survival (C) and caspase-3-like activity (D) were measured for control (Randsi) or Apaf1 (Apafsi) silencing in the presence or absence of Doxo (1 μmol/L) and SVT016426 treatment (SVT; 5 μmol/L) (mean ± SD, $n = 3$). Cell viability was measured by both, trypan blue and MTT assay. Asterisks represent significant differences relative to apoptotic treatment as determined by one-way ANOVA test with Bonferroni's multiple comparison post-test (**$P < 0.05$).

inhibition was close to 20% of the total cell population. This death inhibition ratio, although still modest, could be of relevance finding ways to describe future apoptosis inhibition-based therapies to devastating diseases as neurodegenerative diseases and to decrease secondary effects of cancer chemotherapy.

Apaf1 inhibition permits cell recovery after hypoxia or DNA-damage

We were interested in analyzing not only cell death inhibition but also the possibility of SVT016426-dependent cell recovery. As SVT016426 acts downstream mitochondria we reasoned that only cells in the early phases of apoptosis would be able to recover.

To test this hypothesis we designed a cellular assay to separate populations in different apoptotic stages. We enriched the early apoptotic cell population by applying soft conditions of apoptosis induction. Then we developed a method based in HeLa cells that overexpressed Cyt *c*-GFP cells (HeLa-GFP) that allowed the analysis of cellular Cyt *c*-GFP distribution in living cells. However, we found that due to overexpression, in homeostatic HeLa-GFP cells a fraction of Cyt *c*-GFP was located in the cytoplasm. Then, HeLa-GFP cells were subjected to fluorescence-activated cell sorting (named thereof sorting, st) to select a specific and homogeneous HeLa-GFPst cell population with moderate Cyt *c*-GFP expression levels and absence of Cyt *c*-GFP in cytosol to allow further quantitative procedures. This HeLa-GFPst population allowed setting up a method to analyze Cyt *c* release in living cells subjected to long hypoxia conditions and to sub-lethal Doxo treatment. Cyt *c* released from mitochondria upon an apoptotic stimuli, is rapidly degraded by the proteasome (MacFarlane et al., 2002; Rehm et al., 2003; Ferraro et al., 2008), then Cyt *c*-GFP released from HeLa-GFPst mitochondria will produce a change in fluorescence signal intensity allowing cytofluorimetric measurements without necessity for cell permeabilization. This vital method would allow the cell sorting procedures required for the analysis of cells at different stages of the apoptotic pathway (see below). We initially set-up the method with Doxo-induced apoptotic HeLa-GFPst cells submitted to cytofluorimetric measurements using the 520 nm channel to detect GFP fluorescence. We observed the appearance of a GFP low-intensity peak (Fig. 2A and 2B; peak 2) upon apoptosis induction which was very significantly reduced by lactacystin-induced proteasome inhibition, thus validating the set-up (Fig. 2A and 2B; peak 4) (Rehm et al., 2003). To quantitatively analyze the progression of cells through the cell death pathway according to the cytofluorimetric results, three independent cell populations were defined according to their GFP fluorescence signal intensity and size (Fig. 2C) upon apoptotic induction. The cell population with an average size characteristic of the HeLa-GFPst cell line (as assessed by forward scatter - FS+) and high GFP fluorescence was considered as healthy (H; GFP +/FS+). Cells showing FS+ and diminished GFP fluorescence signal were considered as early apoptotic (EA; GFP-/FS+) and the cell population of reduced size (as assessed by forward scatter - FS-) and low GFP fluorescent signal was considered as late apoptotic (LA; GFP-/FS-). According to such cell population definition, we applied sorting to obtain differentially purified cell populations that were analyzed for GFP

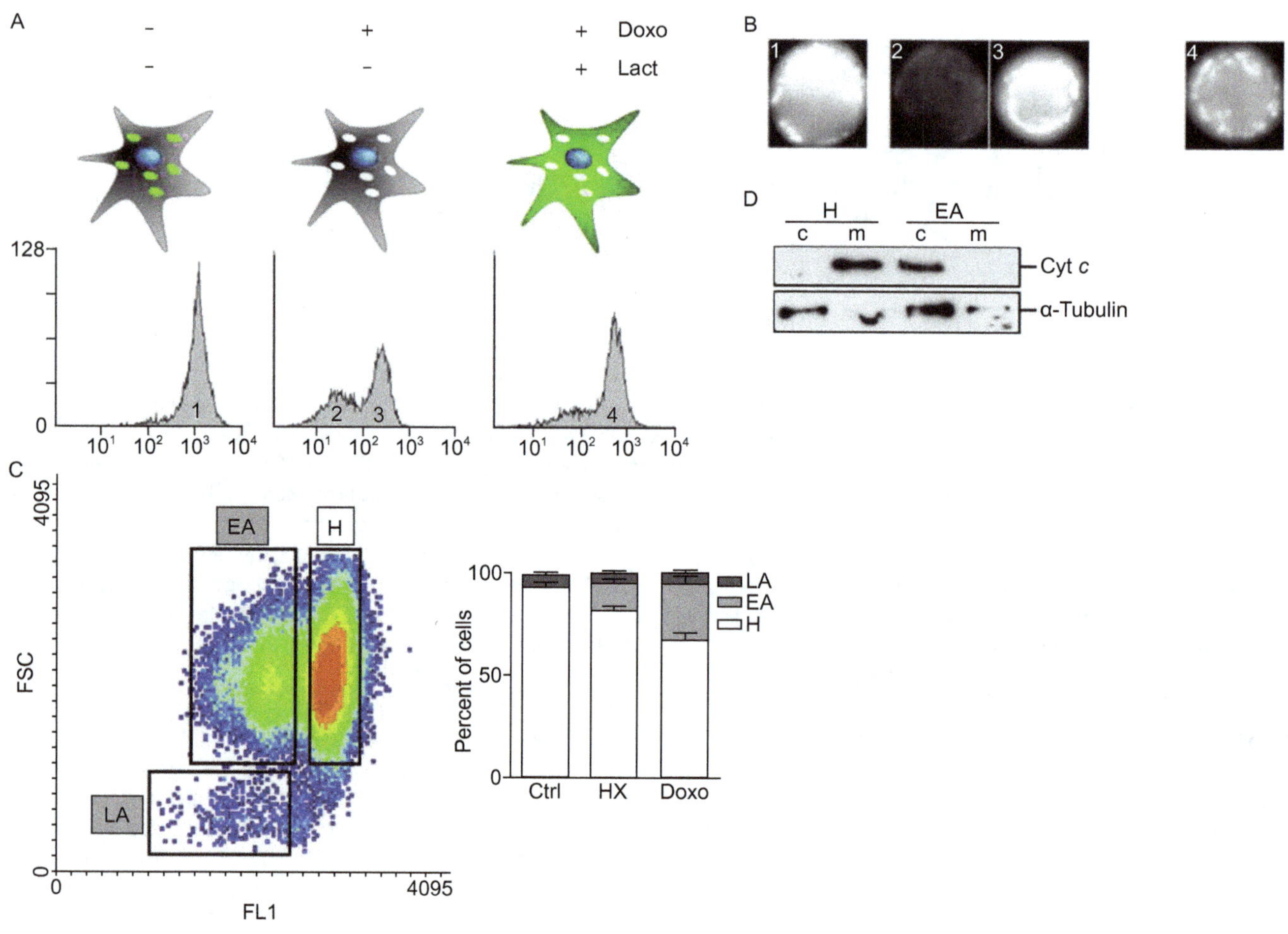

Figure 2. An experimental model to separate cellular populations at different stages of apoptotic cell death based on Cyt *c* localization. (A) Determination of Cyt *c* release in HeLa-GFPst by flow cytometry. HeLa-GFPst cells were induced to apoptosis with 1 μmol/L Doxo for 6 h in the presence or absence of 10 μmol/L of the proteasome inhibitor lactacystin. Samples were subjected to cytofluorometric analysis to detect Cyt *c*-GFP. Two peaks of different intensity can be distinguished. The high intensity peak corresponds to healthy cells (Cyt *c* in the mitochondria). After an apoptotic insult a second peak of low intensity appears (Cyt *c* released and degraded by the proteasome in the cytosol). Inhibition of the proteasome under an apoptotic insult recovers the high intensity peak probing that the degradation of Cyt *c* in the cytoplasm is responsible of the low intensity peak. (B) HeLa-GFP cells treated as described in (A) were submitted to cell sorting. Representative fluorescence microscopy images of the different cellular populations corresponding to peaks in (A) are shown. (C) HeLa-GFPst cells were induced to apoptosis either submitting cells to HX for 4 days or adding 0.25 μmol/L Doxo for 24 h. Three different populations corresponding to healthy (H), early apoptotic (EA) and late apoptotic (LA) could be identified according to the degree of Cyt *c* release and cellular size. Quantification of the different populations observed in the absence (Ctrl) or in the presence of the apoptotic insults (HX; Doxo) are shown in the right panel. (D) Endogenous Cyt *c* localized to mitochondrial (m) and cytosolic (c) fractions in H and EA populations, respectively.

signal under fluorescence microscope (Fig. 2B) and for endogenous Cyt *c* subcellular localization upon cell fractionation (Fig. 2D) to corroborate that our measurements of Cyt *c*-GFP mirrored the general behavior of endogenous Cyt *c*. As expected, H cells appeared intensely stained with GFP bearing a clear hallmark of mitochondrial network in both, untreated population (Fig. 2B, peak 1) as well as in cells isolated from the GFP+/FS+ peak upon Doxo treatment (Fig. 2B, peak 3). In contrast, cells pooled from the GFP- peak that appeared upon Doxo treatment were significantly less intense (Fig. 2B, peak 2) whereas GFP-associated fluorescence intensity of Doxo and lactacystin-treated cells was similar to that of the healthy untreated cells (Fig. 2B, peak 4). Importantly, endogenous Cyt *c* was confined to mitochondria in H cell population while it was found at cytosolic fractions in EA cell population (Fig. 2D). These results suggest that mitochondrial released Cyt *c* is susceptible of a proteasome-dependent degradation that can be partially inhibited by proteasome inhibitors. To further validate our experimental system of purification of cell populations at different stages in the apoptotic pathway, we performed measurements of caspase-3/7 activity. As expected, no caspase-3/7 activity was detected in H or EA cells (results not shown).

Next, we addressed the question if apoptosis induced by different apoptotic insults can be not only prevented by SVT016426-dependent Apaf1 inhibition but also whether or not cells can be recovered from such insults. Experimental cellular models of hypoxia and Doxo-induced apoptosis were then evaluated. HeLa-GFPst cells were incubated in hypoxia/hypercapnea (HX) conditions (1% O_2, 18% CO_2) conditions for 4 days or treated with 0.25 μmol/L Doxo for 24 h and submitted to sorting in order to obtain clean pools of H, EA and LA cells. The cross-contamination among the different pools was thoroughly controlled and was kept as low as less than 1% (0.6%–0.7% in most cases, data not shown). The three different cell pools were then plated on dishes and maintained in HX conditions for 6 additional days or treated further with 0.5 μmol/L Doxo for 24 h in the presence or absence of 5 μmol/L SVT016426 (scheme in Fig. 3A and 3B upper panels). The cytofluorimetric measurements revealed that only 30% of the initial H cells in prolonged HX conditions and 50% of the Doxo-treated H cells remained without entering the cell death pathway. However, in the presence of the Apaf1 inhibitor SVT016426 the number of unaffected H cells increased to 49% ± 2% and to 56% ± 5% for the HX and Doxo treated cells, respectively. This increase was correlated with a 1.6-fold reduction of the number of LA cells (from 38% ± 4% to 23% ± 4%) in the case of the HX population, thus confirming the protective effect that chemical Apaf1 inhibitors exert on cells previous to the apoptotic insult. The EA pool subjected to the HX conditions by the 6 additional days or exposed to Doxo progressed through the cell death pathway as assessed by the presence of 48% ± 7% or 7% ± 1% of LA cells, respectively (Fig. 3A and 3B lower panels; Fig. S1B). In HX conditions, treatment with SVT016426 decreased the number of the LA cells by a 3.3-fold to 15% ± 2% concomitant with a markedly increase in the EA population (37% ± 4% to 58% ± 4% in control and SVT016426 treated, respectively). Moreover, we observed that the presence of H cells from a pool of pure EA cells was substantially increased in those EA cells treated with SVT016426 (from 15% ± 2% to 27% ± 1% on the EA cell population under HX and from 23% ± 5% to 50% ± 6% in the EA cell population treated with Doxo) (Fig. 3A and 3B lower panel white bars; Fig. S1A). These results suggest that cells which were plated as a pure EA population triggered an SVT016426-dependent recovery program. Then, inhibition of Apaf1 not only prevents cells entering the apoptotic program but helps cells to recover from early apoptotic stages. SVT016426 treated cells are able to initiate a normal cell division program as previously demonstrated in colony formation assays (Orzaez et al., 2014). As expected, the LA population did not undergo any changes (data not shown).

Autophagy is implicated in SVT016426 cell recovery

Autophagy is a catabolic cytoprotective mechanism that constitutes a cellular defense against stress induced by starvation (Madeo et al., 2010). It is characterized by the formation of double-membrane vesicles named autophagosomes that engulf cytoplasmic organelles and proteins. The autophagosomes fuse with lysosomes and their luminal content is degraded. Beclin-1 is a protein characterized as essential for the initiation of autophagy. In fact, it has been demonstrated that is the activation of a Beclin-1 dependent autophagy pathway that permits glycolytic-dependent ATP generation allowing recovery from death-stimulus of apoptosome-deficient proneural cells (Ferraro et al., 2008). We were interested in analyzing the role of autophagy in the SVT016426-dependent cell recovery from an early apoptotic to a healthy state. Thus, we analyzed the protein levels of Beclin-1, p62 and the lipidated form of the microtubule-associated protein 1A/1B-light chain 3 alpha type I (LC3-I) (Kabeya et al., 2000; Kabeya et al., 2004) as hallmarks of induced autophagy. Indeed, we observed an overexpression of Beclin-1, a lipidation of the LC3-I (LC3-II) and, a moderate decrease of p62 substrate, in SVT016426-treated when compared to unprotected HeLa cells under HX conditions (Fig. 4A), thus suggesting an increased rate of autophagy (Boya et al., 2005; Pattingre et al., 2005). The presence of autophagosome structures was also observed by transmission electron microscopy (TEM, Fig. 4B and 4C). Similar observations were derived from SVT016426-treated cells when Doxo was used as apoptotic insult (Fig. 5A and 5B). The SVT016426-dependent autophagy induction correlated with an increment in the ATP levels in SVT016426-treated HeLa cells under HX conditions (Fig. 4D) or treated with Doxo (Fig. 5C) when compared with cells not receiving the SVT016426 treatment. Finally, to confirm that cell recovery was mediated by the activation of the autophagy program we silenced Beclin-1 in the Doxo model and cell survival was analyzed. As expected, recovery by SVT016426 is not

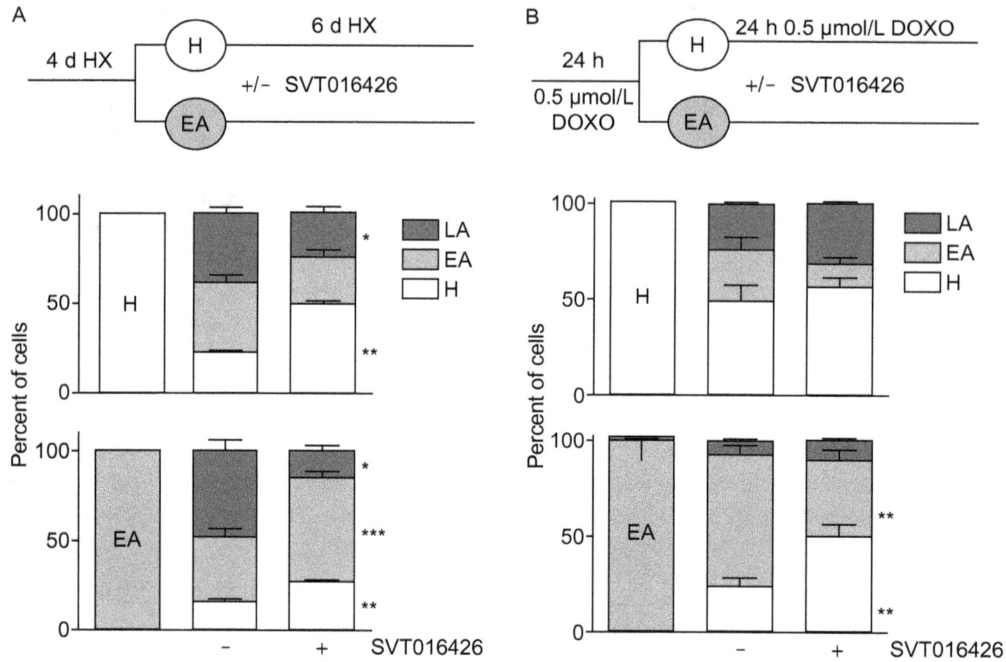

Figure 3. Apaf1 chemical inhibition allows cell recovery after the onset of apoptosis. (A) HeLa-GFPst cells were sorted after 4 days of HX into H, EA and LA (not shown) populations. H and EA cells were reseeded and resubmitted to HX for 6 additional days in the presence or absence of 5 μmol/L SVT016426. The evolution of the different populations at the end of the experiment was quantified by flow cytometry. (B) HeLa-GFPst cells were induced to apoptosis by adding 0.25 μmol/L Doxo. As in (A), cells were sorted into H, EA and LA (not shown) populations. H and EA cells were reseeded and treated with 0.5 μmol/L Doxo for an additional period of 24 h in the presence or absence of 5 μmol/L SVT016426. The evolution of the different populations at the end of the experiment was quantified by flow cytometry. Data represent mean ± SD of three independent experiments.

observed when Beclin-1 is knocked-down thus reinforcing the hypothesis that autophagy could be implicated in cell recovery (Fig. 5D).

DISCUSSION

Here we have described an Apaf1-dependent cell program that permits cells to recover from intrinsic apoptotic insults. The use of selective chemical Apaf1 inhibitors (SVT016426 is a representative member) as chemical biology probes highlighted Apaf1 as a principal regulator not only in cell death but also in cell recovery pathways. As shown here, chemical inhibition of Apaf1 is necessary to allow intrinsic apoptosis-injured cells to have access to survival events that implies the participation of autophagy. The pro-survival adaptive function of autophagy acquires then relevance in this context. Our results suggest that cells at early stages of execution of the apoptotic program, even after Cyt *c* loss, can be recovered to cells with hallmarks of normal (healthy) cells. The cells readapt metabolic conditions and activate autophagy to sustain ATP production. It has been previously shown that Apaf1 genetically deprived proneural cells of embryonic origin can recover from apoptosis when glycolysis is active and the apoptotic stimulus is removed (Ferraro et al., 2008).

Our study combines the use of homogeneous well defined cell populations at different stages of apoptosis with small molecules that allowed the characterization of specific recovery programs. In cells of mammalian origin it has been long observed that the responsiveness to physiological stimuli is not uniform. In particular when a clonal cell population is induced to commit apoptosis some cells die while others survive. Among cells that survive one can probably find cells at different stages of apoptosis as it has been reported when extrinsic apoptosis was induced (Albeck et al., 2008). In this study we have validated a flow cytometry-based method to distinguish and to split the different cell populations originated when an original cell population was subjected to intrinsic apoptosis stimuli. The efficient characterization of three (healthy, early- and late-apoptosis) populations has permitted to unambiguously determine that the Apaf1 inhibitor not only prevents cell death but helps cells to recover from apoptosis. Overall, SVT016426 provides a powerful tool for characterizing the role of Apaf1. Such inhibitors were originally identified while searching for compounds that could prevent the activation

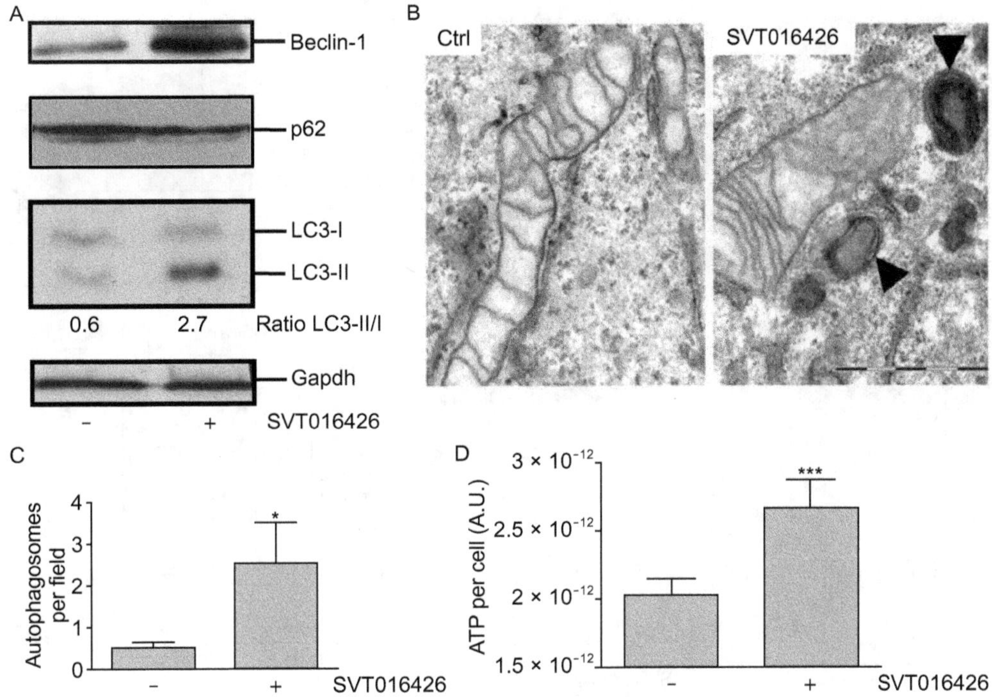

Figure 4. Autophagy is implicated in cell recovery upon Apaf1 inhibition. (A) HeLa cells were treated as in Fig. 3A and expression changes of proteins implicated in autophagy, namely Beclin-1, p62 and LC3-I and -II (the lipidated form of LC3-I), were evaluated by Western blot. Gapdh was used as loading control. (B) Cellular morphology of HX-treated HeLa cells was analyzed by transmission electron microscopy (TEM). Representative images of HX-control and HX-SVT016426 treated cells are shown (Black arrows indicate the presence of autophagosomes). Scale bar equal to 1 µm. (C) Quantification of the number of autophagosomes per field was performed analyzing 40 randomly selected fields per image from two independent experiments. (D) The ATP content per cell in the presence or absence of 5 µmol/L SVT016426 was measured. Results shown are the mean of three independent experiments. Statistical significance was assessed by means of two-tailed Student's *t*-test (*$P < 0.05$; ***$P < 0.001$).

of procaspase-9 when analyzed in an *in vitro* reconstituted apoptosome (Malet et al., 2006). It is now well established that the Apaf1 inhibitors provide cytoprotective effect to cells subjected to apoptosis stimulus in different cellular models (Malet et al., 2006; Mondragon et al., 2008; Santamaria et al., 2009; Orzaez et al., 2014; Sancho et al., 2014b). Furthermore, the Apaf1 inhibitor has been used to confirm the effect of caspase-3 inhibition on GluR1 synaptic distribution in a mouse model of Alzheimer's disease (D'Amelio et al., 2011).

The discovery of candidate drugs that attenuate the effects of unwanted excessive apoptosis will benefit from the knowledge of how after an apoptotic stimulus, cells are able to recover and the molecular mechanisms that use cells to this end. Recent data (Ferraro et al., 2008) suggest that such cellular recovery should use still undiscovered pathways that link intrinsic apoptosis and metabolism networks. Albeck et al. (Albeck et al., 2008) using single-cell analysis approaches demonstrated that upon defined metabolic conditions, cells induced to extrinsic apoptosis can recover. In contrast, the links between the intrinsic apoptosis pathway and cellular pathways that could permit cell recovery still present more questions than answers. Nevertheless current knowledge suggests that the apoptotic machinery should interact functionally with key decision-making components of cellular metabolism (Colell et al., 2007; Ferraro et al., 2008) cell growth and cell division. These components would provide crosstalks between the intricate cellular pathways that define a particular cellular fate.

Early studies pointed to inhibitors of caspases as the most attractive molecules to develop anti-apoptotic drugs; however, chemical inhibition or knocking-out caspases is only able to prevent apoptosis but not cell death (Boya et al., 2005). In these conditions cells die by different processes. Using our system to split cells at different stages of the apoptotic program, we demonstrated that inhibition of Apaf1 enhanced apparition of healthy cells within the early apoptotic population. Similar behavior has been reported in *C. elegans*. Cells that undergo morphological changes accompanying CED-3 activation could recover completely when subjected to weak pro-apoptotic signals if they were not engulfed by phagocytic cells (Hoeppner et al., 2001; Reddien et al., 2001).

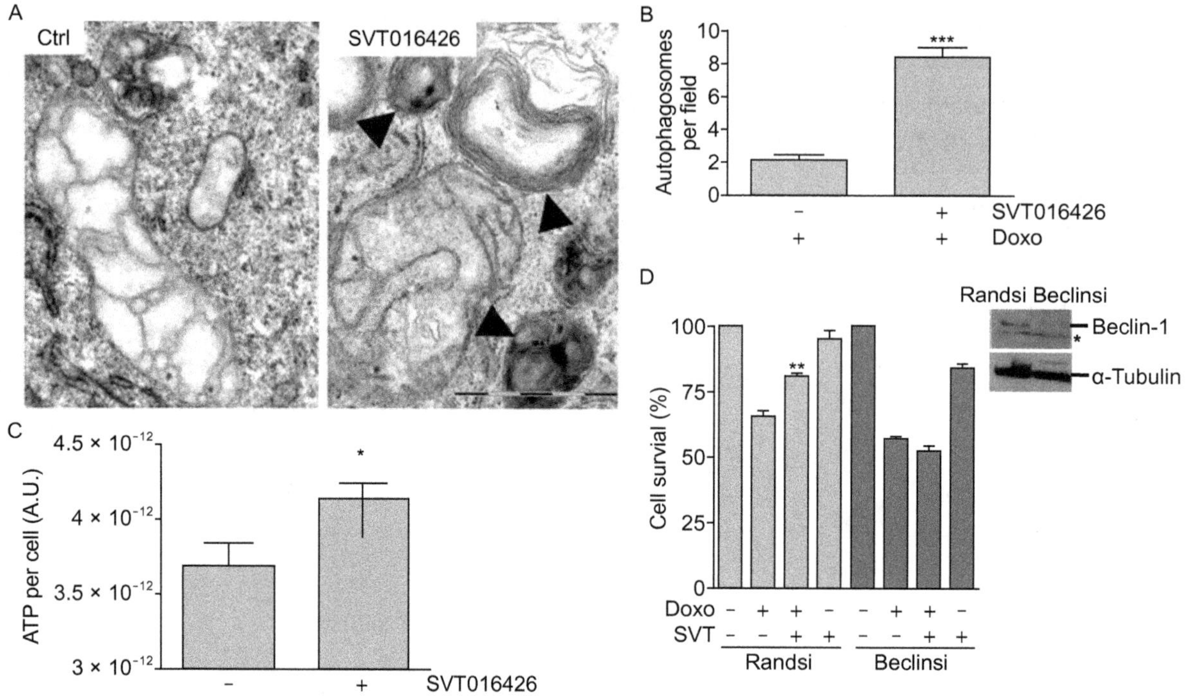

Figure 5. Autophagy is implicated in cell recovery upon Apaf1 inhibition. (A) Representative TEM images of HeLa cells treated with 0.25 µmol/L Doxo for 24 h and maintained for additional 24 h with 0.5 µmol/L Doxo in the presence or absence of 5 µmol/L SVT016426. Black arrows indicate the presence of autophagosomes. Scale bar equal to 1 µm. (B) Quantification of the number of autophagosomes per field was performed analyzing 40 randomly selected fields per image from two independent experiments. (C) Measurement of ATP content per cell in HeLa cells treated as described in (A) in the presence or absence of 5 µmol/L SVT016426. Statistical significance was assessed by means of two-tailed Student's *t*-test (*$P < 0.05$; ***$P < 0.001$). (D) Cell survival was measured by MTT for control (Randsi) or Beclin (Beclinsi) silencing in the presence or absence of Doxo (1 µmol/L) and SVT016426 treatment (SVT; 5 µmol/L) (mean ± SD, $n = 3$). Asterisks represent significant differences relative to apoptotic treatment as determined by one-way ANOVA test with Bonferroni's multiple comparison post-test (**$P < 0.05$). Immunoblotting for Beclin-1 and α-tubulin (loading control) was provided in cell extracts silenced using a random siRNA (Randsi) or Beclin siRNA (Beclinsi). Asterisk marks for unspecific bands recognized by the antibody.

A cell full recovery mechanism must be a complex process that has to overcome irreversible caspase-dependent cleavage of substrates through restraining caspase-3 catalytic activity. This function is played by inhibitor of apoptosis proteins (IAPs) and proteasome-dependent caspase degradation (Albeck et al., 2008). Our data identify Apaf1 as a molecule that should be included in light of its role, when chemically inhibited, as a molecule participating in the process.

Several evidences have demonstrated that differentiated cells can recover after MOMP (Martinou et al., 1999; Deshmukh et al., 2000; Tang et al., 2009) and our study identifies Apaf1 and autophagy induction as part of the molecular mechanism of the recovery process, in particular in cells affected by hypoxic or sublethal doses of intrinsic apoptosis inducers. However, which is the mechanism and the components of the machinery providing cell recovery and whether and how they regulate apoptosis remain to be elucidated.

MATERIALS AND METHODS

Cell culture, reagents and antibodies

HeLa and HeLa-GFP cells were maintained in Dulbecco's Modified Eagle's Medium (DMEM) supplemented with 10% foetal bovine serum (FBS). Media and supplements for cell culture were purchased from Gibco-Invitrogen. Doxorrubicin (Doxo) was obtained from Sigma-Aldrich. Anti LC3 and anti caspase-9 antibodies were purchased from Cell Signaling. Anti Beclin-1 and anti Apaf1 antibodies were purchased from BD Transduction Laboratories, p62 antibody from Enzo Life Sciences and anti α-tubulin from Sigma-Aldrich.

Cell-free caspase activation assay

S100 cytosolic extract from HeLa was obtained as previously described (Dignam et al., 1983; Malet et al., 2006). Extract (0.1 mg/mL) was incubated with different concentrations of SVT016426 (0, 1, 5, 10 and 40 µmol/L) in assay buffer (20 mmol/L HEPES, 10 mmol/L

KCl, 1.5 mmol/L $MgCl_2$, 1 mmol/L EDTA, 1 mmol/L EGTA, 1 mmol/L DTT, 0.1 mmol/L PMSF) plus dATP (10 mmol/L) and Cyt *c* (10 μmol/L) for 40 min at 30°C. Ac-DEVD-afc substrate (20 μmol/L; Enzo Life Sciences) was used to measure caspase-3-like activity using a Victor 2 spectrofluorimeter.

Cell-based caspase activation assay

HeLa cells were seeded in 6-well plate at a cellular density of 1 × 10^5 cells/mL. Then, Lipofectamine™ 2000 (Invitrogen) was used according to the manufacturer's instructions to transfect cells with a control random siRNA (Randsi; 100 nmol/L) and Apaf1 siRNA (Apafsi; 100 nmol/L) from Cell Signaling. 24 h later, cells were pre-treated with SVT016426 (5 μmol/L) for 1 h, then Doxo (1 μmol/L) was added. After 24 h cells were harvested and cytosolic extracts were obtained as described previously. Total protein (50 μg) was mixed with assay buffer containing 20 μmol/L Ac-DEVD-afc substrate. Activity was measured using a Victor 2 spectrofluorimeter.

Trypan blue cell viability assay

HeLa cells seeded and treated as described before, were detached and 0.5% trypan blue dye added in solution. Live cells possess intact cell membranes that exclude the dye, whereas dead cells do not. Unstained (viable) and stained (non-viable) cells were counted separately in a hemacytometer and total number of viable cells in the population was calculated.

MTT assay

Mitochondrial activity was measured using a MTT colorimetric assay. HeLa cells were grown in 24-well plates at a cell density of 1.0 × 10^5 cells/well. After 24 h, cells were transfected with a control random siRNA (Randsi; 100 nmol/L) and Beclin-1 siRNA (Beclinsi; 100 nmol/L) from Cell Signaling. 24 h later, cells were treated with SVT016426 (5 μmol/L) and Doxo (1 μmol/L) for 30 h. Four hours before the end of the treatment 20 μL/well of MTT (5 mg/mL in PBS) was added to each well and the plates were incubated for a further 4 h at 37°C. Finally, the medium was removed and the precipitated formazan crystals were dissolved in optical grade DMSO. The plates were read at 570 nm on a Victor 2 spectrofluorimeter.

Hypoxia treatment and cell sorting

HeLa-GFP cells were seeded at 7.5 × 10^5 cells/mL on a 150 mm sterile dish, allowed to set for 24 h. Cultures were submitted to HX for 4 days in Thermo Electron Corp. incubator. Populations were sorted in a Beckman Coulter MOFLO High Speed Cell Sorter, according to the size and GFP signal intensity as explained in the main text. After sorting, cells were seeded in 6-well plates at a cell density of 1.5 × 10^5 cells/mL, treated or not with 5 μmol/L SVT016426 and placed for additional 6 days in HX. Cell population analysis was performed using a Beckman Coulter cytofluorometer.

Doxorubicin treatment and cell sorting

HeLa-GFP cells were seeded at 7.5 × 10^5 cells/mL on a 150 mm sterile dish, allowed to set for 24 h and treated with 0.25 μmol/L Doxo for additional 24 h. Populations were sorted as described above and treated for 24 h with 0.5 μmol/L Doxo. Cell population analysis was performed in a Beckman Coulter cytofluorometer.

Transmission electron microscopy

HeLa cells were seeded at 3 × 10^4 cells per chamber in a Lab-Tek chamber slides of 4 wells (Nalge Nunc International) and treated as indicated in each case. Then, the cells were fixed for 1 h in 3.5% glutaraldehyde at 37°C and postfixed for 1 h in 2% OsO_4 at room temperature. Cellular staining was performed at 4°C for 2 h in 2% uranyl acetate in the dark. Finally, cells were rinsed in sodium phosphate buffer (0.1 mol/L, pH 7.2), dehydrated in ethanol, and infiltrated overnight in Araldite (Durcupan, Fluka). Following polymerization, embedded cultures were detached from the chamber slide and glued to Araldite blocks. Serial semi-thin (1.5 μm) sections were cut with an Ultracut UC-6 (Leica), mounted onto slides and stained with 1% toluidine blue. Selected semi-thin sections were glued (Super Glue, Loctite) to araldite blocks and detached from the glass slide by repeated freezing (in liquid nitrogen) and thawing. Ultrathin (0.07 μm) sections were prepared with the Ultracut and stained with lead citrate. Finally, photomicrographs were obtained under a transmission electron microscope (FEI Tecnai Spirit G2) using a digital camera (Morada, Soft Imaging System, Olympus).

ATP measurements

HeLa cells were seeded at a density of 5 × 10^4 cells/mL and treated as indicated. ATP measurement was performed in duplicate employing ATPLite kit (Perkin Elmer) according to manufacturer's instructions. The ATP content was normalized per number of cells.

Statistical analysis

Data were analysed using GraphPad software and statistical significance was assessed by means of two-tailed Student's *t*-test ($*P < 0.05$). All data were expressed as the mean ± SD.

ACKNOWLEDGMENTS

We are grateful to Dr. C. Muñoz-Pinedo (IDIBELL, Barcelona) for supplying the HeLa-GFP cell line. We thank E. Sirvent and R. Montava for technical assistance and Dr A. Martínez Romero, D. Gil Casanova and Mario Soriano members of the CIPF flow cytometry core and electronic microscopy facility, respectively, for their assistance. This paper is dedicated to the memory of Professor Enrique Pérez-Payá. This work was supported in part by Laboratorios Salvat, SA, by grants from the Spanish Ministry of Science and Innovation (MICINN-BIO2007-60066, -SAF2010 15512, -SAF2008-00048 and CSD2008-00005C), and by Generalitat Valenciana (GV) Prometeo 2010/005 (funded in part with ERDF) to E.P.-P.

ABBREVIATIONS

AIP, Apaf1 inhibitory protein; Apaf-1, apoptotic protease-activating factor 1; Cyt *c*, cytochrome c; Cyt *c*-GFP, fluorescent GFP-labelled Cyt *c*; dATP, deoxyadenosine triphosphate; Doxo, doxorubicin; HeLa-GFP, human cervix adenocarcinoma (HeLa) cells stably expressing Cyt c-GFP; LC3-I, microtubule-associated protein 1A/1B-light chain 3 alpha type I; LC3-II, microtubule-associated protein 1A/1B-light chain 3 alpha type II; MOMP, mitochondrial outer membrane permeabilization; TEM, transmission electron microscopy; zVAD-fmk, Z-Val-Ala-Asp(OMe)-fluoromethylketone.

COMPLIANCE WITH ETHICS GUIDELINES

Anna Gortat, Mónica Sancho, Laura Mondragón, Àngel Messeguer, Enrique Pérez-Payá and Mar Orzáez declare that they have no conflict of interest. This article does not contain any studies with human or animal subjects performed by the any of the authors.

REFERENCES

Albeck JG, Burke JM, Aldridge BB, Zhang M, Lauffenburger DA, Sorger PK (2008) Quantitative analysis of pathways controlling extrinsic apoptosis in single cells. Mol Cell 30:11–25

Andreu-Fernandez V, Genoves A, Messeguer A, Orzaez M, Sancho M, Perez-Paya E (2013) BH3-mimetics- and cisplatin-induced cell death proceeds through different pathways depending on the availability of death-related cellular components. PLoS One 8: e56881

Boya P, Gonzalez-Polo RA, Casares N, Perfettini JL, Dessen P, Larochette N, Metivier D, Meley D, Souquere S, Yoshimori T et al (2005) Inhibition of macroautophagy triggers apoptosis. Mol Cell Biol 25:1025–1040

Brenner D, Mak TW (2009) Mitochondrial cell death effectors. Curr Opin Cell Biol 21:871–877

Cheng Y, Deshmukh M, D'Costa A, Demaro JA, Gidday JM, Shah A, Sun Y, Jacquin MF, Johnson EM, Holtzman DM (1998) Caspase inhibitor affords neuroprotection with delayed administration in a rat model of neonatal hypoxic-ischemic brain injury. J Clin Invest 101:1992–1999

Colell A, Ricci JE, Tait S, Milasta S, Maurer U, Bouchier-Hayes L, Fitzgerald P, Guio-Carrion A, Waterhouse NJ, Li CW et al (2007) GAPDH and autophagy preserve survival after apoptotic cytochrome c release in the absence of caspase activation. Cell 129:983–997

D'Amelio M, Cavallucci V, Middei S, Marchetti C, Pacioni S, Ferri A, Diamantini A, De Zio D, Carrara P, Battistini L et al (2011) Caspase-3 triggers early synaptic dysfunction in a mouse model of Alzheimer's disease. Nat Neurosci 14:69–76

Debatin KMFS (2006) Apoptosis and cancer therapy. WILEY-VCH, Weinheim

Deshmukh M, Kuida K, Johnson EM Jr (2000) Caspase inhibition extends the commitment to neuronal death beyond cytochrome c release to the point of mitochondrial depolarization. J Cell Biol 150:131–143

Dignam JD, Lebovitz RM, Roeder RG (1983) Accurate transcription initiation by RNA polymerase II in a soluble extract from isolated mammalian nuclei. Nucleic Acids Res 11:1475–1489

Fearnhead HO (2001) Cell-free systems to study apoptosis. Methods Cell Biol 66:167–185

Ferraro E, Pulicati A, Cencioni MT, Cozzolino M, Navoni F, di Martino S, Nardacci R, Carri MT, Cecconi F (2008) Apoptosome-deficient cells lose cytochrome c through proteasomal degradation but survive by autophagy-dependent glycolysis. Mol Biol Cell 19:3576–3588

Gao Y, Liang W, Hu X, Zhang W, Stetler RA, Vosler P, Cao G, Chen J (2009) Neuroprotection against hypoxic-ischemic brain injury by inhibiting the apoptotic protease activating factor-1 pathway. Stroke 41:166–172

Green DR, Kroemer G (2005) Pharmacological manipulation of cell death: clinical applications in sight? J Clin Invest 115:2610–2617

Hoeppner DJ, Hengartner MO, Schnabel R (2001) Engulfment genes cooperate with ced-3 to promote cell death in *Caenorhabditis elegans*. Nature 412:202–206

Hoglen NC, Chen LS, Fisher CD, Hirakawa BP, Groessl T, Contreras PC (2004) Characterization of IDN-6556 (3-[2-(2-tert-butyl-phenylaminooxalyl)-amino]-propionylamino]-4-oxo-5-(2,3,5,6-tetrafluoro-phenoxy)-pentanoic acid): a liver-targeted caspase inhibitor. J Pharmacol Exp Ther 309:634–640

Hotchkiss RS, Karl IE (2003) The pathophysiology and treatment of sepsis. N Engl J Med 348:138–150

Kabeya Y, Mizushima N, Ueno T, Yamamoto A, Kirisako T, Noda T, Kominami E, Ohsumi Y, Yoshimori T (2000) LC3, a mammalian homologue of yeast Apg8p, is localized in autophagosome membranes after processing. Embo J 19:5720–5728

Kabeya Y, Mizushima N, Yamamoto A, Oshitani-Okamoto S, Ohsumi Y, Yoshimori T (2004) LC3, GABARAP and GATE16 localize to autophagosomal membrane depending on form-II formation. J Cell Sci 117:2805–2812

Kunstle G, Leist M, Uhlig S, Revesz L, Feifel R, MacKenzie A, Wendel A (1997) ICE-protease inhibitors block murine liver injury and apoptosis caused by CD95 or by TNF-alpha. Immunol Lett 55:5–10

Linton SD (2005) Caspase inhibitors: a pharmaceutical industry perspective. Curr Top Med Chem 5:1697–1717

MacFarlane M, Merrison W, Bratton SB, Cohen GM (2002) Proteasome-mediated degradation of Smac during apoptosis: XIAP promotes Smac ubiquitination in vitro. J Biol Chem 277:36611–36616

Madeo F, Tavernarakis N, Kroemer G (2010) Can autophagy promote longevity? Nat Cell Biol 12:842–846

Malet G, Martin AG, Orzaez M, Vicent MJ, Masip I, Sanclimens G, Ferrer-Montiel A, Mingarro I, Messeguer A, Fearnhead HO et al (2006) Small molecule inhibitors of Apaf-1-related caspase- 3/-9 activation that control mitochondrial-dependent apoptosis. Cell Death Differ 13:1523–1532

Martinou I, Desagher S, Eskes R, Antonsson B, Andre E, Fakan S, Martinou JC (1999) The release of cytochrome c from mitochondria during apoptosis of NGF-deprived sympathetic neurons is a reversible event. J Cell Biol 144:883–889

Miyazaki K, Yoshida H, Sasaki M, Hara H, Kimura G, Mak TW, Nomoto K (2001) Caspase-independent cell death and mitochondrial disruptions observed in the Apaf1-deficient cells. J Biochem 129:963–969

Mochizuki H, Hayakawa H, Migita M, Shibata M, Tanaka R, Suzuki A, Shimo-Nakanishi Y, Urabe T, Yamada M, Tamayose K et al (2001) An AAV-derived Apaf-1 dominant negative inhibitor prevents MPTP toxicity as antiapoptotic gene therapy for Parkinson's disease. Proc Natl Acad Sci USA 98:10918–10923 **Epub 12001 Sep 10914**

Mondragon L, Orzaez M, Sanclimens G, Moure A, Arminan A, Sepulveda P, Messeguer A, Vicent MJ, Perez-Paya E (2008) Modulation of cellular apoptosis with apoptotic protease-activating factor 1 (apaf-1) inhibitors. J Med Chem 51:521–529

Mondragon L, Galluzzi L, Mouhamad S, Orzaez M, Vicencio JM, Vitale I, Moure A, Messeguer A, Perez-Paya E, Kroemer G (2009) A chemical inhibitor of Apaf-1 exerts mitochondrioprotective functions and interferes with the intra-S-phase DNA damage checkpoint. Apoptosis 14:182–190

Orzaez M, Sancho M, Marchan S, Mondragon L, Montava R, Valero JG, Landeta O, Basanez G, Carbajo RJ, Pineda-Lucena A et al (2014) Apaf-1 inhibitors protect from unwanted cell death in in vivo models of kidney ischemia and chemotherapy induced ototoxicity. PLoS One 9:e110979

Pattingre S, Tassa A, Qu X, Garuti R, Liang XH, Mizushima N, Packer M, Schneider MD, Levine B (2005) Bcl-2 antiapoptotic proteins inhibit Beclin 1-dependent autophagy. Cell 122:927–939

Peter ME, Krammer PH (2003) The CD95(APO-1/Fas) DISC and beyond. Cell Death Differ 10:26–35

Reddien PW, Cameron S, Horvitz HR (2001) Phagocytosis promotes programmed cell death in C. elegans. Nature 412:198–202

Rehm M, Dussmann H, Prehn JH (2003) Real-time single cell analysis of Smac/DIABLO release during apoptosis. J Cell Biol 162:1031–1043

Sancho M, Gortat A, Herrera AE, Andreu-Fernandez V, Ferraro E, Cecconi F, Orzaez M, Perez-Paya E (2014a) Altered mitochondria morphology and cell metabolism in Apaf1-deficient cells. PLoS One 9:e84666

Sancho M, Herrera AE, Orzaez M, Perez-Paya E (2014b) Inactivation of Apaf1 reduces the formation of mutant huntingtin-dependent aggregates and cell death. Neuroscience 262:83–91

Santamaria B, Benito-Martin A, Ucero AC, Aroeira LS, Reyero A, Vicent MJ, Orzaez M, Celdran A, Esteban J, Selgas R et al (2009) A nanoconjugate Apaf-1 inhibitor protects mesothelial cells from cytokine-induced injury. PLoS One 4:e6634

Schotte P, Declercq W, Van Huffel S, Vandenabeele P, Beyaert R (1999) Non-specific effects of methyl ketone peptide inhibitors of caspases. FEBS Lett 442:117–121

Springer JE, Azbill RD, Knapp PE (1999) Activation of the caspase-3 apoptotic cascade in traumatic spinal cord injury. Nat Med 5:943–946

Srinivasula SM, Ahmad M, Fernandes-Alnemri T, Alnemri ES (1998) Autoactivation of procaspase-9 by Apaf-1-mediated oligomerization. Mol Cell 1:949–957

Takemura G, Fujiwara H (2007) Doxorubicin-induced cardiomyopathy from the cardiotoxic mechanisms to management. Prog Cardiovasc Dis 49:330–352

Tang HL, Yuen KL, Tang HM, Fung MC (2009) Reversibility of apoptosis in cancer cells. Br J Cancer 100:118–122

Van Noorden CJ (2001) The history of Z-VAD-FMK, a tool for understanding the significance of caspase inhibition. Acta Histochem 103:241–251

von Coelln R, Kugler S, Bahr M, Weller M, Dichgans J, Schulz JB (2001) Rescue from death but not from functional impairment: caspase inhibition protects dopaminergic cells against 6-hydroxydopamine-induced apoptosis but not against the loss of their terminals. J Neurochem 77:263–273

Yaoita H, Ogawa K, Maehara K, Maruyama Y (2000) Apoptosis in relevant clinical situations: contribution of apoptosis in myocardial infarction. Cardiovasc Res 45:630–641

Yuan J (2009) Neuroprotective strategies targeting apoptotic and necrotic cell death for stroke. Apoptosis 14:469–477

4

Transitions between epithelial and mesenchymal states during cell fate conversions

Xiang Li, Duanqing Pei, Hui Zheng✉

CAS Key Laboratory of Regenerative Biology, Guangdong Provincial Key Laboratory of Stem Cell and Regenerative Medicine, Guangzhou Institutes of Biomedicine and Health, Chinese Academy of Sciences, Guangzhou 510530, China
✉ Correspondence: zheng_hui@gibh.ac.cn (H. Zheng)

ABSTRACT

Cell fate conversion is considered as the changing of one type of cells to another type including somatic cell reprogramming (de-differentiation), differentiation, and trans-differentiation. Epithelial and mesenchymal cells are two major types of cells and the transitions between these two cell states as epithelial-mesenchymal transition (EMT) and mesenchymal-epithelial transition (MET) have been observed during multiple cell fate conversions including embryonic development, tumor progression and somatic cell reprogramming. In addition, MET and sequential EMT-MET during the generation of induced pluripotent stem cells (iPSC) from fibroblasts have been reported recently. Such observation is consistent with multiple rounds of sequential EMT-MET during embryonic development which could be considered as a reversed process of reprogramming at least partially. Therefore in current review, we briefly discussed the potential roles played by EMT, MET, or even sequential EMT-MET during different kinds of cell fate conversions. We also provided some preliminary hypotheses on the mechanisms that connect cell state transitions and cell fate conversions based on results collected from cell cycle, epigenetic regulation, and stemness acquisition.

KEYWORDS EMT, MET, cell states, cell fate conversion, iPSC generation, trans-differentiation, differentiation

INTRODUCTION

As two major types of cells in most animals, epithelial cells are known for their basement membrane, apical-basal axis of polarity, gap junction, immobility, and so on, while the characteristics of mesenchymal cells are almost just opposite, loosely associated, no polarity, and high mobility. Although the two types of cells are so different from each other, the transitions between epithelial and mesenchymal states, epithelial-mesenchymal transition (EMT) and mesenchymal-epithelial transition (MET), have been observed clearly and studied extensively during a variety of biological processes including embryonic development, cancer progression, and somatic cell reprogramming.

The first observation of EMT can be dated back to as early as 1890 when some ductal epithelial cells were described to acquire mesenchymal characteristics in breast tumors progression as reviewed previously (Nieto, 2011). It is suggested that EMT enables the tumor cells to acquire mesenchymal characteristics like loose cell-cell interaction and migratory ability, which further lead to the detachment of cells from tumor mass and the invasion of cells to other tissues. Although the requirement and significance of EMT for cancer progression has been debated for a long time, the migratory and invasive properties induced by EMT are widely accepted to be critical for metastasis (Peinado et al., 2007; Polyak and Weinberg, 2009).

EMT has been observed in multiple biological processes, especially embryonic development. The generation of adult tissues and organs requires multiple rounds of sequential EMT-MET, which is used to refer an EMT followed with its reversed process, MET (Thiery et al., 2009). For example, the formation and migration of mesenchymal cells is essential for the transformation from single-layered blastula into three-layered gastrula in various metazoans (Nakaya and Sheng, 2008). In addition, the development of heart has been considered as a good example for sequential EMT-

MET. The formation of cardiac mesodermal cells during gastrulation and the organization of them into a two-layered epithelium later are considered as the first round of sequential EMT-MET. Another two rounds of sequential EMT-MET are observed during the folding around primitive foregut and the formation of four heart compartments (Nakajima et al., 2000; Thiery et al., 2009).

Since embryonic development can be considered as a combination of multiple differentiation processes from pluripotent or multipotent stem cells to somatic cells, and cancer progression is an acquisition of stem cells properties at least partially, it is reasonable to suggest the critical roles of EMT/MET or sequential EMT-MET during other kinds of cell fate conversions, which was summarized and discussed in the following sections of current review.

EMT/MET DURING DIFFERENT CELL FATE CONVERSIONS

In general, the cell fate conversions are considered as the changes of cells from one type to another, and can be frequently observed during embryonic development when most types of cells are generated. The success of somatic cells to regain pluripotency after nuclear transfer or exogenous expression of four pluripotency-related transcriptional factors makes de-differentiation or iPSC generation as one new type of cell fate conversion (Gurdon, 1962a, b; Takahashi et al., 2007; Takahashi and Yamanaka, 2006). The direct reprogramming from fibroblasts to functional neurons enables cell fate conversion to cover an even wider area including differentiation, de-differentiation, and trans-differentiation (Vierbuchen et al., 2010). Since the combination of de-differentiation and differentiation or trans-differentiation alone can use cells isolated from patient skin or urine to generate patient-specific functional cells (Muller et al., 2009; Wang et al., 2012; Zhou et al., 2012), which have the potential to treat a variety of diseases without facing ethics concerns and immunologic rejections (Grabel, 2012; Nishikawa et al., 2008), a lot of efforts have been put to study the mechanisms underlying cell fate conversions to optimize these technologies for clinic application.

EMT/MET during iPSC generation

In 2006, Dr. Yamanaka's laboratory successfully induced mouse embryonic fibroblasts (MEF) into embryonic stem cells (ESC)-like cells with Oct4, Klf4, c-Myc, and Sox2. The generated cells were named induced pluripotent stem cells (iPSC). iPSC are able to generate tumors containing a variety of tissues from all three germ layers cells after being transplanted into nude mice and viable, fertile live-born progeny by tetraploid complementation (Takahashi and Yamanaka, 2006; Zhao et al., 2009). Because of the potential application of iPSC in clinic, the method to generate iPSC has been improved a lot since 2006 by using somatic cells other than fibroblasts for reprogramming (Aasen et al., 2008; Kim et al., 2008), by generating iPSC in other species like rat and pig (Esteban et al., 2009; Liao et al., 2009; Liu et al., 2008), by identifying new combinations of transcriptional factors (Maekawa et al., 2011; Yu et al., 2007), by developing new strategy to deliver transcriptional factors (Hou et al., 2013; Warren et al., 2010; Yu et al., 2009; Zhou et al., 2009), and by optimizing the culture systems (Chen et al., 2011; Chen et al., 2010a; Esteban et al., 2010).

During the generation of iPSC, multiple changes have been observed with the MEF, including those in gene expression profiles (Brambrink et al., 2008), epigenetic state (Koche et al., 2011; Watanabe et al., 2013), cell morphology (Li et al., 2010; Samavarchi-Tehrani et al., 2010), and cellular metabolism (Folmes et al., 2011). Among these changes, the transition of MEF from mesenchymal state to epithelial state has been recognized as a required step during the early phase of reprogramming (Li et al., 2010; Samavarchi-Tehrani et al., 2010). Inhibiting early MET by inducing EMT with transformation growth factor (TGF)-β or Snail1 prevents iPSC generation. In addition, Sox2 and c-Myc inhibit the expression of Snail1, TGF-β1, and TGF-β receptor 2 to suppress the mesenchymal characteristics of MEF, while Klf4 induces the epithelial properties by up-regulating E-cadherin directly (Li et al., 2010). Promoted MET has been used to explain the beneficial roles played by miR-302, Glis2, and several small molecules during iPSC generation (Chen et al., 2010b; Ichida et al., 2009; Maekawa et al., 2011; Subramanyam et al., 2011). Furthermore, the interactions between Vitamin C which increases the efficiency and quality for iPSC generation (Chen et al., 2012; Esteban and Pei, 2012; Esteban et al., 2010) and Ten-eleven translocation methylcytosine dioxygenase 1 (Tet1) which can replace Oct4 during iPSC generation (Gao et al., 2013) are also related to MET (Chen et al., 2013).

Although MET is necessary for fibroblasts reprogramming, there are still several questions remaining unclear. Firstly, since the process from MEF to iPSC can be considered as a reversed process of embryonic development at least partially, the reversed process of sequential EMT-MET, which is also a sequential EMT-MET, might be observed during iPSC generation. Secondly, the different or even opposite regulatory roles of the four transcriptional factors, like on TGFβ1 and TGFβ receptor 2, make MET induction more complex (Li et al., 2010). These two questions were answered in one of our recent publications (Liu et al., 2013a). Considering the complex functions of the four factors in regulating pluripotency (Kashyap et al., 2009; Pan et al., 2002) and the proper expression level of Oct4 required to maintain pluripotency (Pan et al., 2006), we proposed that the four factors have different functions during iPSC generation and may even have counteractions under certain circumstances or at certain time points. In order to diminish or decrease the possible counteractions, we introduced the four factors into MEF at different time points during iPSC generation.

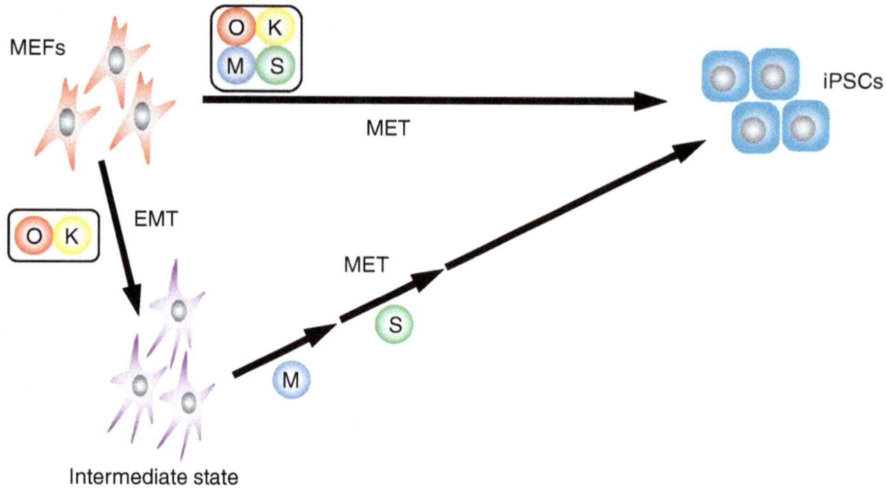

Figure 1. Schematic illustration of EMT/MET during iPSC generation. Simultaneous and time-dependent introduction of the four Yamanaka factor, Oct4 (O), Klf4 (K), c-Myc (M), and Sox2 (S) into MEF resulted in MET and sequential EMT-MET during the processes.

Briefly, we determined reprogramming efficiencies with totally 74 different infection sequences. One particular infection sequence, Oct4 and Klf4 first, c-Myc next and Sox2 last, generated iPSC with the highest efficiency, about 600% of basal level (Liu et al., 2013a) (Fig. 1). After analyzing the differences between simultaneously infection (OKMS method) and the time-dependent infection of the four factors (OK+M+S method), we identified a temporary EMT in the early phase of iPSC generation as demonstrated by an impaired up-regulation of E-cadherin and temporary up-regulation of Snail2. Such temporary EMT was further confirmed with qPCR, Western blotting, FACS analysis, and wound healing assay (Liu et al., 2013a). Consistent with previous reports (Chiou et al., 2010; Liu et al., 2012), the short EMT and delayed MET were suggested to be resulted from the different abilities of Oct4 and Sox2/Klf4 to regulate the expression of E-cadherin and Snail2 (Liu et al., 2013a). In OKMS method, since the four factors were introduced simultaneously, the influences from Sox2 and Klf4 on Snail2 and E-cadherin were be larger than those from Oct4, which led to the consistent down-regulation of Snail2 and rapid up-regulation of E-cadherin. In OK+M+S method, since Sox2 was infected at least, Oct4 would overcome the counteractions from Klf4, which led to temporary up-regulation of Snail2 and impaired up-regulation of E-cadherin.

The reports on MET during iPSC generation in 2010 suggest this temporary EMT should generate more difficulties for later MET and decrease the reprogramming efficiency at the first glance (Li et al., 2010; Samavarchi-Tehrani et al., 2010). However, inducing EMT before MET with short treatment of TGFβ during OKMS reprogramming promoted reprogramming, while blocking EMT before MET with Repsox short treatment during OK+M+S reprogramming impaired reprogramming. Thus introducing a short EMT before MET is beneficial for iPSC generation, especially from MEF.

To explain how short EMT was induced on MEF which are already in mesenchymal state and why sequential EMT-MET promotes iPSC generation, we proposed a new model in previous report (Liu et al., 2013a). As two major cell states, both mesenchymal state and epithelial state are collective concepts and each covers a variety of distinct but similar cell states. We proposed that both of the two states have an optimal or central state within them, which can be named as optimal mesenchymal state and optimal epithelial state. The first hypothesis is to suppose a shortcut pathway between the optimal mesenchymal and epithelial states. When reprogrammed with OK+M+S method, MEF, though in mesenchymal state, would be converted into the optimal mesenchymal state after a short EMT, and took the shortcut pathway to the optimal epithelial state via a delayed MET and to iPSC finally (Fig. 2). When reprogrammed with OKMS method, MEF would not be able to take the advantages of the shortcut pathway and went a longer distance to epithelial state or iPSC, which makes its efficiency lower than that observed with OK+M+S method. This hypothesis was supported by the experiment to identify the optimal cell state for OKMS and OK+M+S reprogramming. When we used TGFβ/RepSox to induce EMT/MET in MEF and determined the ratio of the reprogramming efficiency of OK+M+S method to that of OKMS, we found that the highest ratio is in MEF with 12-h TGFβ treatment, and both longer treatment with TGFβ or Repsox treatment decreased the ratio.

Another available hypothesis is that the optimal mesenchymal state may be more suitable for the cell fate conversions, which is partially supported by the connection between EMT and stem cell characteristics (Hayashida et al., 2011; Mani et al., 2008). OK+M+S method-induced short

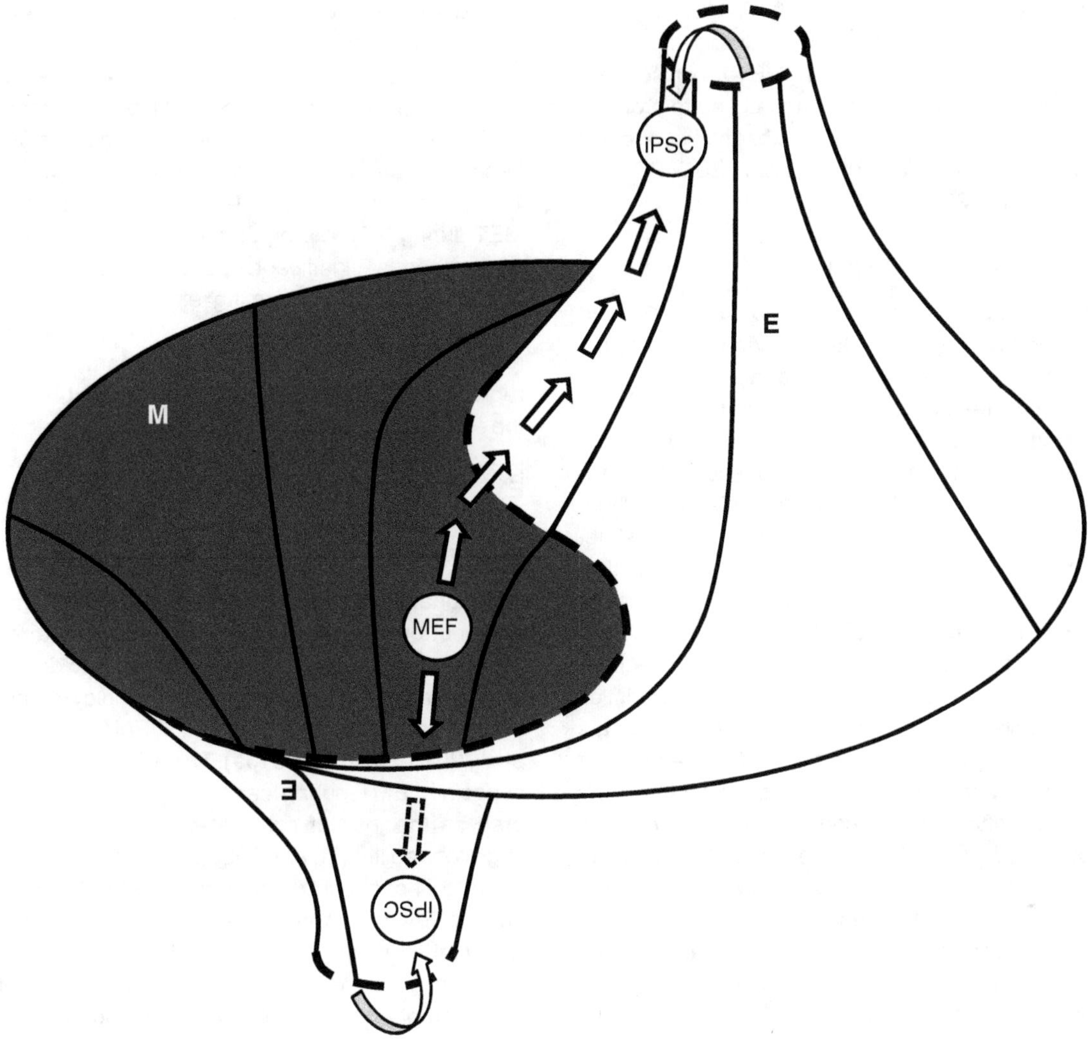

Figure 2. Schematic illustration of the potential shortcut between optimal mesenchymal and epithelial states. Mesenchymal state (M) and epithelial state (E) are collective concepts of two groups of cell states. The optimal mesenchymal state and optimal epithelial state are located in the middle of M and E, respectively. The shortcut was illustrated by putting the two optimal states on the reversed side of each other. MEF can be induced into iPSC by crossing the barriers between the two states. MEF can also be induced into the optimal mesenchymal state first and then to iPSC via optimal epithelial state.

EMT would convert MEF into this optimal mesenchymal state for easier reprogramming, whereas OKMS method could not fulfill this function. Of course, there should be other explanations for the ability of short EMT to promote MEF reprogramming, and how optimal mesenchymal and epithelial state are defined or inter-converted should be investigated in depth even if our hypotheses are right.

EMT/MET during *in vitro* differentiation

Multiple rounds of sequential EMT-MET make embryonic development an excellent model and hot topic for EMT/MET research (Nieto, 2011; Thiery et al., 2009). However, the *in vitro* differentiation of ESC or iPSC is also useful for EMT/MET research because of their similarity to embryonic development and relative simplicity of the system. As a membrane marker for epithelial cells, E-cadherin has also been used as one of the markers for undifferentiated ESC (Li et al., 2012). Loss of E-cadherin expression, which suggests an EMT, can be observed immediately after ESC differentiation (Eastham et al., 2007). If considering EMT as an early step for ESC differentiation, MET should also be observed somewhere during the differentiation of ESC to epithelial cells.

Take the differentiation from iPSC to NSC as an example, immediate up-regulation of N-cadherin, a marker for mesenchymal cells, is essential for the efficient differentiation. However, E-cadherin expression is required to support the self-renewal of NSC (Karpowicz et al., 2009). Thus the expression switches between E-cadherin and N-cadherin, which suggests the transitions between epithelial and mesenchymal states (Gravdal et al., 2007; Maeda et al., 2005), might be observed multiple times during the differentiation from iPSC to NSC. In addition, MET has also been observed during the differentiation of hepatic stem/progenitor cells, suggesting the possibility to observe sequential EMT-MET during the differentiation from ESC/iPSC to hepatic cells (Li et al., 2011).

EMT/MET during trans-differentiation

The successful trans-differentiation of somatic cells into functional neurons (Sheng et al., 2012a; Vierbuchen et al., 2010), NSC (Kim et al., 2011a; Sheng et al., 2012b; Wang et al., 2012), multilineage blood progenitors (Szabo et al., 2010), hepatocyte-like cells (Huang et al., 2011) or cardiomyocytes (Efe et al., 2011; Ieda et al., 2010) suggests a new route to generate target cells for transplantation without employing pluripotent stem cells as an intermediate state. The observation of EMT or MET during these trans-differentiation processes is greatly anticipated, not only because both the mesenchymal cells (fibroblasts) and epithelial cells (cells isolated from urine) have been used for trans-differentiation, but also because of the different characteristics received by the cells after trans-differentiation (Huang et al., 2011; Vierbuchen et al., 2010; Wang et al., 2012). Actually, if the cells were in different cell states (mesenchymal or epithelial) before and after cell fate conversions, EMT or MET should be observed during the conversions. Although the existence of sequential EMT-MET has not been reported yet, complex transitions between mesenchymal and epithelial state should exist during the NSC trans-differentiation for the similar reasons mentioned above and the critical functions of N-cadherin in neuron-neuron interaction (Tan et al., 2010).

THE CONTRIBUTIONS OF EMT/MET TO CELL FATE CONVERSIONS

The observations of EMT/MET during different kinds of cell fate conversions do not enable us to answer the question that EMT/EMT is a by-product or a significant cause for cell fate conversions. Take MET during iPSC generation from MEF for example, MEF and iPSC definitely have the characteristics of mesenchymal and epithelial cells respectively. Thus the successful conversion from MEF to iPSC must be accompanied by a MET process. MET is demonstrated to be required for MEF reprogramming, because reprogramming was greatly impaired when EMT was induced or MET was inhibited (Li et al., 2010). However, this necessity might be explained by that cells will not become iPSC without epithelial characteristics.

One way to answer the question above is to study the reprogramming of epithelial cells. Ciliary body epithelial cells have been reported to have higher reprogramming efficiency to iPSC than fibroblasts (Ni et al., 2013). NSC, which require E-cadherin for self-renewal (Karpowicz et al., 2009), can be reprogrammed into iPSC with only two factors, Oct4 with Klf4 or c-Myc (Kim et al., 2008). In addition, cells isolated from urine have epithelial characteristics and have been considered as one excellent somatic cell types to generate iPSC or NSC not only because of their easy accessibility but also because of the high reprogramming efficiency (Wang et al., 2012; Zhou et al., 2011). These reports suggested that the MET during MEF reprogramming might only be a by-product rather than a significant cause. However, it is difficult to compare the reprogramming efficiencies between different types of somatic cells, because they have their own characteristics and can not be put at the same start point during reprogramming. Thus in the following text, we will discuss the connection between EMT/MET and other biological processes and how the marker events or factors during EMT/MET may contribute to cell fate conversions.

Marker factors and events during EMT/MET

Unfortunately, the studies on MET are fewer than those on EMT not only because MET is regarded as a reversed and following process of EMT during embryonic development but also because both MET and EMT employ same regulatory system and most method to inhibit EMT can induce MET (Li et al., 2010; Liu et al., 2013a). The marker events during EMT, especially EMT during cancer progression, can be considered as the loss of tight cell-cell junction and the gain of migratory ability (Thiery, 2002). Thus the switch from E-cadherin to N-cadherin demonstrated with qPCR, immunoblotting or FACS analysis has been used to evaluate EMT (Gravdal et al., 2007; Liu et al., 2013a; Maeda et al., 2005), because E-cadherin plays critical role in maintaining the adhesion of epithelial cells (van Roy and Berx, 2008). Other membrane markers are also used to characterize epithelial and mesenchymal states, like epithelial cell adhesion molecule, epithelial keratins, Occludin, and Claudin3 for epithelial cells (Li et al., 2010; Litvinov et al., 1997), or Fibronectin and Vimentin for mesenchymal cells (Freire-de-Lima et al., 2011; Zhou et al., 2008). Many transcriptional factors, Snail1/2, Zeb1/2, Twist1/2, and E47, are also involved in EMT as reviewed extensively in previous reports (Nieto, 2011; Thiery et al., 2009). In addition, EMT is twisted with multiple signaling pathways. For example, TGFβ treatment can trigger both cell death and EMT (Massague, 2008), while modulating Snail1 expression can shift the balance between cell death and EMT (Franco et al., 2010). TGFβ molecules and receptors have also been used as markers for EMT (Li et al., 2010). EMT is also regulated by WNT/β-catenin (Li et al., 2013; Wu et al., 2012), FGF (Zhang et al., 2006), Sonic Hedgehog (Panman et al., 2006), mTOR (Lamouille et al., 2012), and other signaling pathways (Wang et al., 2010).

EMT/MET and cell cycle

Cell cycle regulation is an old topic in cell biology and is observed during almost all kinds of cell fate conversions. iPSC have a high proliferation rate and short G_1 phase similar to ESC (Neganova and Lako, 2008; White and Dalton, 2005). MEF gain these cell cycle signatures at the early phase of reprogramming. Proliferation induction and cell cycle arrest respectively promotes and impairs MEF reprogramming (Ruiz et al., 2010). In addition, these is a positive correlation between proliferation rate and reprogramming efficiency, which further confirmed the connection between cell cycle regulation and iPSC generation (Hanna et al., 2009).

On the other hand, EMT and cell cycle regulation are also inter-connected. TGFβ can lead to proliferation arrest and has extensive interaction with p53 (Adorno et al., 2009; Massague, 2008). Forced expression of p53 increased the mRNA levels of TGFβ1/2 (Fujiwara et al., 1994; Kannan et al., 2000). p53 is also essential for the responses to TGFβ signals by promoting the activation of multiple TGFβ target proteins (Cordenonsi et al., 2003; Termen et al., 2013). EMT induced by TGFβ in AML-12 cells is in a cell cycle-dependent manner and takes place at G_1/S phase, where lies the major difference on cell cycle signatures between MEF and iPSC (Yang et al., 2006). In addition, another EMT inducer, Zeb1, also suppresses proliferation (Hugo et al., 2013), while p53 can suppress the expression of Zeb1 directly or via microRAN34/200c (Chang et al., 2011; Kim et al., 2011b). Therefore the decrease in reprogramming efficiency when TGFβ was included in the medium for iPSC generation might be mediated by the impaired proliferation (Li et al., 2010).

To explain the positive correlation between proliferation rate and reprogramming efficiency, Dr. Jaenisch's laboratory has provided a model which suggested the early phase of MEF reprogramming was a stochastic process (Buganim et al., 2012; Hanna et al., 2009). That is each round of cell cycle would enable MEF to choose to enter iPSC cell fate or to stay in MEF cell fate. In addition, the inability of proliferation induction to increase reprogramming efficiency in y-axis when the number of cell cycle rather than the length of reprogramming time was used in x-axis suggests that the increase in the number of cell cycle accounts for almost all the effects of high proliferation rate on reprogramming (Hanna et al., 2009). This hypothesis is strongly supported by the short of observations with iPSC generated from cells in G_0 phase.

The following question is what kind of events or markers during cell cycle promote reprogramming. One possibility is the DNA methylation regulation during cell cycle progression (Bou Kheir and Lund, 2010). There are multiple tasks like growth, expansion, and communication for cells in G_1 phase, whereas cells are focused to DNA replication in S phase. Thus genes that specify the original cell fate would be silenced in S phase, as partially confirmed by the relative higher DNA methylation in S phase than G_1 phase (Brown et al., 2007). The protein coded by these genes would be gradually degraded from S phase until the end of M phase, which resulted in a good opportunity for genes that specify different cell fates to function at the beginning of G_1 phase. Since genes specify original cell fate is in a hypo-methylated state in contrast to the hyper-methylated state of genes specify other cell fates, the cells would keep their original fate without outside stimulants. However, when the expression of genes specifying for other cell fates were increased significantly like during iPSC generation and trans-differentiation, the possibility for cells to switch cell fate will increase significantly. In addition, it is also reasonable to propose that decreasing the overall DNA methylation might facilitate cell fate conversion by reducing the methylation levels on the genes specifying for other cell fates. This hypothesis is supported by the study on Tet1 and lysine-specific demethylase 1 (Chen et al., 2013; Gao et al., 2013; Lin et al., 2011). Over-expressing Tet1 increases the reprogramming efficiency, possibly by reducing DNA methylation.

EMT/MET and histone modifications

As just discussed on DNA methylation, histone modifications are also required to be duplicated and inherited during cell cycle progression (Probst et al., 2009). Thus histone modifications may also be possible to mediate the interaction between cell cycle and cell fate conversions. However, the complex regulation on histone modifications during cell cycle and undetermined functions of different histone modifications make our discussion more difficult.

The histone modifications affect cell fate conversions from a variety of aspects (Papp and Plath, 2012). Increasing H3K4 methylation with lysine-specific demethylase 1 inhibitor, decreasing H3K9 methylation with Setdb1, reducing H3K27 methylation with Utx, and inhibiting H3K36 methylation with Jhdm1a/1b are all reported to benefit iPSC generation (Chen et al., 2012; Mansour et al., 2012; Wang et al., 2011a; Wang et al., 2011b). Thus, by affecting cell cycle, EMT/MET may influence histone modifications and subsequent cell fate conversion. Although difficult to eliminate the participation of cell cycle, there are many reports on the interactions between EMT/MET and histone modifications. Histone deacetylases 3 not only increases H3K4 methylation by interacting with hypoxia-induced WD repeat-containing protein 5 to activate mesenchymal gene expression, but also serves as an essential co-repressor to repress epithelial gene expression (Wu et al., 2011). LSD1 which functions in the demethylation of H3K4 and H3K9 not only serves in NuRD complex to block EMT during breast cancer metastasis (Wang et al., 2009), but also forms complex with Snail1 and Twist to repress E-cadherin expression (Fu et al., 2011; Lin et al., 2010). Considering the extensive interactions between EMT/MET and histone modifications as reviewed (Nieto, 2011), it is reasonable to hypothesize that EMT/MET may contribute to cell fate conversions by modulating histone modifications and other epigenetic properties.

EMT/MET and stemness

As mentioned above, EMT was first observed in breast cancer and plays critical roles during cancer progression

(Thiery et al., 2009). On the other hand, cancer stem cells (CSCs) are characterized as a special group of cancer cells with enhanced tumorigenicity, partial stemness as in self-renewal and differentiation, and $CD44^{+}/CD24^{-/low}$ on cell surface (Al-Hajj et al., 2003; Dalerba et al., 2007). These two concepts, EMT and CSCs, are not well connected until the report by Mani et al. in 2008. The induced EMT in human mammary epithelial cells increased the $CD44^{+}/CD24^{-/low}$ population with properties associated with mammary epithelial stem cells like abilities to form mammosphere and to differentiate into myoepithelial or luminal epithelial cells (Mani et al., 2008). The connection between EMT and CSCs is further established with the common characteristics between cells undergoing EMT and CSCs in TGFβ signaling pathway activation, circulation in blood, and chemo-resistance as reviewed (Hayashida et al., 2011). Thus it seems that EMT may be able to induce stemness in cells at least during cancer progression.

However, the studies on ESC and iPSC are not fully consistent with the hypothesis above. ESC as isolated from inner cell mass are considered to be typical epithelial cells. The epithelial marker, E-cadherin, together with proteins like SSEA1, alkaline phosphatase, Oct4, Nanog, and others have been used to determine the undifferentiated state of ESC (Horie et al., 2010; Redmer et al., 2011). The expression of E-cadherin in undifferentiated ESC decreased immediately after the induction of differentiation (D'Amour et al., 2005; Eastham et al., 2007). In addition, blocking the up-regulation of E-cadherin significantly decreased the efficiency during iPSC generation from MEF (Li et al., 2010). Although the absolute requirement of E-cadherin for pluripotency is still under debate (Eastham et al., 2007; Ullmann et al., 2007), E-cadherin contributes to the survival, self renewal, and pluripotency of ESC at least partially (Li et al., 2012). Thus it seems that EMT may prohibit stemness in cells like iPSC and ESC. The opposite influences of EMT on stemness observed in the two sets of studies above suggest that either EMT has multiple roles or different aspects of EMT function differentially in regulating stemness.

FUTURE DIRECTIONS

We have provided a brief review on the connection between EMT/MET and cell cycle, epigenetic regulation, and stemness to provide possible mechanisms underlying the contributions of EMT/MET to cell fate conversions. Of course, EMT/MET also twists with other biological processes or factors including cell senescence with telomerase reverse transcriptase (Qiao et al., 2012; Yu et al., 2014), hypoxia with hypoxia induced factor 1α (Liu et al., 2013b; Marie-Egyptienne et al., 2013), microRNAs like miR-200 (Ocana and Nieto, 2008; Park et al., 2008), and so on, we did not listed them all in current review. Since the interactions among these biological processes are also complex as those between cell cycle and epigenetic regulation mentioned above, the future study on how EMT/MET contributes to cell fate conversions should not be limited in one or few biological processes.

In our opinion, the well established connections between EMT/MET and large amount of biological processes is the major problem faced by researchers who intend to study the contributions of EMT/MET to cell fate conversions. Firstly, the complex interactions among different biological processes make it difficult to study one aspect at one time. For example, cell cycle, DNA methylation, and tumor progresses have been suggested to function together under certain circumstances (Robertson et al., 2000; Wu et al., 1993). Secondly, these complex interactions lead to opposite observations on how EMT/MET affect cell fate conversions under different conditions. TGFβ treatment in whole process of iPSC generation from MEF decreased reprogramming efficiency, while pre-treatment or short treatment (Day 0–1.5 during reprogramming) with TGFβ decreased the proliferation rate and significantly increased the reprogramming efficiency. (Li et al., 2010; Liu et al., 2013a). The studies on EMT during CSCs and iPSC generation suggested the opposite roles of EMT on stemness as mentioned above. The ability of periphery cells around undifferentiated ESC to have both mesenchymal characteristics and ESC properties makes the correlation between EMT and stemness even more complex (Ullmann et al., 2007). Thirdly but not lastly, different biological processes have distinct preference with different aspects of EMT/MET. For example, the function of E-cadherin during EMT/MET may have a closer connection with cell-cell adhesion and WNT/β-catenin pathway (Tian et al., 2011; van Roy and Berx, 2008), while TGFβ may prefer to interact with p53 (Dupont et al., 2004; McPherson, 1996).

To solve the problem above, one possibility is to put EMT and MET together and study sequential EMT-MET during cell fate conversions. Sequential EMT-MET has been observed during embryonic development and iPSC generation (Liu et al., 2013a; Thiery et al., 2009). In addition, EMT has long been considered important during the initial phase of metastasis, MET is also suggested to be important for the latter phase of metastasis, when cancer cells regain similarities to primary tumors at the secondary sites (Chaffer et al., 2007). Although sequential EMT-MET has not been reported during differentiation and trans-differentiation, it is reasonable to suggest that inducing sequential EMT-MET in proper manner may facilitate different cell fate conversions.

To explain why temporary EMT before EMT promotes iPSC generation from MEF, we proposed a hypothesized model with optimal mesenchymal and optimal epithelial states as above. Considering the sequential EMT-MET, the preliminary hypothesis that an optimal mesenchymal may function as intermediate for efficient cell fate conversions might be more reasonable. This hypothesis is able to explain the beneficial effects of temporary EMT during iPSC generation and the opposite roles of EMT on stemness regulation. The hypothesis is supported by the chromosomal instability during EMT and in mesenchymal stem cells (Miura et al., 2006; Roschke et al., 2008) and challenged by the short of evidences during a variety of cell fate conversions.

ACKNOWLEDGEMENTS

This review work has been supported in part by the Strategic Priority Research Program of the Chinese Academy of Sciences (XDA01020302, XDA01020401), the Guangzhou International Science and Technology Cooperation Projects from the Bureau of Science and Information Technology of the Guangzhou Municipal Government (2012J5100007), the Guangdong Natural Science Foundation (S2012010010087), and the National Natural Science Foundation of China (Grant No. 31100773).

ABBREVIATIONS

EMT, epithelial-mesenchymal transition; ESC, embryonic stem cells; iPSC, induced pluripotent stem cells; MEF, mouse embryonic fibroblasts; MET, mesenchymal-epithelial transition; OKMS, Oct4, Klf4, c-Myc and Sox2 were introduced into MEF simultaneously during reprogramming; OK+M+S; Oct4 and Klf4 were first, c-Myc was next, and Sox2 was last introduced into MEF during reprogramming; Tet1, Ten-eleven translocation methylcytosine dioxygenase 1.

COMPLIANCE WITH ETHICS GUIDELINES

Xiang Li, Duanqing Pei and Hui Zheng declare that they have no conflict of interest.

This article does not contain any studies with human or animal subjects performed by the any of the authors.

REFERENCES

Aasen T, Raya A, Barrero MJ, Garreta E, Consiglio A, Gonzalez F, Vassena R, Bilic J, Pekarik V, Tiscornia G et al (2008) Efficient and rapid generation of induced pluripotent stem cells from human keratinocytes. Nat Biotechnol 26:1276–1284

Adorno M, Cordenonsi M, Montagner M, Dupont S, Wong C, Hann B, Solari A, Bobisse S, Rondina MB, Guzzardo V et al (2009) A Mutant-p53/Smad complex opposes p63 to empower TGFbeta-induced metastasis. Cell 137:87–98

Al-Hajj M, Wicha MS, Benito-Hernandez A, Morrison SJ, Clarke MF (2003) Prospective identification of tumorigenic breast cancer cells. Proc Natl Acad Sci USA 100:3983–3988

Bou Kheir T, Lund AH (2010) Epigenetic dynamics across the cell cycle. Essays Biochem 48:107–120

Brambrink T, Foreman R, Welstead GG, Lengner CJ, Wernig M, Suh H, Jaenisch R (2008) Sequential expression of pluripotency markers during direct reprogramming of mouse somatic cells. Cell Stem Cell 2:151–159

Brown SE, Fraga MF, Weaver IC, Berdasco M, Szyf M (2007) Variations in DNA methylation patterns during the cell cycle of HeLa cells. Epigenetics 2:54–65

Buganim Y, Faddah DA, Cheng AW, Itskovich E, Markoulaki S, Ganz K, Klemm SL, van Oudenaarden A, Jaenisch R (2012) Single-cell expression analyses during cellular reprogramming reveal an early stochastic and a late hierarchic phase. Cell 150:1209–1222

Chaffer CL, Thompson EW, Williams ED (2007) Mesenchymal to epithelial transition in development and disease. Cells Tissues Organs 185:7–19

Chang CJ, Chao CH, Xia W, Yang JY, Xiong Y, Li CW, Yu WH, Rehman SK, Hsu JL, Lee HH et al (2011) p53 regulates epithelial-mesenchymal transition and stem cell properties through modulating miRNAs. Nat Cell Biol 13:317–323

Chen J, Liu J, Han Q, Qin D, Xu J, Chen Y, Yang J, Song H, Yang D, Peng M et al (2010a) Towards an optimized culture medium for the generation of mouse induced pluripotent stem cells. J Biol Chem 285:31066–31072

Chen T, Yuan D, Wei B, Jiang J, Kang J, Ling K, Gu Y, Li J, Xiao L, Pei G (2010b) E-cadherin-mediated cell-cell contact is critical for induced pluripotent stem cell generation. Stem Cells 28:1315–1325

Chen J, Liu J, Chen Y, Yang J, Chen J, Liu H, Zhao X, Mo K, Song H, Guo L et al (2011) Rational optimization of reprogramming culture conditions for the generation of induced pluripotent stem cells with ultra-high efficiency and fast kinetics. Cell Res 21:884–894

Chen J, Liu H, Liu J, Qi J, Wei B, Yang J, Liang H, Chen Y, Chen J, Wu Y et al (2012) H3K9 methylation is a barrier during somatic cell reprogramming into iPSCs. Nature Genet 45:34–42

Chen J, Guo L, Zhang L, Wu H, Yang J, Liu H, Wang X, Hu X, Gu T, Zhou Z et al (2013) Vitamin C modulates TET1 function during somatic cell reprogramming. Nature Genet 45:1504–1509

Chiou SH, Wang ML, Chou YT, Chen CJ, Hong CF, Hsieh WJ, Chang HT, Chen YS, Lin TW, Hsu HS et al (2010) Coexpression of Oct4 and Nanog enhances malignancy in lung adenocarcinoma by inducing cancer stem cell-like properties and epithelial-mesenchymal transdifferentiation. Cancer Res 70:10433–10444

Cordenonsi M, Dupont S, Maretto S, Insinga A, Imbriano C, Piccolo S (2003) Links between tumor suppressors: p53 is required for TGF-beta gene responses by cooperating with Smads. Cell 113:301–314

Dalerba P, Cho RW, Clarke MF (2007) Cancer stem cells: models and concepts. Ann Rev Med 58:267–284

D'Amour KA, Agulnick AD, Eliazer S, Kelly OG, Kroon E, Baetge EE (2005) Efficient differentiation of human embryonic stem cells to definitive endoderm. Nat Biotechnol 23:1534–1541

Dupont S, Zacchigna L, Adorno M, Soligo S, Volpin D, Piccolo S, Cordenonsi M (2004) Convergence of p53 and TGF-beta signaling networks. Cancer Lett 213:129–138

Eastham AM, Spencer H, Soncin F, Ritson S, Merry CL, Stern PL, Ward CM (2007) Epithelial-mesenchymal transition events during human embryonic stem cell differentiation. Cancer Res 67:11254–11262

Efe JA, Hilcove S, Kim J, Zhou H, Ouyang K, Wang G, Chen J, Ding S (2011) Conversion of mouse fibroblasts into cardiomyocytes using a direct reprogramming strategy. Nat Cell Biol 13:215–222

Esteban MA, Pei D (2012) Vitamin C improves the quality of somatic cell reprogramming. Nature Genet 44:366–367

Esteban MA, Xu J, Yang J, Peng M, Qin D, Li W, Jiang Z, Chen J, Deng K, Zhong M et al (2009) Generation of induced pluripotent stem cell lines from Tibetan miniature pig. J Biol Chem 284:17634–17640

Esteban MA, Wang T, Qin B, Yang J, Qin D, Cai J, Li W, Weng Z, Chen J, Ni S et al (2010) Vitamin C enhances the generation of mouse and human induced pluripotent stem cells. Cell Stem Cell 6:71–79

Folmes CD, Nelson TJ, Martinez-Fernandez A, Arrell DK, Lindor JZ, Dzeja PP, Ikeda Y, Perez-Terzic C, Terzic A (2011) Somatic oxidative bioenergetics transitions into pluripotency-dependent glycolysis to facilitate nuclear reprogramming. Cell Metabol 14:264–271

Franco DL, Mainez J, Vega S, Sancho P, Murillo MM, de Frutos CA, Del Castillo G, Lopez-Blau C, Fabregat I, Nieto MA (2010) Snail1 suppresses TGF-beta-induced apoptosis and is sufficient to trigger EMT in hepatocytes. J Cell Sci 123:3467–3477

Freire-de-Lima L, Gelfenbeyn K, Ding Y, Mandel U, Clausen H, Handa K, Hakomori SI (2011) Involvement of O-glycosylation defining oncofetal fibronectin in epithelial-mesenchymal transition process. Proc Natl Acad Sci USA 108:17690–17695

Fu J, Qin L, He T, Qin J, Hong J, Wong J, Liao L, Xu J (2011) The TWIST/Mi2/NuRD protein complex and its essential role in cancer metastasis. Cell Res 21:275–289

Fujiwara T, Mukhopadhyay T, Cai DW, Morris DK, Roth JA, Grimm EA (1994) Retroviral-mediated transduction of p53 gene increases TGF-beta expression in a human glioblastoma cell line. Int J Cancer J Int du Cancer 56:834–839

Gao Y, Chen J, Li K, Wu T, Huang B, Liu W, Kou X, Zhang Y, Huang H, Jiang Y et al (2013) Replacement of Oct4 by Tet1 during iPSC induction reveals an important role of DNA methylation and hydroxymethylation in reprogramming. Cell Stem Cell 12:453–469

Grabel L (2012) Prospects for pluripotent stem cell therapies: into the clinic and back to the bench. J Cell Biochem 113:381–387

Gravdal K, Halvorsen OJ, Haukaas SA, Akslen LA (2007) A switch from E-cadherin to N-cadherin expression indicates epithelial to mesenchymal transition and is of strong and independent importance for the progress of prostate cancer. Clin Cancer Res 13:7003–7011

Gurdon JB (1962a) Adult frogs derived from the nuclei of single somatic cells. Dev Biol 4:256–273

Gurdon JB (1962b) The developmental capacity of nuclei taken from intestinal epithelium cells of feeding tadpoles. J Embryol Exper Morphol 10:622–640

Hanna J, Saha K, Pando B, van Zon J, Lengner CJ, Creyghton MP, van Oudenaarden A, Jaenisch R (2009) Direct cell reprogramming is a stochastic process amenable to acceleration. Nature 462:595–601

Hayashida T, Jinno H, Kitagawa Y, Kitajima M (2011) Cooperation of cancer stem cell properties and epithelial-mesenchymal transition in the establishment of breast cancer metastasis. J Oncol 2011:591427

Horie M, Ito A, Kiyohara T, Kawabe Y, Kamihira M (2010) E-cadherin gene-engineered feeder systems for supporting undifferentiated growth of mouse embryonic stem cells. J Biosci Bioeng 110:582–587

Hou P, Li Y, Zhang X, Liu C, Guan J, Li H, Zhao T, Ye J, Yang W, Liu K et al (2013) Pluripotent stem cells induced from mouse somatic cells by small-molecule compounds. Science 341:651–654

Huang P, He Z, Ji S, Sun H, Xiang D, Liu C, Hu Y, Wang X, Hui L (2011) Induction of functional hepatocyte-like cells from mouse fibroblasts by defined factors. Nature 475:386–389

Hugo HJ, Pereira L, Suryadinata R, Drabsch Y, Gonda TJ, Gunasinghe NP, Pinto C, Soo ET, van Denderen B, Hill P et al (2013) Direct repression of MYB by ZEB1 suppresses proliferation and epithelial gene expression during epithelial-to-mesenchymal transition of breast cancer cells. Breast Cancer Res 15:R113

Ichida JK, Blanchard J, Lam K, Son EY, Chung JE, Egli D, Loh KM, Carter AC, Di Giorgio FP, Koszka K et al (2009) A small-molecule inhibitor of tgf-Beta signaling replaces sox2 in reprogramming by inducing nanog. Cell Stem Cell 5:491–503

Ieda M, Fu JD, Delgado-Olguin P, Vedantham V, Hayashi Y, Bruneau BG, Srivastava D (2010) Direct reprogramming of fibroblasts into functional cardiomyocytes by defined factors. Cell 142:375–386

Kannan K, Amariglio N, Rechavi G, Givol D (2000) Profile of gene expression regulated by induced p53: connection to the TGF-beta family. FEBS Lett 470:77–82

Karpowicz P, Willaime-Morawek S, Balenci L, DeVeale B, Inoue T, van der Kooy D (2009) E-Cadherin regulates neural stem cell self-renewal. J Neurosci 29:3885–3896

Kashyap V, Rezende NC, Scotland KB, Shaffer SM, Persson JL, Gudas LJ, Mongan NP (2009) Regulation of stem cell pluripotency and differentiation involves a mutual regulatory circuit of the NANOG, OCT4, and SOX2 pluripotency transcription factors with polycomb repressive complexes and stem cell microRNAs. Stem Cells Devel 18:1093–1108

Kim JB, Zaehres H, Wu G, Gentile L, Ko K, Sebastiano V, Arauzo-Bravo MJ, Ruau D, Han DW, Zenke M et al (2008) Pluripotent stem cells induced from adult neural stem cells by reprogramming with two factors. Nature 454:646–650

Kim J, Efe JA, Zhu S, Talantova M, Yuan X, Wang S, Lipton SA, Zhang K, Ding S (2011a) Direct reprogramming of mouse fibroblasts to neural progenitors. Proc Natl Acad Sci USA 108:7838–7843

Kim T, Veronese A, Pichiorri F, Lee TJ, Jeon YJ, Volinia S, Pineau P, Marchio A, Palatini J, Suh SS et al (2011b) p53 regulates epithelial-mesenchymal transition through microRNAs targeting ZEB1 and ZEB2.J Exper Med 208:875–883

Koche RP, Smith ZD, Adli M, Gu H, Ku M, Gnirke A, Bernstein BE, Meissner A (2011) Reprogramming factor expression initiates widespread targeted chromatin remodeling. Cell Stem Cell 8:96–105

Lamouille S, Connolly E, Smyth JW, Akhurst RJ, Derynck R (2012) TGF-beta-induced activation of mTOR complex 2 drives epithelial-mesenchymal transition and cell invasion. J Cell Sci 125:1259–1273

Li R, Liang J, Ni S, Zhou T, Qing X, Li H, He W, Chen J, Li F, Zhuang Q et al (2010) A mesenchymal-to-epithelial transition initiates and is required for the nuclear reprogramming of mouse fibroblasts. Cell Stem Cell 7:51–63

Li B, Zheng YW, Sano Y, Taniguchi H (2011) Evidence for mesenchymal-epithelial transition associated with mouse hepatic stem cell differentiation. PloS one 6:e17092

Li L, Bennett SA, Wang L (2012) Role of E-cadherin and other cell adhesion molecules in survival and differentiation of human pluripotent stem cells. Cell Adhesion Migr 6:59–70

Li X, Xu Y, Chen Y, Chen S, Jia X, Sun T, Liu Y, Li X, Xiang R, Li N (2013) SOX2 promotes tumor metastasis by stimulating epithelial-to-mesenchymal transition via regulation of WNT/beta-catenin signal network. Cancer Lett 336:379–389

Liao J, Cui C, Chen S, Ren J, Chen J, Gao Y, Li H, Jia N, Cheng L, Xiao H et al (2009) Generation of induced pluripotent stem cell lines from adult rat cells. Cell Stem Cell 4:11–15

Lin Y, Wu Y, Li J, Dong C, Ye X, Chi YI, Evers BM, Zhou BP (2010) The SNAG domain of Snail1 functions as a molecular hook for recruiting lysine-specific demethylase 1. EMBO J 29:1803–1816

Lin SL, Chang DC, Lin CH, Ying SY, Leu D, Wu DT (2011) Regulation of somatic cell reprogramming through inducible mir-302 expression. Nucleic Acids Res 39:1054–1065

Litvinov SV, Balzar M, Winter MJ, Bakker HA, Briaire-de Bruijn IH, Prins F, Fleuren GJ, Warnaar SO (1997) Epithelial cell adhesion molecule (Ep-CAM) modulates cell-cell interactions mediated by classic cadherins. J Cell Biol 139:1337–1348

Liu H, Zhu F, Yong J, Zhang P, Hou P, Li H, Jiang W, Cai J, Liu M, Cui K et al (2008) Generation of induced pluripotent stem cells from adult rhesus monkey fibroblasts. Cell Stem Cell 3:587–590

Liu YN, Abou-Kheir W, Yin JJ, Fang L, Hynes P, Casey O, Hu D, Wan Y, Seng V, Sheppard-Tillman H et al (2012) Critical and reciprocal regulation of KLF4 and SLUG in transforming growth factor beta-initiated prostate cancer epithelial-mesenchymal transition. Mol Cell Biol 32:941–953

Liu X, Sun H, Qi J, Wang L, He S, Liu J, Feng C, Chen C, Li W, Guo Y et al (2013a) Sequential introduction of reprogramming factors reveals a time-sensitive requirement for individual factors and a sequential EMT-MET mechanism for optimal reprogramming. Nat Cell Biol 15:829–838

Liu Y, Mukhopadhyay P, Pisano MM, Lu X, Huang L, Lu Q, Dean DC (2013b) Repression of Zeb1 and hypoxia cause sequential mesenchymal-to-epithelial transition and induction of aid, Oct4, and Dnmt1, leading to immortalization and multipotential reprogramming of fibroblasts in spheres. Stem Cells 31:1350–1362

Maeda M, Johnson KR, Wheelock MJ (2005) Cadherin switching: essential for behavioral but not morphological changes during an epithelium-to-mesenchyme transition. J Cell Sci 118:873–887

Maekawa M, Yamaguchi K, Nakamura T, Shibukawa R, Kodanaka I, Ichisaka T, Kawamura Y, Mochizuki H, Goshima N, Yamanaka S (2011) Direct reprogramming of somatic cells is promoted by maternal transcription factor Glis1. Nature 474:225–229

Mani SA, Guo W, Liao MJ, Eaton EN, Ayyanan A, Zhou AY, Brooks M, Reinhard F, Zhang CC, Shipitsin M et al (2008) The epithelial-mesenchymal transition generates cells with properties of stem cells. Cell 133:704–715

Mansour AA, Gafni O, Weinberger L, Zviran A, Ayyash M, Rais Y, Krupalnik V, Zerbib M, Amann-Zalcenstein D, Maza I et al (2012) The H3K27 demethylase Utx regulates somatic and germ cell epigenetic reprogramming. Nature 488:409–413

Marie-Egyptienne DT, Lohse I, Hill RP (2013) Cancer stem cells, the epithelial to mesenchymal transition (EMT) and radioresistance: potential role of hypoxia. Cancer Lett 341:63–72

Massague J (2008) TGFbeta in Cancer. Cell 134:215–230

McPherson JM (1996) p53/TGF-beta/cancer: an intriguing connection. Cytokine Growth Factor Rev 7:295

Miura M, Miura Y, Padilla-Nash HM, Molinolo AA, Fu B, Patel V, Seo BM, Sonoyama W, Zheng JJ, Baker CC et al (2006) Accumulated chromosomal instability in murine bone marrow mesenchymal stem cells leads to malignant transformation. Stem Cells 24:1095–1103

Muller LU, Daley GQ, Williams DA (2009) Upping the ante: recent advances in direct reprogramming. Mol Ther 17:947–953

Nakajima Y, Yamagishi T, Hokari S, Nakamura H (2000) Mechanisms involved in valvuloseptal endocardial cushion formation in early cardiogenesis: roles of transforming growth factor (TGF)-beta and bone morphogenetic protein (BMP). Anat Rec 258:119–127

Nakaya Y, Sheng G (2008) Epithelial to mesenchymal transition during gastrulation: an embryological view. Devel Growth Differ 50:755–766

Neganova I, Lako M (2008) G1 to S phase cell cycle transition in somatic and embryonic stem cells. J Anat 213:30–44

Ni A, Wu MJ, Nakanishi Y, Chavala SH (2013) Facile and efficient reprogramming of ciliary body epithelial cells into induced pluripotent stem cells. Stem cells and development 22:2543–2550

Nieto MA (2011) The ins and outs of the epithelial to mesenchymal transition in health and disease. Ann Rev Cell Devel Biol 27:347–376

Nishikawa S, Goldstein RA, Nierras CR (2008) The promise of human induced pluripotent stem cells for research and therapy. Nat Rev Mol Cell Biol 9:725–729

Ocana OH, Nieto MA (2008) A new regulatory loop in cancer-cell invasion. EMBO Rep 9:521–522

Pan GJ, Chang ZY, Scholer HR, Pei D (2002) Stem cell pluripotency and transcription factor Oct4. Cell Res 12:321–329

Pan G, Li J, Zhou Y, Zheng H, Pei D (2006) A negative feedback loop of transcription factors that controls stem cell pluripotency and self-renewal. FASEB J 20:1730–1732

Panman L, Galli A, Lagarde N, Michos O, Soete G, Zuniga A, Zeller R (2006) Differential regulation of gene expression in the digit forming area of the mouse limb bud by SHH and gremlin 1/FGF-mediated epithelial-mesenchymal signalling. Development 133:3419–3428

Papp B, Plath K (2012) Reprogramming to pluripotency: stepwise resetting of the epigenetic landscape. Cell Res 21:486–501

Park SM, Gaur AB, Lengyel E, Peter ME (2008) The miR-200 family determines the epithelial phenotype of cancer cells by targeting the E-cadherin repressors ZEB1 and ZEB2. Genes Dev 22:894–907

Peinado H, Olmeda D, Cano A (2007) Snail, Zeb and bHLH factors in tumour progression: an alliance against the epithelial phenotype? Nature Rev Cancer 7:415–428

Polyak K, Weinberg RA (2009) Transitions between epithelial and mesenchymal states: acquisition of malignant and stem cell traits. Nature Rev Cancer 9:265–273

Probst AV, Dunleavy E, Almouzni G (2009) Epigenetic inheritance during the cell cycle. Nat Rev Mol Cell Biol 10:192–206

Qiao B, Gopalan V, Chen Z, Smith RA, Tao Q, Lam AK (2012) Epithelial-mesenchymal transition and mesenchymal-epithelial transition are essential for the acquisition of stem cell properties in hTERT-immortalised oral epithelial cells. Biol Cell 104:476–489

Redmer T, Diecke S, Grigoryan T, Quiroga-Negreira A, Birchmeier W, Besser D (2011) E-cadherin is crucial for embryonic stem cell pluripotency and can replace OCT4 during somatic cell reprogramming. EMBO Rep 12:720–726

Robertson KD, Keyomarsi K, Gonzales FA, Velicescu M, Jones PA (2000) Differential mRNA expression of the human DNA methyltransferases (DNMTs) 1, 3a and 3b during the G(0)/G(1) to S phase transition in normal and tumor cells. Nucleic Acids Res 28:2108–2113

Roschke AV, Glebov OK, Lababidi S, Gehlhaus KS, Weinstein JN, Kirsch IR (2008) Chromosomal instability is associated with higher expression of genes implicated in epithelial-mesenchymal transition, cancer invasiveness, and metastasis and with lower expression of genes involved in cell cycle checkpoints, DNA repair, and chromatin maintenance. Neoplasia 10:1222–1230

Ruiz S, Panopoulos AD, Herrerias A, Bissig KD, Lutz M, Berggren WT, Verma IM, Izpisua Belmonte JC (2010) A high proliferation rate is required for cell reprogramming and maintenance of human embryonic stem cell identity. Curr Biol 21:45–52

Samavarchi-Tehrani P, Golipour A, David L, Sung HK, Beyer TA, Datti A, Woltjen K, Nagy A, Wrana JL (2010) Functional genomics reveals a BMP-driven mesenchymal-to-epithelial transition in the initiation of somatic cell reprogramming. Cell Stem Cell 7:64–77

Sheng C, Zheng Q, Wu J, Xu Z, Sang L, Wang L, Guo C, Zhu W, Tong M, Liu L et al (2012a). Generation of dopaminergic neurons directly from mouse fibroblasts and fibroblast-derived neural progenitors. Cell Res

Sheng C, Zheng Q, Wu J, Xu Z, Wang L, Li W, Zhang H, Zhao XY, Liu L, Wang Z et al (2012b) Direct reprogramming of Sertoli cells into multipotent neural stem cells by defined factors. Cell Res 22:208–218

Subramanyam D, Lamouille S, Judson RL, Liu JY, Bucay N, Derynck R, Blelloch R (2011) Multiple targets of miR-302 and miR-372 promote reprogramming of human fibroblasts to induced pluripotent stem cells. Nat Biotechnol 29:443–448

Szabo E, Rampalli S, Risueno RM, Schnerch A, Mitchell R, Fiebig-Comyn A, Levadoux-Martin M, Bhatia M (2010) Direct conversion of human fibroblasts to multilineage blood progenitors. Nature 468:521–526

Takahashi K, Yamanaka S (2006) Induction of pluripotent stem cells from mouse embryonic and adult fibroblast cultures by defined factors. Cell 126:663–676

Takahashi K, Tanabe K, Ohnuki M, Narita M, Ichisaka T, Tomoda K, Yamanaka S (2007) Induction of pluripotent stem cells from adult human fibroblasts by defined factors. Cell 131:861–872

Tan ZJ, Peng Y, Song HL, Zheng JJ, Yu X (2010) N-cadherin-dependent neuron-neuron interaction is required for the maintenance of activity-induced dendrite growth. Proc Natl Acad Sci USA 107:9873–9878

Termen S, Tan EJ, Heldin CH, Moustakas A (2013) p53 regulates epithelial-mesenchymal transition induced by transforming growth factor beta. J Cell Physiol 228:801–813

Thiery JP (2002) Epithelial-mesenchymal transitions in tumour progression. Nature Rev Cancer 2:442–454

Thiery JP, Acloque H, Huang RY, Nieto MA (2009) Epithelial-mesenchymal transitions in development and disease. Cell 139:871–890

Tian X, Liu Z, Niu B, Zhang J, Tan TK, Lee SR, Zhao Y, Harris DC, Zheng G (2011) E-cadherin/beta-catenin complex and the epithelial barrier. J Biomed Biotechnol 2011:567305

Ullmann U, In't Veld P, Gilles C, Sermon K, De Rycke M, Van de Velde H, Van Steirteghem A, Liebaers I (2007) Epithelial-mesenchymal transition process in human embryonic stem cells cultured in feeder-free conditions. Molecular Human Rep 13:21–32

van Roy F, Berx G (2008) The cell-cell adhesion molecule E-cadherin. Cell Mol Life Sci 65:3756–3788

Vierbuchen T, Ostermeier A, Pang ZP, Kokubu Y, Sudhof TC, Wernig M (2010) Direct conversion of fibroblasts to functional neurons by defined factors. Nature 463:1035–1041

Wang Y, Zhang H, Chen Y, Sun Y, Yang F, Yu W, Liang J, Sun L, Yang X, Shi L et al (2009) LSD1 is a subunit of the NuRD complex and targets the metastasis programs in breast cancer. Cell 138:660–672

Wang Z, Li Y, Kong D, Sarkar FH (2010) The role of Notch signaling pathway in epithelial-mesenchymal transition (EMT) during development and tumor aggressiveness. Curr Drug Targets 11:745–751

Wang Q, Xu X, Li J, Liu J, Gu H, Zhang R, Chen J, Kuang Y, Fei J, Jiang C et al (2011a) Lithium, an anti-psychotic drug, greatly enhances the generation of induced pluripotent stem cells. Cell Res 21:1424–1435

Wang T, Chen K, Zeng X, Yang J, Wu Y, Shi X, Qin B, Zeng L, Esteban MA, Pan G et al (2011b) The histone demethylases Jhdm1a/1b enhance somatic cell reprogramming in a vitamin-C-dependent manner. Cell Stem Cell 9:575–587

Wang L, Wang L, Huang W, Su H, Xue Y, Su Z, Liao B, Wang H, Bao X, Qin D et al (2012) Generation of integration-free neural progenitor cells from cells in human urine. Nature Methods 10:84–89

Warren L, Manos PD, Ahfeldt T, Loh YH, Li H, Lau F, Ebina W, Mandal PK, Smith ZD, Meissner A et al (2010) Highly efficient reprogramming to pluripotency and directed differentiation of human cells with synthetic modified mRNA. Cell Stem Cell 7:618–630

Watanabe A, Yamada Y, Yamanaka S (2013) Epigenetic regulation in pluripotent stem cells: a key to breaking the epigenetic barrier. Philos Trans Royal Soc London Ser B Biol Sci 368:20120292

White J, Dalton S (2005) Cell cycle control of embryonic stem cells. Stem Cell Rev 1:131–138

Wu J, Issa JP, Herman J, Bassett DE Jr, Nelkin BD, Baylin SB (1993) Expression of an exogenous eukaryotic DNA methyltransferase gene induces transformation of NIH 3T3 cells. Proc Natl Acad Sci USA 90:8891–8895

Wu MZ, Tsai YP, Yang MH, Huang CH, Chang SY, Chang CC, Teng SC, Wu KJ (2011) Interplay between HDAC3 and WDR5 is essential for hypoxia-induced epithelial-mesenchymal transition. Mol Cell 43:811–822

Wu ZQ, Li XY, Hu CY, Ford M, Kleer CG, Weiss SJ (2012) Canonical Wnt signaling regulates Slug activity and links epithelial-mesenchymal transition with epigenetic Breast Cancer 1, Early Onset (BRCA1) repression. Proc Natl Acad Sci USA 109:16654–16659

Yang Y, Pan X, Lei W, Wang J, Song J (2006) Transforming growth factor-beta1 induces epithelial-to-mesenchymal transition and apoptosis via a cell cycle-dependent mechanism. Oncogene 25:7235–7244

Yu J, Vodyanik MA, Smuga-Otto K, Antosiewicz-Bourget J, Frane JL, Tian S, Nie J, Jonsdottir GA, Ruotti V, Stewart R et al (2007)

Induced pluripotent stem cell lines derived from human somatic cells. Science 318:1917–1920

Yu J, Hu K, Smuga-Otto K, Tian S, Stewart R, Slukvin II, Thomson JA (2009) Human induced pluripotent stem cells free of vector and transgene sequences. Science 324:797–801

Yu L, Liu S, Guo W, Zhang C, Zhang B, Yan H, Wu Z (2014) hTERT promoter activity identifies osteosarcoma cells with increased EMT characteristics. Oncol Lett 7:239–244

Zhang X, Stappenbeck TS, White AC, Lavine KJ, Gordon JI, Ornitz DM (2006) Reciprocal epithelial-mesenchymal FGF signaling is required for cecal development. Development 133:173–180

Zhao XY, Li W, Lv Z, Liu L, Tong M, Hai T, Hao J, Guo CL, Ma QW, Wang L et al (2009) iPS cells produce viable mice through tetraploid complementation. Nature 461:86–90

Zhou C, Nitschke AM, Xiong W, Zhang Q, Tang Y, Bloch M, Elliott S, Zhu Y, Bazzone L, Yu D et al (2008) Proteomic analysis of tumor necrosis factor-alpha resistant human breast cancer cells reveals a MEK5/Erk5-mediated epithelial-mesenchymal transition phenotype. Breast Cancer Res BCR 10:R105

Zhou H, Wu S, Joo JY, Zhu S, Han DW, Lin T, Trauger S, Bien G, Yao S, Zhu Y et al (2009) Generation of induced pluripotent stem cells using recombinant proteins. Cell Stem Cell 4:381–384

Zhou T, Benda C, Duzinger S, Huang Y, Li X, Li Y, Guo X, Cao G, Chen S, Hao L et al (2011) Generation of Induced Pluripotent Stem Cells from Urine. J Am Soc Nephrol 22:1221–1228

Zhou T, Benda C, Dunzinger S, Huang Y, Ho JC, Yang J, Wang Y, Zhang Y, Zhuang Q, Li Y et al (2012) Generation of human induced pluripotent stem cells from urine samples. Nature Protoc 7:2080–2089

5

Differential regulation of H3S10 phosphorylation, mitosis progression and cell fate by Aurora Kinase B and C in mouse preimplantation embryos

Wenzhi Li[1], Peizhe Wang[1], Bingjie Zhang[2], Jing Zhang[1], Jia Ming[1], Wei Xie[2], Jie Na[1]✉

[1] Center for Stem Cell Biology and Regenerative Medicine, School of Medicine, Tsinghua University, Beijing 100084, China
[2] Center for Stem Cell Biology and Regenerative Medicine, MOE Key Laboratory of Bioinformatics, THU-PKU Center for Life Sciences, School of Life Sciences, Tsinghua University, Beijing 100084, China
✉ Correspondence: jie.na@tsinghua.edu.cn (J. Na)

ABSTRACT

Coordination of cell division and cell fate is crucial for the successful development of mammalian early embryos. Aurora kinases are evolutionarily conserved serine/threonine kinases and key regulators of mitosis. Aurora kinase B (AurkB) is ubiquitously expressed while Aurora kinase C (AurkC) is specifically expressed in gametes and preimplantation embryos. We found that increasing AurkC level in one blastomere of the 2-cell embryo accelerated cell division and decreasing AurkC level slowed down mitosis. Changing AurkB level had the opposite effect. The kinase domains of AurkB and AurkC were responsible for their different ability to phosphorylate Histone H3 Serine 10 (H3S10P) and regulate metaphase timing. Using an Oct4-photoactivatable GFP fusion protein (Oct4-paGFP) and fluorescence decay after photoactivation assay, we found that AurkB overexpression reduced Oct4 retention in the nucleus. Finally, we show that blastomeres with higher AurkC level elevated pluripotency gene expression, which were inclined to enter the inner cell mass lineage and subsequently contributed to the embryo proper. Collectively, our results are the first demonstration that the activity of mitotic kinases can influence cell fate decisions in mammalian preimplantation embryos and have important implications to assisted reproduction.

KEYWORDS Aurora kinase, mouse preimplantation embryo, cell fate, development, mitosis

INTRODUCTION

Precise and timely segregation of sister chromatids during cell division is essential for early embryo development. In mammalian preimplantation embryos, chromosome segregation errors happen frequently, this can lead to deleterious downstream events, such as chromosome number and structure alteration, DNA damage, cytokinesis failure, which eventually lead to embryo arrest or severe developmental defects (Bolton et al., 2016; Vanneste et al., 2009; Wong et al., 2010). Global gene expression profiling studies discovered that many cell division kinases and regulators are highly enriched in preimplantation embryos (Xue et al., 2013). The high rate of cell division error implies that the regulation of mitosis, particularly the activity of cell division checkpoints, might be different from that in somatic cells. Spindle assembly checkpoint (SAC) is a surveillance mechanism safeguards the accurate separation of sister chromatids (Musacchio, 2015). It remains active until all kinetochores attach to spindle microtubules correctly (Musacchio, 2015). When SAC is active, Aurora kinase B (AurkB) phosphorylates and inactivates CDC20, to prevent the activation of anaphase-promoting complex/cyclosome (APC/C) (Hagting et al., 2002; Nasmyth, 2005; Ruchaud et al., 2007). Upon the inactivation of SAC, or reduction in AurkB activity, APC/C becomes active and ubiquitinates securin which leads to its degradation (Hagting et al., 2002). Separase, which is repressed by securin, becomes activated and cleaves the components of cohesin to permit the segregation of sister chromatids (Schvartzman et al., 2010). Thus, securin protein level negatively correlates with the activity of SAC, and the rate of its degradation reflects the dynamics of SAC inactivation during mitosis (Hagting et al., 2002). Due to the difficulties of manipulating the mouse preimplantation embryo, there has not been any study using securin degradation to study SAC activity in this system.

AurkC is a mammalian specific Aurora kinase closely related to AurkB (Carmena and Earnshaw, 2003; Lin et al., 2014; Sasai et al., 2004). It is specially highly expressed in mammalian gamete and required for spermatocyte meiosis (Dieterich et al., 2007). During oocytes meiosis I, AurkC displayed distinct function from AurkB (Balboula and Schindler, 2014; Sharif et al., 2010). It regulates localized chromosome passenger complex (CPC) activity and kinetochore microtubule (K-MT) attachments during the metaphase of meiosis I, thus promoting APC/C activation and securin degradation, while too much AurkC resulted in cytokinesis failure in meiotic I (Balboula and Schindler, 2014; Sharif et al., 2010). After fertilization, a significant number of AurkC$^{-/-}$ embryos showed division defect which results in subfertility, while AurkB seemed not required in cleavage stage embryos but become important in blastocysts (Fernandez-Miranda et al., 2011; Schindler et al., 2012). In human, both AurkB and AurkC are expressed in preimplantation embryos, and AurkC level remained high until 8-cell stage (Avo Santos et al., 2011). Above studies highlighted the importance of AurkB and C function during preimplantation development, however, a direct comparative analysis of their mitotic roles in preimplantation embryos has been lacking and whether their activity may affect cell fate decision is another intriguing question.

In mammalian preimplantation embryos, cell division timing appeared to associate with cell fate choices (Chazaud and Yamanaka, 2016; Zernicka-Goetz, 2006). At 8-cell stage, the earlier dividing cells were more likely to locate to the inside of the morula, while the later dividing cells staying at outside (Tabansky et al., 2013; Zernicka-Goetz, 2006; Zernicka-Goetz et al., 2009). Subsequently, the inside cells gained advantage to enter the inner cell mass (ICM) lineage and outside cells become the trophecoderm (TE). Thus, the prior cell division advantage was accumulated through the first three cell divisions after fertilization (Zernicka-Goetz, 2006).

Here by using mRNA overexpression, siRNA knocking-down and live imaging, we show that the speed of mitosis and the degradation rate of securin in mouse preimplantation embryos can be controlled by the level of AurkB and AurkC. Further, AurkB and AurkC had different influence on the nuclear dynamics of Oct4 as shown by an Oct4-photoactivatable GFP fusion protein (Oct4-paGFP). Increasing AurkC level promoted pluripotency gene expression. Finally, we show that cells with more AurkC protein tend to locate to ICM and contribute to the embryo proper. Our results demonstrated that difference in the activity of cell division kinases has significant impact on cell fate decision.

Figure 1. AurkB and AurkC have different effect on Securin degradation and sister chromatids separation. (A) Q-PCR analysis showing AurkB and AurkC mRNA expression level during mouse preimplantation development, each sample were normalized by AurkB level at zygote stage, the bar and whiskers indicate means and SEM, $^{**}P < 0.01$, $^{***}P < 0.001$. (B) AurkC localization in a dividing blastomere. DNA (blue), AurkC (green), Tubulin (red). Scale bar, 40 μm. (C) The localization of AurkB and AurkC (green) during interphase, DNA (blue). Scale bars, 20 μm. (D) Q-PCR analysis showing AurkB and AurkC expression level in the daughter cells from earlier and later dividing cells, the level of AurkB in the later dividing cells was considered "1", the bar and whiskers indicate means and SEM, $^{**}P < 0.01$. (E) Schematic view of live imaging and Securin-mCherry degradation assay. (F) Selected frames from time lapse imaging movies of Securin-mCherry degradation in Control, AurkB-OE, AurkC-OE, siAurkB, and siAurkC groups embryos. Securin-mCherry (red), H2B-GFP (green), time interval, 15 min. Scale bars, 20 μm. (G) The Securin-mCherry degradation analysis in control (n = 40), AurkB-OE (n = 24), AurkC-OE (n = 52), siAurkB (n = 50), and siAurkC (n = 42) groups, from NEBD to chromosome separation. Y-axis, relative fluorescence intensity; X-axis, time in minutes (min). The continuous line and whiskers indicate means and SEM. (H) Quantification of the cell division time (from NEBD to chromatin separation, time/min, Y-axis) in control (n = 40), AurkB-OE (n = 24), AurkC-OE (n = 52), siAurkB (n = 50), and siAurkC (n = 42) groups (X-axis), each dot represents one embryo in the group, and the bar and whiskers indicate means and SD, $^{*}P < 0.05$, $^{**}P < 0.01$, $^{***}P < 0.001$.

RESULTS

Aurora kinase B and C differentially regulate mitosis progression in mouse preimplantation embryos

Previously, we showed that AurkB and AurkC differentially regulated securin degradation during mouse oocyte meiosis (Sharif et al., 2010). However, it is unclear whether this is also the case during embryo mitosis. We first examined AurkB and AurkC expression in mouse preimplantation embryos. Q-PCR analysis showed that AurkC was highly expressed throughout preimplantation stages, and AurkB was hardly detectable before 2-cell stage. From 2-cell to morula stage, although AurkB was transcribed, its level was still lower than that of AurkC (Fig. 1A). Next we used HA tagged AurkB and AurkC to reveal their subcellular distribution (Fig. 1B and 1C). During mitosis, AurkC associated with the chromosome, and translocated to the midbody between the two daughter cells during anaphase (Fig. 1B).

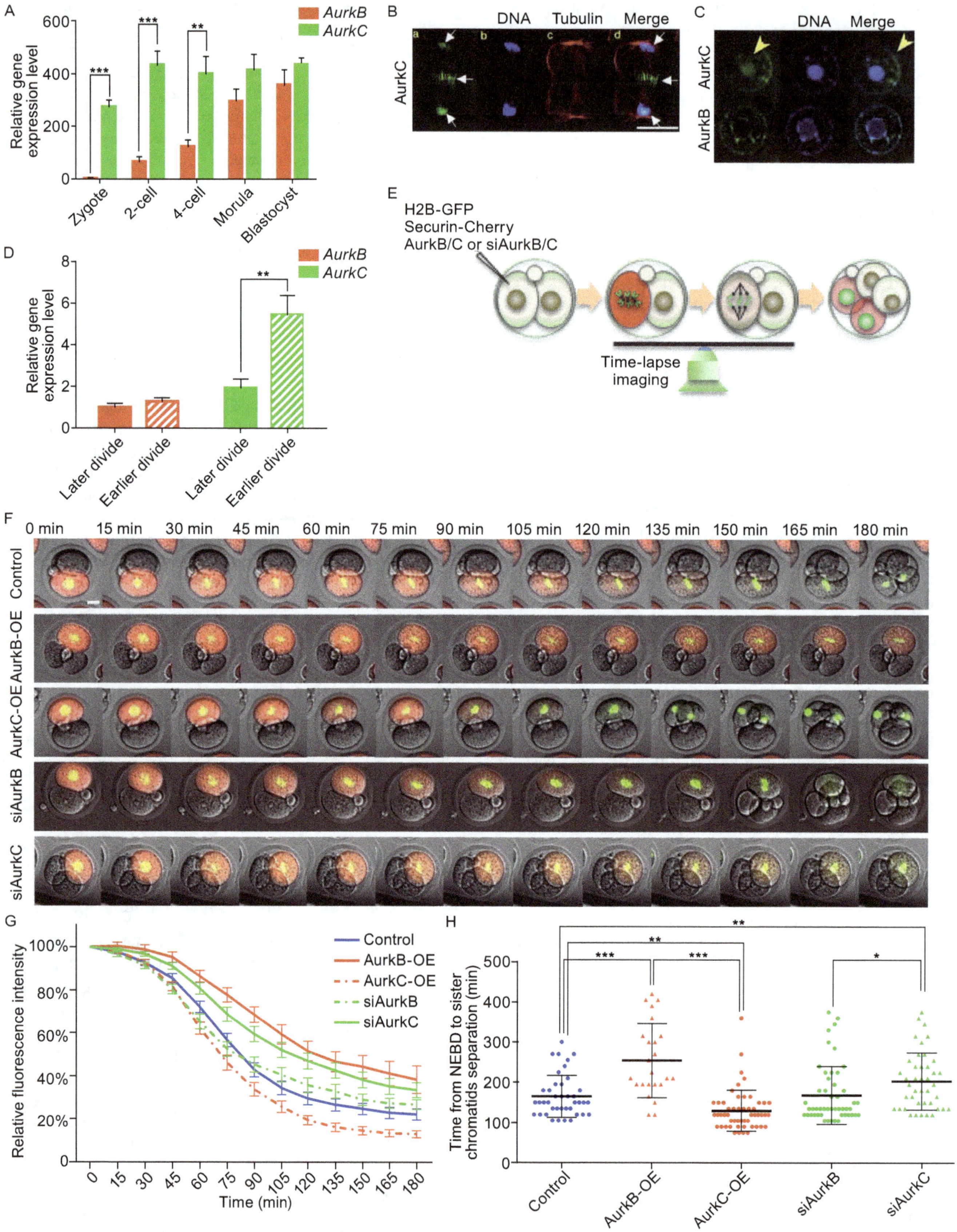
A
AurkB
AurkC
Relative gene expression level
***
***
**
Zygote
2-cell
4-cell
Morula
Blastocyst
B
DNA
Tubulin
Merge
AurkC
C
DNA
Merge
AurkC
AurkB
D
Relative gene expression level
Later divide
Earlier divide
E
H2B-GFP
Securin-Cherry
AurkB/C or siAurkB/C
Time-lapse imaging
F
0 min
15 min
30 min
45 min
60 min
75 min
90 min
105 min
120 min
135 min
150 min
165 min
180 min
Control
AurkB-OE
AurkC-OE
siAurkB
siAurkC
G
Relative fluorescence intensity
Time (min)
Control
AurkB-OE
AurkC-OE
siAurkB
siAurkC
H
Time from NEBD to sister chromatids separation (min)
Control
AurkB-OE
AurkC-OE
siAurkB
siAurkC

During interphase, both AurkB and AurkC showed nuclear localization, while AurkC but not AurkB was seen in the nucleolus (Fig. 1C). During preimplantation experiment, the two blastomeres of the 2-cell embryo often divide asynchronously. We also compared AurkB and AurkC expression level in earlier and later dividing cells at early 4-cell stage. Mouse embryos were first cultured to very late 2-cell stage and then monitored every hour. Embryos with 3 cells (one blastomere has divided) were picked and dissociated. The daughter cells from the early and late dividing blastomere were separated and cultured to early 4-cell stage. Q-PCR analysis showed that earlier dividing cells had higher AurkC level but similar AurkB level compared with late dividing cells (Fig. 1D).

Next we asked whether cell division kinetics in preimplantation embryos can be altered by manipulating AurkB and AurkC levels. AurkB or AurkC mRNA or siRNA were co-injected with Securin-mCherry and H2B-GFP mRNAs into one blastomere of 2-cell embryos and time-lapse microscopy was used to monitor the speed of Securin degradation (Fig. 1E). The specificity of siRNA targeting AurkB and AurkC, and overexpression level of mRNAs were confirmed (Fig. S1B and S1C). During interphase, Securin-mCherry showed diffused cytoplasmic distribution and maintained a relatively stable level. The intensity of the red fluorescence represented Securin protein level, its value just before the nuclear envelope break down (NEBD) was considered as 100%. In the control blastomeres injected with Securin-mCherry and H2B-GFP mRNAs, 90 min after NEBD, Securin level decreased to 50% of the initial level. After 180 min, when the red fluorescence reached about 20%–30% of the initial level, the sister chromatids separated (Fig. 1F and Movie S1A–E). Overexpression of AurkB (AurkB-OE) significantly increased the time required for Securin-mCherry degradation. It took almost 150 min for Securin to reduce to 50%, and after 180 min, Securin was still at 40% level (Fig. 1G). During the prolonged metaphase, the chromosomes remain condensed (Fig. 1F), while sister chromatids stayed at the equatorial plan of the cell and failed to separate (Fig. 1F). On the other hand, overexpression of AurkC (AurkC-OE) significantly accelerated Securin-mCherry degradation, the red fluorescence reduced 50% within 75 min (Fig. 1G). Moreover, it reached less than 20%, even lower level than in control cells (Fig. 1G). Conversely, knocking down of AurkC through siRNA (siAurkC) in blastomeres delayed 50% Securin-mCherry degradation time to 120 min (Fig. 1G), similar to the timing in AurkB overexpression blastomeres. Sister chromatid separation was also delayed in AurkC knock-down cells (Fig. 1G). AurkB knock-down (siAurkB) cells had similar 50% Securin-degradation time as control cells. Consequently, the time required to complete cell division was also affected by altering AurkB and AurkC levels (Fig. 1H). The control group and siAurkB group had similar cell division time, 166.1 ± 8.240 min (n = 40) and 169.2 ± 10.16 min (n = 50) respectively. The cell division time was the longest in AurkB-OE group, 255.0 ± 17.96 min (n = 25). SiAurkC group had the second longest division time, 204.3 ± 10.93 min (n = 42). AurkC-OE group had the shortest division time, 130.8 ± 6.838 min (n = 54). Consequently, AurkC-OE cells tend to divide earlier than sister uninjected cells, and AurkB-OE and siAurkC cells divided later than uninjected cells (Fig. S1D). The above results suggested that in cleavage stage embryos, the timing of division can be controlled by varying AurkB and AurkC levels. AurkC appeared to be the major Aurora kinase regulating the mitosis speed, while too much AurkB seemed not compatible with the fast cell cycle progression rate in preimplantation development.

The kinase domains of AurkB and C were responsible for their distinct functions

AurkB and AurkC shared similar domain structure. To better understand how AurkB and AurkC differentially regulate mitosis progression, we constructed kinase domain swapped AurkB/C chimeric proteins, namely AurkBCB and AurkCBC (Fig. 2A). We also generated a kinase dead AurkC (AurkC-KD) which carried T171A and T175A mutations in the cAMP-dependent protein kinase phosphorylation sites, the double mutants could abolish the activity of AurkC (Chen and Tang, 2002). The mRNAs of mutant Aurora kinases were mixed with Securin-mCherry and H2B-GFP mRNAs and microinjected into one blastomere of 2-cell stage embryos, followed by time-lapse imaging during 2 to 4-cell division as described in Fig. 1E. Quantification of Securin-mCherry degradation rate revealed that AurkCBC, which had the kinase domain from AurkB, showed delayed Securin-mCherry degradation similar with AurkB-OE. Cells took about 120 min (n = 21) to reach 50% of the Securin-mCherry fluorescence (Fig. 2B). While AurkBCB, which contained the AurkC kinase domain, showed accelerated degradation of Securin-mCherry. Cells expressing AurkBCB took about 75 min (n = 22) to reach 50% Securin level, which was similar to the degradation curve of AurkC-OE cells (Fig. 2B). Interestingly, the slope of the degradation curve in AurkCBC-OE cells was slightly steeper than that from AurkB-OE cells (Fig. 2B), suggesting that the AurkC domains flanking the kinase domain may also play some roles in dampening the checkpoint activity. In AurkBCB-OE cells, although the degradation of Securin-mCherry was initially faster than that in AurkC-OE cells, it did not reach as low level as in AurkC-OE cells (Fig. 2B), suggesting that the N- and C-terminus non-kinase domain of AurkB can elevate the AurkC kinase activity. The overexpression of AurkC-KD strongly inhibited Securin-mCherry degradation which reduced only 40% after 180 min (Fig. 2B). Consequently, nearly 90% (n = 26) of the cells that overexpressed AurkC-KD failed to complete cytokinesis (Movie S2). We examined the cellular localization of mutant Aurora kinases. Interestingly, AurkBCB showed similar location as AurkC during metaphase, it distributed along the chromosome (Fig. 2C),

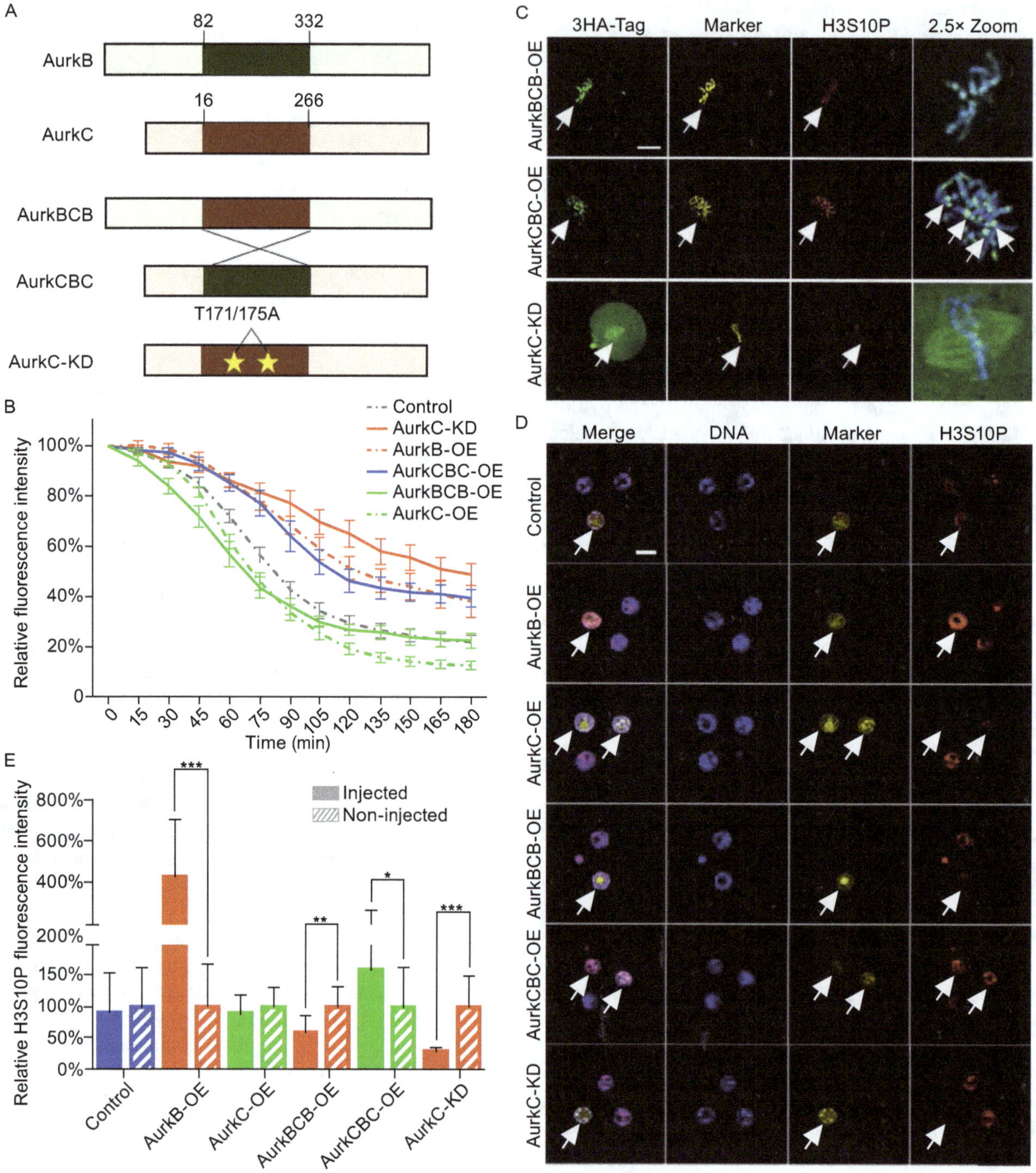

Figure 2. The kinase domain of AurkB and AurkC determined their biological activity. (A) Schematic view of the domain structure of AurkB, C, CBC, BCB, and kinase deficient AurkC-KD. (B) The Securin-mCherry fluorescence (Y-axis) degradation curve in AurkBCB-OE (*n* = 21), AurkCBC-OE (*n* = 22), and AurkC-KD (*n* = 26) groups, from NEBD to sister chromatids separation. Y-axis, fluorescence intensity; X-axis, time (min). The continuous line and whiskers indicate means and SEM. (C) 3HA-AurkBCB, CBC, and AurkC-KD localization during 2–4 cell stage division. DNA (blue), 3HA-tag (green), H2B-GFP (yellow), H3S10P (red). Scale bar, 20 µm. (D) H3S10P immunostaining of 4-cell stage embryos in the control (*n* = 26), AurkB-OE (*n* = 28), AurkC-OE (*n* = 24), AurkBCB-OE (*n* = 30), AurkCBC-OE (*n* = 22), and AurkC-KD (*n* = 24) groups. DNA (blue), H2B-GFP (marker, yellow), H3S10P (red). Scale bars, 20 µm. (E) Relative fluorescence intensity of H3S10P in injected and no injected cells. Control (*n* = 26), AurkB-OE (*n* = 28), AurkC-OE (*n* = 24), AurkBCB-OE (*n* = 30), AurkCBC-OE (*n* = 22), and AurkC-KD (*n* = 24) groups. The bar and whiskers indicate means and SD, *$P < 0.05$, **$P < 0.01$, ***$P < 0.001$.

while AurkCBC appeared to concentrate more on the kinetochore/centromere region (Fig. 2C-b, arrow heads). In contrast, significant amount of AurkC-KD diffusely distributed in the cytoplasm, some of it associated with spindle microtubules and chromosomes (Fig. 2C-c). Above results suggest that the kinase domain was key to the unique function of AurkB and AurkC, namely, the specific chromosome location and differential regulation of SAC activity. As AurkC-KD failed to localize properly, it may sequester other key chromosome passenger proteins that required for the degradation of securin-mCherry thus lead to failure in sister chromatids separation.

In preimplantation embryos, low level of H3S10P was present during interphase. As H3S10 is a well-known substrate of AurkB phosphorylation during mitosis in somatic cells (Crosio et al., 2002). We also examined the effect of AurkB and AurkC towards interphase H3S10P. Immunostaining revealed that the overexpression of AurkB significantly up-regulated H3S10P level during interphase (Fig. 2D-b). Quantification showed more than 4-fold increase of H3S10P fluorescence intensity in AurkB-OE cells (n = 28), whereas AurkC-OE (n = 24) did not significantly affect H3S10P level at this stage (Fig. 2D-c and 2E). In contrary, AurkC-KD significantly reduced interphase H3S10P (Fig. 2D-f). The H3S10P intensity in AurkC-KD cells were reduced to about 70% of that in control cells (n = 24) (Fig. 2E). The level of interphase H3S10P seemed to positively correlate with the presence of AurkB kinase domain, as the chimeric AurkBCB reduced H3S10P level by about 40% (n = 30), while AurkCBC increased H3S10P level by 60% (n = 22) (Fig. 2D-a, 2D-b, and 2E). To discover whether AurkB/C may affect other important histone modifications in preimplantation embryos indirectly through differentially regulation of H3S10P level, we performed immunostaining of H3K9me3 and H3K4me3, which are histone marks of repressed and active chromatin respectively. They did not appear to be significantly affected by AurkB or AurkC (Figs. S2 and S3). Although H3K9me3 staining (Fig. S2A and S2B) showed a negative correlation with H3S10P, a recent paper reported that the affinity of H3K9me3 antibody can be affected by the adjacent H3S10 phosphorylation (Rothbart et al., 2015). Hirota et al. reported that H3S10 phosphorylation by AurkB could lead to HP1 dissociation from the heterochromatin (Hirota et al., 2005). However, HP1 expression pattern was similar in AurkB or AurkC overexpression cells (Fig. S2A).

Aurora kinase B and C level affected Oct4 nuclear kinetics in cleavage stage embryos

We next tested whether the level of AurkB/C may affect the biological activity of key pluripotency factors. To this end, we constructed a photoactivable GFP fused to the C-terminus of Oct4 (Oct4-paGFP) similar to the one described by Plachta et al. (2011). We injected one blastomere of the late 2-cell stage embryo with a mRNA mixture of Oct4-paGFP, H2B-mCherry, and AurkB or AurkC. Embryos were cultured to mid-4-cell stage in the incubator, the two daughter cells of the injected cell can be recognized by the H2B-mCherry expression. We then performed the fluorescence decay after photoactivation assay (FDAP) using a confocal microscope (Fig. 3A). Upon targeted photoactivation of Oct4-paGFP by a 40% intensity of 405 nm laser for 3 s, the nucleus in the focus plan, which labelled with H2B-mCherry, showed bright green fluorescence, indicating successful conversion of the photoactivatable protein (Fig. S4A and S4B). Due to different focus plan, only the nucleus in focus appeared with fluorescence (Fig. S4C). Z-plans (2 μm/section, 31 sections in total) were acquired every 10 min for 4 h. The fluorescence intensity of Oct4-paGFP was quantified over time (Fig. 3B, 3C, and Movie S3A–E). The delay rate of Oct4-paGFP appeared to fall into two clusters: cluster 1, slower decay, and cluster 2, faster decay, which were likely to represent higher and lower affinity of Oct4 binding to DNA (Fig. 3C-a, control). This was similar to the previous report by Plachta et al. (2011). Interestingly, upon alteration of AurkB and C levels, cells from injected 4-cell embryos still had cluster 2 Oct4-paGFP decay. However, in AurkB-OE and siAurkC groups, fewer cells had cluster 1 decay or the decay rate of cluster 1 shifted towards cluster 2 (Fig. 3C-b and 3C-e). This suggests that increasing AurkB level or decreasing AurkC level may affect the kinetics of Oct4-DNA binding.

Figure 3. Aurora kinase B and C level affected Oct4 kinetics in the nucleus. (A) Schematic view of Oct4-paGFP fluorescence decay after photoactivation (FDAP) assay. (B) Selected frames from time lapse imaging movies of Oct4-paGFP degradation in control, AurkB-OE, AurkC-OE, siAurkB, and siAurkC groups embryos. H2B-mCherry (red), Oct4-paGFP (green), time interval, 10 min. Scale bars, 20 μm. (C) The Oct4-paGFP relative fluorescence (Y-axis) degradation analysis in control (n = 23), AurkB-OE (n = 26), AurkC-OE (n = 22), siAurkB (n = 20), and siAurkC (n = 23) groups, time interval 10 min, every symbol line represents an embryo in the group and the continuous line and whiskers in merged curve indicate means and SEM.

Cells with higher AurkC level tend to enter the ICM lineage

To investigate whether different levels of AurkB and AurkC may influence cell fate, we dissociated early 8-cell stage embryos into single cells and compared the pluripotency genes expression in the daughter cells divided from injected cells. In AurkB-OE and siAurkC embryos which had delayed cell cycle, the levels of pluripotency genes *Oct4*, *Nanog*, *Klf4*, *Prdm14*, and *Nr5a2* were significantly reduced (Fig. 4A). While in AurkC-OE and siAurkB cells, which had

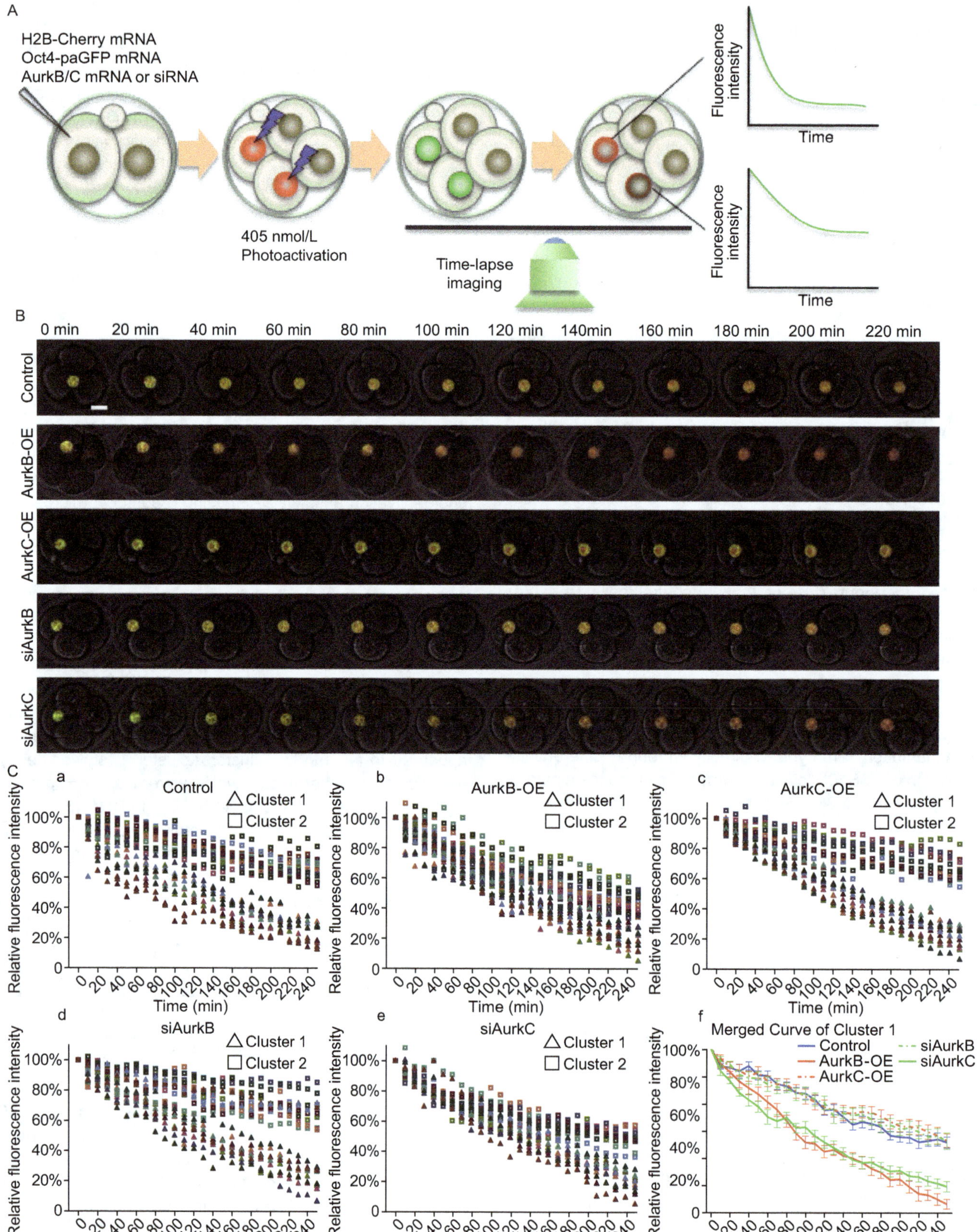
A
H2B-Cherry mRNA
Oct4-paGFP mRNA
AurkB/C mRNA or siRNA
405 nmol/L
Photoactivation
Time-lapse
imaging
Fluorescence intensity
Time
Fluorescence intensity
Time
B
0 min
20 min
40 min
60 min
80 min
100 min
120 min
140min
160 min
180 min
200 min
220 min
Control
AurkB-OE
AurkC-OE
siAurkB
siAurkC
C
a
Control
△Cluster 1
□Cluster 2
b
AurkB-OE
c
AurkC-OE
d
siAurkB
e
siAurkC
f
Merged Curve of Cluster 1
Control
AurkB-OE
AurkC-OE
siAurkB
siAurkC
Relative fluorescence intensity
Time (min)

accelerated mitosis, the expression levels of above genes were similar with the control group (Fig. 4A).

To find out whether changing AurkB/C level may affect cell fate at blastocyst stage, we injected AurkB or AurkC mRNA/siRNA (with H2B-GFP as a marker) into one cell of 2-cell stage embryos and cultured them to blastocyst stage (E4.0). They were fixed and immunostained for Oct4 and Cdx2 proteins, followed by confocal microscopy analysis (Fig. 4B). We found that the total number of H2B-GFP positive cells was not significantly different in AurkB/C-OE or siAurkB/C groups, the average positive cell number was 26.54 ± 1.047 in control group (n = 41), 25.90 ± 0.9521 in AurkB-OE group (n = 30), 24.91 ± 0.8429 in AurkC-OE group (n = 32), 24.95 ± 0.8951 in siAurkB group (n = 40), and 24.94 ± 1.067 in siAurkC group (n = 31) (Fig. 4C). Interestingly, in AurkB-OE embryos, the percentage of Oct4 positive cells among H2B-GFP cells was clearly reduced (20.86% ± 1.13, n = 30) compared to that in control embryos (31.23% ± 1.21, n = 41) (Fig. 4D). While higher percentage of daughter cells from AurkC-OE cells had Oct4 expression (36.18% ± 1.45, n = 32). AurkB knock-down did not affect the ratio of Oct4 positive cells (31.92% ± 1.61, n = 40), but AurkC knock-down lead to fewer Oct4 positive cells amongst H2B-GFP cells (24.76% ± 1.30, n = 31), which was similar to AurkB-OE.

Next, we used ROSA26Sor$^{tm4(ACTB\text{-}tdTomato,\ \text{-}EGFP)\ Luo}$ (ROSA) transgenic mice to trace the lineage of AurkB/C-OE cells. H2B-GFP, AurkB or AurkC mRNAs were co-injected with CRE mRNA into one cell of 4-cell stage embryos from ROSA female mice crossed with F1 male mice. Upon CRE protein expression, it will excise the membrane-bound Tomato (mTomato) cassette to allow the expression of the downstream membrane-bound GFP (mGFP), thus the progeny of the cell where CRE was expressed will display mGFP fluorescence, while the rest of the embryo will have mTomato red fluorescence. After injection, embryos were transferred into the oviduct of pseudopregnant ICR female mice. At E13.5, embryos were dissected for observation. Three types of GFP cell distribution can be observed: located only to the embryo body, or only to the placenta, or to both embryo body and placenta (Fig. 5A). Progenies derived from cells with transient AurkB overexpression were more likely to contribute to the placenta (5/11) and less likely to the embryo body (2/11), while cell descendent from transient AurkC overexpression blastomere preferred to contribute to embryo body (6/9) (Fig. 5B). These results suggested that blastomeres with higher AurkC level indeed were more likely to take the pluripotent cell fate and contribute to the embryo development.

Taken together, our results suggested that the levels of AurkB and C in cleavage stage embryos may influence cell division speed as well as histone H3S10 phosphorylation level, and subsequently affect cell fate choice during development.

Figure 4. Aurora kinase B and C affected pluripotency genes expression and cell fate during early morula stage. (A) Relative genes expression analysis (control, AurkB-OE, AurkC-OE, siAurkB, siAurkC) of early morula stage (8-cell stage) embryos. Each sample were normalized by control, the bar and whiskers indicate means and SEM, $^{*}P < 0.05$, $^{**}P < 0.01$, $^{***}P < 0.001$. (B) Immunostaining of Oct4 and Cdx2 in late blastocyst stage of the control (n = 41), AurkB-OE (n = 30), AurkC-OE (n = 32), siAurkB (n = 40), and siAurkC (n = 31) groups. DNA (blue), H2B-GFP (marker, yellow), Oct4 (red), Cdx2 (green), 3D projections (max. proj.) of merged images are shown in the right panels. Scale bars, 20 μm. (C) Average cell number of H2B-GFP positive cell in the control (n = 41), AurkB-OE (n = 30), AurkC-OE (n = 32), siAurkB (n = 40), and siAurkC (n = 31) groups, each point represents data from one embryo, the bar and whiskers indicate means and SD, H2B-GFP positive cells are the daughter cells of the injected blastomere. (D) Percentage of H2B-GFP positive cells also expressed Oct4 in control (n = 41), AurkB-OE (n = 30), AurkC-OE (n = 32), siAurkB (n = 40), and siAurkC (n = 31) groups, each dot represents data from one embryo, the bar and whiskers indicate means and SD, $^{*}P < 0.05$, $^{***}P < 0.001$.

DISCUSSION

In this study, by modulation of AurkB or AurkC levels in part of mouse preimplantation embryo, combined with live cell imaging and lineage tracing experiments, we showed that changing key mitotic kinase activities can significantly impact the cell division dynamics and cell fate decision.

Previous research demonstrated that in both mouse and human cleavage stage embryos, AurkC was present at much higher level than AurkB, a ubiquitously expressed mitotic kinase (Avo Santos et al., 2011; Fernandez-Miranda et al., 2011; Schindler et al., 2012; Sharif et al., 2010). Despite their similarities in the protein structure and cellular localization at chromosome passenger complex, AurkC displayed unique characteristics different from AurkB. Lin et al. showed that in cancer cells, overexpression of AurkC could displace AurkB from centromere and impaired SAC function (Lin et al., 2014). Our previous results also showed that high level AurkC in mouse oocytes accelerated securin degradation, caused chromosome mis-alignment and cytokinesis failure (Sharif et al., 2010). The mitosis in cleavage mammalian embryos are particularly error prone, consequently, a high proportion of chromosomal mosaicism was observed (Avo Santos et al., 2011; Vanneste et al., 2009). Curiously, AurkC but not AurkB was higher expressed during this time window, and AurkC expression level was higher in the daughter cells from the earlier dividing blastomere of the 2-cell embryo. Using a live-imaging based

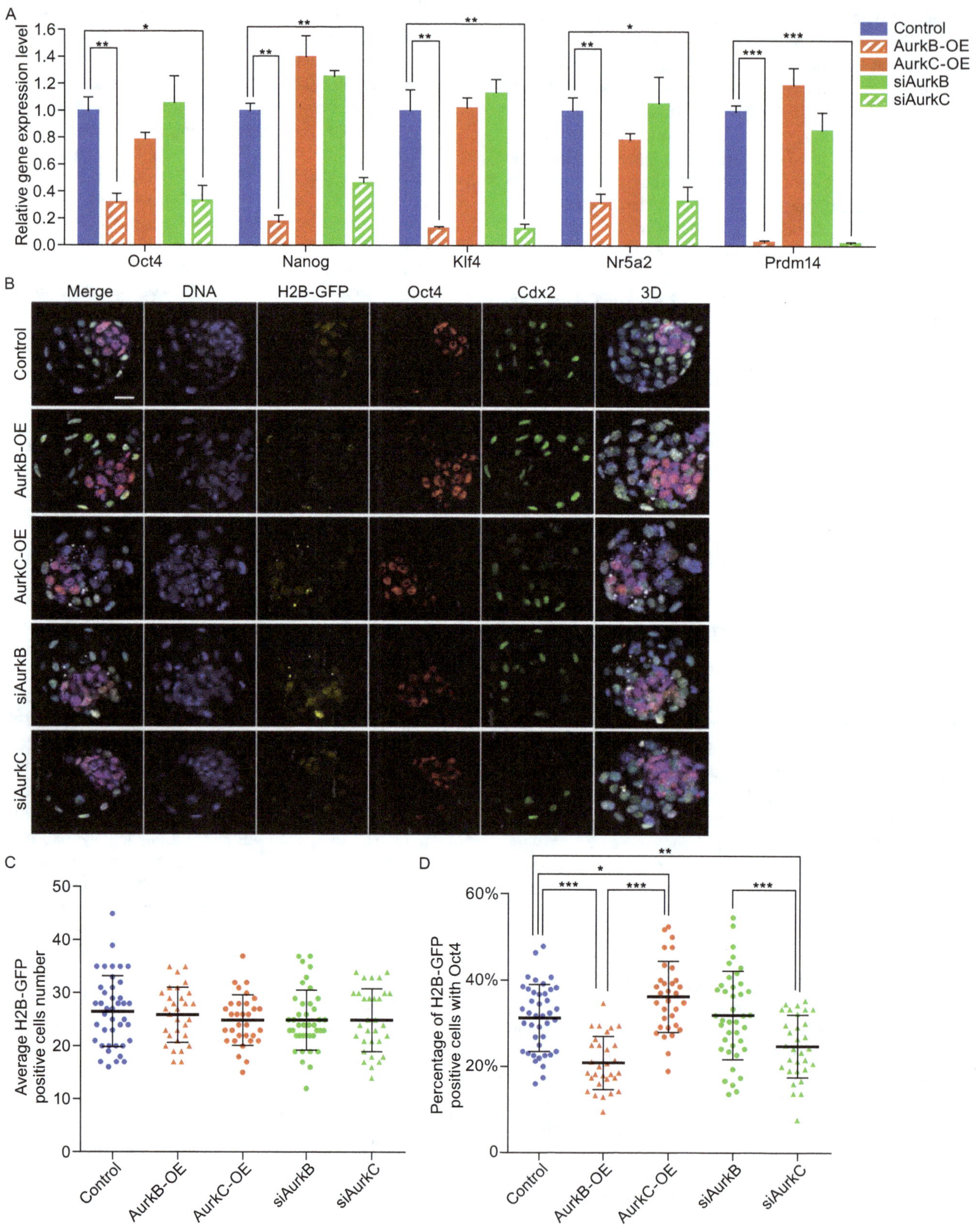
A
Relative gene expression level
1.6
1.4
1.2
1.0
0.8
0.6
0.4
0.2
0.0
Oct4
Nanog
Klf4
Nr5a2
Prdm14
Control
AurkB-OE
AurkC-OE
siAurkB
siAurkC
B
Merge
DNA
H2B-GFP
Oct4
Cdx2
3D
Control
AurkB-OE
AurkC-OE
siAurkB
siAurkC
C
Average H2B-GFP positive cells number
D
Percentage of H2B-GFP positive cells with Oct4
60%
40%
20%
0

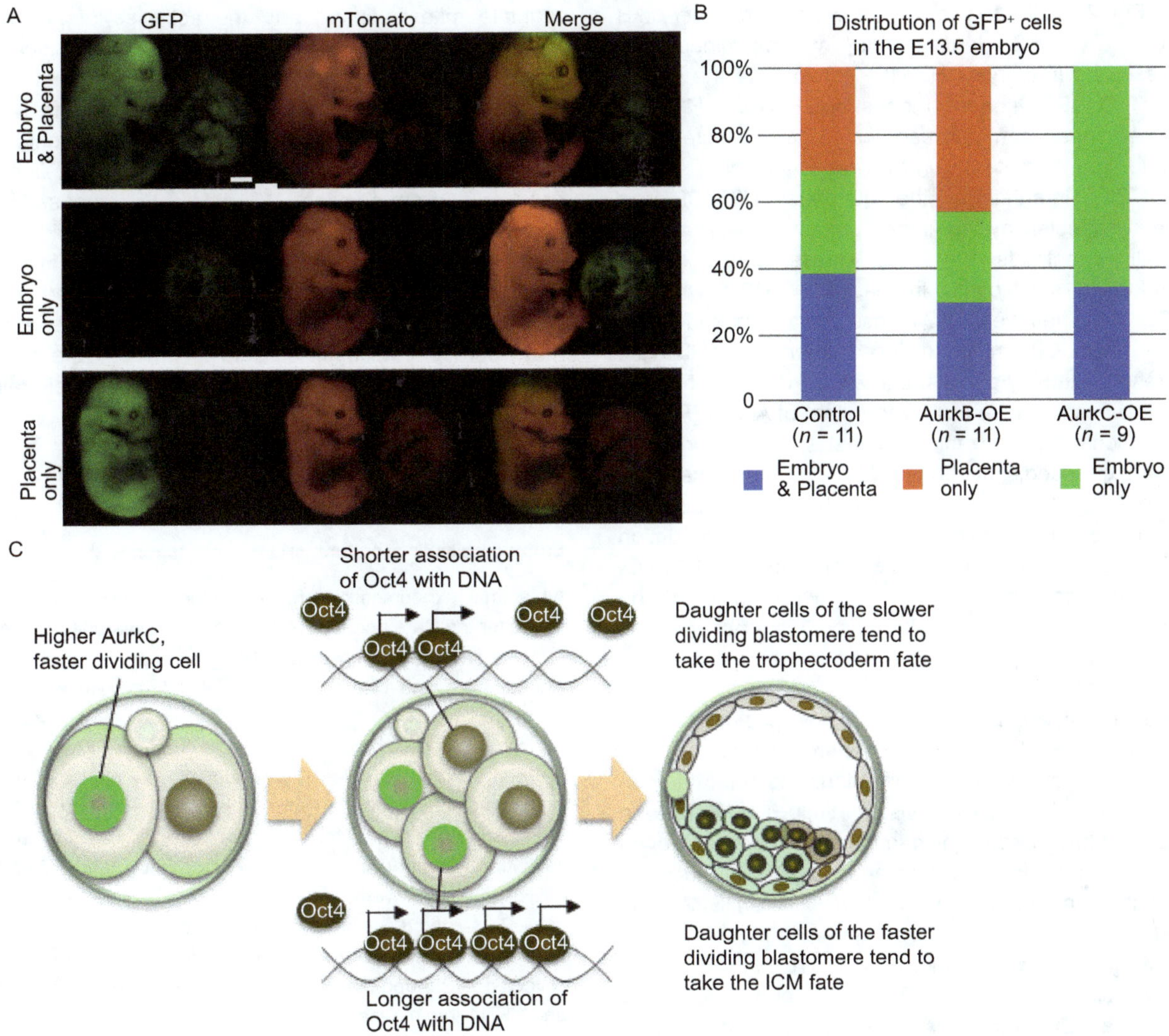

Figure 5. Long term effect of AurkB and AurkC overexpression during preimplantation embryo development. (A) Images of GFP positive cells in E13.5 Rosa embryos and placenta. The distribution was listed at the side of images. Scale bars, 2 mm. (B) Bar graph quantification of the percentage of embryo or placenta or both. Control (n = 11), AurkB-OE (n = 11), AurkC-OE (n = 9) groups. (C) Model of how cell division may affect cell fate decision in preimplantation mouse embryos.

securin degradation assay, we showed that increasing AurkC level advanced securin destruction, while knocking down AurkC had the opposite effect. In contrast, more AurkB protein correlated with slower securin degradation and vice versa. The degradation rate of securin can be considered as a readout of SAC and APC/C activity (Hagting et al., 2002). Thus, our results imply that the relatively high level of AurkC may create a more "relaxed" SAC state that adapts to the cleavage stage embryos. While AurkB represents a "stricter" SAC which is normally present in somatic cells. It has been proposed that after fertilization, it may take several cell cycles to transform all the components of cell division from meiosis to mitosis (Clift and Schuh, 2013). Moreover, upon fertilization, there's dramatic epigenetic reprogramming to create an open chromatin state that permits the robust zygotic genome activation during the first few mitotic cell cycles (Clift and Schuh, 2013). The special chromatin modifications may be recognized as abnormal by cell cycle checkpoints and activate stress response pathway. Therefore, we speculate it may be necessary for cleavage stage embryos to evolve a special cell division checkpoint mechanism, AurkC may be an important member of such mechanism.

It has been shown that the kinase activity is essential for the function of AurkB (Krenn and Musacchio, 2015). To have a deeper understanding of the molecular basis that leads to distinct functions of AurkB and C, we generated series of mutant Aurora kinases: AurkBCB, AurkCBC, and the kinase dead AurkC-KD. We found that the degradation rate of securin was largely determined by the AurkB/C kinase domain (Fig. 2B). H3S10 is a well-known substrate of AurkB (Crosio et al., 2002). In this study, we examined both metaphase and interphase H3S10P to evaluate the kinase activity of AurkB/C series of constructs. Strong H3S10P during metaphase can be observed in cells injected with either wild type or mutant Aurora kinases including AurkC-KD (Fig. 2C). Interestingly, the interphase H3S10P was significantly affected by the kinase domain overexpressed. Specifically, constructs with AurkB kinase domain (AurkB and AurkCBC) significantly increased H3S10P, whereas AurkBCB which has the AurkC kinase domain flanked by the AurkB N- and C-terminus reduced interphase H3S10P. These results strongly suggest that kinase domain of AurkC may have weaker kinase activity than that of AurkB (Fig. 2D and 2E). Consistent with this hypothesis, when AurkC-KD was overexpressed, interphase H3S10P was markedly diminished (Fig. 2D and 2E). The kinase dead mutations also lead to significantly slower Securin degradation (Fig. 2B), this was likely due to severely mis-localized AurkC-KD protein during mitosis (Fig. 2C) which inhibited the function of APC/C. Similar phenotype was observed in mouse oocytes ectopically expressed kinase dead AurkC (Yang et al., 2010).

The ability of AurkB/C to regulate interphase H3S10P leads us to ask whether they have different impact on transcription factor activity, gene expression, and cell fate decisions. Indeed, both cell division timing and histone modifications have been shown to affect cell fate decision in preimplantation embryos (Zernicka-Goetz, 2006). Recently, AurkB was reported to phosphorylate Oct4 on S229 and reduced its binding ability to target genes in mouse embryonic stem cells (mESCs) (Shin et al., 2016). In our experiments, we utilized a Oct4-paGFP construct and FDAP assay (Plachta et al., 2011) to investigate how AurkB and AurkC variation may affect Oct4 kinetics. We found that raising AurkB level caused faster Oct4-paGFP decay in the nucleus which is likely to reflect reduced affinity of Oct4 binding to DNA (Fig. 3C). Gene expression analysis also confirmed that the expression of several Oct4 target genes were decreased when AurkB level was high (Fig. 4A). Although we didn't observe that AurkC altered the kinetics of Oct4-paGFP decay in 4-cell embryos, knocking-down AurkC by siRNA significantly reduced the expression of pluripotency marker genes (Fig. 4A). Moreover, more daughter cells from the AurkC-OE cell expressed Oct4 and preferably occupied ICM position (Fig. 4B and 4D). In contrast, significant fewer daughter cells from AurkB-OE or siAurkC cell expressed Oct4 and entered ICM lineage (Fig. 4B and 4D). Lineage tracing experiments also confirmed that progeny cells derived from the AurkC-OE blastomere at 4-cell stage tend to contribute to the embryo proper (Fig. 5A and 5B).

In summary, our results demonstrated that it is possible to control cell fate by transiently modulating the level of AurkB and C in preimplantation embryos. We proposed a model as depicted in Fig. 5C, during preimplantation development, key cell division kinases such as AurkB and C may differentially regulate mitosis timing and the activities of pluripotency transcription factors thus to influence cell position, gene expression and these effects cumulate to impact cell fate decisions. Our study also leads to more questions, what are the factors that affect the activity of cell division regulators in preimplantation embryos, is it just the stochastic fluctuation or inherited from oocytes? Will different interphase H3S10P level lead to additional change in histone modification? As chromosome abnormality caused by erroneous mitosis can lead to severe consequences, mRNA mediated transient modulation of cell division checkpoints or epigenetic state may represent a possible route to improve the quality of *in vitro* fertilized mammalian embryos without alteration of the embryo genome.

MATERIALS AND METHODS

Embryo collection, culture, and microinjection

All animal experiments were conducted in accordance with the Guide for the Care and Use of Animals for Research Purposes. The protocol for mouse embryo isolation was approved by Institutional Animal Care and Use Committee and Internal Review Board of Tsinghua University.

Oocytes and embryos were collected from wild type F1 (C57BL/6xDBA) females (Charles River) aspreviously described (Na and Zernicka-Goetz, 2006). ROSA26Sor$^{tm4(ACTB-tdTomato, -EGFP)}$ Luo transgenic mice (JAX stock number 007676) were obtained from Jackson laboratory and maintained as homozygotes. Zygotes for mRNA injections were collected from female mice 25–26 h post-hCG. 2-, 4-, and 8-cell embryos were collected from female mice 46, 56 or 64 h post-hCG, respectively. Morula and blastocysts were collected at 2.5 dpc or 4 dpc, respectively. Microinjection of mRNA and siRNA into mouse preimplantation embryos were performed on a Leica DMI3000B microscope equipped with a Leica micromanipulator as previously described (Na and Zernicka-Goetz, 2006) at desired stages.

Plasmid construction, mRNA synthesis, and siRNA preparation

HA tagged (N-terminus) AurkB and AurkC, Securin-mCherry, H2B-GFP or mCherry and Oct4-paGFP were engineered and subcloned into RN3P vector for *in vitro* transcription of mRNA. Capped mRNAs were generated using a T3 mMESSAGE mMACHINE Kit according to the manufacture's instructions (AM1348, Ambion/Thermo Fisher Scientific). SiRNA targeting Aurora B and C were designed and purchased from siRNA Design Service (Sigma). SiRNA with scrambled sequence was used as the control.

Embryo fixation and immunostaining

For immunostaining, mouse preimplantation embryos were first treated with Acidic Tyrode solution to remove the zona pellucida. Then the embryos were fixed with 1% PFA in PBS in 4°C overnight. After fixation, embryos were permeabilized with 0.25% Triton X-100 at room temperature for 20 min and blocked with 3% BSA in PBS at 4°C overnight. Primary antibodies incubation was carried out in 4°C overnight. The primary antibodies include: monoclonal mouse anti-

HA (ab130275, Abcam), monoclonal rat anti-αTubulin (sc-53029, Santa Cruz), monoclonal rabbit anti-H3S10P (#9701S, CST), polyclonal rabbit anti-Oct4 (ab19857, Abcam), monoclonal mouse anti-Cdx2 (CDX2-88, Biogenex). Then the samples were incubated with DyLight 488/549/633 conjugated Goat anti mouse or rabbit IgG (H + L) antibodies (#35502, #35557, #35512, Thermo) at 4°C overnight, and nucleus were stained with DAPI. After staining, embryos were mounted on coverslips in Vectashield mounting medium (H-1000, Vectashield) and imaged on Nikon A1 confocal microscope.

Q-PCR

Embryos were first lysed in 0.2% Triton X-100 solution. Then reverse transcription and amplification of cDNA were performed using the Smart-seq2 protocol (Picelli et al., 2014). Q-PCR reactions and analysis were performed with GoTaq Q-PCR Master Mix (A6002, Promega) on a Bio-rad CFX96 Q-PCR Amplifier. The primer sequences and annealing temperature are listed in Table S1.

Time-lapse imaging experiments

Securin-mCherry degradation experiment was performed on a Leica AF6000 inverted microscope with a 20× objective (NA = 0.7). The microscope was equipped with a PECON environmental control chamber, fully motorized stage (VETXA) with plate heating insert (PECON). Embryos were cultured to 37°C and 5% CO_2 atmosphere. Oct4-paGFP FDAP assay were performed on a Nikon A1 confocal microscope equipped with TOKAI HIT atmospheric chamber using a method modified from Plachta et al. (2011). In brief, at 4-cell stage, 2 daughter cells of the injected blastomere from the 2-cell embryo were identified based on the expression of H2B-mCherry. Their nuclear regions were selectively photoactivated using the 405 nm laser within 5 min. Then z-series scan (2 μm per section) was performed every 10 min for up to 4 h across each nuclear region.

Embryo transfer into foster female mice and postimplantation embryo dissection

Foster female mice (F1) were crossed with ligated male mice 0.5 day before embryo transfer operation. Successful mating was determined by the presence of vaginal plug the next morning. At 1.5 dpc, plugged pseudo-pregnant female mice were anesthetized, CRE and AurkB/C mRNA injected ROSA 4-cell stage embryos were transferred into the oviduct using glass transfer pipette and the mice were nursed until vivification. The operated mice were sacrificed after 12 days, post-implantation embryos with placenta (13.5 dpc equivalent) were isolated from the uterus and stripped off the extraembryonic membranes. The embryo proper with the placenta were imaged on a Leica M165 FC integrated microscope.

ACKNOWLEDGMENTS

This work was supported by the National Natural Science Foundation of China (Grant No. 31171381), NSFC-MRC China-UK collaborative project grant 81261130320, the Beijing Natural Science Foundation grant 20151100084 (J.N.), the National Basic Research Program (973 Program) (Nos. 2016YFC0900301 and 2015CB856201), the Youth Thousand Scholar Program of China (W. X.), and the funding from the Tsinghua-Peking Center for Life Sciences (W.X., J.N.). We thank the animal facility, SLSTH-Nikon Biological Imaging Center of Tsinghua University for assistance with fluorescence imaging and Dr. Richard de Grijs for helpful comments and English language editing of the manuscript. All authors have contributed to, read, and approved the manuscript for submission.

ABBREVIATIONS

APC/C, anaphase-promoting complex/cyclosome; AurkB, Aurora kinase B; AurkC, Aurora kinase C; CPC, chromosome passenger complex; ICM, inner cell mass; K-MT, kinetochore microtubule; NEBD, nuclear envelope break down; Oct4-paGFP, Oct4-photoactivatable GFP fusion protein; SAC, Spindle assembly checkpoint; TE, trophecoderm

COMPLIANCE WITH ETHICAL GUIDELINES

This article does not contain any studies with human subjects performed by any of the authors. All institutional and national guidelines for the care and use of laboratory animals were followed. Wenzhi Li, Peizhe Wang, Bingjie Zhang, Jing Zhang, Jia Ming, Wei Xie, and Jie Na declare that they have no conflicts of interest.

AUTHORS' CONTRIBUTIONS

Wenzhi Li and Jie Na conceived the study and designed the experiments; Wenzhi Li, Peizhe Wang, Jing Zhang, Jia Ming and Jie Na generated plasmids used in this study, performed microinjection of mouse preimplantation embryos and all functional studies; Bingjie Zhang and Wei Xie contributed to histone modification analysis; Wenzhi Li and Jie Na wrote and revised the manuscript. All authors read and approved the final version of the manuscript.

REFERENCES

Santos MA, van de Werken C, de Vries M, Jahr H, Vromans MJ, Laven JS, Fauser BC, Kops GJ, Lens SM, Baart EB (2011) A role for Aurora C in the chromosomal passenger complex during human preimplantation embryo development. Hum Reprod 26:1868–1881

Balboula AZ, Schindler K (2014) Selective disruption of aurora C kinase reveals distinct functions from aurora B kinase during meiosis in mouse oocytes. PLoS Genet 10:e1004194

Bolton H, Graham SJ, Van der Aa N, Kumar P, Theunis K, Fernandez Gallardo E, Voet T, Zernicka-Goetz M (2016) Mouse model of chromosome mosaicism reveals lineage-specific depletion of aneuploid cells and normal developmental potential. Nat Commun 7:11165

Carmena M, Earnshaw WC (2003) The cellular geography of aurora kinases. Nat Rev Mol Cell Biol 4:842–854

Chazaud C, Yamanaka Y (2016) Lineage specification in the mouse preimplantation embryo. Development 143:1063–1074

Chen SH, Tang TK (2002) Mutational analysis of the phosphorylation sites of the Aie1 (Aurora-C) kinase in vitro. DNA Cell Biol 21:41–46

Clift D, Schuh M (2013) Restarting life: fertilization and the transition from meiosis to mitosis. Nat Rev Mol Cell Biol 14:549–562

Crosio C, Fimia GM, Loury R, Kimura M, Okano Y, Zhou H, Sen S, Allis CD, Sassone-Corsi P (2002) Mitotic phosphorylation of histone H3: spatio-temporal regulation by mammalian aurora kinases. Molecular and Cellular Biology 22:874–885

Dieterich K, Soto Rifo R, Faure AK, Hennebicq S, Ben Amar B, Zahi M, Perrin J, Martinez D, Sele B, Jouk PS, Ohlmann T, Rousseaux S, Lunardi J, Ray PF (2007) Homozygous mutation of AURKC yields large-headed polyploid spermatozoa and causes male infertility. Nat Genet 39:661–665

Fernandez-Miranda G, Trakala M, Martin J, Escobar B, Gonzalez A, Ghyselinck NB, Ortega S, Canamero M, Perez de Castro I, Malumbres M (2011) Genetic disruption of aurora B uncovers an essential role for aurora C during early mammalian development. Development 138:2661–2672

Hagting A, Den Elzen N, Vodermaier HC, Waizenegger IC, Peters JM, Pines J (2002) Human securin proteolysis is controlled by the spindle checkpoint and reveals when the APC/C switches from activation by Cdc20 to Cdh1. J Cell Biol 157:1125–1137

Hirota T, Lipp JJ, Toh BH, Peters JM (2005) Histone H3 serine 10 phosphorylation by Aurora B causes HP1 dissociation from heterochromatin. Nature 438:1176–1180

Krenn V, Musacchio A (2015) The aurora B kinase in chromosome bi-orientation and spindle checkpoint signaling. Front Oncol 5:225

Lin BW, Wang YC, Chang-Liao PY, Lin YJ, Yang ST, Tsou JH, Chang KC, Liu YW, Tseng JT, Lee CT, Lee JC, Hung LY (2014) Overexpression of Aurora-C interferes with the spindle checkpoint by promoting the degradation of Aurora-B. Cell Death Dis 5: e1106

Musacchio A (2015) The molecular biology of spindle assembly checkpoint signaling dynamics. Curr Biol 25:R1002–R1018

Na J, Zernicka-Goetz M (2006) Asymmetric positioning and organization of the meiotic spindle of mouse oocytes requires CDC42 function. Curr Biol 16:1249–1254

Nasmyth K (2005) How do so few control so many? Cell 120:739–746

Picelli S, Faridani OR, Bjorklund AK, Winberg G, Sagasser S, Sandberg R (2014) Full-length RNA-seq from single cells using Smart-seq2. Nat Protoc 9:171–181

Plachta N, Bollenbach T, Pease S, Fraser SE, Pantazis P (2011) Oct4 kinetics predict cell lineage patterning in the early mammalian embryo. Nat Cell Biol 13:117–123

Rothbart SB, Dickson BM, Raab JR, Grzybowski AT, Krajewski K, Guo AH, Shanle EK, Josefowicz SZ, Fuchs SM, Allis CD, Magnuson TR, Ruthenburg AJ, Strahl BD (2015) An interactive database for the assessment of histone antibody specificity. Mol Cell 59:502–511

Ruchaud S, Carmena M, Earnshaw WC (2007) Chromosomal passengers: conducting cell division. Nat Rev Mol Cell Biol 8:798–812

Sasai K, Katayama H, Stenoien DL, Fujii S, Honda R, Kimura M, Okano Y, Tatsuka M, Suzuki F, Nigg EA, Earnshaw WC, Brinkley WR, Sen S (2004) Aurora-C kinase is a novel chromosomal passenger protein that can complement Aurora-B kinase function in mitotic cells. Cell Motil Cytoskeleton 59:249–263

Schindler K, Davydenko O, Fram B, Lampson MA, Schultz RM (2012) Maternally recruited Aurora C kinase is more stable than Aurora B to support mouse oocyte maturation and early development. Proc Natl Acad Sci USA 109:E2215–E2222

Schvartzman JM, Sotillo R, Benezra R (2010) Mitotic chromosomal instability and cancer: mouse modelling of the human disease. Nat Rev Cancer 10:102–115

Sharif B, Na J, Lykke-Hartmann K, McLaughlin SH, Laue E, Glover DM, Zernicka-Goetz M (2010) The chromosome passenger complex is required for fidelity of chromosome transmission and cytokinesis in meiosis of mouse oocytes. J Cell Sci 123:4292–4300

Shin J, Kim TW, Kim H, Kim HJ, Suh MY, Lee S, Lee HT, Kwak S, Lee SE, Lee JH, Jang H, Cho EJ, Youn HD (2016) Aurkb/PP1-mediated resetting of Oct4 during the cell cycle determines the identity of embryonic stem cells. Elife 5:e10877

Tabansky I, Lenarcic A, Draft RW, Loulier K, Keskin DB, Rosains J, Rivera-Feliciano J, Lichtman JW, Livet J, Stern JN, Sanes JR, Eggan K (2013) Developmental bias in cleavage-stage mouse blastomeres. Curr Biol 23:21–31

Vanneste E, Voet T, Le Caignec C, Ampe M, Konings P, Melotte C, Debrock S, Amyere M, Vikkula M, Schuit F, Fryns JP, Verbeke G, D'Hooghe T, Moreau Y, Vermeesch JR (2009) Chromosome instability is common in human cleavage-stage embryos. Nat Med 15:577–583

Wong CC, Loewke KE, Bossert NL, Behr B, De Jonge CJ, Baer TM, Reijo Pera RA (2010) Non-invasive imaging of human embryos before embryonic genome activation predicts development to the blastocyst stage. Nat Biotechnol 28:1115–1121

Xue Z, Huang K, Cai C, Cai L, Jiang CY, Feng Y, Liu Z, Zeng Q, Cheng L, Sun YE, Liu JY, Horvath S, Fan G (2013) Genetic programs in human and mouse early embryos revealed by single-cell RNA sequencing. Nature 500:593–597

Yang KT, Li SK, Chang CC, Tang CJ, Lin YN, Lee SC, Tang TK (2010) Aurora-C kinase deficiency causes cytokinesis failure in meiosis I and production of large polyploid oocytes in mice. Mol Biol Cell 21:2371–2383

Zernicka-Goetz M (2006) The first cell-fate decisions in the mouse embryo: destiny is a matter of both chance and choice. Curr Opin Genet Dev 16:406–412

Zernicka-Goetz M, Morris SA, Bruce AW (2009) Making a firm decision: multifaceted regulation of cell fate in the early mouse embryo. Nat Rev Genet 10:467–477

6

The role of mitochondria in osteogenic, adipogenic and chondrogenic differentiation of mesenchymal stem cells

Qianqian Li[1,2], Zewen Gao[1,2], Ye Chen[1,2]✉, Min-Xin Guan[1,2]

[1] Division of Clinical Genetics and Genomics, The Children's Hospital, Zhejiang University School of Medicine, Hangzhou 310058, China
[2] Institute of Genetics, Zhejiang University and Department of Genetics, Zhejiang University School of Medicine, Hangzhou 310058, China
✉ Correspondence: yechency@zju.edu.cn (Y. Chen)

ABSTRACT

Mesenchymal stem cells (MSCs) are progenitors of connective tissues, which have emerged as important tools for tissue engineering due to their differentiation potential along various cell types. In recent years, accumulating evidence has suggested that the regulation of mitochondria dynamics and function is essential for successful differentiation of MSCs. In this paper, we review and provide an integrated view on the role of mitochondria in MSC differentiation. The mitochondria are maintained at a relatively low activity level in MSCs, and upon induction, mtDNA copy number, protein levels of respiratory enzymes, the oxygen consumption rate, mRNA levels of mitochondrial biogenesis-associated genes, and intracellular ATP content are increased. The regulated level of mitochondrial ROS is found not only to influence differentiation but also to contribute to the direction determination of differentiation. Understanding the roles of mitochondrial dynamics during MSC differentiation will facilitate the optimization of differentiation protocols by adjusting biochemical properties, such as energy production or the redox status of stem cells, and ultimately, benefit the development of new pharmacologic strategies in regenerative medicine.

KEYWORDS mesenchymal stem cells, mitochondria, differentiation

INTRODUCTION

Mesenchymal stem cells (MSCs) are multipotent cells originally found in the bone marrow that give rise to various specialized cells, including osteoblasts, adipocytes, and chondrocytes. MSC studies have progressed rapidly since the initial report of human MSCs isolation. These cells exhibit considerable promise for application in rebuilding damaged mesenchymal tissues (Chen et al., 2008b; Parekkadan and Milwid, 2010; Savkovic et al., 2014). So far, most MSC studies have focused on the activity of the nuclear genome; characteristics of mitochondrial activities have largely lagged behind. Mitochondria are essential organelles inside eukaryotic cells that are responsible for cellular energy production. Interestingly, several recent studies have shown variations in the abundances, morphology, and functions of mitochondria in different cell types that adapt to environmental and cellular cues (Collu-Marchese et al., 2015; Forni et al., 2016; Zhang et al., 2013). Here we summarize the current knowledge about the involvement of mitochondria in the stemness and differentiation progress of MSCs.

MITOCHONDRIAL PROPERTIES ARE MODIFIED DURING MSC DIFFERENTIATION

Mitochondria are a distinguishing feature of eukaryotic cells, with an essential role in organismal longevity given their critical function in energy metabolism and cell death. In particular, the mitochondrion is one of the most sophisticated and dynamic responsive sensing systems in the cell. In addition, recent studies have proven that balanced mitochondrial dynamics and morphology are crucial to the maintenance of tissue homeostasis: mitochondrial dynamics regulates the fate of stem cells.

The significance of the perinuclear arrangement of mitochondria, which may be a characteristic feature of stem cells, is generally accepted. Taking MSCs as an example, the mitochondrial are mainly gathered around the nucleus in undifferentiated cells, whereas they are more uniformly distributed throughout the cytoplasm of differentiated cells (Hofmann et al., 2012; Quinn et al., 2013). In addition, the mitochondrial-to-cytoplasm area ratio increases during differentiation relative to that of undifferentiated cells (Forni et al., 2016; Lambertini et al., 2015). Researchers have confirmed that the area of individual mitochondria generally remains constant, which means that the increase in the mitochondrial-to-cytoplasm area ratio is not the result of mitochondrial swelling. Indeed, increased mitochondrial biogenesis is observed during both adipogenesis and osteogenesis, leading to an increased abundance of mitochondria in differentiated cells. It was shown that the level of the mitochondrial outer membrane protein TOM20 was greatly improved during adipogenic differentiation, and enhanced MitoTracker Green Staining further confirmed the increase in mitochondrial mass (Zhang et al., 2013). Correspondingly, mtDNA content was increased in differentiated cells (Wanet et al., 2014), and this increase was more striking in adipogenic cell differentiation (Forni et al., 2016). During osteogenesis, a dynamic change was revealed by qRT-PCR, with an initial decline upon induction, followed by a subsequent increase (Chen et al., 2008a). Furthermore, the morphology of mitochondria gradually becomes slender after the process of differentiation,and it has been shown that osteogenic differentiation of MSCs is accompanied by cristae development.

Studies have proven that the conditions of metabolic activity are different in MSCs from those of differentiated offspring cells: MSCs are more dependent on glycolysis, whereas differentiated cells depend more on oxidative metabolism (Hofmann et al., 2012; Hsu et al., 2016). Long-term cultured MSCs have been shown to exhibit a significantly lower proportion of cells with a perinuclear mitochondrial distribution, and these cells exhibit a higher ATP content and tendency to differentiate compared with cells from an earlier passage. Upon initiation of the differentiation process, mitochondria in MSCs are activated via a yet unknown mechanism and oxidative phosphorylation becomes the major source of ATP. This bioenergetic switch is especially important for osteogenic differentiation of MSCs as reported by Kowaltowski et al. (Tahara et al., 2009), and the ATP content of maturing cultures continuously increases along with nodule morphogenesis. Like osteogenic differentiation, oxygen consumption and the activities of respiratory enzyme complexes increase significantly during adipogenesis. In addition, Pietila et al. showed that red fluorescence of the probe JC-1 gradually became green along with the process of differentiation, indicating that mitochondrial membrane potential is reduced during the differentiation process (Pietila et al., 2012).

THE TIGHT REGULATION OF MSC FATE BY ROS

Reactive oxygen species (ROS) are oxygen-derived small molecules, that react readily with various chemical structures including proteins, lipids, sugars, and nucleic acids. Mitochondria are the most important source of ROS within mammalian cells. For a long time, ROS have been considered to induce cellular dysfunction and organismal death via the destructive oxidation of intra-cellular components. In recent years, there has been an accumulating understanding of their role as signaling molecules (Atashi et al., 2015). Most research groups now believe that only unregulated levels of ROS are harmful, and a regulated basal level of ROS is necessary and advantageous for maintenance of cell functions, such as proliferation, differentiation, and survival (Sart et al., 2015; Wang et al., 2015). During the last decade, the impact of ROS on MSC differentiation has generated a great deal of interest due to its potential application in clinic.

Investigations have suggested the existence of a link between oxidative stress and impaired skeletal integrity. Increased intra-cellular levels of ROS were concluded to underline the observation that elder donor-derived MSCs exhibit a reduced potential for osteogenic differentiation (Tan et al., 2015), which is a pivotal pathogenetic mechanism of age-related bone and bone strength loss. Meanwhile, there is accumulating *in vitro* evidence suggesting that excess ROS impairs osteogenic differentiation. Studies have revealed that exogenous H_2O_2 reduced the activity of alkaline phosphatase, a marker of osteogenic differentiation, in culture (Lee et al., 2006; Tahara et al., 2009). Similarly, Chen et al. observed that the antioxidant enzymes SOD2 and catalase were significantly upregulated upon osteogenic differentiation, which led to a dramatic decrease in the intracellular ROS level (Chen et al., 2008a). In addition, by using cell viability assay, the researchers showed that differentiated cells were more resistant to exogenous ROS stress than undifferentiated cells. More recently, Kim et al. demonstrated that upregulation of ROS inhibits osteogenic differentiation of MSCs, in part through inhibition of the Hedgehog (Hh) signaling pathway, which is essential for bone development and maintenance (Kim et al., 2012).

Many research groups have focused on the osteogenic and adipogenic potential of aged MSCs. In general, MSCs from aged donors shift the balance in favor of adipocyte differentiation at the expense of osteoblast differentiation (Geissler et al., 2012). These results were consistent with the observations that ROS generation was increased during adipogenic differentiation. However, it is still controversial whether increased ROS is essential for adipocyte differentiation or it is a byproduct of the differentiation process. Yasunari et al. found that differentiation-inducing agents induced ROS generation in MSCs by an H_2DCF assay, and the application of N-acetyl-L-cysteine (NAC) blocked adipogenic differentiation (Kanda et al., 2011). Tormos et al. revealed that ROS generated from mitochondrial complex III is required to initiate adipocyte differentiation by genetical manipulation of the complex (Tormos et al., 2011). In addition, these studies have shown that mitochondrial-targeted antioxidants could inhibit adipocyte differentiation, which was rescued by the addition of exogenous hydrogen peroxide.

These results implied that increased ROS generation is not simply a consequence of adipocyte differentiation.

Although intensive research has been carried out on chondrogenic differentiation of MSCs, the role of ROS in chondrogenesis is less well characterized. Investigators have noticed that the ROS level was specifically high in developing chondrocytes of the embryonic limb, and chondrocyte maturation was accompanied by a progressive decrease in catalase activity in growing cartilage (Salas-Vidal et al., 1998; Schnabel et al., 2006). Regarding *in vitro* chondrogenic differentiation, ROS generation was increased during induction (Heywood and Lee, 2016; Jallali et al., 2007; Morita et al., 2007). Moreover, the differentiation markers of chondrocyte are upregulated by addition of H_2O_2, and chondrogenesis was inhibited by administration of antioxidant NAC, suggesting that ROS plays a critical role in chondrogenesis. Consistent with this, Kim et al. proved that ROS generated by Nox2 or Nox4 is essential for survival and differentiation in the early stage of chondrogenesis (Kim et al., 2010).

MOLECULAR LINKS OF MITOCHONDRIAL FUNCTION AND DYNAMICS WITH MSC DIFFERENTIATION

There is a substantial body of evidence indicating that mitochondrial morphology and function are modulated during MSC differentiation. However, the molecular mechanisms linking mitochondrial dynamics with the regulation of differentiation are still poorly understood. It is believed that an orchestrated series of events are involved in MSC fate determination, for example, osteoblast differentiation requires the activation of many osteoblastgenic transcription factors (including Runx2, Osterix, and β-catenin) and inhibition of adipogenic transcription factors (PPARγ and CEBPα) (Chen et al., 2016). Growing evidence supports the bifunctional role of many transcription factors in the control of both nuclear and mitochondrial gene expression (Fig. 1).

The nuclear receptor peroxisome proliferator-activated receptor γ (PPARγ) is a crucial cellular and metabolic switch (Patel et al., 2014; Wan, 2010). Studies have shown that activation of PPARγ shifts the balance of MSCs fate by favoring adipocyte differentiation and inhibiting osteogenesis (Stechschulte et al., 2016; Wan, 2010). *In vitro* suppression of PPARγ enhanced osteogenesis and reduced adipogenesis. Consequently, $PPAR\gamma^{+/-}$ mice display a significantly higher bone mass due to increased bone generation (Akune et al., 2004). Huang et al. overexpressed PPARγ coactivator 1α (PGC-1α), which is the master regulator of mitochondrial biogenesis, in MSCs and observed a significant increase in genes related to mitochondrial functions and lipid metabolism (Huang et al., 2011). The over dose of PGC-1α prevented MSC differentiation into osteocytes upon induction; on the other hand, the adipogenic potential of the MSCs was increased. In addition, they found that PGC-1α knockdown inhibited adipocyte differentiation. Studies on PGC-1α have energized mitochondria research in stem cells biology, and have increased the chances that strategies to improve mitochondrial activity will be discovered.

Studies have demonstrated that mitochondria are maintained at a much lower activity state in MSCs compared with differentiated cells (Sanchez-Arago et al., 2013). Upon osteogenic induction, mitochondrial functions are upregulated to fulfill greater energy demand or facilitate other biochemical reactions that take place in cells. Wei et al. demonstrated that, upon osteogenic induction of MSCs, the levels of proteins involved in mitochondrial biogenesis are increased, including PGC-1α, mtTFA, DNA polymerase γ, enzymes of the TCA cycle, and protein subunits of respiratory enzymes (Hsu et al., 2016). Lambertini et al. found that nuclear factor of activated T cell complex 1 (NFATc1) is recruited to mtDNA and acts as a negative regulator of mtDNA transcription, which reveals the involvement of NFATc1 in the mineralization process (Lambertini et al., 2015). Sirtuins are protein deacetylases that are thought to play evolutionarily conserved roles in lifespan extension (Denu and Hematti, 2016). Recent studies have revealed that they also play roles in MSCs differentiation. SIRT1 is the most widely studied sirtuin, which deacetylates a number of substrates such as PGC-1α. It has been proven to be crucial for stem cell maintenance and differentiation (Min-Wen et al., 2016). The activation of SIRT1 impairs adipogenesis and enhances MSC osteogenesis. SIRT2 was found to regulate adipocyte differentiation through inhibition of PPARγ. Sirt3, Sirt5, and Sirt7 were also reported to be involved in mitochondrial biogenesis and the activation of mitochondrial function during adipogenic differentiation (Denu and Hematti, 2016). SIRT6 deficiency results in the impairment of differentiation potential of hMSCs into bone and cartilage (Pan et al., 2016).

Mitochondrial dynamics and their regulatory processes are supposed to be modulated during differentiation, leading to altered bioenergetic profiles. In the process of adipogenic differentiation and osteogenic differentiation, the content of citrate synthase increased obviously (Forni et al., 2016). Then, the expression of glycolytic enzymes and the yield of lactate were reduced during the osteogenic induction (Zhang et al., 2013). This suggests that the three carboxylic acid cycle process that occurs in the mitochondria is enhanced. Recently, Forni et al. showed that mitochondrial fusion proteins Mfn1 and Mfn2 were upregulated in the early stages of adipogenesis and osteogenesis, along with mitochondrial elongation (Forni et al., 2016). Regarding chondrogenesis, there was increased expression of fission proteins Drp1, Fis1, and Fis2. Knocking down of these genes resulted in a loss of differentiation ability. Moreover, enhanced mitophagy was observed during chondrogenesis. The authors claimed that the activation of mitophagy led to higher mitochondrial turnover during early chondrogenesis. Nuschke et al. found that accumulation of cleaved type II light chain 3 (LC3-II) protein, a marker for active autophagosomes, was correlated with osteogenic differentiation, indicating activation of

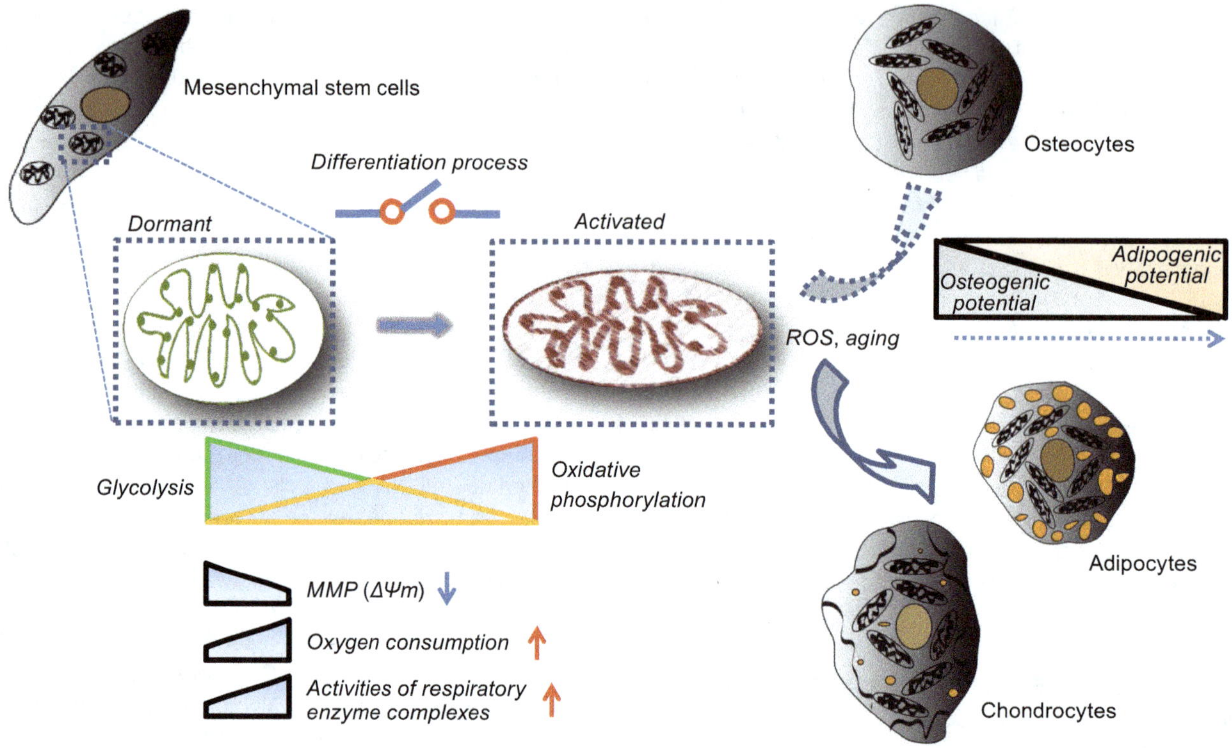

Figure 1. The regulation of mitochondria dynamics and function is essential for successful differentiation of MSCs. Mitochondria are mainly seen gathered around the nucleus in MSCs, whereas the mitochondria are more uniformly distributed in the cytoplasm of differentiated cells. Along with the process of differentiation, the morphology of mitochondria gradually becomes slender, and the number of mitochondria increases. The metabolic pattern has changed from glycolysis to oxidative phosphorylation; therefore, increased oxygen consumption and respiratory enzyme complex activation becomes logical. Notably, the membrane potential appears to be reduced in differentiated cells. The advancement of age and the high level of ROS can promote MSCs to adipocytes, whereas a low level of ROS can promote osteogenesis.

autophagy upon stimulation of differentiation (Nuschke et al., 2014). Consistent with this conclusion, Song et al. recently reported that the adipogenic differentiation of MSCs was also promoted through the activation of autophagy (Song et al., 2015). Moreover, adipogenic differentiation could be blocked by the addition of autophagy inhibitors. Nonetheless, further studies are needed to better characterize the regulation of mitophagy during MSC differentiation.

HYPOXIA INFLUENCES MSC DIFFERENTIATION

In the bone marrow, MSCs reside in a unique microenvironment (MSC niche). Recent advances have identified significant metabolic changes in the mitochondria that are regulated by environmental stimuli. Hypoxia, an important feature of MSC niche, has been proved to play an important role in maintaining stem cell fate, self-renewal, and multipotency during the last decade. It was observed that there is an enhanced transcription and synthesis of glycolytic pathway enzymes and reduction of synthesis of proteins involved in mitochondrial catabolism in hypoxic cells. This holds true for MSCs as their proliferation, differentiation, and survival are affected by culture in low O_2 tension. In mammalian cell, oxygen participates in reactions of aerobic energy synthesis as a substrate for cytochrome oxidase, the terminal enzyme of mitochondrial respiratory chain. In hypoxia, decreased mitochondrial size and reduced mitochondrial mean velocity have been observed (Varela-Rey et al., 2009). Mitochondrial morphology is associated with HIF-1α stabilization. Moreover, mitochondria act as O_2 sensors, and contribute to the cell redox potential, ion homeostasis, and energy production. Specifically, hypoxia could reduce mitochondrial fusion by impairing mitochondrial membrane potential, which in turn could induce supercomplexes disassembly, increasing ROS production.

Evaluation of adipocyte lineage-specific transcripts and osteocyte lineage-specific transcripts showed that the expression of ALPL in MSCs in severe hypoxia is higher than in normoxia, indicating MSCs in hypoxia are more prone to osteogenic differentiation than in normoxia (Ejtehadifar et al., 2015). The expression of osteogenesis-related genes, such as alkaline phosphatase, Type I collagen, and osteocalcin was significantly increased under hypoxia (Boyette et al., 2014). Hypoxia suppressed adipogenesis and associated HIF1α and PPARG gene expression in hMSCs and enhanced osteogenesis and associated HIF1A and RUNX2 gene expression (Wagegg et al., 2012; Zhang et al., 2013). Moreover, shRNA-mediated knockdown of HIF-1α suppressed hypoxia-induced osteogenesis (Wagegg et al., 2012). The mechanism was demonstrated that HIF-1α can suppress oxidative metabolism through inhibition of pyruvate dehydrogenase (PDH) by PDH kinase (PDK) and activate the expression of glycolytic enzymes (Papandreou et al., 2006). On the other hand, the mitochondrial electron transport chain deficiency made the expression of HIF-1α severely reduced, and HIF-1 DNA binding was diminished, which suggested that electron transport chain activity is required for activation of HIF-1 (Agani et al., 2000).

So far, most hypoxia studies showed that long term culture of MSCs in hypoxia promotes a genetic program maintaining their undifferentiated and multipotent status. However, the degree and duration of hypoxia described in the literatures vary greatly and may result in very different effects on the proliferation and differentiation capacities of MSC. Future work will continue to explore hypoxia-induced effects and help to position mitochondrial function, dynamics, and signaling within MSC differentiation. And, more remarkable, we cannot draw conclusions that *in vitro* culture in hypoxia mimics the niche until we understand the *in vivo* signature of MSCs.

PERSPECTIVES

Throughout the last decade, rapid progress has been made in the area of stem cells and metabolism. Various studies suggest that mitochondrial activity or dormancy plays a role in maintaining the undifferentiated state of MSCs, whereas proper activation is essential for successful differentiation. In addition, mitochondrial ROS are reported to not only regulate the differentiation process but also contribute to the direction determination of differentiation. However, the understanding of the role played by mitochondria during stem cell differentiation is still in its early stages, and a deeper knowledge of the interaction between mitochondria dynamics and MSC differentiation is needed. Indeed, a series of studies are working on these important open questions: What are the mechanisms restricting mitochondrial biogenesis and favoring a glycolytic metabolism in MSCs? How can activated mitochondrial biogenesis favor various differentiation processes? What are the molecular players linking mitochondrial dynamics and the differentiation processes? Do noncoding RNAs decode the interplay between mitochondria and cell differentiation? Taken together, the data we have summarized here clearly indicate a critical role for mitochondria dynamics during MSCs differentiation. We believe this rapidly moving field promise many more to come, and progress in this research will facilitate the optimization of *in vitro* differentiation protocols, which can finally benefit the better design of cell engineering using MSCs.

ACKNOWLEDGEMENTS

The authors were supported by the National Basic Research Program (973 Program) (No. 2014CB541702), the National Natural Science Foundation of China (Grant No. 31671305), and the Natural Science Foundation of Zhejiang province (R15H080001).

ABBREVIATIONS

ATP, adenosine triphosphate; HIF, hypoxia inducible factor; LC3-II, cleaved type II light chain 3; MSCs, mesenchymal stem cells; NFATc1, nuclear factor of activated T cell complex 1; PDH, pyruvate dehydrogenase; PGC-1α, PPARγ coactivator 1α; PPARγ, peroxisome proliferator-activated receptor γ; ROS, reactive oxygen species.

COMPLIANCE WITH ETHICS GUIDELINES

Qianqian Li, Zewen Gao, Ye Chen, and Min-Xin Guan declare that they have no conflict of interest with respect to research, authorship, and/or publication of this article.

REFERENCES

Agani FH, Pichiule P, Chavez JC, LaManna JC (2000) The role of mitochondria in the regulation of hypoxia-inducible factor 1 expression during hypoxia. J Biol Chem 275:35863–35867

Akune T, Ohba S, Kamekura S, Yamaguchi M, Chung UI, Kubota N, Terauchi Y, Harada Y, Azuma Y, Nakamura K et al (2004) PPARgamma insufficiency enhances osteogenesis through osteoblast formation from bone marrow progenitors. J Clin Investig 113:846–855

Atashi F, Modarressi A, Pepper MS (2015) The role of reactive oxygen species in mesenchymal stem cell adipogenic and osteogenic differentiation: a review. Stem cells and development 24:1150–1163

Boyette LB, Creasey OA, Guzik L, Lozito T, Tuan RS (2014) Human bone marrow-derived mesenchymal stem cells display enhanced clonogenicity but impaired differentiation with hypoxic preconditioning. Stem cells Transl Med 3:241–254

Chen CT, Shih YR, Kuo TK, Lee OK, Wei YH (2008a) Coordinated changes of mitochondrial biogenesis and antioxidant enzymes during osteogenic differentiation of human mesenchymal stem cells. Stem Cells 26:960–968

Chen Y, Shao JZ, Xiang LX, Dong XJ, Zhang GR (2008b) Mesenchymal stem cells: a promising candidate in regenerative medicine. Int J Biochem Cell Biol 40:815–820

Chen Q, Shou P, Zheng C, Jiang M, Cao G, Yang Q, Cao J, Xie N, Velletri T, Zhang X et al (2016) Fate decision of mesenchymal stem cells: adipocytes or osteoblasts? Cell Death Differ 23:1128–1139

Collu-Marchese M, Shuen M, Pauly M, Saleem A, Hood DA (2015) The regulation of mitochondrial transcription factor A (Tfam) expression during skeletal muscle cell differentiation. Biosci Rep 35:e00221

Denu RA, Hematti P (2016) Effects of Oxidative Stress on Mesenchymal Stem Cell Biology. Oxidative Med Cell Longev 2016:2989076

Ejtehadifar M, Shamsasenjan K, Movassaghpour A, Akbarzadehlaleh P, Dehdilani N, Abbasi P, Molaeipour Z, Saleh M (2015) The effect of hypoxia on mesenchymal stem cell biology. Adv Pharm Bull 5:141–149

Forni MF, Peloggia J, Trudeau K, Shirihai O, Kowaltowski AJ (2016) Murine mesenchymal stem cell commitment to differentiation is regulated by mitochondrial dynamics. Stem Cells 34:743–755

Geissler S, Textor M, Kuhnisch J, Konnig D, Klein O, Ode A, Pfitzner T, Adjaye J, Kasper G, Duda GN (2012) Functional comparison of chronological and in vitro aging: differential role of the cytoskeleton and mitochondria in mesenchymal stromal cells. PLoS ONE 7:e52700

Heywood HK, Lee DA (2016) Bioenergetic reprogramming of articular chondrocytes by exposure to exogenous and endogenous reactive oxygen species and its role in the anabolic response to low oxygen. J Tissue Eng Regen Med 10:1–9

Hofmann AD, Beyer M, Krause-Buchholz U, Wobus M, Bornhauser M, Rodel G (2012) OXPHOS supercomplexes as a hallmark of the mitochondrial phenotype of adipogenic differentiated human MSCs. PLoS ONE 7:e35160

Hsu YC, Wu YT, Yu TH, Wei YH (2016) Mitochondria in mesenchymal stem cell biology and cell therapy: From cellular differentiation to mitochondrial transfer. Semin Cell Dev Biol 52:119–131

Huang PI, Chen YC, Chen LH, Juan CC, Ku HH, Wang ST, Chiou SH, Chiou GY, Chi CW, Hsu CC et al (2011) PGC-1alpha mediates differentiation of mesenchymal stem cells to brown adipose cells. J Atheroscler Thromb 18:966–980

Jallali N, Ridha H, Thrasivoulou C, Butler P, Cowen T (2007) Modulation of intracellular reactive oxygen species level in chondrocytes by IGF-1, FGF, and TGF-beta1. Connect Tissue Res 48:149–158

Kanda Y, Hinata T, Kang SW, Watanabe Y (2011) Reactive oxygen species mediate adipocyte differentiation in mesenchymal stem cells. Life Sci 89:250–258

Kim KS, Choi HW, Yoon HE, Kim IY (2010) Reactive oxygen species generated by NADPH oxidase 2 and 4 are required for chondrogenic differentiation. J Biol Chem 285:40294–40302

Kim M, Kim C, Choi YS, Kim M, Park C, Suh Y (2012) Age-related alterations in mesenchymal stem cells related to shift in differentiation from osteogenic to adipogenic potential: implication to age-associated bone diseases and defects. Mech Ageing Dev 133:215–225

Lambertini E, Penolazzi L, Morganti C, Lisignoli G, Zini N, Angelozzi M, Bonora M, Ferroni L, Pinton P, Zavan B et al (2015) Osteogenic differentiation of human MSCs: Specific occupancy of the mitochondrial DNA by NFATc1 transcription factor. Int J Biochem Cell Biol 64:212–219

Lee DH, Lim BS, Lee YK, Yang HC (2006) Effects of hydrogen peroxide (H2O2) on alkaline phosphatase activity and matrix mineralization of odontoblast and osteoblast cell lines. Cell Biol Toxicol 22:39–46

Min-Wen JC, Jun-Hao ET, Shyh-Chang N (2016) Stem cell mitochondria during aging. Semin Cell Dev Biol 52:110–118

Morita K, Miyamoto T, Fujita N, Kubota Y, Ito K, Takubo K, Miyamoto K, Ninomiya K, Suzuki T, Iwasaki R et al (2007) Reactive oxygen species induce chondrocyte hypertrophy in endochondral ossification. J Exp Med 204:1613–1623

Nuschke A, Rodrigues M, Stolz DB, Chu CT, Griffith L, Wells A (2014) Human mesenchymal stem cells/multipotent stromal cells consume accumulated autophagosomes early in differentiation. Stem cell Res Ther 5:140

Pan H, Guan D, Liu X, Li J, Wang L, Wu J, Zhou J, Zhang W, Ren R, Zhang W et al (2016) SIRT6 safeguards human mesenchymal stem cells from oxidative stress by coactivating NRF2. Cell Res 26:190–205

Papandreou I, Cairns RA, Fontana L, Lim AL, Denko NC (2006) HIF-1 mediates adaptation to hypoxia by actively downregulating mitochondrial oxygen consumption. Cell Metab 3:187–197

Parekkadan B, Milwid JM (2010) Mesenchymal stem cells as therapeutics. Annu Rev Biomed Eng 12:87–117

Patel JJ, Butters OR, Arnett TR (2014) PPAR agonists stimulate adipogenesis at the expense of osteoblast differentiation while inhibiting osteoclast formation and activity. Cell Biochem Funct 32:368–377

Pietila M, Palomaki S, Lehtonen S, Ritamo I, Valmu L, Nystedt J, Laitinen S, Leskela HV, Sormunen R, Pesala J et al (2012) Mitochondrial function and energy metabolism in umbilical cord blood- and bone marrow-derived mesenchymal stem cells. Stem Cells Dev 21:575–588

Quinn KP, Sridharan GV, Hayden RS, Kaplan DL, Lee K, Georgakoudi I (2013) Quantitative metabolic imaging using endogenous fluorescence to detect stem cell differentiation. Sci Rep 3:3432

Salas-Vidal E, Lomeli H, Castro-Obregon S, Cuervo R, Escalante-Alcalde D, Covarrubias L (1998) Reactive oxygen species participate in the control of mouse embryonic cell death. Exp Cell Res 238:136–147

Sanchez-Arago M, Garcia-Bermudez J, Martinez-Reyes I, Santacatterina F, Cuezva JM (2013) Degradation of IF1 controls energy metabolism during osteogenic differentiation of stem cells. EMBO Rep 14:638–644

Sart S, Song L, Li Y (2015) Controlling redox status for stem cell survival, expansion, and differentiation. Oxidative Med Cell Longev 2015:105135

Savkovic V, Li H, Seon JK, Hacker M, Franz S, Simon JC (2014) Mesenchymal stem cells in cartilage regeneration. Curr Stem Cell Res Ther 9:469–488

Schnabel D, Salas-Vidal E, Narvaez V, Sanchez-Carbente Mdel R, Hernandez-Garcia D, Cuervo R, Covarrubias L (2006) Expression and regulation of antioxidant enzymes in the developing limb support a function of ROS in interdigital cell death. Dev Biol 291:291–299

Song BQ, Chi Y, Li X, Du WJ, Han ZB, Tian JJ, Li JJ, Chen F, Wu HH, Han LX et al (2015) Inhibition of notch signaling promotes the adipogenic differentiation of mesenchymal stem cells through autophagy activation and PTEN-PI3K/AKT/mTOR pathway. Cell Physiol Biochem 36:1991–2002

Stechschulte LA, Czernik PJ, Rotter ZC, Tausif FN, Corzo CA, Marciano DP, Asteian A, Zheng J, Bruning JB, Kamenecka TM et al (2016) PPARG post-translational modifications regulate bone formation and bone resorption. EBioMedicine 10:174–184

Tahara EB, Navarete FD, Kowaltowski AJ (2009) Tissue-, substrate-, and site-specific characteristics of mitochondrial reactive oxygen species generation. Free Radic Biol Med 46:1283–1297

Tan J, Xu X, Tong Z, Lin J, Yu Q, Lin Y, Kuang W (2015) Decreased osteogenesis of adult mesenchymal stem cells by reactive oxygen species under cyclic stretch: a possible mechanism of age related osteoporosis. Bone Res 3:15003

Tormos KV, Anso E, Hamanaka RB, Eisenbart J, Joseph J, Kalyanaraman B, Chandel NS (2011) Mitochondrial complex III ROS regulate adipocyte differentiation. Cell Metab 14:537–544

Varela-Rey M, Embade N, Ariz U, Lu SC, Mato JM, Martinez-Chantar ML (2009) Non-alcoholic steatohepatitis and animal models: understanding the human disease. Int J Biochem Cell Biol 41:969–976

Wagegg M, Gaber T, Lohanatha FL, Hahne M, Strehl C, Fangradt M, Tran CL, Schonbeck K, Hoff P, Ode A et al (2012) Hypoxia promotes osteogenesis but suppresses adipogenesis of human mesenchymal stromal cells in a hypoxia-inducible factor-1 dependent manner. PLoS ONE 7:e46483

Wan Y (2010) PPARgamma in bone homeostasis. Trends Endocrinol metab 21:722–728

Wanet A, Remacle N, Najar M, Sokal E, Arnould T, Najimi M, Renard P (2014) Mitochondrial remodeling in hepatic differentiation and dedifferentiation. Int J Biochem Cell Biol 54:174–185

Wang W, Zhang Y, Lu W, Liu K (2015) Mitochondrial reactive oxygen species regulate adipocyte differentiation of mesenchymal stem cells in hematopoietic stress induced by arabinosylcytosine. PLoS ONE 10:e0120629

Zhang Y, Marsboom G, Toth PT, Rehman J (2013) Mitochondrial respiration regulates adipogenic differentiation of human mesenchymal stem cells. PLoS ONE 8:e77077

7

C. elegans-based screen identifies lysosome-damaging alkaloids that induce STAT3-dependent lysosomal cell death

Yang Li[1,3]✉, Yu Zhang[2], Qiwen Gan[3], Meng Xu[3], Xiao Ding[2], Guihua Tang[2], Jingjing Liang[3], Kai Liu[3], Xuezhao Liu[3], Xin Wang[5], Lingli Guo[2], Zhiyang Gao[3], Xiaojiang Hao[2,4]✉, Chonglin Yang[3,5]✉

[1] Department of Pharmacology, Key Laboratory of Metabolism and Molecular Medicine (The Ministry of Education), School of Basic Medical Science, Fudan University, Shanghai 200032, China
[2] State Key Laboratory of Phytochemistry and Plant Resources in Western China, Kunming Institute of Botany, Chinese Academy of Sciences, Kunming 650021, China
[3] State Key Laboratory of Molecular Developmental Biology, Institute of Genetics and Developmental Biology, Chinese Academy of Sciences, Beijing 100101, China
[4] The Key Laboratory of Chemistry for Natural Product of Guizhou Province, Chinese Academy of Science, Guiyang 550002, China
[5] State Key Laboratory for Conservation and Utilization of Bio-Resources in Yunnan, Center for Life Sciences, School of Life Sciences, Yunnan University, Kunming 650091, China
✉ Correspondence: oceanyangli@fudan.edu.cn (Y. Li), haoxj@mail.kib.ac.cn (X. Hao), clyang@genetics.ac.cn (C. Yang)

ABSTRACT

Lysosomes are degradation and signaling centers within the cell, and their dysfunction impairs a wide variety of cellular processes. To understand the cellular effect of lysosome damage, we screened natural small-molecule compounds that induce lysosomal abnormality using *Caenorhabditis elegans* (*C. elegans*) as a model system. A group of vobasinyl-ibogan type bisindole alkaloids (ervachinines A–D) were identified that caused lysosome enlargement in *C. elegans* macrophage-like cells. Intriguingly, these compounds triggered cell death in the germ line independently of the canonical apoptosis pathway. In mammalian cells, ervachinines A–D induced lysosomal enlargement and damage, leading to leakage of cathepsin proteases, inhibition of autophagosome degradation and necrotic cell death. Further analysis revealed that this ervachinine-induced lysosome damage and lysosomal cell death depended on STAT3 signaling, but not RIP1 or RIP3 signaling. These findings suggest that lysosome-damaging compounds are promising reagents for dissecting signaling mechanisms underlying lysosome homeostasis and lysosome-related human disorders.

Yang Li and Yu Zhang have contributed equally to this work; leading contact: oceanyangli@fudan.edu.cn

KEYWORDS lysosome, alkaloids, lysosomal cell death, STAT3, *Caenorhabditis elegans*

INTRODUCTION

Lysosomes are acidic single-membrane organelles that function as the major sites for digesting cargoes received from several pathways, including endocytosis, phagocytosis and autophagy. The lysosome contains >60 different hydrolytic enzymes, many of which are activated at low pH. The acidity of the lysosomal lumen is generated and maintained by v-ATPase, an ATP-dependent proton pump. The lysosome contains over 150 membrane proteins that are required for the integrity and homeostasis of the organelle (Saftig and Klumperman, 2009). Impairment of lysosomal function is responsible for more than 70 lysosomal storage diseases (LSDs) (Macauley, 2016) and contributes to many other human diseases, such as neurodegenerative disorders and cancers (Nixon, 2013; Perera et al., 2015). Lysosomes also participate in several types of cell death including apoptosis and necroptosis (Taniguchi et al., 2015; Kreuzaler et al., 2011). In particular, lysosomal damage leads to lysosomal cell death (LCD) under specific physiological or pathological conditions, for instance, mammary gland involution after lactation, neutrophil aging, and bacterial infection (Sargeant et al., 2014; Loison et al., 2014; Prince et al., 2008).

LCD is characterized by lysosomal membrane permeabilization (LMP) and release of cathepsin proteases into the cytoplasm (Boya and Kroemer, 2008). In the cytoplasm, cathepsins act as executioners of cell death by mechanisms that are not well understood. Interestingly, while LCD is known to occur independently of caspases, cytoplasmic cathepsins can cleave Bid, promoting mitochondrial translocation of the proapoptotic proteins Bax and Bak, which in turn induce mitochondrial membrane permeabilization and caspase-dependent apoptosis (Oberle et al., 2010). Cathepsins can also promote the degradation of antiapoptotic proteins such as XIAP to facilitate apoptosis (Oberle et al., 2010). In addition, cathepsins promote LMP and thus amplify LCD signals (Oberle et al., 2010). The caspase-independent feature of LCD offers an important alternative for designing therapeutic strategies for cancer treatment. Because cancer cells generally carry mutations in proapoptotic factors or overexpression of antiapoptotic factors, they are usually resistant to apoptosis. However, it has been found that LCD can be induced in such apoptosis-resistant cells (Gonzalez et al., 2012). Induction of LCD was also found to restrict propagation of invading bacterial pathogens (Almaguel et al., 2010; Zhu et al., 2015). Thus, identifying potent LCD-inducing compounds may provide valuable reagents both for dissecting mechanisms underlying LCD and for treating lysosome-related human diseases.

In this study, we took advantage of the unique endolysosome system in macrophage-like cells in *Caenorhabditis elegans* (*C. elegans*) to screen for natural compounds that induced lysosomal abnormality. Our screen identified a group of vobasinyl-ibogan type bisindole alkaloids (ervachinines A–D) that caused abnormal lysosome enlargement in both *C. elegans* and mammalian cells. We further found that these lysosome-targeting natural compounds induced LMP and LCD in a STAT3-dependent manner. These findings suggest that ervachinines A–D are promising candidates for dissecting the signals underlying lysosome homeostasis and for developing therapeutic reagents for human disorders resulting from defective apoptosis.

RESULTS

Using *C. elegans* as a model to screen for natural compounds that induce lysosomal abnormality

C. elegans has 6 specialized macrophage-like cells, namely coelomocytes, which are highly active in fluid-phase endocytosis (Sato et al., 2016). Coelomocytes contain endosomes and lysosomes that are easily distinguished with differential interference contrast (DIC) optics or fluorescent markers (Fig. 1A). These features make *C. elegans* an ideal organism for screening small-molecule compounds that can affect endosome-lysosome trafficking. To identify compounds that induce endosomal or lysosomal abnormalities, we carried out a screen by treating larval stage 4 (L4) worms cultured in liquid medium with individual natural compounds at several concentrations and then observed the change in organelle morphology under DIC optics. A group of bisindole alkaloids isolated from *Ervatamia chinensis* (Meschini et al., 2008; Guo et al., 2012), named as HEC-19 (ervachinine A), HEC-20 (ervachinine C), HEC-21 (ervachinine D) and HEC-23 (ervachinine B), induced vacuolar enlargement in coelomocytes (Fig. 1B–D and Table S1). Among them, HEC-23 had the strongest effect (Fig. 1D), and it induced vacuolar enlargement in time- and dose-dependent manners (Fig. 1E and 1F).

To determine the identities of the enlarged vacuoles induced by HEC-23, we treated worms expressing endosome- or lysosome-specific proteins tagged with fluorescent proteins. HEC-23-enlarged vacuoles were positive for mCherry::CUP-5 (lysosomal calcium channel), LMP-1::GFP (lysosomal membrane protein) and ASP-1::dsRed (lysosomal hydrolase) (Fig. 1G–J). However, HEC-23 did not change the sizes of early endosomes labeled by 2xFYVE::GFP, an indicator of early endosome-specific phosphatidylinositol 3-phosphate (PI3P) (Fig. 1G). These results indicate that HEC-23 specifically enlarged lysosomes in coelomocytes.

HEC-23 impairs lysosomal degradation and increases the number of cell corpses in the germline

Next, we investigated whether HEC-23 affects the delivery of endocytic cargoes to the lysosome by injecting Texas-Red BSA (TR-BSA) into the body cavity of HEC-23-treated worms and monitoring its appearance in lysosomes in coelomocytes (Liu et al., 2016). Following injection, TR-BSA similarly appeared in lysosomes labeled with LMP-1::GFP in control animals and the enlarged LMP-1::GFP-positive lysosomes in HEC-23-treated animals, suggesting that HEC-23-induced lysosomal enlargement does not affect lysosomal cargo delivery (Fig. 2A). We then used *arIs36* animals to examine whether lysosomal degradation capacity is compromised in the enlarged lysosomes induced by HEC-23. These animals express in the body cavity a secreted soluble GFP (ssGFP) driven by a heat-shock promoter, which is taken up by coelomocytes and degraded in lysosomes (Fares and Greenwald, 2001). We treated *arIs36* animals with HEC-23 and performed time-course monitoring of ssGFP signals in coelomocytes following heat shock. While ssGFP was similarly taken up into coelomocytes in control animals and HEC-23-treated animals, the ssGFP persisted much longer in HEC-23-treated coelomocytes than in control

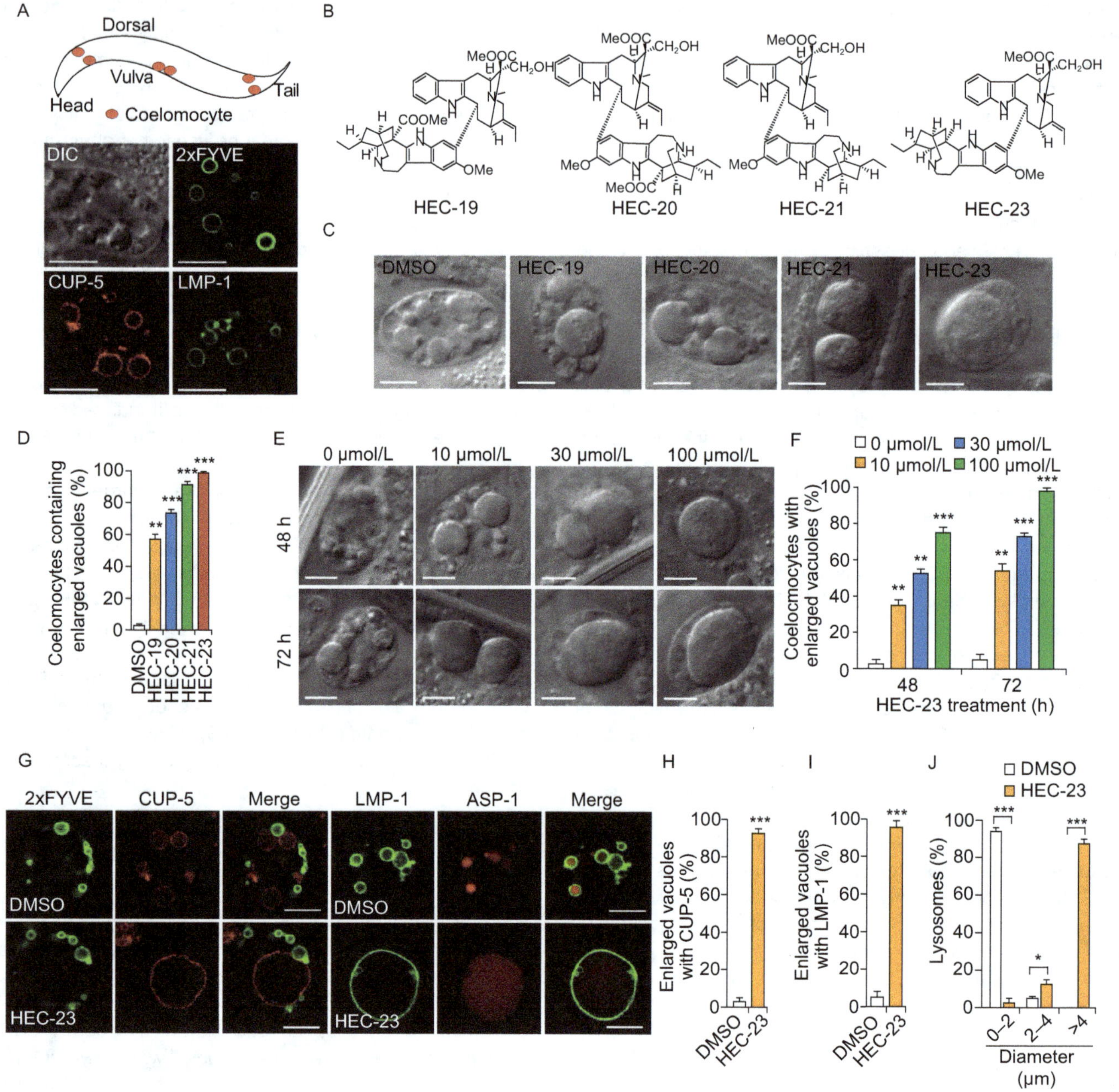

Figure 1. HEC-23 induces lysosomal enlargement in coelomocytes. (A) Representative images of endosomes and lysosomes in *C. elegans* coelomocytes. The top panel shows a schematic depiction of 3 pairs of coelomocytes (in red) in *C. elegans.* The bottom panels show a DIC image of a coelomocyte and images of 2xFYVE::GFP-labeled early endosomes, mCherry::CUP-5-labeled lysosomes, and LMP-1::GFP-labeled lysosomes. Scale bars, 10 μm. (B) Structures of HEC family compounds. (C and D) HEC family compounds induce enlargement of vacuoles in coelomocytes. Worms were treated with indicated HEC compounds at 100 μmol/L for 48 h. DIC images (C) are shown for the vacuoles and quantifications are shown in (D). (E and F) Representative DIC images (E) and quantification (F) of vacuole enlargement induced by HEC-23. (G) Effect of HEC-23 on vacuoles positive for 2xFYVE::GFP, mCherry::CUP-5, LMP-1::GFP and ASP-1::dsRed. Scale bars, 10 μm. (H and I) Quantification of vacuoles labeled with mCherry::CUP-5 (H) and LMP-1::GFP (I) in animals treated with HEC-23. (J) Quantification of lysosome sizes in worms treated with HEC-23 (100 μmol/L, 48 h). Data (mean ± SEM) were from 3 independent experiments. $^{**}P < 0.01$, $^{***}P < 0.001$.

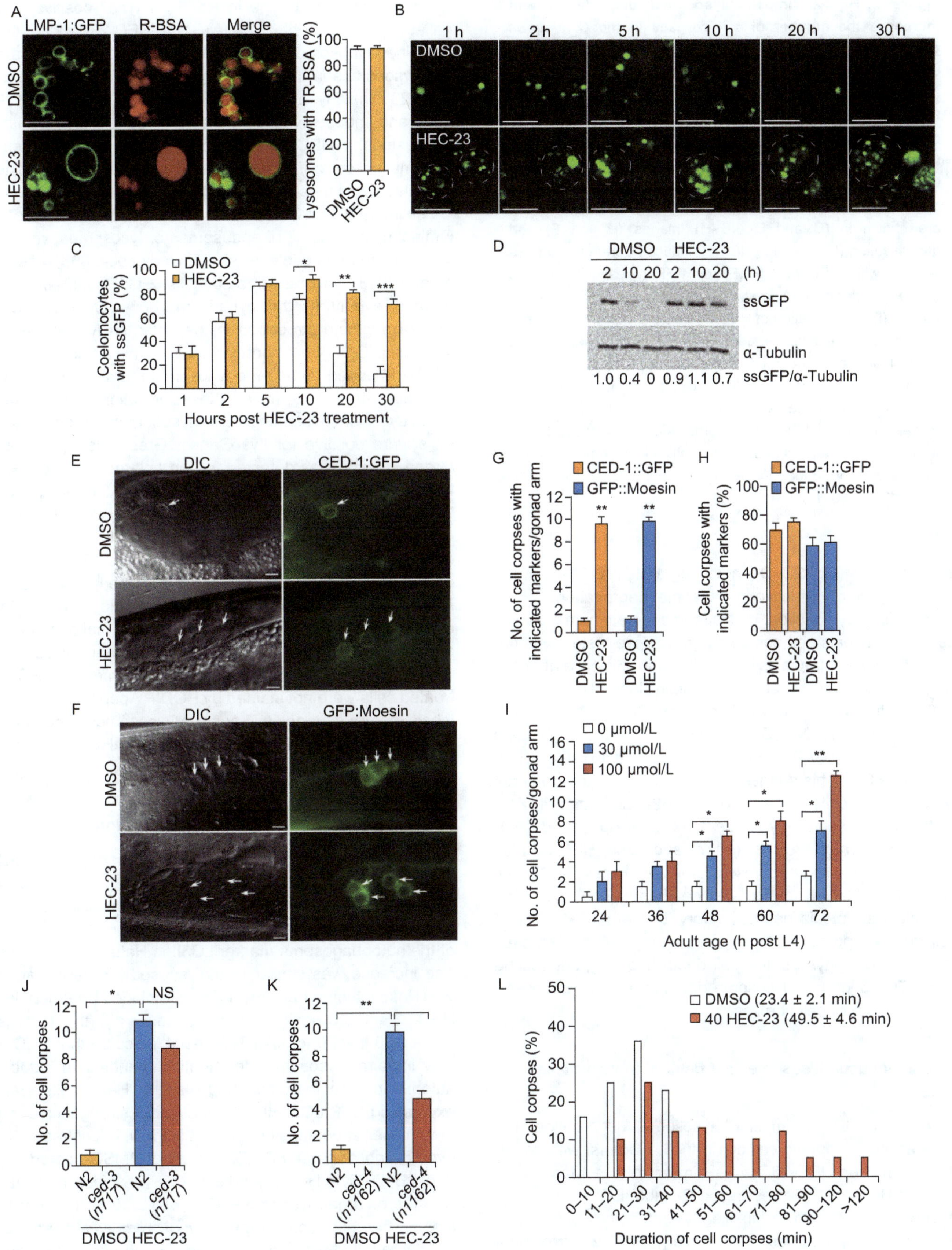
A
LMP-1:GFP
R-BSA
Merge
DMSO
HEC-23
Lysosomes with TR-BSA (%)
B
1 h
2 h
5 h
10 h
20 h
30 h
C
Coelomocytes with ssGFP (%)
Hours post HEC-23 treatment
D
DMSO
HEC-23
2 10 20 2 10 20 (h)
ssGFP
α-Tubulin
1.0 0.4 0 0.9 1.1 0.7 ssGFP/α-Tubulin
E
DIC
CED-1:GFP
F
GFP:Moesin
G
CED-1::GFP
GFP::Moesin
No. of cell corpses with indicated markers/gonad arm
H
Cell corpses with indicated markers (%)
I
0 μmol/L
30 μmol/L
100 μmol/L
No. of cell corpses/gonad arm
Adult age (h post L4)
J
NS
No. of cell corpses
N2
ced-3 (n717)
K
ced-4 (n1162)
L
DMSO (23.4 ± 2.1 min)
40 HEC-23 (49.5 ± 4.6 min)
Cell corpses (%)
Duration of cell corpses (min)

Figure 2. HEC-23 impairs lysosomal degradation and increases the number of cell corpses in the *C. elegans* germline. (A) Representative images of TR-BSA localization in LMP-1::GFP-positive lysosomes following HEC-23 treatment. (B–D) Time-course analysis of ssGFP signals in DMSO- and HEC-23-treated coelomocytes. Expression of ssGFP under the control of a heat-shock promoter was induced at 33°C for 30 min, and the uptake and degradation of ssGFP in coelomocytes were monitored at the indicated time points (B). The dashed circles indicate HEC-23-enlarged lysosomes. Quantifications are shown in (C and D). (E–H) Images (E and F) and quantification (G and H) of HEC-23-induced germ cell corpses labeled with CED-1::GFP (E) or GFP::moesin (F) in the *C. elegans* germline. Arrows indicate cell corpses. Scale bars, 20 μm. (I) Quantification of HEC-23-induced germ cell corpses in animals at the indicated adult ages. 30 animals were scored for each time point. (J and K) Quantification of HEC-23-induced germ cell corpses in *ced-3* (I) and *ced-4* (J) loss-of-function mutants. (L) Durations of germ cell corpses in DMSO- and HEC-23 (100 μmol/L)-treated worms. Cell corpses from >30 worms were analyzed. For all quantifications, data (mean ± SEM) were from 3 independent experiments. **P* < 0.05, ***P* < 0.01, ****P* < 0.001. NS, not significant.

coelomocytes (Fig. 2B–D), indicating that HEC-23-enlarged lysosomes were defective in lysosomal degradation.

Because dysfunction of lysosomes contributes to cell death and affects cell corpse clearance, we tested whether HEC-23 affects apoptosis in germlines. In wild-type animals, HEC-23 treatment caused a significant increase in button-like structures that were encircled by GFP-tagged engulfment receptor CED-1 (CED-1::GFP) or the F-actin-binding protein moesin (GFP::moesin) (Xu et al., 2014) (Fig. 2E–I, and Table S1). This suggests that the button-like structures were dying cells. The *ced-4(n1162)* and *ced-3(n717)* mutants are respectively deficient in CED-4/Apaf1 and CED-3/Caspase, which are required for apoptosis, HEC-23 also induced button-like cell corpses in these mutants (Fig. 2J, K). Thus, HEC-23 likely induced cell death independently of the canonical apoptosis pathway (Horvitz et al., 1994). These cell corpses persisted much longer than the cell corpses resulting from physiological cell death (Fig. 2L), suggesting that HEC-23-induced lysosome damage also compromised the clearance of cell corpses.

HEC-23 induces lysosome enlargement in mammalian cells

Given that HEC-23 impaired lysosomal degradation and induced non-apoptotic cell death in *C. elegans*, we investigated whether HEC-23 had a similar role in mammalian cells. In HeLa cells expressing RFP-Rab5, EGFP-Rab7 and mCherry-LAMP1, which label early endosomes, late endosomes and lysosomes, respectively, HEC-23 treatment caused a strong enlargement of LAMP1-positive lysosomes, while no obvious change in Rab5- or Rab7-positive organelles was detected (Fig. 3A). The HEC-23-treated cells contained lysosomes with diameters up to 2 μm, compared with lysosomes in diameter ≤1 μm in control cells (Fig. 3B). HEC-23 similarly enlarged lysosomes in human HEK293 cells and mouse NIH3T3 cells (Fig. 3C). These findings suggested that HEC-23 had a specific effect on lysosomes in mammalian cells. To investigate this further, we examined the integrity of these enlarged lysosomes using GFP-fused Galectin 3 (EGFP-Gal3), which binds to β-galactoside on luminal glycoproteins of endosomes or lysosomes with ruptured membranes (Liu et al., 1995; Ono et al., 2003; Maejima et al., 2013). In mock-treated cells, EGFP-Gal3 was distributed evenly in the cytoplasm; however, EGFP-Gal3 formed punctate structures in the enlarged lysosomes following HEC-23 treatment (Fig. 3D). This suggested that HEC-23 caused damage to lysosomal membranes. To corroborate this, we examined the acidification of lysosomes using LysoSensor staining. While lysosomes in mock-treated cells were positive for LysoSensor Green, no LysoSensor Green staining was detected for the enlarged lysosomes in HEC-23-treated cells (Fig. 3E). Thus, the acidification of HEC-23-enlarged lysosomes was impaired. Furthermore, we examined the localization of cathepsin L, a lysosomal cathepsin protease. In DMSO-treated cells, EGFP-tagged cathepsin L localized well in LAMP1-positive lysosomes, but HEC-23-induced enlarged lysosomes did not contain cathepsin L (Fig. 3F), suggesting that lysosomal damage led to leakage of cathepsins from the enlarged lysosomes. Consistent with these findings, the lysosomes in HEC-23-treated cells were not stained by BODIPY pepstatin A, which labels mature (cathepsin D-positive) lysosomes (Chen et al., 2000) (Fig. 3G). This result indicates that HEC-23 impairs lysosomal maturation.

HEC-23 inhibits autophagosome degradation

Because HEC-23 induced lysosome damage, we investigated if HEC-23 has an effect on autophagy, a lysosome-dependent process for degradation of cellular contents. HEC-23 treatment strongly increased the endogenous level of the autophagosome marker LC3II in HeLa cells (Fig. 4A). The increase was similar to that caused by the lysosomal v-ATPase inhibitor bafilomycin A1. Combined treatment with HEC-23 and bafilomycin A1 did not cause a further increase in the LC3 level (Fig. 4A). This result indicated that HEC-23 only impaired lysosomal degradation, rather than inducing autophagosome formation (Fig. 4A). In HeLa cells stably expressing CFP-LC3, HEC-23 induced a strong increase in the number and intensity of CFP-LC3 foci, in contrast to the even distribution of CFP-LC3 in control DMSO-treated cells (Fig. 4B–D). The HEC-23-induced CFP-LC3 foci colocalized with the enlarged lysosomes (Fig. 4E), suggesting that the fusion of autophagosomes with lysosomes was normal but the degradation was compromised. In HeLa cells transiently expressing RFP-GFP-LC3, HEC-23 induced formation of

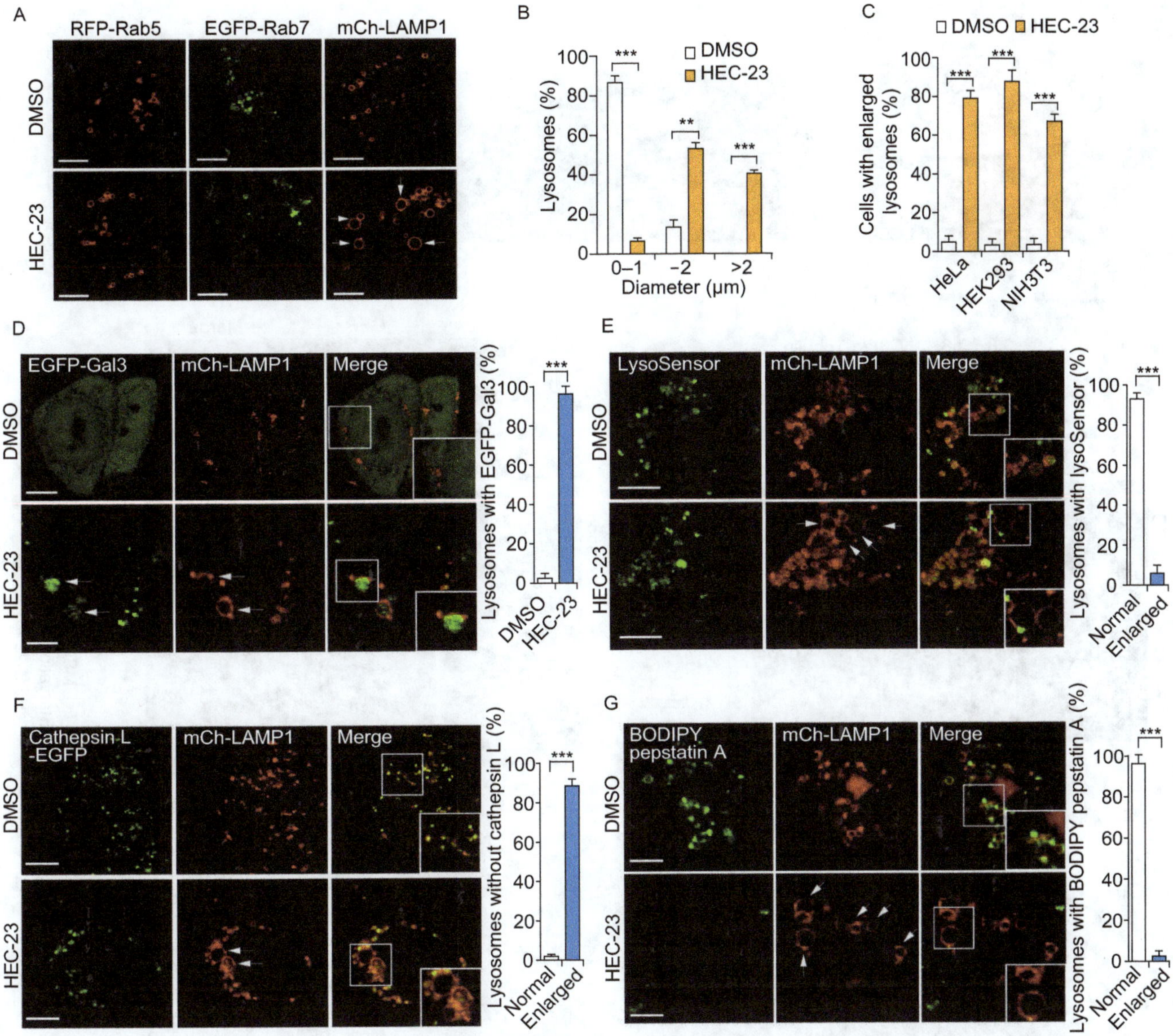

Figure 3. HEC-23 induces lysosomal damage in mammalian cells. (A and B) Images (A) and quantification (B) of HEC-23-induced enlargement of LAMP1-positive lysosomes in HeLa cells. (C) Quantification of HEC-23-induced lysosomal enlargement in the indicated cell types. (D) Representative images (left) and quantification (right) of EGFP-Gal3 in mCh-LAMP1-positive lysosomes in HeLa cells treated with HEC-23. (E) Representative images (left) and quantification (right) of LysoSensor Green staining in mCh-LAMP1-positive lysosomes in HeLa cells treated with HEC-23. (F) Representative images (left) and quantification (right) of cathepsin L-EGFP in mCh-LAMP1-positive lysosomes in HeLa cells treated with HEC-23. (G) Representative images (left) and quantification (right) of BODIPY-Pepstatin A in mCh-LAMP1-positive lysosomes in HeLa cells treated with HEC-23. In (D–G), boxed regions in the merged images are magnified and shown in the bottom right corners. Cells were treated with HEC-23 at 10 μmol/L for 3 h. Data (mean ± SEM) were from 3 independent experiments. $^{**}P < 0.01$, $^{***}P < 0.001$. Bars represent 10 μm in all images.

LC3 foci positive for both GFP and RFP (Fig. 4F), indicating that HEC-23 caused accumulation of autophagosomes by inhibiting their degradation. In contrast, starvation-induced LC3 foci were mostly positive for RFP, owing to the quenching of GFP signals following autolysosome formation (Fig. 4F). Using transmission electron microscopy (TEM), we confirmed that there was indeed an increase in the number of autophagosomes/autolysosomes with accumulated substrates in the lumen (Fig. 4G). Thus, HEC-23-induced lysosome damage inhibited degradation of autophagosomes.

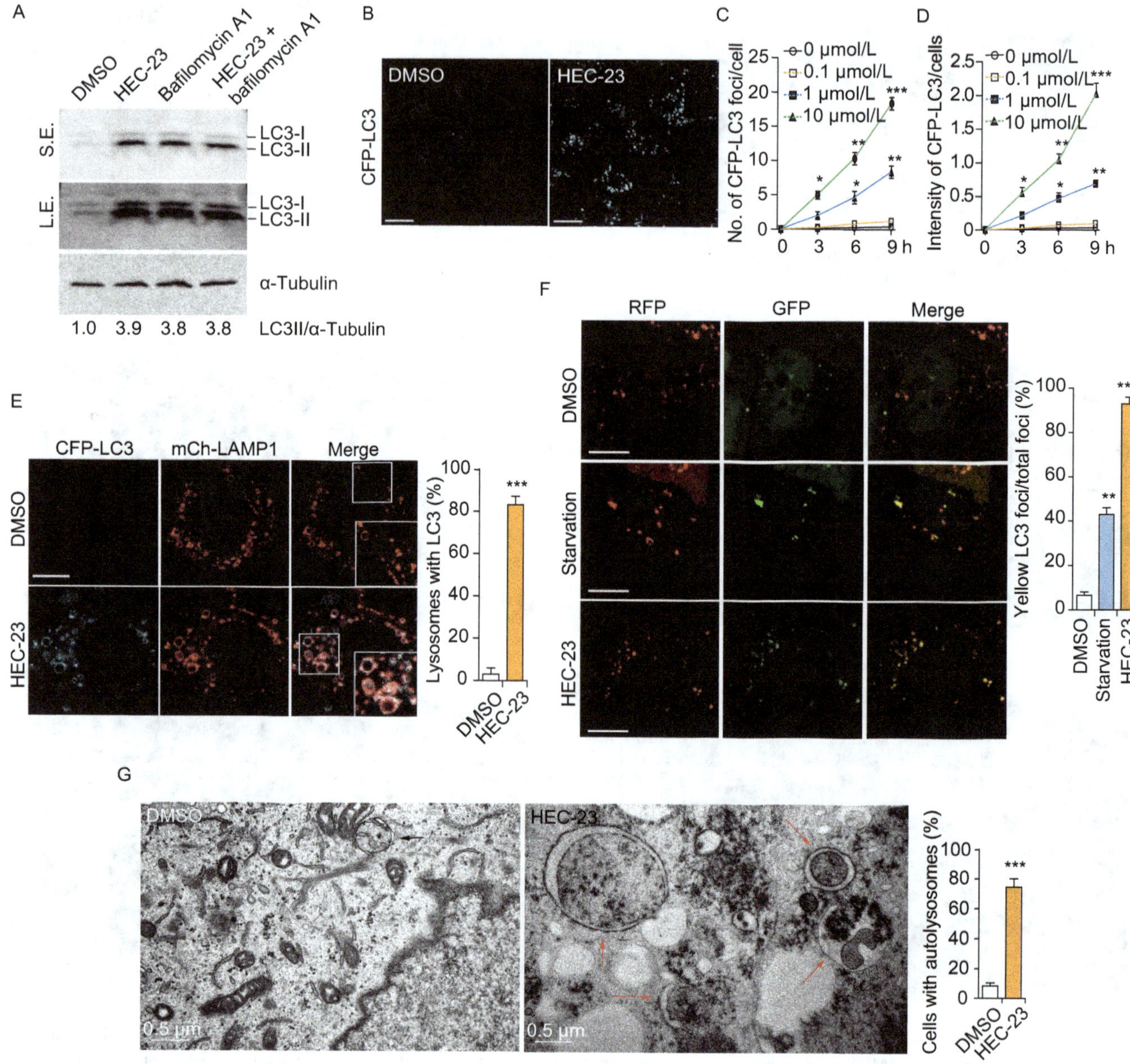

Figure 4. HEC-23 impairs autophagosome degradation. (A) Immuno-blotting of LC3 in HeLa cells treated with HEC-23 (10 μmol/L, 6 h) and bafilomycin A1 (0.4 μmol/L, 6 h). Relative fold changes of LC3-II intensity are indicated at the bottom. (B–D) Representative images (B) and quantifications (C and D) of CFP-LC3 foci in HeLa cells treated with HEC-23. (E) Representative images (left) and quantification (right) of the colocalization of stably expressed CFP-LC3 and mCh-LAMP1 in cells treated without or with HEC-23. Boxed regions in the merged images are magnified and shown in the bottom right corners. (F) Representative images of RFP-GFP-LC3 foci in cells treated with DMSO, starvation (EBSS, 3 h), or HEC-23 (10 μmol/L, 3 h). (G) Transmission electron microscopy images of cargo-containing autolysosomes in HeLa cells without or with HEC-23 treatment (10 μmol/L, 6 h). For quantifications, data (mean ± SEM) were from 3 independent experiments. *$P < 0.05$, **$P < 0.01$, ***$P < 0.001$. Bars represent 10 μm in all images except (G).

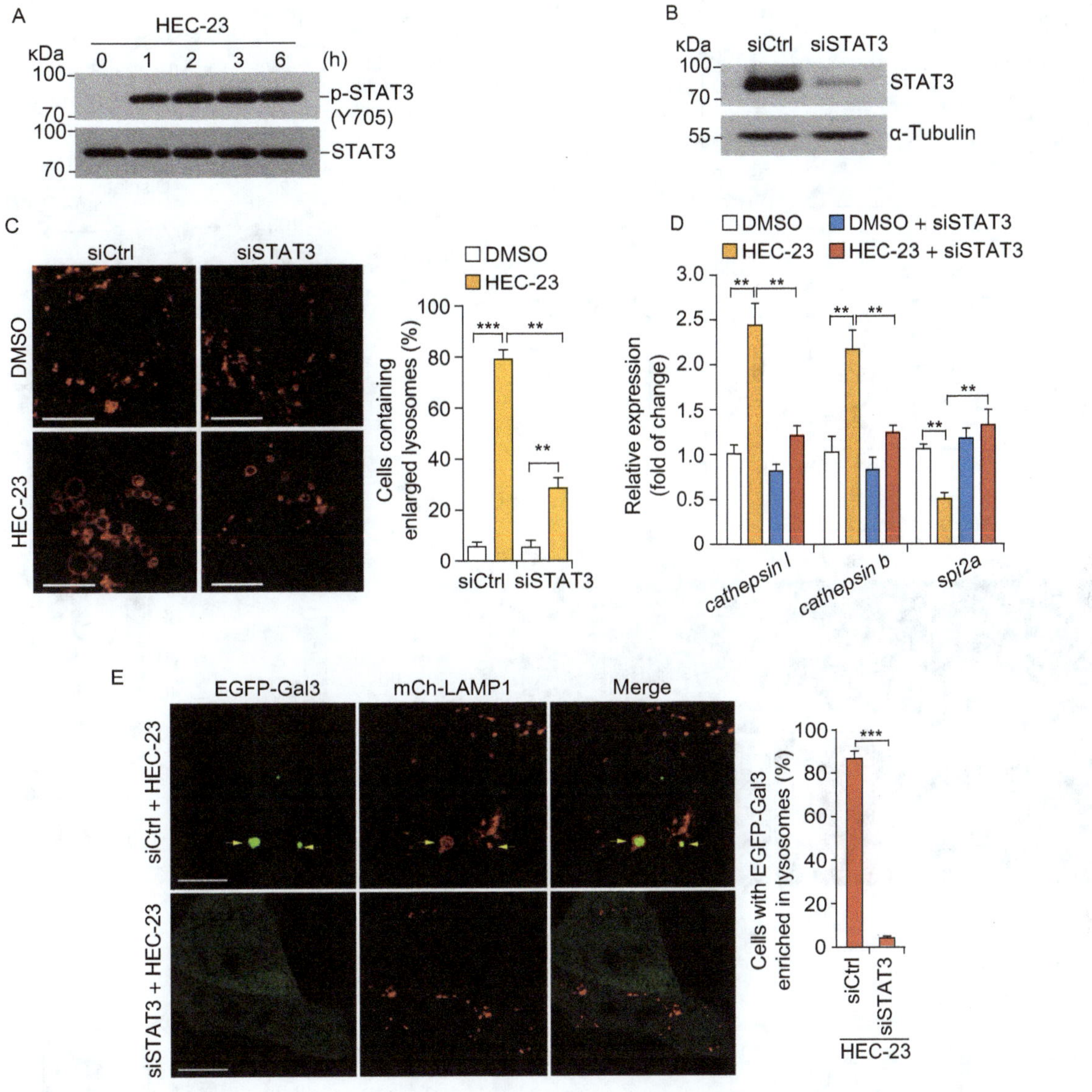

Figure 5. HEC-23-induced enlargement of lysosomes depends on STAT3 activation. (A) Immuno-blotting of STAT3 phosphorylation at Y705 in HeLa cells following HEC-23 treatment. (B) Immuno-blotting of STAT3 in HeLa cells treated with siSTAT3. (C) Representative images (left) and quantification (right) of lysosomes labeled with mCherry-LAMP1 in HEC-23-treated HeLa cells pretreated with control siRNA (siCtrl) and STAT3 siRNA (siSTAT3). (D) qPCR-analysis of the expression of STAT3 target genes in siCtrl or siSTAT3 cells treated with HEC-23. (E) Representative images (left) and quantification (right) of lysosome damage in HEC-23-treated HeLa cells pretreated with control siRNA (siCtrl) and STAT3 siRNA (siSTAT3). The concentration of HEC-23 was 10 μmol/L in all treatments. Data representing mean ± SEM were derived from 3 independent experiments. $^{**}P < 0.01$, $^{***}P < 0.001$ (Scale bars, 10 μm).

HEC-23-induced lysosomal enlargement depends on STAT3 activation

Recently, it was found that STAT3 controls cell death during mammary gland involution by regulating LMP (Kreuzaler et al., 2011; Sargeant et al., 2014). During this process, STAT3 is activated, which up-regulates expression of cathepsins B and L but down-regulates their endogenous inhibitor Spi2A, causing LMP-mediated necrosis in mammary epithelial cells. We thus investigated if STAT3 signaling is involved in HEC-23-induced lysosomal enlargement. HEC-23 strongly increased the phosphorylation of STAT3 at tyrosine 705 (Y705) (Fig. 5A). Importantly, knocking down *STAT3* significantly suppressed HEC-23-induced lysosomal enlargement and damage (Fig. 5B, 5C and 5E). HEC-23 treatment resulted in elevated expression of *cathepsin l* and *cathepsin b*, but decreased the expression of *spi2a*, all of which are STAT3 target genes (Fig. 5D). Furthermore, knocking down *STAT3* reversed the HEC-23-induced

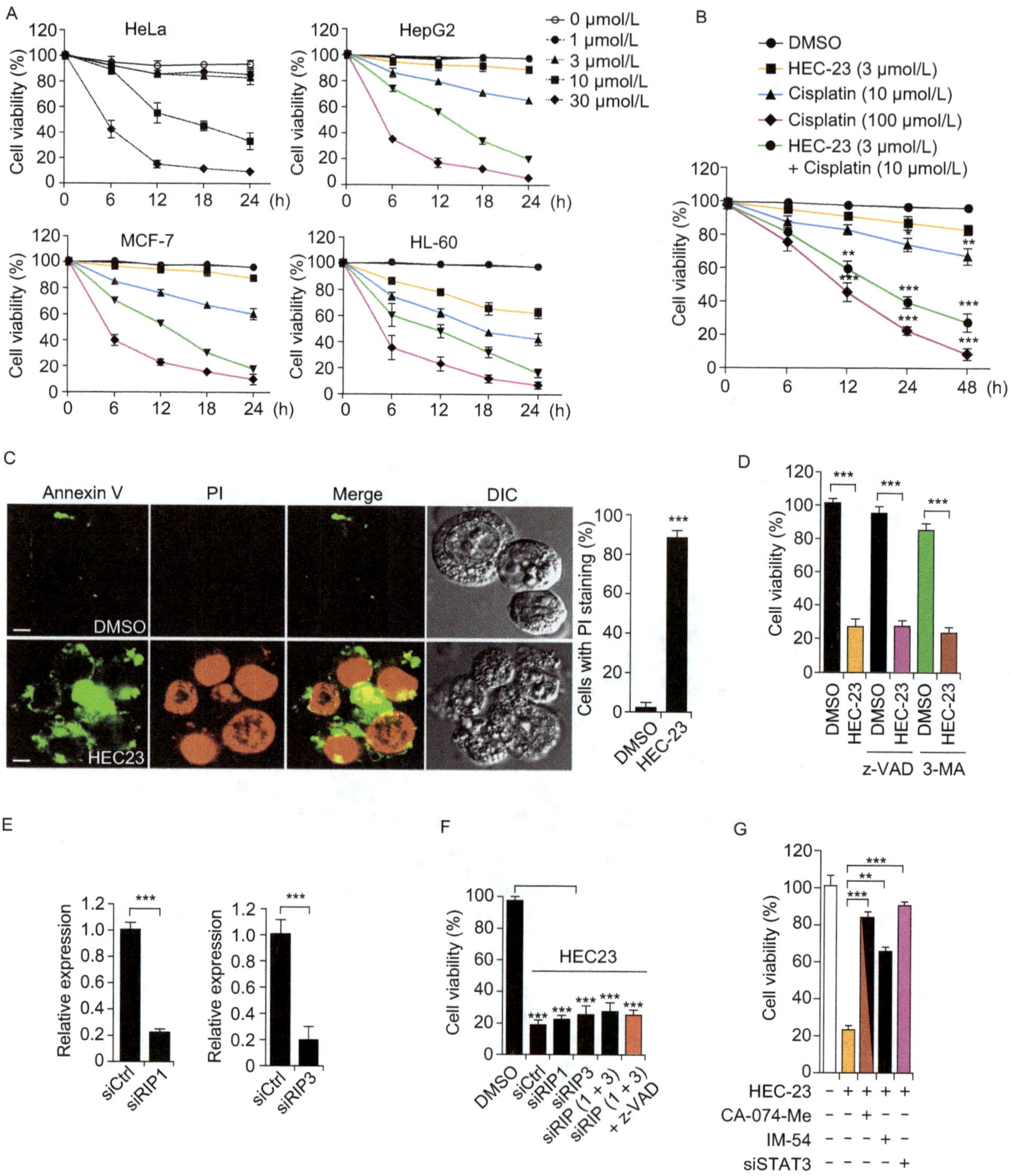

Figure 6. HEC23-induced cell death is dependent on the STAT3-cathepsin pathway. (A) Quantifications of cell death induced by HEC-23 in the indicated types of carcinoma cells. (B) Quantification of cell death induced by HEC-23 together with or without cisplatin in HeLa cells. (C) Representative images (left) and flow cytometry quantification (right) of necrosis induced by HEC-23 (10 μmol/L, 12 h). (D) Quantification of cell death induced by HEC-23 (10 μmol/L, 24 h) in HeLa cells pre-treated with z-VAD (20 μmol/L) or 3-MA (1 μmol/L) for 30 min. (E) qPCR evaluation of the knockdown efficiency of siRIP1 and siRIP3. (F) Quantification of the viability of HeLa cells treated with HEC-23 and siRIP1, siRIP3 or both in the presence or absence of z-VAD. (G) Quantification of cell viability in siCtrl or siSTAT3 HeLa cells treated with HEC-23, CA-074-Me or IM-54. HeLa cells were transfected with siRNA for 48 h, followed by treatment with CA-074-Me (20 μmol/L) or IM-54 (2 μmol/L) for 30 min. The cells were then treated with HEC-23 for 24 h. MTT assays were performed to examine the viability of the cells. The concentration of HEC-23 was 10 μmol/L in all treatments. Data representing mean ± SEM were derived from 3 independent experiments. ***P* < 0.01, ****P* < 0.001 (Scale bars, 10 μm).

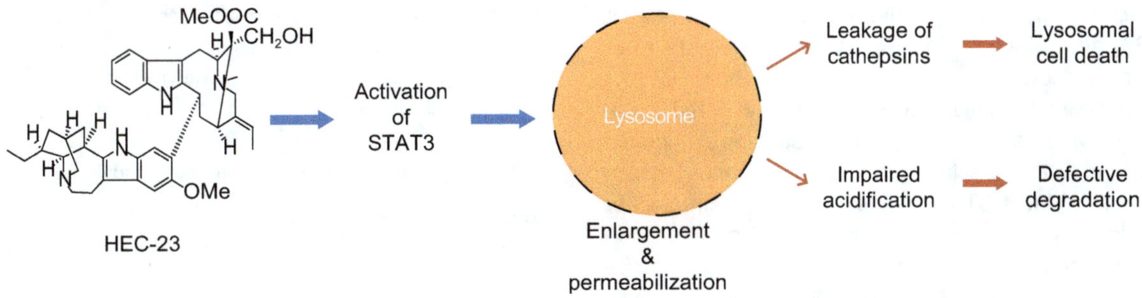

Figure 7. Schematic diagram of HEC23-induced lysosomal enlargement and cell death.

changes in the expression of these genes (Fig. 5D). Altogether, these findings suggest that STAT3 signaling is required for HEC-23-induced lysosomal enlargement and damage.

HEC-23 induces cathepsin-dependent necrosis through STAT3 signaling

In carcinoma cells, including HeLa, HepG2, MCF-7 and HL-60, HEC-23 induced cell death in a concentration-dependent manner (Fig. 6A). In addition, HEC-23 had a synergistic effect on cell death with cisplatin, a well-established anti-tumor therapeutic drug (Fruh et al., 2016) (Fig. 6B). Further analysis indicated that >90% of the dying HEC-23-treated cells were positive for propidium iodide (PI) staining, indicating that HEC-23 likely induced non-apoptotic cell death (Fig. 6C). In support of this hypothesis, pretreatment with the pan-caspase inhibitor z-VAD or the autophagic inhibitor 3-MA (3-methyladenine) did not inhibit the cell death induced by HEC-23 (Fig. 6D).

To determine which signaling pathway is required for HEC-23-induced cell death, we performed siRNA (small interfering RNA) to knock down RIP1 and RIP3, the key regulators of necroptosis (Sun and Wang, 2014). Knock-down of RIP1, RIP3, or both, did not affect HEC-23-induced cell death (Fig. 6E, F), suggesting that HEC-23 does not act through these two factors. In contrast, treatment of cells with *STAT3* siRNA, the cathepsin inhibitor CA-074-Me or the necrosis inhibitor IM-54 strongly suppressed HEC-23-induced cell death (Fig. 6G). These findings indicated that HEC23-induced STAT3 activation not only upregulated gene expression of cathepsins (Fig. 5D), but also resulted in LMP (Fig. 5C, E) and subsequent leakage of cathepsins (Fig. 3F) and necrosis (Fig. 6G). Altogether, these results indicate that STAT3 signaling is required for HEC-23-induced lysosomal dysfunction and LCD (Fig. 7).

DISCUSSION

In this study, we present *C. elegans* as a model for screening natural compounds that target lysosomes. We found that a group of alkaloids, named HECs, enlarge the size and impair the integrity, acidification and digestion capacity of lysosomes, and these effects are conserved in different species. HEC-induced lysosomal impairment results in the accumulation of cell corpses and non-apoptotic cell death in *C. elegans* and lysosomal cell death in mammalian cells.

The use of *C. elegans* to identify lysosome-targeting small-molecule compounds has many advantages. Firstly, the macrophage-like coelomocytes are active in fluid-phase endocytosis and pinocytosis, which facilitate the entry of compounds into the cell. Secondly, lysosomes are easily identified in a living animal using DIC optics or fluorescently labeled makers. Thirdly, the powerful genetic approaches and the availability of numerous mutant alleles make it possible to dissect the mechanisms underlying the function of the compounds.

Apoptosis and necrosis have been extensively studied during the development of organisms and the pathogenesis of diseases. In the present study, we find that mutation of neither *ced-3* nor *ced-4* could totally block the HEC-23-induced elevation of cell corpses. This demonstrates that HEC-23 induces non-apoptotic cell death in *C. elegans*. We also performed knock-down of RIP1 and RIP3, which did not reverse HEC-23-induced necrosis in mammalian cells. These data demonstrate that HEC-23-induced necrosis is independent of the RIP1/3 pathway.

Intriguingly, LMP also plays a crucial physiological role in regression of the mammary gland. Conditional deletion of STAT3 causes an obviously delay in involution of the mammary gland and reduces the level of cell death (Kreuzaler et al., 2011; Sargeant et al., 2014). In the present study, we found that HEC-23 induces the activation of STAT3, up-regulation of cathepsins B and L and down-regulation of Spi2A. Importantly, knock-down of STAT3 abolished lysosomal enlargement and necrosis induced by HEC-23 treatment. Furthermore, inhibition of cathepsins reverses the HEC-23-induced reduction in cell viability. These findings indicate that HEC23-induced necrosis is dependent on the STAT3-cathepsin pathway.

We found that HEC-23 treatment significantly promotes the phosphorylation of STAT3 at Y705. Unfortunately, we failed to obtain functional biotin-labeled compounds related to HEC-23, since they all lost the ability to enlarge lysosomes in cells. However, we can still propose how HEC-23 may interact with potential targets in the JAK-STAT3 signaling pathways. Firstly, HEC-23 may function as an agonist to directly bind with and activate JAK (Janus kinases), which in turn phosphorylates STAT3 at the Y705 site (Vainchenker and Constantinescu, 2013). Secondly, HEC-23 may function as an antagonist to bind with Src homology region 2 domain-containing phosphatase-1 (SHP-1) or suppressor of cytokine signaling (SOCS), which inhibits JAK activity in the cytoplasm (Vainchenker and Constantinescu, 2013). Thirdly, HEC-23 may function as an antagonist to bind with E3 SUMO-protein ligases (PIASs) or protein tyrosine phosphatases (PTPs), which inhibit STAT3 transcription activity in the nucleus (Vainchenker and Constantinescu, 2013).

Usually, LCD remains functional in apoptosis-resistant cancer cells. The fact that HEC-23 induced non-apoptotic cell death in several different cancer cell lines suggests that HEC-23 family compounds can potentially be used to develop therapeutic reagents for cancers that are resistant to apoptosis-based therapy. Notably, the antimalarial agent mefloquine was found to induce LMP and release of cathepsins into the cytoplasm of human acute myeloid leukemia (AML) cells, which provides a novel and promising therapeutic strategy for AML (Sukhai et al., 2013). Further studies are needed to investigate the application of HEC-23 and STAT3 activation to the therapeutic treatment of AML and other cancers.

MATERIALS AND METHODS

C. elegans strains and genetics

The Bristol strain N2 is used as wild type. The mutant alleles used in this study are *vps-18*(*tm1125*), *arl-8*(*tm2388*), *ced-3*(*n717*) and *ced-4* (*n1162*). The integrated arrays are: *smIs34*($P_{ced\text{-}1}$*ced-1::gfp*), *yqIs121*($P_{ced\text{-}1}$*gfp::moesin*), *cdIs85*($P_{unc\text{-}122}$*2xfyve::gfp*), *cdIs97* ($P_{unc\text{-}122}$*mCherry::cup-5*), *cdIs131*($P_{unc\text{-}122}$*gfp::rab-5*), *pwIs50*($P_{lmp\text{-}1}$*lmp-1::gfp*), and *tmIs225*($P_{asp\text{-}1}$*asp-1::dsRed*). *C. elegans* cultures and genetic crosses were performed according to standard procedures.

Screen for natural compounds that induce enlargement of lysosomes

All compounds used in this study were isolated from plants and compound structures were determined by means of 1D NMR, 2D NMR and MS as previously reported (Li et al., 2016). In total, 257 natural compounds (35 alkaloids, 23 triterpenes, 46 diterpenes, 22 sesquiterpenes, 17 monoterpenes, 18 tetranortriterpenoids, 12 flavonoids, 11 coumarins, 25 sterols, 27 lignanoids, 11 saponins, and 10 cyclic peptides) were tested individually for their ability to induce lysosome enlargement. Worms were cultured in M9 solution (1 L contains 3 g KH_2PO_4, 6 g Na_2HPO_4, 5 g NaCl and 1 mmol/L $MgSO_4$) with X1666. 20–30 worms at L4 stage were transferred into liquid culture with or without different natural compounds in 24-well plates in a shaker (120 rpm, 20°C). 48 h or 72 h later, worms were examined for lysosomal changes in coelomocytes under a fluorescence microscope with DIC optics. The investigators were blinded to compound identities during the screen.

Microscopy and trafficking experiments in coelomocytes

Adult worms were immobilized with 2.5 mmol/L levamisole in M9 solution and mounted on 2% agarose pads for imaging. DIC pictures were captured by using an AxioImager M1 (Carl Zeiss). Fluorescence images were obtained by using an inverted FV1000 confocal microscope system (IX81; Olympus). TR-BSA trafficking assays in *C. elegans* coelomocytes were performed as previously described[29]. In brief, TR-BSA (Sigma-Aldrich; 1 mg/mL in water) was injected into the body cavity of adult worms. Then worms were cultured on NGM plates seeded with *Escherichia coli* OP50. Worms were imaged by confocal microscopy at different time points after injection. For each time point, similar results were obtained in more than 30 coelomocytes from 15–20 different worms.

Germ cell corpse analysis

Animals synchronized to different adult stages were scored for germ cell corpses under Nomarski optics. Animals were grown in liquid culture at 20°C unless otherwise indicated. For cell corpse analysis in HEC-23-treated worms, animals were scored for germ cell corpses after 12 h, 24 h, 36 h, 48 h and 72 h. At each time point, germ cell corpses in the meiotic region of one gonad arm were counted for every animal, and ≥30 animals were analyzed.

Cell culture, transfection and reagents

All cell lines were cultured at 37°C with 5% CO_2 in Dulbecco's modified Eagle's medium (DMEM) supplemented with 10% fetal bovine serum (FBS) (HyClone), 100 U/mL penicillin and 100 mg/mL streptomycin. No cell lines used in this study were found in the database of commonly misidentified cell lines that is maintained by ICLAC and NCBI Biosample. All cell lines were from American type culture collection (ATCC). Transient transfections were performed with Lipofectamine 2000 (Invitrogen, Carlsbad, CA) following the manufacturers' instructions. TR-BSA and CA-074-Me were from Tocris Bioscience (Bristol, UK). bafilomycin A1, 3-MA, z-VAD, necrosis inhibitor IM-54 and cisplatin were from Calbiochem (Darmstadt, Germany). The mature lysosome dye BODIPY-pepstatin A, LysoSensor, Annexin V and PI were purchased from Invitrogen Life Technologies (Carlsbad, CA). The antibody against LC3 was from MBL. Antibodies against p-STAT3 and STAT3 were from Cell Signaling Technology. Mouse monoclonal antibodies for α-tubulin were purchased from Sigma-Aldrich (St. Louis, MO). HRP-, Cy3- and FITC-conjugated secondary antibodies were obtained from Jackson ImmunoResearch Laboratories (West Grove, PA).

Western blot

Cells were lysed in ice-cold RIPA buffer (20 mmol/L Tris-HCl pH 7.5, 100 mmol/L NaCl, 0.1% SDS, 0.5% sodium deoxycholate, 1 mmol/L PMSF) containing Complete Protease Inhibitor Cocktail (1 tablet in 50 mL RIPA buffer) and Phosphatase Inhibitor Cocktail Tablets (1 tablet in 10 mL RIPA buffer) (Roche, Basel, Switzerland). Cell lysates were spun down at 12,000 rpm for 10 min at 4°C. 20 μg of each supernatant were resolved by sodium dodecyl sulfate polyacrylamide gel electrophoresis (SDS-PAGE) and probed with the indicated antibodies. α-Tubulin was used as the internal control.

Small interfering RNA (siRNA)

RNA oligos used for siRNA in this study were:

Human gene	Oligo
STAT3	5′-UGUAAUGCAUGACAGCCUGTT-3′ 5′-AUUGUCUUUCUUCUGCCGCTT-3′ 5′-UUGAUGUUGAACCUUCGUCTT-3′
RIP1	5′-AUCCCUGCUCUCUUCAGUGTT-3′ 5′-AGGGCUGCUUUCCUUGGCCTT-3′ 5′-UCAUCAGCCUGGAGUCCAGTT-3′
RIP3	5′-AUUUGAAUGUAAAGGACUCTT-3′ 5′-AGUAACAAAUUCAUGGCACTT-3′ 5′-AUUUCAUACAACAGGACGCTT-3′
Control siRNA	5′-UUCUCCGAACGUGUCACGUTT-3′

Cells were transfected with 100 pmol RNA oligos twice (at 0 h and 24 h) using Lipofectamine 2000 in 6-well plates or confocal culture dishes. The efficiency of siRNAs was evaluated by Western blot or qRT-PCR.

Quantitative real-time reverse-transcription PCR (qPCR)

RNA was isolated from cells by using TRIzol reagent (Invitrogen) as recommended by the manufacturer. A reverse transcription kit (Promega) was used to reverse transcribe RNA (1 μg) in a 20 μL-reaction mixture. Quantification of gene expression was performed using a real-time PCR system (7900HT Fast; Applied Biosystems) in triplicate. Amplification of the sequence of interest was normalized with the reference endogenous *GAPDH* gene.

Expression vectors

The mammalian expression vector pEGFP-N2-cathepsin L was constructed by inserting the cDNA of cathepsin L between the *Hind*III and *Kpn*I sites of the pEGFP-N2 vector using standard protocols confirmed by sequencing.

The following vectors were kindly provided by other scientists: mCherry-LAMP1 (Dr. Li Yu, Tsinghua University, China), EGFP-Galectin3 (Dr. Tamotsu Yoshimori, Osaka University, Japan), RFP-GFP-LC3 (Dr. Hong Zhang, Institute of Biophysics, CAS). All expression constructs were confirmed by DNA sequencing.

Transmission electron microscopy

HeLa cells were cultured on plastic cover slices with DMEM containing 10% FBS. After HEC-23 treatment for 3 h, the cells were fixed in fixation buffer (2.5% glutaraldehyde, 1% paraformaldehyde in PBS) on ice for 1 h. The samples were then post-fixed by 1% OsO_4 for 2 h, followed by dehydration in a graded ethanol series (30%, 50%, 70%, 90% and 100%). After rinsing with propylene oxide (100%) for 3 times, the samples were infiltrated stepwise in increasing concentrations of embed 812 resin (propylene oxide:resin 2:1 for 3 h and 1:1 for 5 h). Then, the samples were incubated in 100% fresh resin twice for 8 h, and transferred into fresh resin in an embedding mold and polymerized in a 60°C oven for 3 days. Ultrathin sections (70 nm) were generated with a diamond knife (Diatome) on an ultramicrotome (Ultracut UCT; Leica Microsystems), and collected on copper grids (EMS). The slices on copper grids were stained with 2% UAc and 1% citric acid for 10 min. Then, the samples were visualized with a JEM-1400 TEM at 80 kV. Pictures were recorded with a Gatan832 (4k × 2.7k) CCD camera.

MTT assay for cell viability

Cells were cultured in 96-well plates with DMEM containing 10% FBS. After HEC-23 treatment for 24 h, 15 μL of dye solution was added into each well. Then the plates were incubated at 37°C for 2 h in a humidified CO_2 incubator. 100 μL of stop solution was added to each well, and the absorbance was recorded at 570 nm using a 96-well plate reader. A reference wavelength at 630 nm was used. The MTT kit was purchased from Promega (Cat# G4002).

Cell death quantification by flow cytometry

After treatment with HEC-23 for 24 h, the cells were harvested and washed with PBS. Cells were stained in binding buffer containing Annexin V (5 μL) and PI (10 μL) for 15–30 min in the dark. After extensive washing, cells were suspended in PBS and transferred into tubes for quantification of cell death by flow cytometry using a FACS AriaII machine (BD Biosciences). Data were analyzed by using FlowJo software (FLOWJO, LLC).

Statistics and reproducibility

Data were analyzed with Prism (GraphPad Software) to generate curves and bar charts. Statistical analyses were performed using *t*-tests or ANOVA. $P < 0.05$, indicated with *, was considered statistically significant. $P < 0.01$, indicated with **, was considered significant. $P < 0.001$, indicated with ***, was considered extremely significant. $P > 0.05$ was considered not significant (NS).

ACKNOWLEDGEMENTS

This research was supported by grants 31230043 (to C.Yang), 21432010 (to X. Hao and C. Yang), and 81473122 (to Y. Zhang) from the National Natural Science Foundation of China, 2013CB910102 from the National Basic Research Program of China, and the CAS Interdisciplinary Innovation Team (to C. Yang), the Youth Innovation Promotion Association of CAS (2015323), CAS "Light of West China" Program (to Y. Zhang), the Young Academic and Technical Leader Raising Foundation of Yunnan Province (to Y. Zhang), the Technological Leading Talent Project of Yunnan Province (to X. Hao), the Startup Funding of Fudan University (to Y. Li), and Funding for Construction of Outstanding Universities in Shanghai (to Y. Li). We also appreciate Dr. Isabel Hanson to proofread this manuscript.

ABBREVIATIONS

AML, acute myeloid leukemia; *C. elegans*, *Caenorhabditis elegans*; DIC, differential interference contrast; FBS, fetal bovine serum; LCD, lysosomal cell death; LMP, lysosomal membrane permeabilization; LSDs, lysosomal storage diseases; PI, propidium iodide; PTPs, protein tyrosine phosphatases; SHP-1, Src homology region 2 domain-containing phosphatase-1; SOCS, suppressor of cytokine signaling; TEM, transmission electron microscopy

AUTHOR CONTRIBUTIONS

Y Li, C. Yang, Y. Zhang and X. Hao designed the experiments and wrote the paper. Q. Gan and M. Xu contributed to the experiments shown on Fig. 2E–H and J–L. X. Ding and G. Tang contributed to compound screening. K. Liu, X. Liu, L. Guo and Z. Gao contributed materials. J. Liang and X. Wang performed transmission electron microscopy. Y. Li and Y. Zhang analyzed results. All authors reviewed and approved the final version of the manuscript.

COMPLIANCE WITH ETHICS GUIDELINES

Yang Li, Yu Zhang, Qiwen Gan, Meng Xu, Xiao Ding, Guihua Tang, Jingjing Liang, Kai Liu, Xuezhao Liu, Xin Wang, Lingli Guo, Zhiyang Gao, Xiaojiang Hao, and Chonglin Yang declare that they have no conflict of interest. All institutional and national guidelines for the care and use of laboratory animals were followed.

REFERENCES

Almaguel F, Liu J, Pacheco F, De Leon D, Casiano C, De Leon M (2010) Lipotoxicity-mediated cell dysfunction and death involve lysosomal membrane permeabilization and cathepsin L activity. Brain Res 1318:133–143

Boya P, Kroemer G (2008) Lysosomal membrane permeabilization in cell death. Oncogene 27(50):6434–6451

Chen C, Chen W, Zhou M, Arttamangkul S, Haugland RP (2000) Probing the cathepsin D using a BODIPY FL-pepstatin A: applications in fluorescence polarization and microscopy. J Biochem Biophys Methods 42(3):137–151

Fares H, Greenwald I (2001) Genetic analysis of endocytosis in *Caenorhabditis elegans*: coelomocyte uptake defective mutants. Genetics 159(1):133–145

Fruh M, Ris HB, Xyrafas A, Peters S, Mirimanoff RO, Gautschi O, Pless M, Stupp R (2016) Preoperative chemoradiotherapy with cisplatin and docetaxel for stage IIIB non-small-cell lung cancer: 10-year follow-up of the SAKK 16/01 trial. Ann Oncol 27 (10):1971–1973

Gonzalez P, Mader I, Tchoghandjian A, Enzenmuller S, Cristofanon S, Basit F, Debatin KM, Fulda S (2012) Impairment of lysosomal integrity by B10, a glycosylated derivative of betulinic acid, leads to lysosomal cell death and converts autophagy into a detrimental process. Cell Death Differ 19(8):1337–1346

Guo L, He H, Di Y, Li S, Cheng Y, Yang W, Li Y, Yu J, Zhang Y, Hao X (2012) Indole alkaloids from *Ervatamia chinensis*. Phytochemistry 74:140–145

Horvitz HR, Shaham S, Hengartner MO (1994) The genetics of programmed cell death in the nematode *Caenorhabditis elegans*. Cold Spring Harb Symp Quant Biol 59:377–385

Kreuzaler P, Staniszewska A, Li W, Omidvar N, Kedjouar B, Turkson J, Poli V, Flavell R, Clarkson R, Watson C (2011) Stat3 controls lysosomal-mediated cell death in vivo. Nat Cell Biol 13(3):303–309

Li Y, Xu M, Ding X, Yan C, Song Z, Chen L, Huang X, Wang X, Jian Y, Tang G, Tang C, Di Y, Mu S, Liu X, Liu K, Li T, Wang Y, Miao L, Guo W, Hao X, Yang C (2016) Protein kinase C controls lysosome biogenesis independently of mTORC1. Nat Cell Biol 18 (10):1065–1077

Liu FT, Hsu DK, Zuberi RI, Kuwabara I, Chi EY, Henderson WR Jr (1995) Expression and function of galectin-3, a beta-galactoside-binding lectin, in human monocytes and macrophages. Am J Pathol 147(4):1016–1028

Liu K, Jian Y, Sun X, Gao Z, Zhang Z, Liu X, Li Y, Xu J, Jing Y, Mitani S, He S, Yang C (2016) Negative regulation of phosphatidylinositol 3-phosphate levels in early-to-late endosome conversion. J Cell Biol 212(2):181–198

Loison F, Zhu H, Karatepe K, Kasorn A, Liu P, Ye K, Zhou J, Cao S, Gong H, Jenne D, Remold-O'Donnell E, Xu Y, Luo HR (2014) Proteinase 3-dependent caspase-3 cleavage modulates neutrophil death and inflammation. J Clin Invest 124(10):4445–4458

Macauley S (2016) Combination therapies for lysosomal storage diseases: a complex answer to a simple problem. Pediatr Endocrinol Rev 13(Suppl 1):639–648

Maejima I, Takahashi A, Omori H, Kimura T, Takabatake Y, Saitoh T, Yamamoto A, Hamasaki M, Noda T, Isaka Y, Yoshimori T (2013) Autophagy sequesters damaged lysosomes to control lysosomal biogenesis and kidney injury. EMBO J 32(17):2336–2347

Meschini S, Condello M, Calcabrini A, Marra M, Formisano G, Lista P, De Milito A, Federici E, Arancia G (2008) The plant alkaloid voacamine induces apoptosis-independent autophagic cell death on both sensitive and multidrug resistant human osteosarcoma cells. Autophagy 4(8):1020–1033

Nixon R (2013) The role of autophagy in neurodegenerative disease. Nat Med 19(8):983–997

Oberle C, Huai J, Reinheckel T, Tacke M, Rassner M, Ekert PG, Buellesbach J, Borner C (2010) Lysosomal membrane permeabilization and cathepsin release is a Bax/Bak-dependent, amplifying event of apoptosis in fibroblasts and monocytes. Cell Death Differ 17(7):1167–1178

Ono K, Kim SO, Han J (2003) Susceptibility of lysosomes to rupture is a determinant for plasma membrane disruption in tumor necrosis factor alpha-induced cell death. Mol Cell Biol 23(2):665–676

Perera R, Stoykova S, Nicolay B, Ross K, Fitamant J, Boukhali M, Lengrand J, Deshpande V, Selig M, Ferrone C, Settleman J, Stephanopoulos G, Dyson N, Zoncu R, Ramaswamy S, Haas W, Bardeesy N (2015) Transcriptional control of autophagy-lysosome function drives pancreatic cancer metabolism. Nature 524 (7565):361–365

Prince L, Bianchi S, Vaughan K, Bewley M, Marriott H, Walmsley S, Taylor G, Buttle D, Sabroe I, Dockrell D, Whyte M (2008) Subversion of a lysosomal pathway regulating neutrophil apoptosis by a major bacterial toxin, pyocyanin. J Immunol 180 (5):3502–3511

Saftig P, Klumperman J (2009) Lysosome biogenesis and lysosomal membrane proteins: trafficking meets function. Nat Rev Mol Cell Biol 10(9):623–635

Sargeant T, Lloyd-Lewis B, Resemann H, Ramos-Montoya A, Skepper J, Watson C (2014) Stat3 controls cell death during mammary gland involution by regulating uptake of milk fat globules and lysosomal membrane permeabilization. Nat Cell Biol 16(11):1057–1068

Sato K, Norris A, Sato M, Grant B (2016) *C. elegans* as a model for membrane traffic. Wormbook. https://doi.org/10.1895/wormbook.1.77.2

Sukhai MA, Prabha S, Hurren R, Rutledge AC, Lee AY, Sriskanthadevan S, Sun H, Wang X, Skrtic M, Seneviratne A, Cusimano M, Jhas B, Gronda M, MacLean N, Cho EE, Spagnuolo PA, Sharmeen S, Gebbia M, Urbanus M, Eppert K, Dissanayake D, Jonet A, Dassonville-Klimpt A, Li X, Datti A, Ohashi PS, Wrana J, Rogers I, Sonnet P, Ellis WY, Corey SJ, Eaves C, Minden MD, Wang JC, Dick JE, Nislow C, Giaever G, Schimmer AD (2013) Lysosomal disruption preferentially targets acute myeloid leukemia cells and progenitors. J Clin Invest 123(1):315–328

Sun L, Wang X (2014) A new kind of cell suicide: mechanisms and functions of programmed necrosis. Trends Biochem Sci 39 (12):587–593

Taniguchi M, Ogiso H, Takeuchi T, Kitatani K, Umehara H, Okazaki T (2015) Lysosomal ceramide generated by acid sphingomyelinase triggers cytosolic cathepsin B-mediated degradation of X-linked inhibitor of apoptosis protein in natural killer/T lymphoma cell apoptosis. Cell Death Dis 6:e1717

Vainchenker W, Constantinescu SN (2013) JAK/STAT signaling in hematological malignancies. Oncogene 32(21):2601–2613

Xu M, Liu Y, Zhao L, Gan Q, Wang X, Yang C (2014) The lysosomal cathepsin protease CPL-1 plays a leading role in phagosomal degradation of apoptotic cells in *Caenorhabditis elegans. Mo.l.* Biol Cell 25(13):2071–2083

Zhu W, Tao L, Quick M, Joyce A, Qu J, Luo Z (2015) Sensing cytosolic RpsL by macrophages induces lysosomal cell death and termination of bacterial infection. PLoS Pathog 11(3): e1004704

8

Characterization of human αβTCR repertoire and discovery of D-D fusion in TCRβ chains

Peipei Liu[1,2], Di Liu[3], Xi Yang[1,2], Jing Gao[1,2], Yan Chen[1], Xue Xiao[1], Fei Liu[1], Jing Zou[1], Jun Wu[1], Juncai Ma[3], Fangqing Zhao[5], Xuyu Zhou[1,2]✉, George F. Gao[1,2,4,5]✉, Baoli Zhu[1,2]✉

[1] CAS Key Laboratory of Pathogenic Microbiology and Immunology, Institute of Microbiology, Chinese Academy of Sciences, Beijing 100101, China
[2] University of Chinese Academy of Sciences, Beijing 100049, China
[3] Network Information Center, Institute of Microbiology, Chinese Academy of Sciences, Beijing 100101, China
[4] Chinese Center for Disease Control and Prevention (China CDC), Beijing 102206, China
[5] Research Network of Immunity and Health (RNIH), Beijing Institutes of Life Science, Chinese Academy of Sciences, Beijing 100101, China
✉ Correspondence: zhubaoli@im.ac.cn (B. Zhu), zhouxy@im.ac.cn (X. Zhou), gaof@im.ac.cn (G. F. Gao)

ABSTRACT

The characterization of the human T-cell receptor (TCR) repertoire has made remarkable progress, with most of the work focusing on the TCRβ chains. Here, we analyzed the diversity and complexity of both the TCRα and TCRβ repertoires of three healthy donors. We found that the diversity of the TCRα repertoire is higher than that of the TCRβ repertoire, whereas the usages of the V and J genes tended to be preferential with similar TRAV and TRAJ patterns in all three donors. The V-J pairings, like the V and J gene usages, were slightly preferential. We also found that the TRDV1 gene rearranges with the majority of TRAJ genes, suggesting that TRDV1 is a shared TRAV/DV gene (TRAV42/DV1). Moreover, we uncovered the presence of tandem TRBD (TRB D gene) usage in ~2% of the productive human TCRβ CDR3 sequences.

KEYWORDS TCR repertoire, next-generation sequencing, V/J usage, V-J pairing, CDR3, D-D fusion

INTRODUCTION

The T lymphocytes that are engaged in cellular immunity are mainly αβ T cells. The T-cell receptors (TCRs) located on the T cell surface play key roles in recognition of the vast number of antigenic peptides presented by the major histocompatibility complex (p-MHC). Heterodimeric TCRs are composed of two polypeptide chains, the α and the β chains, each containing one variable domain and one constant domain (Murphy et al., 2007). The antigenic specificity of T lymphocytes is primarily determined by the amino acid sequences of the hypervariable complementarity determining region 3 (CDR3). The nucleotide sequences of CDR3 are generated by somatic recombination of segregated germline variable (V), diversity (D), and joining (J) gene segments for the TCR β chain (TRB), and V and J gene segments for the TCR α chain (TRA).

Aside from αβ T lymphocytes, humans harbor a small subset of T cells called γδ T lymphocytes that possess distinct TCRs on their surface. These TCRs are made up of one γ chain and one δ chain. Resembling the TRB, the nucleotide sequences encoding the TCR δ chain (TRD) variable region are also formed by somatic recombination of the V, D, and J gene segments in the TRD gene locus. This locus is nestled in the TRA locus between the TRAV and TRAJ genes on chromosome 14 (Loh et al., 1988; Takihara et al., 1988). Generally, the TRD genes should only encode TRD (Lefranc, 2001). However, some TRDV gene segments form joints with TRAJ gene and can therefore be used to synthesize TRA. Furthermore, some specific $CD8^+$ T cell clones with $TRDV1^+$ TRA recognize and kill HIV-infected target cells (Ueno et al., 2003), while other $TRDV1^+$ $CD8^+$ T cell clones are believed to contribute to chronic neutropenia (Bank et al., 1992; Bank et al., 2003). In total, five out of eight TRDV genes, including TRDV4 to TRDV8, have been classified into this group (Lefranc, 2001), and their names are designated in the form of TRAV/DV plus the corresponding sequential numbers. The remaining TRDV2 and TRDV3 are hypothesized to exclusively rearrange with TRDD genes (Lefranc, 2001). For TRDV1, some reports show that it, resembling the five shared TRAV/DV gene segments, can also rearrange with a small fraction of TRAJ gene segments to synthesize TRA (Miossec, 1990; Miossec et al., 1991).

Somatic recombination begins with lymphoid-specific recognition of conserved recombination signal sequences (RSSs). An RSS that flanks the V, D, and J gene segments consists of a conserved block of seven nucleotides, "the heptamer", which is always contiguous with the coding sequence, followed by a non-conserved region known as "the spacer", which is either 12 or 23 bp long, followed by a second conserved block of nine nucleotides, "the nonamer" (Bassing et al., 2002). The recombination occurs between gene segments that are flanked by RSSs of unequal spacer lengths, which is called the 12/23 rule. For immunoglobulin, the D gene segments are flanked by the same RSS containing a 12-bp spacer. It is now confirmed that even though it violates the 12/23 rule, direct joining of one D gene segment to another can occur in most species. In human B cells, D-D fusions are found in approximately 5% of genes that encode antibodies and are the major mechanism accounting for the unusually long CDR3 loops in heavy chains (Larimore et al., 2012; Meek et al., 1989; Murphy et al., 2007). For TRB, inactivation of the TRB enhancer results in non-standard TRBD1-TRBD2 joints in double positive thymocytes in mice (Hempel et al., 1998; Ryu et al., 2003). In humans, the D-D fusions fulfill the 12/23 rule but have only been reported in the TRD (Olaru et al., 2005) and not in the TRB.

Due to the huge diversity of the TCR repertoire, which is estimated to be 2.5×10^7 per individual (Arstila et al., 1999) and theoretically calculated to be much greater (Nikolich-Zugich et al., 2004), it is daunting to precisely and efficiently characterize the repertoire. Classical DNA cloning and Sanger sequencing techniques are laborious and generally have limited throughput (Arstila et al., 1999; Gorski et al., 1994; Pannetier et al., 1993). However, deep sequencing now permits interrogation of complex sequencing targets including the TCR repertoire, at unprecedented depth and reasonable cost (Mardis, 2008; Shendure and Ji, 2008). The diversity, complexity, and specificity of the human TRB repertoire in healthy donors and patients has recently been explored (Klarenbeek et al., 2012; Klarenbeek et al., 2010; Li et al., 2012; Neller et al., 2013; Quigley et al., 2010; Robins et al., 2009; Robins et al., 2010; Sherwood et al., 2011; Venturi et al., 2011; Wang et al., 2010; Warren et al., 2011). However, there is still no comprehensive and systematic study of the TRA repertoire.

Here, we characterized the TRA and TRB repertoires of three healthy donors with distinct human leukocyte antigen (HLA) types and ethnic backgrounds (Table 1). We found that the diversity of the TRA repertoire is higher than that of the TRB repertoire, and the usage of V and J gene segments was preferential for both chains, though the overall frequency pattern was conserved among different individuals. In contrast, for each donor, productive V-J pairings were not equally distributed. Strikingly, the TRDV1 gene was used to the same extent as other normal TRAV gene segments when rearranging with TRAJ segments, indicating that TRDV1 is also a V gene segment shared by the TRA and TRD loci. Furthermore, TRB CDR3 sequence composition analysis uncovered D-D fusions in the TRB, which is likely a key mechanism contributing to the formation of long CDR3 sequences.

RESULTS

Workflow and error correction

We used similar approach as previously reported (Freeman et al., 2009) to profile TCR repertoires derived from 50 mL peripheral blood mononuclear cells (PBMCs). Given that error correction is essential for accurate repertoire profiling, we utilized a series of measures to eliminate errors. First, we used 5′ rapid amplification of cDNA ends (5′RACE) to obtain the TCR V domain transcript sequences. The 5′RACE approach avoids the potential bias associated with the use of the multiple primer sets required to amplify all V region sequences and takes advantage of the conserved sequences offered by TRAC for TRA and TRBC1 and TRBC2 (96% nucleotide sequence identity) for TRB1. Then, two rounds of nested PCR were performed with nested C gene primers to enhance PCR production specificity, ultimately yielding the target sequences with a low level of background amplification. Second, we used high-fidelity reverse transcriptase and DNA polymerase to perform the experiments. For reverse transcription (RT), the reverse transcriptase Superscript II has an error rate of 1/15,000. Subsequently, in the two rounds of PCR, we utilized Plantinum Pfx DNA polymerase with an error rate of 1/632,900. Third, we used the quality score to evaluate sequencing errors. If there were any ambiguous nucleotides within the CDR3 region, that sequence was discarded.

However, for accurate identification of CDR3 sequences, we established bioinformatics tools to specifically deal with the deep sequencing data of the TCR repertoires. To validate the bioinformatics tools, we randomly sampled a subset of the total data and manually dissected the raw reads one by one to identify the CDR3 sequences. This comprised the reference set. The prediction of the reference set by our bioinformatics tools demonstrated an average of 93.02% sensitivity and 98.18% precision (Table 2). We also used the IMGT/HighV-QUEST tool (Giudicelli et al., 2011) to analyze our reference set, and obtained an average of 41.65% sensitivity and 96.15% precision (Table 3). Therefore, our bioinformatics tools can accurately and efficiently identify CDR3 sequences.

Table 1. Donor characteristics.

Subject	Sex	Ethnic origin	HLA-A	HLA-B	HLA-Cw
Donor 1	M	Caucasian	*0201/*2501	*0702/*4427	*0501/*0702
Donor 2	F	Mongolian	*0201/*0210	*0801/*4006	*0304/*0702
Donor 3	M	Mongolian	*1101/*2402	*4001/*4001	*0702/*0702

Table 2. Estimation of the precision and sensitivity of the CDR3-identifying method.

Sample set	No. of reads pairs	Predicted positive segments	True positives	False positives	False negatives	Precision [TP/(TP + FP)]	Sensitivity [TP/(TP + FN)]
TRA-D1*	100,000	8476	8400	478	76	99.10%	94.60%
TRA-D2	10,000	1035	1007	30	28	97.30%	97.10%
TRA-D3	10,000	630	619	69	11	98.30%	90.00%
TRB-D1	10,000	640	625	74	15	97.70%	89.40%
TRB-D2	10,000	880	859	71	21	97.60%	92.40%
TRB-D3	10,000	633	627	36	6	99.10%	94.60%
						98.18%	93.02%

* D1, D2, and D3 represent Donor 1, Donor 2, and Donor 3, respectively.

Table 3. Estimation of the precision and sensitivity of the IMGT/HighV-QUEST.

Chain Type	No. of reads pairs	Predicted positive segments	True positives	False positives	False negatives	Precision [TP/(TP + FP)]	Sensitivity [TP/(TP + FN)]
TRA	2860	55	53	2	85	96.40%	38.40%
TRB	2610	73	70	3	86	95.90%	44.90%

Table 4. TCR sequence statistics.

Subject	Chain type	Raw reads	Total CDR3 sequence	Unique CDR3 (nt)	Unique CDR3 (aa)	Unique V(D)J TCR sequences (nt)
Donor 1	α	59,845,380	2,488,327	244,524	179,395	300,935
	β	60,088,204	1,913,167	135,614	102,142	154,851
Donor 2	α	57,261,780	2,991,318	258,163	189,457	318,919
	β	56,050,648	2,414,344	158,174	118,560	182,532
Donor 3	α	60,783,874	1,928,533	164,669	122,348	199,577
	β	56,432,586	1,939,500	121,763	90,426	140,270

As shown in Table 4, approximately 60 million 100-nt long paired-end raw reads were generated for each chain from the three donors. Therefore, the quantity of the raw data was kept at a similar level for all three donors. For instance, we identified at least 1.9 million productive TRA CDR3 sequences from Donor 3, with >164,000 unique TRA CDR3 clonotypes. When the flanking V and J gene segments were taken into account, we characterized an average of 2.7×10^5 and 1.6×10^5 distinct TRA and TRB clonotypes, respectively, for all three donors. Notably, we observed that the TRA repertoire was more diverse than the TRB repertoire in all three donors from the sampled PBMCs, regardless of their disparate HLA backgrounds.

TCRα and TCRβ CDR3 distribution

Figure 1 shows the overall distribution of the TRA and TRB clonotypes and that the abundance of distinct clonotypes can vary several thousand-fold. With respect to the abundance of clonotypes, we classified the TCR clonotypes for both the TRA and TRB repertoires into two groups: group 1 consisted of clonotypes with copy numbers ≤300; and group

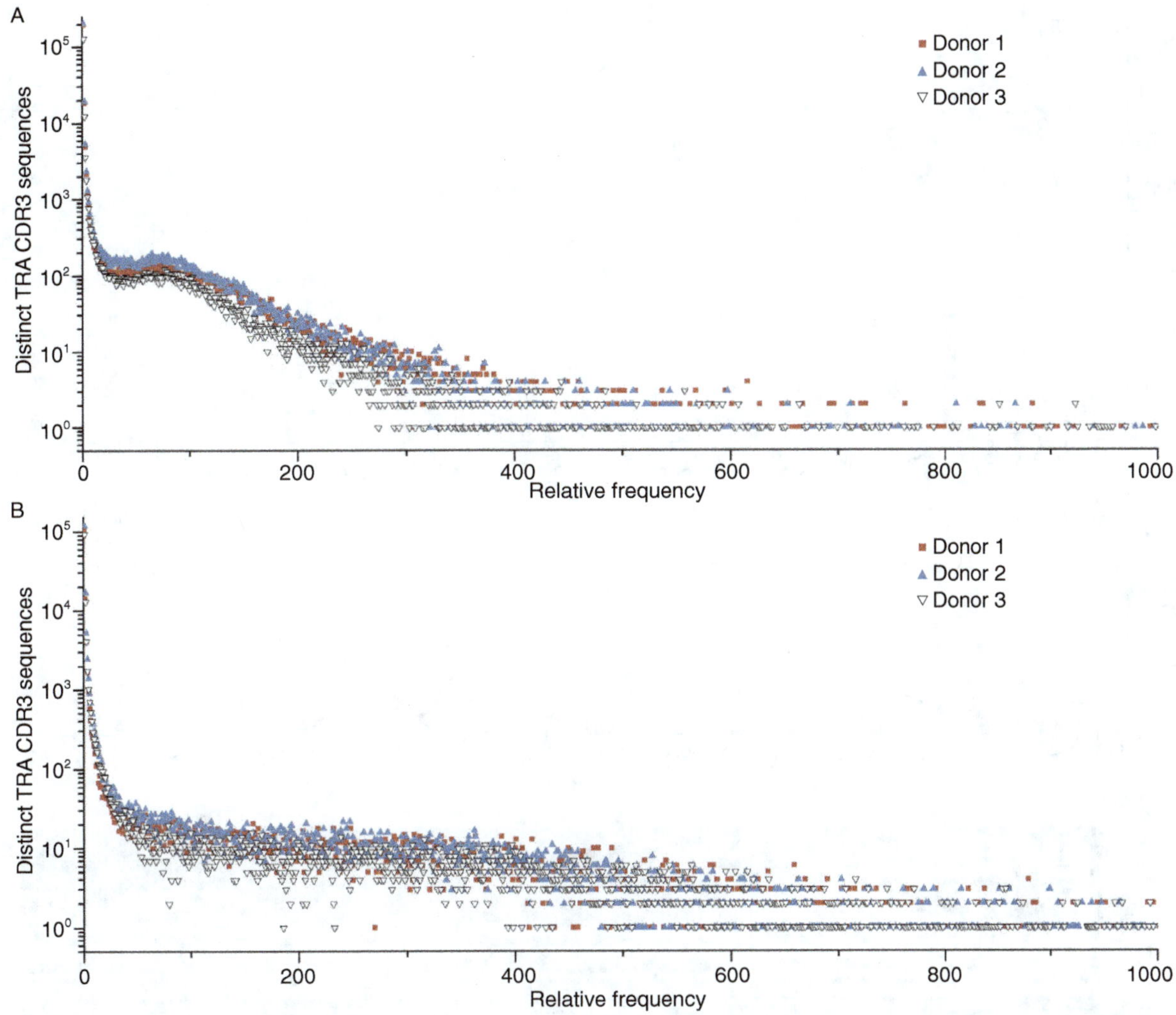

Figure 1. The distribution of the CDR3 sequences derived from 50 mL PBMC of three donors. The CDR3 sequence copy number of the TRA (A) and the TRB (B) observed in three donors was represented by the abscissa, while the ordinate represents the number of unique CDR3 clonotypes. For example, the square 300, 10 means that 10 unique CDR3 clonotypes were each observed 300 times. A small set of sequences found with very large copy number (>1000, copies) are not displayed.

2 consisted of clonotypes with copy numbers >300. For group 1, there were 221,835 TRA clonotypes on average (Fig. 1A), which is much greater than that of the TRB clonotypes (i.e., 136,535) (Fig. 1B). For group 2, there were only 616 TRA clonotypes (Fig. 1A) and 1985 TRB clonotypes (Fig. 1B) on average for the three donors.

The characterization of the TRB repertoire has been systematically reported. In this study, we characterized the TRB repertoire in parallel with the TRA repertoire to validate our approaches. As shown in Fig. S1, the TRBV and TRBJ utilization in the productive CDR3 sequences was highly non-uniform in a given individual. Furthermore, inter-donor conservation of TRBV and TRBJ usage (Fig. S1A and S1B, average Pearson correlation coefficients = 0.83 and 0.85) was observed. The TRBV-TRBJ pairings (Fig. S1C) were biased toward specific TRBV-TRBJ pairs, while striking similarity in the pairing frequency among individuals was observed (average Pearson coefficients = 0.65 ($P < 0.001$)). In aggregate, these results are consistent with previous reports (Freeman et al., 2009; Robins et al., 2009; Warren et al., 2011), which validates our approaches.

TRAV/TRAJ usage and TRAV-TRAJ pairing pattern in healthy donors

When examining the frequency of the TRAV segments listed according to their chromosomal locations, we found that the TRAV usage was notably biased in a given individual (Fig. 2A). Some TRAV segments such as TRAV8-1, 13-1, 20, 27, and 38-2 are preferentially used in comparison with those like 8-3, 8-6, and 8-7, which are almost undetectable. Furthermore, the frequencies of the TRAV segments that are most proximal to the TRAJ cluster were not the highest, while those most distal to the TRAJ cluster were not the

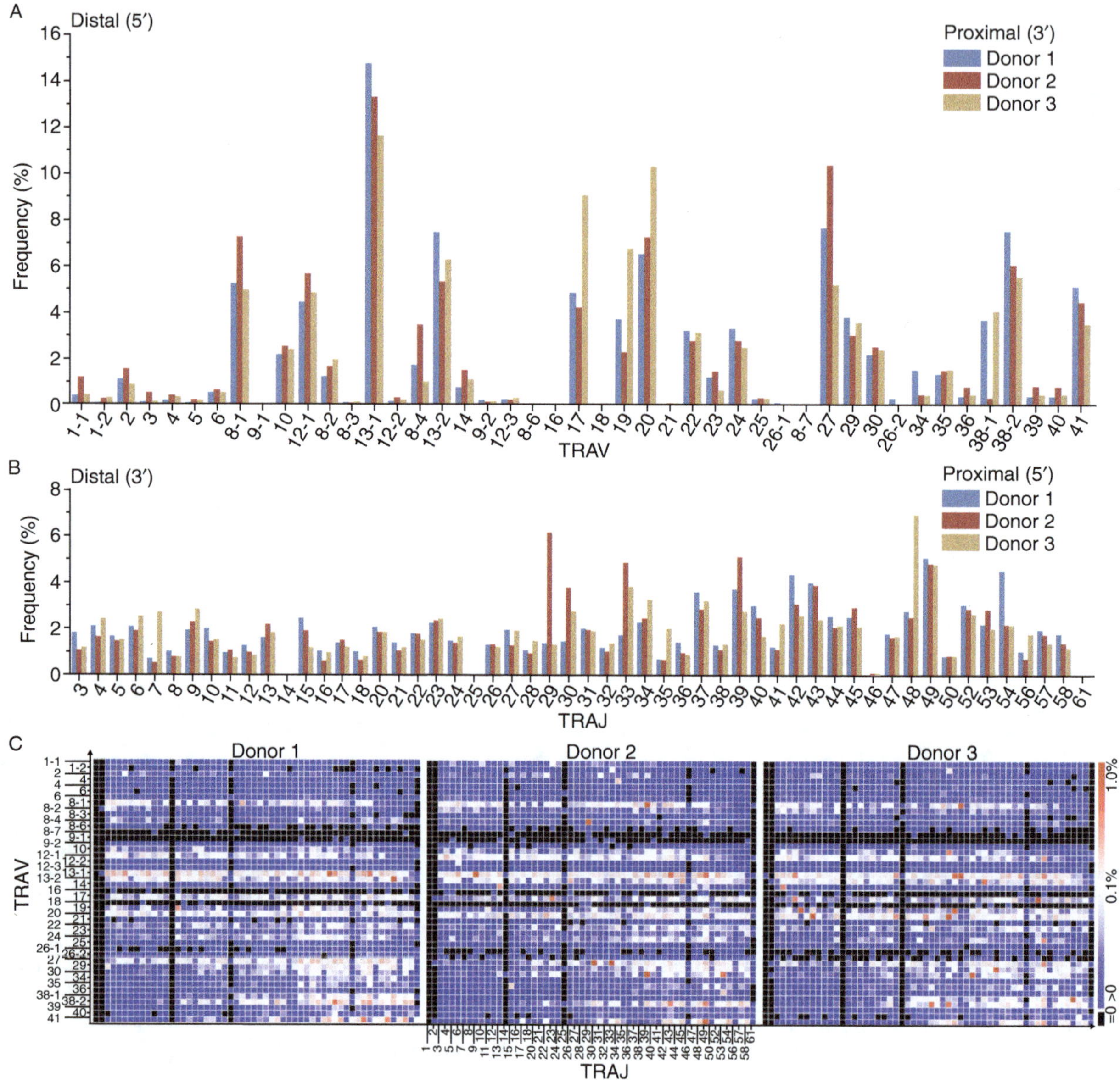

Figure 2. TRAV, TRAJ gene usage and TRAV-TRAJ pairing are highly correlated among donors. Relative frequency of TRAV and TRAJ segments are listed in (A) and (B), respectively, according to their chromosome locations. (C) The heat map of the TRAV and TRAJ pairings for three donors are shown.

rarest. These data indicate that TRAV segments were selected irrespective of distance from TRAJ gene segments. Most intriguingly, pairwise comparisons of TRAV usage between donors produced a Pearson correlation coefficient of 0.90 ± 0.04 (mean ± SD), indicating marked similarity in the TRAV frequency among individuals.

Likewise, TRAJ usage was also non-uniform in a given donor (Fig. 2B). Furthermore, the TRAJ usage patterns observed in the three healthy donors were quantitatively similar to each other, with an average Pearson correlation coefficient of 0.66 ± 0.04 (mean ± SD).

As illustrated in the heat map (Fig. 2C), the abundance of TRAV-TRAJ pairings was strongly correlated among individuals. The average Pearson coefficient of TRAV-TRAJ pairing was 0.334 ($P < 0.001$). Though the TRAV and TRAJ gene segments usage was strikingly quantitatively similar among donors, the extent of TRAV-TRAJ pairing similarity was somewhat reduced. Especially when focusing on the most abundant TRAV-TRAJ parings, we found that they were unique for each individual. Moreover, as the TRAV and TRAJ gene segments are shown according to their chromosomal positions (5′ to 3′ direction), we determined that the TRAV-TRAJ pairing in humans is not compatible with the sequential coordinate gene recombination hypothesis, which means 5′ to 3′ polarized utilization of the TRAJ library may be coordinated with a 3′ to 5′ polarized

A

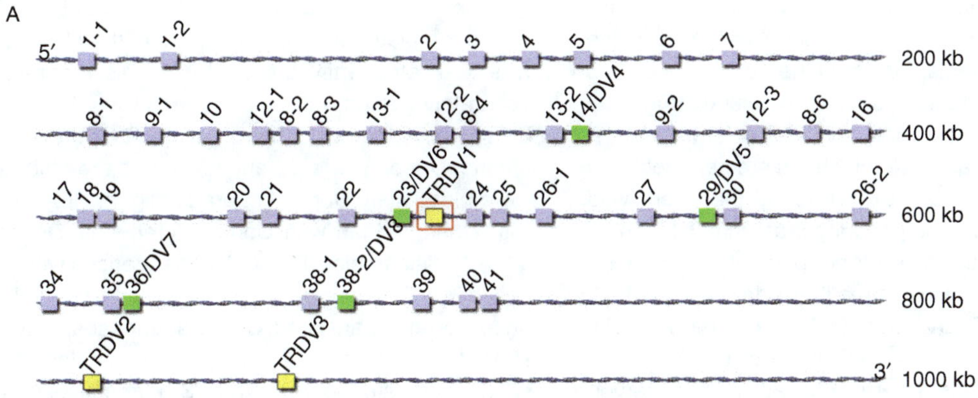

B

Donor	TRDV1	TRDV2	TRDV3	TRAV14/ DV4	TRAV29/ DV5	TRAV23/ DV6	TRAV36/ DV7	TRAV38-2/ DV8	TRAV22	TRAV24
Donor 1	1757	0	0	712	3225	1541	826	4200	678	4178
Donor 2	1307	0	0	918	2668	1608	748	3802	611	2598
Donor 3	927	0	0	2102	2413	564	665	1816	623	5980

C

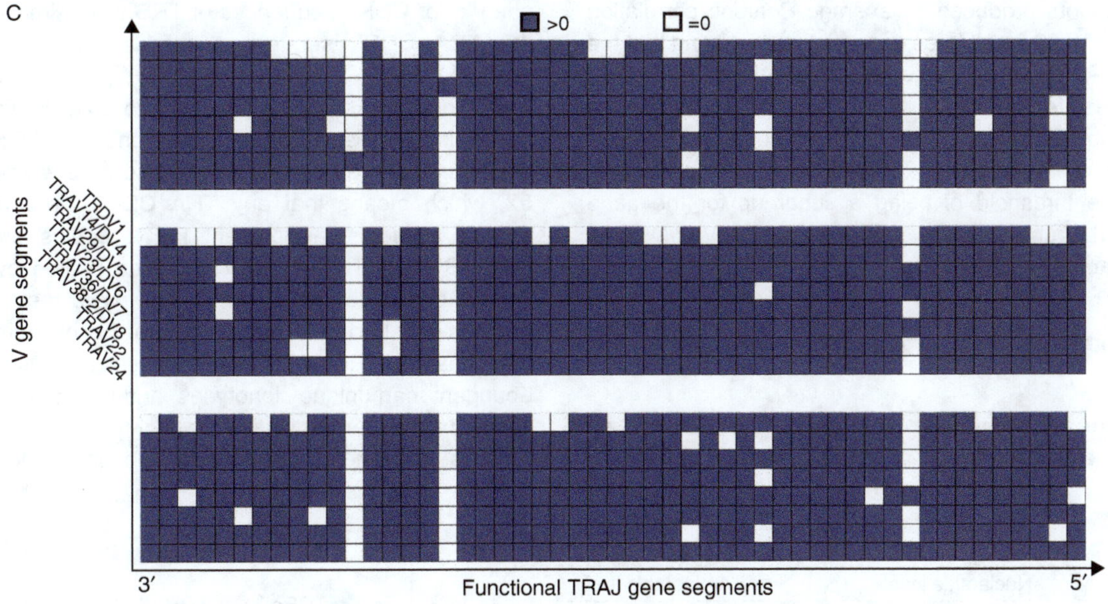

Figure 3. TRDV1 is also a shared TRAV/DV gene segments. (A) TRDV gene segments dispersed in the TRA locus in chromosome 14 (14q11.2). TRDV1, TRDV2, and TRDV3 (yellow squares) were not shared with TRA in previous report, while the other five TRDV segments (green squares) also have their corresponding TRAV names. (B) Reads containing TRDV gene segments and two TRAV segments closest to TRDV1 in chromosome were identified. The counts of the reads were listed. (C) The patterns of related TRAJ segments paired with TRDV1 in three donors were shown. The TRAJ gene segments were listed along 3′ to 5′ direction in the chromosome.

utilization of the library of the TRAV gene segments (Fuschiotti et al., 2007; Huang and Kanagawa, 2001; Krangel, 2009; Pasqual et al., 2002; Roth et al., 1991). This situation is analogous to a recent report focusing on TRA in mice (Genolet et al., 2012).

TRDV1 is used as a common TRAV gene segment: TRAV42/DV1

The TRD locus spans 60 kb on chromosome 14 at 14q11.2 and is nested within the TRA locus (Fig. 3A). The TRD locus is composed of a cluster of one TRDV gene (TRDV2), three TRDD genes, and four TRDJ genes, upstream of the unique TRDC gene (Lefranc, 2001). Another TRDV gene (TRDV3) is located downstream of the TRDC gene, in inverted transcription orientation (Lefranc and Rabbitts, 1990). Resembling the five TRAV/DV gene segments, TRDV1 is dispersed in the TRAV cluster rather than TRDV2 and TRDV3 (Lefranc, 2001). Hence, TRDV1 has a high potential to be a shared TRAV/DV segment.

As illustrated in Fig. 3B, TRDV1 had at least 927 copies, while both TRDV2 and TRDV3, which are believed to exclusively join to TRDD segments (Lefranc, 2001), had zero copies in all donors after noisy sequence elimination. Furthermore, TRDV1 had the same usage level as the other five shared TRAV/DV segments, for which the number of copies ranged from several hundred to several thousands (Fig. 3B).

The TRA gene locus contains 61 TRAJ gene segments, of which 50 are functional, while the others are pseudogenes or open reading frames (ORFs). We found that there was absolutely no signal from the pseudogenes and ORFs in all three donors. For simplicity, we only present the data for genes referred to as functional in the IMGT database. In fact, the rearrangement of TRDV1 with TRAJ is comparable with that of all of the other TRAVs (Fig. 3C). For the other five shared AV/DV segments referred to above, conserved results were also observed for their V-J pairing pattern (Fig. 3C).

The frequency of specific TRDV1-TRAJ recombination within any given donor varied approximately 1000-fold (Table S1). Pairwise comparisons of TRDV1-TRAJ pairings between donors produced an average Pearson correlation coefficient of 0.99 for all three donors, demonstrating that the TRDV1-TRAJ pairing patterns are constant among donors.

These results on TRDV1 usage and TRDV1-TRAJ pairing patterns demonstrated that TRDV1 is also used as a normal TRAV gene segment. With 75% identity at the nucleotide level as the threshold of being a subgroup for the gene segments (Lefranc, 2011), we designated TRAV42/DV1 as the new name for TRDV1.

Shared CDR3 sequences of TRA are the dominant species

To compare repertoire diversity among the donors, we obtained the same amount of PBMCs from all three with informed consent. Furthermore, the amount of mRNA used as the template for all three donors in the 5′RACE reactions was kept stringently the same. In this manner, we could accurately compare the repertoire among donors. As shown in Fig. 4A, at the nucleotide level, each repertoire was mostly unique. Donor 1 was a Caucasian, and the other two donors were Mongolian. Donor 1 shared a higher HLA type similarity with Donor 2 than with Donor 3 (Table 1). Donor 1 shared 5.8% of his sampled TRA CDR3 sequences with Donor 2 but only 2% with Donor 3. We also performed a similar analysis based on the translated CDR3 sequences, where Donor 1 shared 10.4% and 4.7% of his sampled TRA CDR3 sequences with Donors 2 and 3, respectively. This is consistent with the common concept that there are preferred amino acid sequences reactive with any given antigenic epitope. Furthermore, these preferred amino acid sequences are coded by a larger diversity of nucleotide sequences due to the degenerated genetic code. In addition, the patterns of sharing for CDR3 sequences of TRB were similar to that of the TRA CDR3 sequences between the three donors (Fig. 4A).

We next classified sequences into shared subsets versus unique subsets for a given donor. The TRA CDR3 sequences shared by two donors had an average depth of 52, which means that any TRA CDR3 clonotype in the shared subsets was present in 52 copies on average (Fig. 4B). The TRA CDR3 clonotypes shared by all three donors had a deeper depth of 230, while the unique clonotypes only had an average depth of 7. This demonstrated that shared TRA CDR3 sequences are usually more abundant than unique clonotypes, and the more people that share the sequences, the more highly abundant they are. This is consistent with the notion that infection with

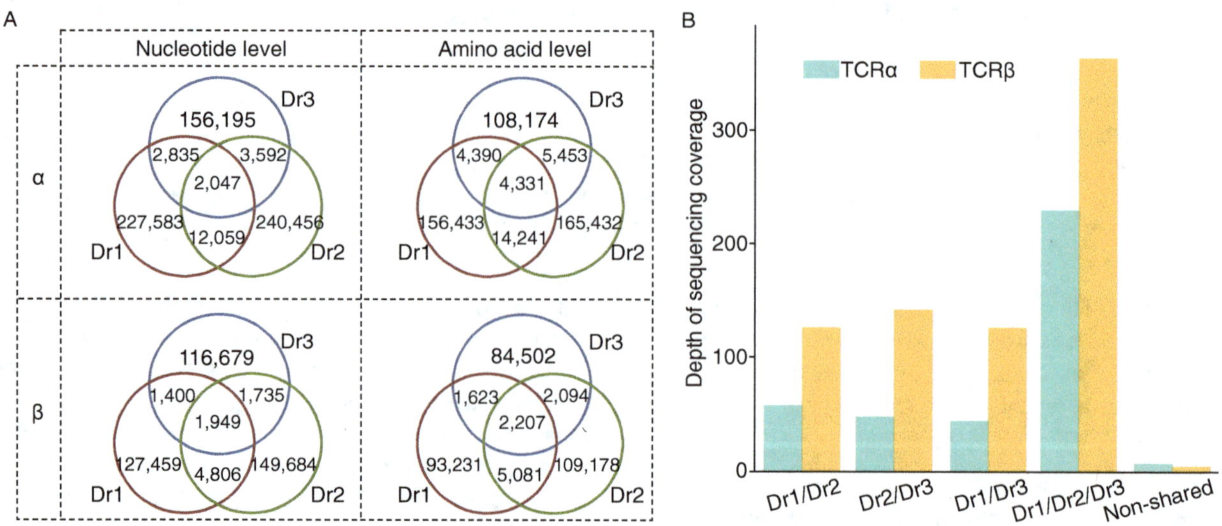

Figure 4. Shared CDR3 sequences are dominant species. (A) For TRA and TRB, the number of nucleotide and amino acid clonotypes of CDR3 sequences that are shared by two or three donors are illustrated in a pie. (B) Comparison of the average depth of the shared CDR3 sequences with non-shared ones. Dr1, Dr2, and Dr3 represent Donor 1, 2 and 3 respectively.

common public pathogens can result in dominant public TCR clonotypes. Our data also demonstrate the consistent sharing pattern of the CDR3 sequences in the TRB repertoire (Fig. 4B).

D-D fusion is found in TRB

While dealing with the CDR3 sequence dataset, we found the existence of long CDR3 sequences (longer than the average 15 amino acids) in the TRB repertoire. We dissected these longer CDR3 sequences to investigate their composition. There are two TRBD gene segments involved in TRB somatic recombination. Due to nucleotide nibbling at both ends of D gene segments during recombination, TRBD gene identification is challenging for a mature CDR3 sequence. Here, we used more stringent criteria than the Junction Analysis tool (which is specially designed to annotate the TCR CDR3 sequences by IMGT) to identify the TRBD gene segments. We restricted our analysis to include only those sequences with: i) at least six nucleotides identical to the intact germline TRBD1 or TRBD2 sequence; and ii) two continuous nucleotides at each ends matching the known germline TRBD gene segments (see MATERIAL AND METHODS). Junction Analysis requires that only two nucleotides of the TRBD should be compared with the user sequences (Yousfi Monod et al., 2004). Not surprisingly, we found tandem D gene usage in some of these long CDR3 segments.

We randomly select an 18-amino acid-long raw read, which had been identified as containing D-D fusions by our approach, and then analyzed it with the IMGT/Junction Analysis program. The program delimited the CDR3 sequence and showed all of the potential rearranged TRBD gene segments (Fig. 5A). Both TRBD1 and TRBD2 were identified with high scores (nine and 10 nucleotides were identical to germline TRBD1 and TRBD2, respectively). We also collected a panel of reads containing D-D fusions in our dataset and placed them in the supplementary material.

As shown in Fig. 5B, 1.75%, 3.45%, and 1.73% of the productive CDR3 clonotypes were identified to contain D-D fusions in Donors 1, 2, and 3, respectively. However, we could not identify any D gene sequences in 17.68%, 15.01%, and 22.86% of the productive CDR3 clonotypes in Donors 1, 2, and 3, respectively. This may be the result of the very short D sequences produced by nucleotide trimming during somatic recombination. Furthermore, the fit of the length distribution of the CDR3 sequences with D-D fusions has a right shift in the abscissa compared to that of the length distribution of the CDR3 sequences with only one D gene segment (Fig. 5C), and those CDR3 sequences in which TRBD could not unambiguously be identified are shortest in average. These results suggest that D-D fusion was a key

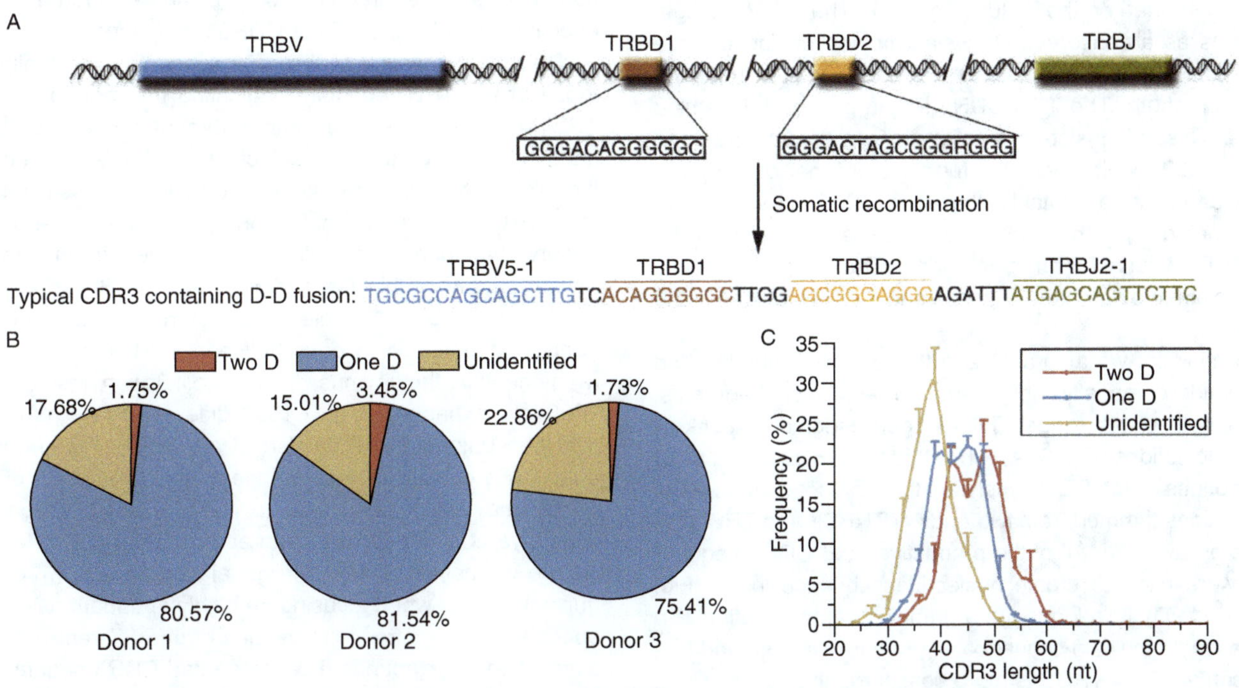

Figure 5. D-D fusions are found in TRB in human beings. (A) A typical CDR3 sequence containing D-D fusion was shown. (B) The percentage of the productive CDR3 sequences containing D-D fusions in three donors. (C) The CDR3 length distributions for the productive TRB CDR3 containing two D segments, one D segment and no D segment. The mean frequency of the three donors is shown, with standard deviations as error bars.

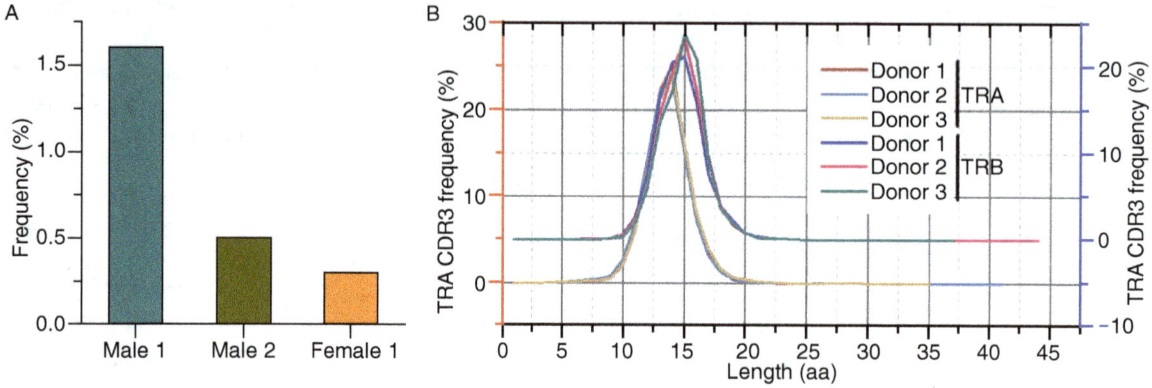

Figure 6. The confirmation of D-D fusion in the TRB. (A) Proportion of the TRB CDR3 clonotypes containing D-D fusions from the data published by Warren and his coworkers (Warren et al., 2011). (B) The lengths of CDR3s in both TRA and TRB sequences appear a Gaussian distribution. CDR3 length distribution curves of TRA for three donors all peak at 14 amino acids while that of TRB peak at 15 amino acids.

mechanism that preferentially contributed to the development of the long TRB CDR3 loops.

To confirm our findings, we randomly selected approximately 10% of the raw reads published by Holt's group (Warren et al., 2011) to annotate their CDR3 sequences using our bioinformatics pipeline. As illustrated in Fig. 6A, we also identified 1.6%, 0.5%, and 0.3% of their CDR3 clonotypes containing D-D fusions for their three samples. Furthermore, we compared the CDR3 length distribution of the TRA with that of the TRB (Fig. 6B). The CDR3 length appears as a conserved Gaussian distribution for all three individuals, and the TRA CDR3 sequences peak at 14 amino acids in length. The TRB CDR3 length peaks at 15 amino acids. This is consistent with a previous report (Freeman et al., 2009), which provides further proof that our approaches are reliable. Notably, the TRB CDR3 curve has an appreciable right shift along the abscissa compared to the TRA CDR3 curve for any given donor. The curve shift should be explained by both the D genes and the D-D fusions in the TRB.

In addition, we also analyzed the length distribution and nucleotide composition of the rearranged TRBD segments. The length of rearranged TRBD1 gene segments peak at nine nucleotides (Fig. S2A), while that of TRBD2 peaks at 13 nucleotides (Fig. S2B), implying that the total number of nucleotides trimmed from two ends of TRBD1 and TRBD2 is three on average. All of the productive TRB CDR3 sequences were used to create nucleotide sequence logos (Fig. S2C and 2D). The logos are a graphical representation of a nucleic acid multiple sequence alignment. We observed that the center of the TRBD is more conserved than the flanking regions. This could be explained by nucleotide nibbling (Murphy et al. 2007), though the bias for calling TRBD gene segments cannot be fully ruled out. Regardless, this is consistent with previous reports (Freeman et al., 2009; Quigley et al., 2010).

DISCUSSION

We used experimental approaches similar to those developed by Freeman and coworkers, and the bioinformatics pipeline we adopted proved to be efficient and accurate in identifying CDR3 sequences from TCR repertoire datasets generated by deep sequencing (Tables 2 and 3). Furthermore, our data on TRB repertoires, including V/J usage, V-J pairing, and CDR3 sharing pattern, is consistent with previous reports. In aggregate, using a similar approach, we effectively identified the TRA CDR3 sequences.

The characteristics of the TRA repertoire are similar to those of the TRB repertoire, indicating that both TRA and TRB experienced similar rearrangement processes regulated by recombination or selection mechanisms. Though the TRA repertoire was estimated to be less diversified than the TCRβ repertoire in whole blood (Arstila et al., 1999), we observed a higher diversity in the TRA repertoire than in the TRB repertoire within the sampled 50 mL PBMCs for any given donor (Table 1). This is consistent with the fact that during the period of 5 days between TRB and TRA rearrangement in the thymus, T cell proliferation results in a 1000-fold expansion, i.e., ~1000 different TRAs may pair with one unique TCRβ (Arstila et al., 1999). Furthermore, dual TCRs caused by less stringent allelic exclusion of the TCRα may also contribute to the TRA-to-TRB pairing (Borgulya et al., 1992; Padovan et al., 1993).

The V-J pairing of each individual displays its own distinguishable traits when focusing on highly abundant pairs. We speculate these differences resulted from different antigen-stimulation backgrounds. For the shared CDR3 sequences, we found that most of them are dominant clonotypes, likely public ones to ward off common pathogens like influenza virus, Epstein-Barr virus, and Cytomegalovirus.

There are contradictory results on the affiliation of TRDV1. Some groups report that TRDV1 can rearrange with some TRAJ gene segments (Bank et al., 1992; Bank et al., 2003; Castelli et al., 1992; Miossec, 1990; Miossec et al., 1991; Ueno et al., 2003), while others believed that it just appears as a normal TRDV gene segment (Lefranc, 2001) that can only be engaged in TRD chain formation. In this study, our approaches could detect transcripts only if they contained the TRAC segment. If TRDV1 can actually rearrange with TRAJ segments, these TRDV1$^+$ TRA sequences would also be captured. Indeed, we did find TRDV1$^+$ TRA reads in the three donors, and TRDV1 had a similar usage level with shared TRAV/DV segments and the two representative TRAV segments. In addition, nearly all of the TRAJ segments could form joints with TRDV1, which is also solid evidence for TRDV being a shared TRAV/DV gene segment. Interestingly, we noticed that TRDV1-TRAJ20 rearrangement was used 80% of the time. However, whether this is constitutive or antigen-selected is still unknown. The ratio of naïve versus memory T cells deserves more examination.

Somatic recombination in TCR and BCR genes generally fulfills the 12/23 rule (Murphy et al., 2007), but in some cases, it violates this rule. For example, in antibody heavy chain gene recombination, D-D fusions are found in approximately 5% of antibodies (Murphy et al., 2007). The D gene segments of both TRB and TRD are flanked by 12- and 23-bp spacers, and the D-D fusion fulfills the 12/23 rule. The D-D fusion has been identified in nearly all functional TCRδ genes (Olaru et al., 2005), but D-D fusion in TCRβ chains has never been reported (Olaru et al., 2005). While identifying D gene segments, we restricted our analysis to include only those sequences with at least six nucleotides and two or more continuous nucleotides at both ends matching a given germline D gene. At this level of stringency, we identified D-D fusion in ~2% of the productive TCRβ rearrangements. We assume that the actually percentage of D-D fusions should be higher than this because some D gene segments might be excluded by our stringent standards during CDR3 identification. Therefore, 2% is likely an underestimate of the overall rate of human TRB D-D rearrangement.

Larimore and coworkers demonstrate that D-D fusion is a key mechanism resulting in the formation of the long CDRH3 in broadly and potent HIV-1-neutralizing antibodies (bNAbs). They identified D-D rearrangements in ~3% of productive IgH rearrangements. For TCRβ, we obtained similar results, i.e., ~2% of the productive TCRβ were shown to contain D-D fusions. These chains also have a longer CDR3 sequences, but we do not know whether they can recognize non-protein epitopes or they are used to bind inaccessible epitopes like the long CDRH3 in bNAbs. This issue deserves further study. We speculate that the use of two D segments may greatly increase the variability of the β chain because extra N-region nucleotides can be added at the junction between the two D gene segments, as well as at the V-D and D-J junctions. Under these conditions, the potentially huge diversity of the αβTCR repertoire would be further enlarged.

MATERIALS AND METHODS

Cell isolation and mRNA extraction

With written informed consent, we draw 50 mL peripheral blood by venipuncture from two males and one female with disparate HLA backgrounds (Table 1). Then 4.7×10^7, 4.6×10^7 and 4.3×10^7 peripheral blood mononuclear cells (PBMC) were isolated immediately by flotation on Ficoll-Hypaque. Using TRIzol (Invitrogen) reagent, the total RNA was extracted from all of the cells according to the manufacturer's instructions. After that, mRNA was purified from total RNA using the Oligotex mRNA Midi Kit (Qiagen) according to the manufacturer's specifications.

5′RACE

Sequences of variable regions of both TRA and TRB were obtained by 5′RACE, which is consisted of three continuous polymerase chain reactions (PCR), i.e. a reverse transcription PCR (RT-PCR) plus two nested PCRs.

PCRs of α and β chain were performed separately but the conditions were the same. First-strand cDNA was primed using a TCRC gene-specific primer (GSP1), and a target-switching primer (Oligo-dG) was also added as the 5′ template. Each RT-PCR contained 100 ng mRNA, 20 pmol oligonucleotides, 40 nmol DTT, 20 nmol each dNTP, 0.5 μmol Tris-HCl pH 8.3, 0.7 μmol KCl, 30 nmol $MgCl_2$ (Invitrogen), and 400 units of Superscript II Reverse Transcriptase (Invitrogen) in a 20-μL volume. Extension was 90 min at 42°C followed by inactivation for 7 min at 72°C. Using an adaptor primer (AP-1) and another C gene specific primer annealing to the 3′ of GSP1, the first nested PCR was carried out with 0.5 μL of the first-strand reactions added as the template. Reaction conditions were as follows: 1 U Platinum Pfx polymerase, 2× Pfx buffer, 1 mmol/L $MgSO_4$, primers 0.3 mmol/L each, and 0.3 mmol/L each dNTP in a 50-μL volume. For cycling, a 2 min denaturation at 94°C followed by 30 cycles of 30 s at 94°C, 30 s at 55°C, and 45 s at 68°C, plus a final extension for 5 min at 68°C. The second nested PCR was conducted to enhance the specificity by using another adaptor primer (AP-2) and another gene-specific primer (GSP-3) annealing to a site located upstream of GSP2 in C gene. PCR reaction conditions were the same with the 1st nested PCR except the template was changed into 0.1 μL product of the former nested PCR. Cycling parameters were as follows: a 2 min denaturation at 94°C followed by 20 cycles of 30 s at 94°C and 75 s at 68°C, with an additional extension for 5 min at 68°C. The adaptor primers i.e., the 5′ forward primers in RT, 1st-PCR and 2nd-PCR were depicted earlier (Ozawa et al., 2006). All of the primers were listed in Supplementary Table S1.

Sonication

Eight aliquots of the 2nd nested PCR reactions were pooled and purified using QIAquick PCR Purification Kit (Qiagen). After 1 h sonication, the eluates were sheared into fragments ranged from 100 bp to 350 bp by bioruptor sonicator (Diagenode). The sample was loaded on an 8% polyacrylamide gel, and the fraction from 150 bp to 300 bp was excised, purified.

Library construction

The double-stranded DNA fragments were comprised of 3′ or 5′ overhangs. T4 DNA polymerase and Klenow enzyme were then used to convert the overhangs into blunt ends. An 'A' base was added to the 3′ end of the blunt phosphorylated DNA fragments, which was ligated with adapters on both ends. The correctly ligated products were purified by agarose gel electrophoresis followed by the QIAquick gel extraction kit (Qiagen). The DNA fragments with adapter molecules on both ends were selected and amplified. The PCR was performed with two primers that anneal to the ends of the adapters. The number of PCR cycles was minimized to avoid skewing the representation of the library. The PCR products were checked and purified by agarose gel electrophoresis. The fragments size and molar concentration of library were respectively determined by the 2100 Bioanalyzer (Agilent) and Real-Time PCR System (ABI).

Cluster Generation

The Cluster Generation is used to hybridize samples onto a flow cell and amplify them for subsequent sequencing. The major steps include: (1) to hybridize library DNA to flow cell; (2) to extend hybridized templates and perform bridge amplification; (3) to block the flow cell; and (4) to hybridize sequencing primers. The Cluster Generation process was based on Illumina Cluster Station system (cBot), which would automatically amplify a single DNA molecule into clonal clusters for the following sequencing process.

Sequencing

The sequencing method was based on Illumina HiSeq 2000 platform and it was a fully automated protocol: the clustered template DNA was sequenced by four-color DNA Sequencing-By-Synthesis (SBS) technology with reversible terminators and removable fluorescence. Raw image files were processed by Illumina pipeline for base-calling with default parameters and the sequences of each individual were generated as 100-bp paired-end reads. The sequencing data were subjected to a strict QC test before bioinformatics analysis.

Identification of full length CDR3 from Assembled reads pair (ARP)

To facilitate the interpretation from tens of millions of reads, we implemented a bioinformatics workflow, including the determination of CDR3, V and J, and tested on human dataset. Manual validation suggests the reliability and implies the applicability for TCR repertoire annotation. (1) Assembled reads pair (ARP) ranges from 140 bp to 174 bp. (2) Built up V peptide dataset from protein sequences of FR3-IMGT region. (3) Built up J peptide dataset from protein sequences of FR4-IMGT region, namely "J^" in our analysis. (4) BLASTx each ARP (with the default parameters) against the above V/J dataset for the finding of FR3-signal and J^-signal.

Identification of full length CDR3 from raw reads

(1) For 90-bp-long raw reads without assembly, BLASTn (the parameter "word size" was set to 7) for finding V-signal and J-signal by constructing two nucleotide dataset: V-segment dataset comprised all known V exon sequences and J-segment dataset comprised all J exons. (2) Removing false positive reads according to CDR3 length and frame direction.

BLASTx (the parameter "word size" was set to 1) for determining the boundaries of CDR3 by constructing CDR3-V and CDR3-J nucleotide dataset. (1) CDR3-V dataset was extracted from the 6 amino acids in V-segment downstream the conserved "C"; CDR3-J dataset was extracted from 5 amino acids upstream the "FGXG" pattern. (2) In the above two analysis routes, the full-length CDR3 region must be assigned from the conserved "C" to the "F" position of "FGXG" pattern, and have no frameshift and stop codon.

Combination of the redundant CDR3 segments and V/J determination

(1) Then we combine above results and discard those redundant CDR3 segments according to sequence ID. (2) Quality check of CDR3-containing raw reads. Those sequences with ambiguous nucleotides (i.e. the base-calling result is N) are deleted from the filtered dataset. (3) Determination of the specific V and J segment from CDR3-identified reads.

D-D fusion identification

First, we identified TRBD1 and TRBD2 from the sequences localized between the TRBV and TRBJ segments. These sequences must be compared to the TRBD1 and TRBD2 gene segments and the alignments with most identical nucleotides are kept as TRBD candidates. If 6 or more nucleotides in a candidate TRBD sequence are identical to the corresponding TRBD segments, it was retained to be analyzed further.

To avoid random matches at both ends of the sequence, 2 or more identical nucleotides with reference sequences at both ends of the candidate TRBD sequences are needed. While TRBD1 and TRBD2 are identified to co-exist in a given CDR3 sequence, and there in no overlap between the two rearranged TRBD segments, the CDR3 was annotated as containing the D-D fusion.

ACKNOWLEDGEMENTS

We thank Dr. Christopher J. Vavrickafor and Boris Tefsen for their critical reading and revision of the manuscript and Dr. Miles P. Davenport for his inspiring discussions. This work is supported by the National Natural Science Foundation of China (NSFC, Grant No. 31030030), the National Basic Research Program (973 Program) (No. 2013CB531500) and the National Natural Science Foundation of China (Grant No. 81373141). G.F.G. is a leading principal investigator of the NSFC Innovative Research Group (Grant No. 81321063).

COMPLIANCE WITH ETHICS GUIDELINES

Peipei Liu, Di Liu, Xi Yang, Jing Gao, Yan Chen, Xue Xiao, Fei Liu, Jing Zou, Jun Wu, Juncai Ma, Fangqing Zhao, Xuyu Zhou, George F. Gao and Baoli Zhu declare that they have no conflict of interest.

This article does not contain any studies with human or animal subjects performed by the any of the authors.

ABBREVIATIONS

CDR3, complementary determining region 3; RSS, recombination signal sequence; TRA, TCR α chain; TRB, TCR β chain; TRD, TCR δ chain.

REFERENCES

Arstila TP, Casrouge A, Baron V, Even J, Kanellopoulos J, Kourilsky P (1999) A direct estimate of the human alphabeta T cell receptor diversity. Science 286:958–961

Bank I, Book M, Cohen L, Kneller A, Rosental E, Pras M, Bassat IB, Ben-Nun A (1992) Expansion of a unique subpopulation of cytotoxic T cells that express a C alpha V delta 1 T-cell receptor gene in a patient with severe persistent neutropenia. Blood 80:3157–3163

Bank I, Cohen L, Kneller A, de Rosbo NK, Book M, Ben-Nun A (2003) Aberrant T-Cell Receptor Signalling of Interferon-γ- and Tumour Necrosis Factor-α-Producing Cytotoxic CD8+Vδ1/Vβ16 T Cells in a Patient with Chronic Neutropenia. Scand J Immunol 58:89–98

Bassing CH, Swat W, Alt FW (2002) The mechanism and regulation of chromosomal V(D)J recombination. Cell 109(Suppl):S45–S55

Borgulya P, Kishi H, Uematsu Y, von Boehmer H (1992) Exclusion and inclusion of alpha and beta T cell receptor alleles. Cell 69:529–537

Castelli C, Mazzocchi A, Salvi S, Anichini A, Sensi M (1992) Use of the V delta 1 variable region in the functional T-cell receptor alpha chain of a WT31+ cytotoxic T lymphocyte clone which specifically recognizes HLA-A2 molecule. Scand J Immunol 35:487–494

Freeman JD, Warren RL, Webb JR, Nelson BH, Holt RA (2009) Profiling the T-cell receptor beta-chain repertoire by massively parallel sequencing. Genome Res 19:1817–1824

Fuschiotti P, Pasqual N, Hierle V, Borel E, London J, Marche PN, Jouvin-Marche E (2007) Analysis of the TCR alpha-chain rearrangement profile in human T lymphocytes. Mol Immunol 44:3380–3388

Genolet R, Stevenson BJ, Farinelli L, Osteras M, Luescher IF (2012) Highly diverse TCRalpha chain repertoire of pre-immune CD8(+) T cells reveals new insights in gene recombination. EMBO J 31:1666–1678

Gorski J, Yassai M, Zhu X, Kissela B, Kissella B, Keever C, Flomenberg N (1994) Circulating T cell repertoire complexity in normal individuals and bone marrow recipients analyzed by CDR3 size spectratyping. Correlation with immune status. J Immunol 152:5109–5119

Hempel WM, Stanhope-Baker P, Mathieu N, Huang F, Schlissel MS, Ferrier P (1998) Enhancer control of V(D)Jrecombination at the TCRβ locus: differential effects on DNA cleavage and joining. Genes & Development 12:2305–2317

Huang C, Kanagawa O (2001) Ordered and coordinated rearrangement of the TCR alpha locus: role of secondary rearrangement in thymic selection. J Immunol 166:2597–2601

Klarenbeek PL, Tak PP, van Schaik BD, Zwinderman AH, Jakobs ME, Zhang Z, van Kampen AH, van Lier RA, Baas F, de Vries N (2010) Human T-cell memory consists mainly of unexpanded clones. Immunol Lett 133:42–48

Klarenbeek PL, Remmerswaal EB, ten Berge IJ, Doorenspleet ME, van Schaik BD, Esveldt RE, Koch SD, ten Brinke A, van Kampen AH, Bemelman FJ et al (2012) Deep sequencing of antiviral T-cell responses to HCMV and EBV in humans reveals a stable repertoire that is maintained for many years. PLoS Pathog 8: e1002889

Krangel MS (2009) Mechanics of T cell receptor gene rearrangement. Curr Opin Immunol 21:133–139

Larimore K, McCormick MW, Robins HS, Greenberg PD (2012) Shaping of human germline IgH repertoires revealed by deep sequencing. J Immunol 189:3221–3230

Lefranc MP (2001) Nomenclature of the human T cell receptor genes. Curr Protoc Immunol Appendix 1, Appendix 1O

Lefranc MP (2011) From IMGT-ONTOLOGY CLASSIFICATION Axiom to IMGT standardized gene and allele nomenclature: for immunoglobulins (IG) and T cell receptors (TR). Cold Spring Harb Protoc 2011:627–632

Lefranc MP, Rabbitts TH (1990) Genetic organization of the human T-cell receptor gamma and delta loci. Res Immunol 141:565–577

Li H, Ye C, Ji G, Wu X, Xiang Z, Li Y, Cao Y, Liu X, Douek DC, Price DA et al (2012) Recombinatorial biases and convergent recombination determine interindividual TCRbeta sharing in murine thymocytes. J Immunol 189:2404–2413

Loh EY, Cwirla S, Serafini AT, Phillips JH, Lanier LL (1988) Human T-cell-receptor delta chain: genomic organization, diversity, and expression in populations of cells. Proc Natl Acad Sci USA 85:9714–9718

Mardis ER (2008) Next-generation DNA sequencing methods. Annu Rev Genomics Hum Genet 9:387–402

Meek KD, Hasemann CA, Capra JD (1989) Novel rearrangements at the immunoglobulin D locus. Inversions and fusions add to IgH somatic diversity. J Exp Med 170:39–57

Miossec C (1990) Further analysis of the T cell receptor gamma/delta+ peripheral lymphocyte subset. The V delta 1 gene segment is expressed with either C alpha or C delta. J Exp Med 171:1171–1188

Miossec C, Caignard A, Ferradini L, Roman-Roman S, Faure F, Michalaki H, Triebel F, Hercend T (1991) Molecular characterization of human T cell receptor alpha chains including a V delta 1-encoded variable segment. Eur J Immunol 21:1061–1064

Murphy KM, Travers P, Walport M (2007) Janeway's Immunobiology. UK, Garland Science, London

Neller MA, Burrows JM, Rist MJ, Miles JJ, Burrows SR (2013) High frequency of herpesvirus-specific clonotypes in the human T cell repertoire can remain stable over decades with minimal turnover. J Virol 87:697–700

Nikolich-Zugich J, Slifka MK, Messaoudi I (2004) The many important facets of T-cell repertoire diversity. Nat Rev Immunol 4:123–132

Olaru A, Petrie HT, Livak F (2005) Beyond the 12/23 rule of VDJ recombination independent of the Rag proteins. J Immunol 174:6220–6226

Ozawa T, Kishi H, Muraguchi A (2006) Amplification and analysis of cDNA generated from a single cell by 5′-RACE: application to isolation of antibody heavy and light chain variable gene sequences from single B cells. Biotechniques 40:469–478

Padovan E, Casorati G, Dellabona P, Meyer S, Brockhaus M, Lanzavecchia A (1993) Expression of two T cell receptor alpha chains: dual receptor T cells. Science 262:422–424

Pannetier C, Cochet M, Darche S, Casrouge A, Zoller M, Kourilsky P (1993) The Sizes of the Cdr3 Hypervariable Regions of the Murine T-Cell Receptor Beta-Chains Vary as a Function of the Recombined Germ-Line Segments. Proc Natl Acad Sci USA 90:4319–4323

Pasqual N, Gallagher M, Aude-Garcia C, Loiodice M, Thuderoz F, Demongeot J, Ceredig R, Marche PN, Jouvin-Marche E (2002) Quantitative and Qualitative Changes in V-J Rearrangements During Mouse Thymocytes Differentiation: Implication For a Limited T Cell Receptor Chain Repertoire. J Exp Med 196:1163–1174

Quigley MF, Greenaway HY, Venturi V, Lindsay R, Quinn KM, Seder RA, Douek DC, Davenport MP, Price DA (2010) Convergent recombination shapes the clonotypic landscape of the naive T-cell repertoire. Proc Natl Acad Sci U S A 107:19414–19419

Robins HS, Campregher PV, Srivastava SK, Wacher A, Turtle CJ, Kahsai O, Riddell SR, Warren EH, Carlson CS (2009) Comprehensive assessment of T-cell receptor beta-chain diversity in alphabeta T cells. Blood 114:4099–4107

Robins HS, Srivastava SK, Campregher PV, Turtle CJ, Andriesen J, Riddell SR, Carlson CS, Warren EH (2010) Overlap and effective size of the human CD8+ T cell receptor repertoire. Sci Transl Med 2, 47ra64

Roth ME, Holman PO, Kranz DM (1991) Nonrandom use of J alpha gene segments. Influence of V alpha and J alpha gene location. J Immunol 147:1075–1081

Ryu CJ, Haines BB, Draganov DD, Kang YH, Whitehurst CE, Schmidt T, Hong HJ, Chen J (2003) The T cell receptor beta enhancer promotes access and pairing of Dbeta and Jbeta gene segments during V(D)J recombination. Proc Natl Acad Sci USA 100:13465–13470

Shendure J, Ji H (2008) Next-generation DNA sequencing. Nat Biotechnol 26:1135–1145

Sherwood AM, Desmarais C, Livingston RJ, Andriesen J, Haussler M, Carlson CS, Robins H (2011). Deep sequencing of the human TCRgamma and TCRbeta repertoires suggests that TCRbeta rearranges after alphabeta and gammadelta T cell commitment. Sci Transl Med 3, 90ra61

Takihara Y, Tkachuk D, Michalopoulos E, Champagne E, Reimann J, Minden M, Mak TW (1988) Sequence and organization of the diversity, joining, and constant region genes of the human T-cell delta-chain locus. Proc Natl Acad Sci U S A 85:6097–6101

Ueno T, Tomiyama H, Fujiwara M, Oka S, Takiguchi M (2003) HLA class I-restricted recognition of an HIV-derived epitope peptide by a human T cell receptor alpha chain having a Vdelta1 variable segment. Eur J Immunol 33:2910–2916

Venturi V, Quigley MF, Greenaway HY, Ng PC, Ende ZS, McIntosh T, Asher TE, Almeida JR, Levy S, Price DA et al (2011) A mechanism for TCR sharing between T cell subsets and individuals revealed by pyrosequencing. J Immunol 186:4285–4294

Wang C, Sanders CM, Yang Q, Schroeder HW Jr, Wang E, Babrzadeh F, Gharizadeh B, Myers RM, Hudson JR Jr, Davis RW et al (2010) High throughput sequencing reveals a complex pattern of dynamic interrelationships among human T cell subsets. Proc Natl Acad Sci U S A 107:1518–1523

Warren RL, Freeman JD, Zeng T, Choe G, Munro S, Moore R, Webb JR, Holt RA (2011) Exhaustive T-cell repertoire sequencing of human peripheral blood samples reveals signatures of antigen selection and a directly measured repertoire size of at least 1 million clonotypes. Genome Res 21:790–797

Yousfi Monod M, Giudicelli V, Chaume D, Lefranc MP (2004) IMGT/JunctionAnalysis: the first tool for the analysis of the immunoglobulin and T cell receptor complex V-J and V-D-J JUNCTIONs. Bioinformatics 20(Suppl 1):i379–i385

9

Structural insights into the recognition of phosphorylated FUNDC1 by LC3B in mitophagy

Mengqi Lv, Chongyuan Wang, Fudong Li, Junhui Peng, Bin Wen, Qingguo Gong Yunyu Shi, Yajun Tang✉

Hefei National Laboratory for Physical Sciences at Microscale and School of Life Sciences, University of Science and Technology of China, Hefei 230027, China

✉ Correspondence: tangyj@ustc.edu.cn (Y. Tang)

ABSTRACT

Mitophagy is an essential intracellular process that eliminates dysfunctional mitochondria and maintains cellular homeostasis. Mitophagy is regulated by the post-translational modification of mitophagy receptors. Fun14 domain-containing protein 1 (FUNDC1) was reported to be a new receptor for hypoxia-induced mitophagy in mammalian cells and interact with microtubule-associated protein light chain 3 beta (LC3B) through its LC3 interaction region (LIR). Moreover, the phosphorylation modification of FUNDC1 affects its binding affinity for LC3B and regulates selective mitophagy. However, the structural basis of this regulation mechanism remains unclear. Here, we present the crystal structure of LC3B in complex with a FUNDC1 LIR peptide phosphorylated at Ser17 (pS_{17}), demonstrating the key residues of LC3B for the specific recognition of the phosphorylated or dephosphorylated FUNDC1. Intriguingly, the side chain of LC3B Lys49 shifts remarkably and forms a hydrogen bond and electrostatic interaction with the phosphate group of FUNDC1 pS_{17}. Alternatively, phosphorylated Tyr18 (pY_{18}) and Ser13 (pS_{13}) in FUNDC1 significantly obstruct their interaction with the hydrophobic pocket and Arg10 of LC3B, respectively. Structural observations are further validated by mutation and isothermal titration calorimetry (ITC) assays. Therefore, our structural and biochemical results reveal a working model for the specific recognition of FUNDC1 by LC3B and imply that the reversible phosphorylation modification of mitophagy receptors may be a switch for selective mitophagy.

KEYWORDS microtubule-associated protein light chain 3 beta (LC3B), fun14 domain-containing protein 1 (FUNDC1), phosphorylation, selective mitophagy

INTRODUCTION

As the source of ATP, mitochondria play a central role in cellular metabolism, stress responses and cell death (Galluzzi et al., 2012). Mitochondrial dysfunction results in a large number of reactive oxygen species (ROS), which damage the mitochondrial and cellular DNA and proteins (Lemasters, 2005; Wallace, 2005; Murphy, 2013). The quality control of mitochondria is crucial for the normal physiological function of cells (Rugarli and Langer, 2012; Javadov and Kuznetsov, 2013). Cells employ mitophagy as a major strategy to eliminate the dysfunctional mitochondria produced under stress (Batlevi and La Spada, 2011; Kanki et al., 2011). The main process of mitophagy in mammals is lysosome-dependent and includes the selective removal of abnormal mitochondria with low membrane potential (Bampton et al., 2005; Levine and Kroemer, 2008; Narendra et al., 2008). A considerable number of studies suggested that mitophagy is regulated through two pathways, including receptor-mediated mitophagy and Parkin-dependent mitophagy (Kim et al., 2007; Narendra et al., 2008; Okamoto et al., 2009; Novak et al., 2010; Liu et al., 2012). Many neurodegenerative diseases, such as Parkinson's disease and Alzheimer's disease, and cancers, such as hepatocellular carcinoma, are related to the dysfunction of mitophagy (Chu et al., 2007; Mizushima et al., 2008; Dikic et al., 2010; Deas et al., 2011; Ding et al., 2011).

Recently, several mitophagy receptors have been identified, such as ATG32 in yeast (Kanki et al., 2009; Okamoto et al., 2009), and NIX/BNIP3L (Novak et al., 2010), BNIP3 (Hanna et al., 2012), FUNDC1 (Liu et al., 2012) in mammalian cells. One of the common characteristics of mammalian mitophagy receptors is their conserved LC3 interaction region (LIR) with a W/Y/FxxL/I/V motif, which can interact with LC3, a bio-marker protein of the autophagosome in mammalian cells (Pankiv et al., 2007; Noda et al., 2008; Novak et al., 2010; Liu et al., 2012). The post-translational modifications of the mitophagy receptors, especially phosphorylation, play the key roles in regulating mitochondria homeostasis (Egan et al., 2011; Liu et al., 2014). In line with these studies, the phosphorylation of ATG32 promotes its interactions with both ATG11 and ATG8 (Farre et al., 2013) and subsequently triggers mitophagy. Moreover, the selective mitophagy receptor BNIP3 enhances its interaction with LC3B and causes an increased mitophagy through the serine phosphorylation of LIR (Zhu et al., 2013).

As a lately reported receptor for hypoxia-induced mitophagy in mammalian cells, FUNDC1, which is localized at the outer membrane of mammalian mitochondria, forms a high hydrophobicity transmembrane domain with three-helices and exposes its N-terminus to the cytoplasm (Liu et al., 2012). Interestingly, several reports have revealed that the outer membrane region of FUNDC1 interacts with LC3B and regulates hypoxia-induced selective mitophagy through the reversible phosphorylation at several critical sites (Liu et al., 2012; Chen et al., 2014; Wu et al., 2014). As previously reported, FUNDC1 interacts with LC3B through its classical LIR - $Y^{18}xxL^{21}$, while the phosphorylation at Tyr18 by Src kinase remarkably reduces the FUNDC1-mediated mitophagy. Under hypoxic stress, Tyr18 is dephosphorylated to promote the interaction between FUNDC1 and LC3B and mitophagy is triggered (Liu et al., 2012). In this context, Ser/Thr kinase CK2 phosphorylates the Ser13 of FUNDC1 in normal cells, while PGAM5 phosphatase in mitochondria dephosphorylates Ser13 under hypoxia stimulation. The dephosphorylation of Ser13 results in the enhanced interaction of FUNDC1 with LC3B, which further leads to the selective removal of dysfunctional mitochondria (Chen et al., 2014). On the other hand, Ser17 of FUNDC1 is phosphorylated by ULK1 kinase under hypoxia or mitochondrial uncouplers stimulation to increase the binding affinity for LC3B and promote mitophagy (Wu et al., 2014) (Fig. 1A). Although various intracellular experiments have confirmed that the reversible phosphorylation of FUNDC1 plays the key role in the regulation of mitophagy (Liu et al., 2012; Chen et al., 2014; Wu et al., 2014), the precise working mechanism remains unclear and needs to be elucidated.

To gain structural insights into the interactions between FUNDC1 and LC3B involved in hypoxia-induced selective mitophagy, it is essential to solve the structure of LC3B in complex with FUNDC1, including LIR and phosphorylated critical residues, at atomic resolution. Although the apo LC3B structure and a set of LC3B-LIR complex structures have been reported (Ichimura et al., 2008; Rogov et al., 2013; Suzuki et al., 2014; McEwan et al., 2015), the structural basis for the phosphorylation-regulated interaction between FUNDC1 LIR peptide and LC3B is still largely unknown. Here, we present the crystal structure of LC3B in complex with a FUNDC1 LIR peptide phosphorylated at Ser17 (pS_{17}). Through the structural analyses, we identified the key residues of LC3B responsible for the specific recognition of the phosphorylated or dephosphorylated FUNDC1. We found that the phosphate group of FUNDC1 pS_{17} binds to LC3B Lys49 and enhances the binding affinity, while the phosphorylation of FUNDC1 Tyr18 may conflicts with the hydrophobic pocket of LC3B and disrupt the interaction. In addition, using the High Ambiguity Driven protein-protein Docking (HADDOCK) and ITC, we were able to show that LC3B Arg10 interacts with FUNDC1 Ser13, but the phosphorylation of FUNDC1 Ser13 may generate steric hindrance for LC3B binding. Furthermore, mutation and ITC assays were performed to validate our observations from the crystal structure. Our structural and *in vitro* interaction analyses provide a detailed elucidation of the specific recognition of FUNDC1 by LC3B and facilitate a deep understanding of how mitophagy receptors utilize the post-translational modification to sense environmental stress and elaborately regulate the selective mitophagy.

RESULTS

The interaction between LC3B and FUNDC1 is dramatically affected by the phosphorylation states of FUNDC1

Three key residues, Ser13, Ser17 and Tyr18, in the outer membrane region of FUNDC1 and their phosphorylation states have been reported to play essential roles in affecting the binding affinity for LC3B and influence the FUNDC1-mediated selective mitophagy (Liu et al., 2012; Chen et al., 2014; Wu et al., 2014). In line with these results, a series of FUNDC1 peptides were first synthesized and tested for their binding affinities for LC3B. These FUNDC1 peptides all include LIR and flanking residues (10–25: DYESDDDSYEVLDLTE), with/without the phosphorylation at different positions. To investigate the LC3B-FUNDC1 interaction quantitatively, we employed ITC to measure the binding affinities of LC3B to the FUNDC1 peptides. First, we measured the binding affinity of LC3B to the FUNDC1 peptide without any phosphorylation modification and the K_D value was fitted to 1.78 ± 0.16 μmol/L. We then measured the binding affinity of LC3B with three peptides containing pS_{13}, pS_{17} and pY_{18}, respectively. The ITC results show that FUNDC1 pS_{17} peptide binds to LC3B ~3-fold stronger than the unphosphorylated FUNDC1 peptide, while the FUNDC1

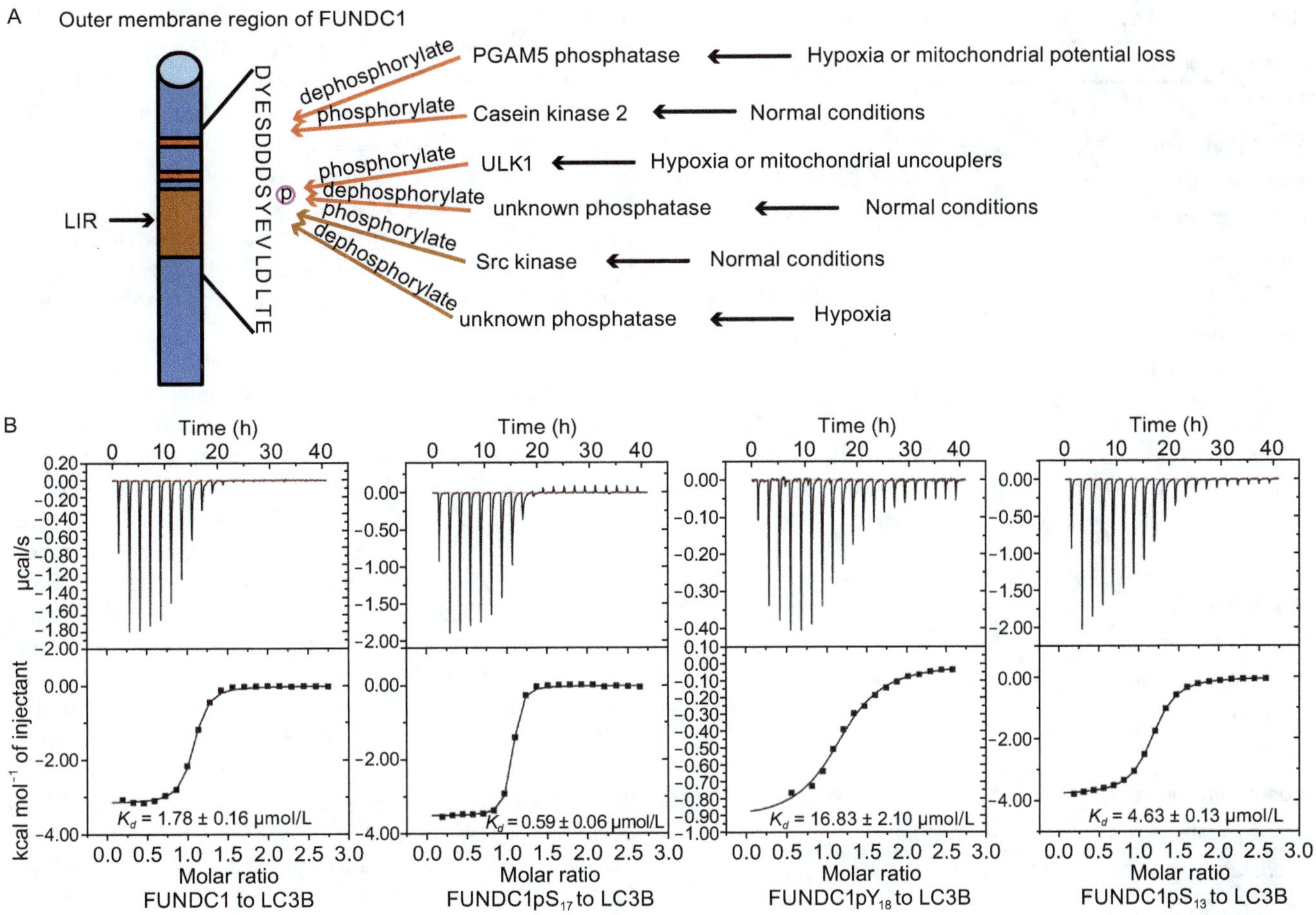

Fig. 1. The signaling pathways of FUNDC1-mediated mitophagy are regulated by phosphorylation modification. (A) Schematic representation of reversible phosphorylation at critical sites in the FUNDC1 mitochondrial outer membrane region. (B) The ITC fitting results of LC3B with unphosphorylated FUNDC1 peptide and FUNDC1 peptides phosphorylated at different positions.

pS_{13} and FUNDC1 pY_{18} peptides bind to LC3B ~3-fold and ~10-fold weaker than the unphosphorylated peptide, respectively (Fig. 1B). Taken together, our results demonstrate that phosphorylation of FUNDC1 Ser17 enhances the interaction between FUNDC1 and LC3B, while the phosphorylation of FUNDC1 Ser13 and Tyr18 reduce the binding affinities, consistent with the previously reported experiment results *in vivo*.(Liu et al., 2012; Chen et al., 2014; Wu et al., 2014)

Overall structure of the LC3B-FUNDC1 pS_{17} peptide complex

To provide structural insights into the interaction between LC3B and FUNDC1, X-ray crystallography was employed to study the complex structure. We chose a FUNDC1 LIR peptide$^{10-25}$ with pS_{17}, which simulates the possible physiological state upon the induction of mitophagy and exhibits a strong binding affinity for LC3B, to co-crystallize with full-length LC3B$^{1-125}$. The crystal structure of the LC3B-FUNDC1 complex was subsequently refined to a resolution of 2.25 Å in space group P 2_1. The crystal structure was solved by molecular replacement using the structure of apo LC3B (PDB ID: 3VTU) as the search model. Finally, the R_{work} and R_{free} of the LC3B-FUNDC1 complex structure were refined to 20.57% and 27.22%, respectively. The detailed crystallographic statistics are summarized in Table 1.

In the final model, the LC3B-FUNDC1 complex includes two molecules in an asymmetric unit both with observable electronic densities for 120 amino acid residues of LC3B$^{5-124}$ and a main part of the FUNDC1 pS_{17} peptide$^{16-23}$ (D-pS_{17}-YEVLDL). The two molecules display a root-mean-square (r.m.s.) deviation for Cα atoms of 0.248 Å, revealing only slight conformational differences between the two molecules in the flexible loops. The overall structure of the LC3B-FUNDC1 complex is depicted in Fig. 2A and 2B. The LC3B in complex with FUNDC1 exhibits four-stranded anti-parallel β sheets

Table 1. Data collection and refinement statistics

Data collection	LC3B-FUNDC1 pS_{17}
Space group	$P2_1$
Wavelength (Å)	0.9792
Resolution (Å)	37.88–2.25 (2.33–2.25)
Cell dimensions	
a, b, c (Å)	40.54, 86.85, 40.54
α, β, γ (°)	90.00, 110.86, 90.00
Unique reflections	12034 (1167)
Completeness (%)	96.8
I/σI	7.5 (3.8)
R_{merge} (%)	7.6 (31.4)
R_{meas} (%)	9.8 (40.6)
$CC1/2^b$	0.988
Refinement	
R_{work} (%)	20.57
R_{free} (%)	27.22
Average B factors ($Å^2$)	
Protein	33.46
H_2O	34.77
Root mean square deviations	
Bond lengths (Å)	0.011
Bond angles (°)	1.450
Ramachandran plot	
Favored (%)	99.2
Allowed (%)	0.8
Disallowed	0

(β1–β4) separated by five α-helices (α1–α5) which is highly similar to the structure of apo LC3B (Rogov et al., 2013) (with r.m.s. deviations for Cα atoms of 0.695 Å). The FUNDC1 pS_{17} peptide binds to one side of the LC3B surface and lies on the β2 of LC3B in an extended conformation (Fig. 2A).

The interactions between LC3B and FUNDC1 pS_{17} peptide[16–23] are shown in Fig. 2C, including hydrogen bonding interactions and hydrophobic interactions. The side chains of FUNDC1 Tyr18 and Leu21 insert into two deep hydrophobic pockets of LC3B, resulting in specific recognitions between LC3B and FUNDC1 (Fig. 2A and 2B). The pocket accommodating the side chain of Tyr18 is mainly composed of several hydrophobic residues including Ile23, Lys51 and Leu53 of LC3B, and forms a hydrogen bond between the carbonyl oxygen of LC3B Asp19 and the hydroxyl group of FUNDC1 Tyr18 (Fig. 3A). To validate the importance of these conserved residues of LC3B involved in FUNDC1 Tyr18 recognition, we introduced alanine mutations into these residues to partially disrupt the hydrophobic pocket and performed ITC assays to measure their binding affinities for the FUNDC1 pS_{17} peptide. As shown in Table 2, all LC3B mutants weaken the interactions of LC3B with the FUNDC1 pS_{17} peptide, particularly I23A and K51A, which exhibit more than 5-fold reductions in binding affinities (K_D = 4.85 ± 0.25 μmol/L and 5.92 ± 0.75 μmol/L, respectively), compared with 0.59 ± 0.06 μmol/L of wild-type LC3B (Fig. 3B). Similar to the mutations effect on the hydrophobic pocket of LC3B, the phosphorylation of FUNDC1 Tyr18 (PTR18) leads to a more extended side chain, which may conflict with the hydrophobic pocket and disrupt the interaction (Fig. 3C).

The interaction pattern between LC3B and FUNDC1 LIR is quite similar to those between LC3B and other autophagy-related LIR in complex structures (Figs. 4A and Fig. S1). Likewise, the LC3B residues involved in hydrophobic interactions with different W/Y/FxxL/I/V LIR motifs are identical (Figs. 4A and S1). To further compare our LC3B-FUNDC1 complex with other LC3B-autophagy receptor complexes, we superimposed our complex structure on the LC3B-p62 complex (PDB code 2ZJD), which is involved in the ubiquitin-dependent selective autophagy (Ichimura et al., 2008) (Fig. 4A). LC3B and the peptides in the two complexes share highly similar conformations, respectively. Both YxxL motif of FUNDC1 and WxxL motif of p62 exhibit a residue with a bulky aromatic side chain (Y or W), which deeply insert into the hydrophobic pockets of LC3B, sharing the similar interactions (Fig. 4A and 4B). Nevertheless, Glu19 in the FUNDC1 LIR forms a distinct interaction with LC3B Arg70 from its corresponding residue in p62, Thr340 (Fig. 4B and 4C). As shown in Fig. 4C, the side chain of FUNDC1 Glu19 is longer than p62 Thr340, posing a steric clash with LC3B Arg70 to induce the torsion of Arg70 side chain. Moreover, unlike p62 Thr340, FUNDC1 Glu19 forms additional salt bonds with the side chain of LC3B Arg70 to stabilize the interaction (Fig. 4D). The neutral and opposite mutations of Arg70 (R70A and R70E) both decrease the binding affinity between LC3B and FUNDC1 pS_{17} peptide significantly (K_D = 15.75 ± 1.50 μmol/L and 60.98 ± 3.44 μmol/L, respectively) (Fig. 4E and Table 2). We also compared our structure with other LC3B-LIR complexes and found the interaction between LC3B Arg70 and FUNDC1 Glu19 is distinctive (Fig. S1). Taken together, these results imply that the interaction between LC3B Arg70 and FUNDC1 Glu19 is essential in the specific recognition between LC3B and FUNDC1.

Lys49 senses the phosphorylation of FUNDC1 Ser17, enhancing binding affinity

As we have mentioned previously, ULK1 kinase promotes the phosphorylation of FUNDC1 Ser17, enhancing the interaction with LC3B during mitophagy (Wu et al., 2014). Our ITC experiments also confirmed an enhanced interaction between the FUNDC1 peptide and LC3B by the phosphorylation of FUNDC1 Ser17 (Fig. 1B). Structural analysis

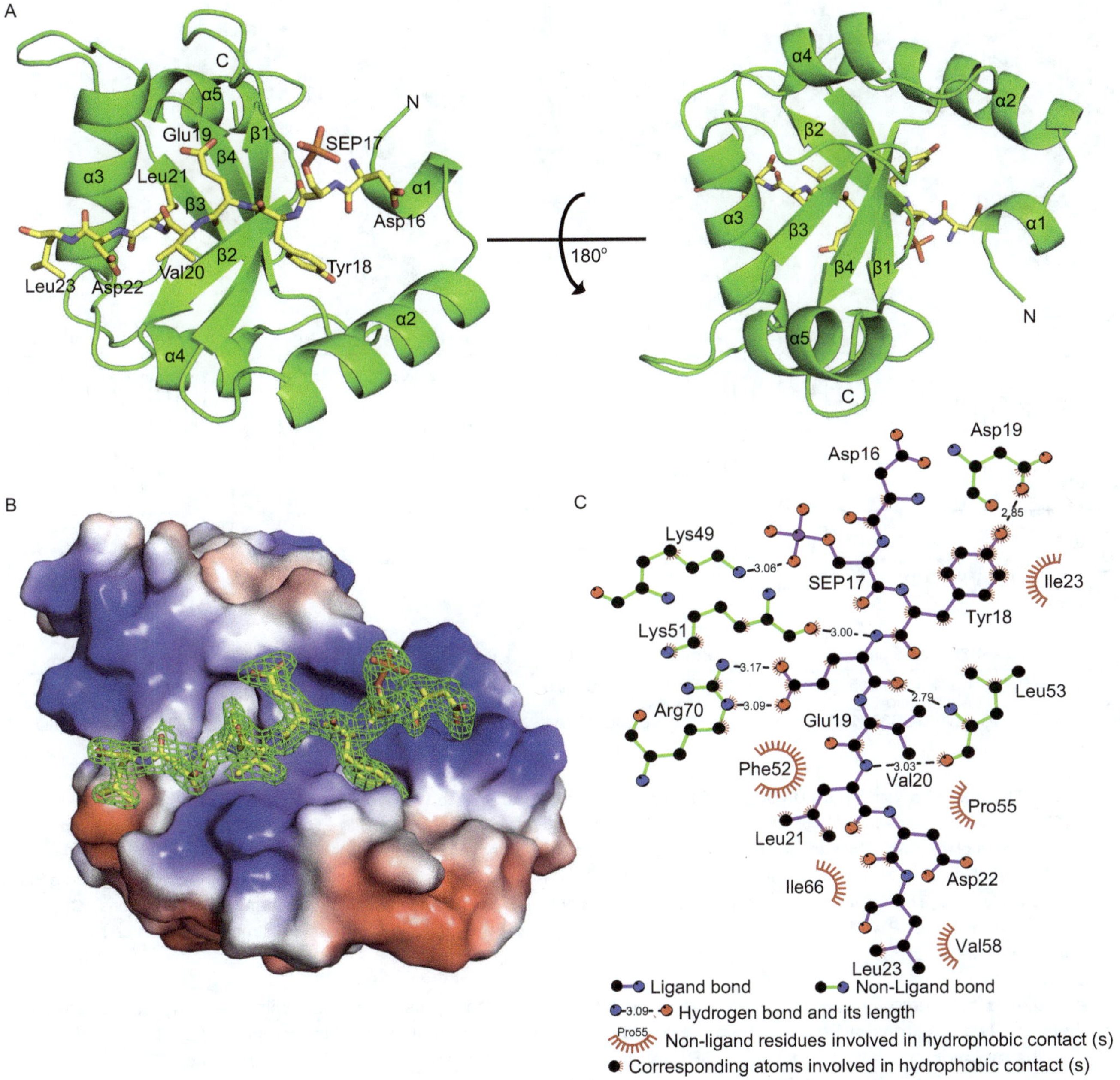

Fig. 2. Crystal structure of LC3B-FUNDC1 pS$_{17}$ peptide complex. (A) Cartoon representation of LC3B (green) in complex with the FUNDC1 pS$_{17}$ peptide (yellow). In the peptide, the phosphorylated Ser17 is labeled as SEP17. (B) The electrostatic potential of the LC3B-FUNDC1 pS$_{17}$ peptide complex surface, in which positively charged, negatively charged and neutral areas are represented in blue, red and white, respectively. The 2Fo-Fc omit density map of the FUNDC1 pS$_{17}$ peptide was contoured at 1.0 σ (green). (C) Schematic representations of the recognition of FUNDC1 pS$_{17}$ peptide (colored purple and labeled in red) by LC3B (colored green and labeled in black).

shows that the enhancement of binding affinity induced by FUNDC1 pS$_{17}$ is due to an additional hydrogen bond formed between LC3B Lys49 and the phosphate group of FUNDC1 pS$_{17}$ (Fig. 5A). As shown in Fig. 5B, the alanine substitution of LC3B Lys49 (K49A), which prevents the formation of this hydrogen bond, results in a ~6-fold decrease in the binding affinity of LC3B for FUNDC1 pS$_{17}$ peptide (K_D = 3.56 ± 0.21 μmol/L). Furthermore, the phosphorylation of Ser17 leads to an elongated and electronegative side chain, allowing it easily approach the positively charged side chain of Lys49. Mutation of Lys49 to glutamine (K49E) results in a >200-fold reduction in binding affinity

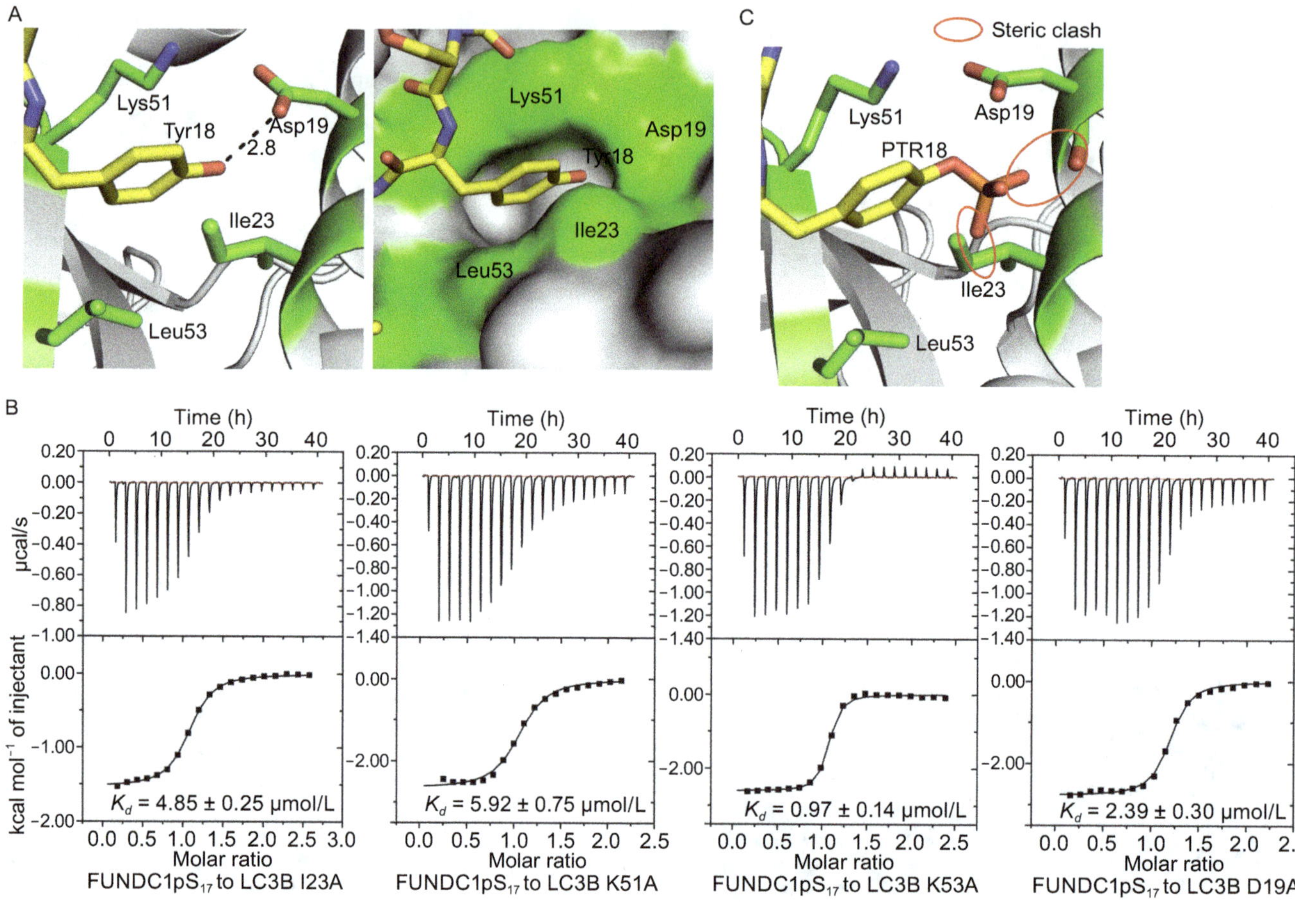

Fig. 3. The phosphorylation of FUNDC1 Tyr18 disrupts the interaction with the hydrophobic pocket of LC3B. (A) FUNDC1 Tyr18 inserts into the deep hydrophobic pocket of LC3B and forms a hydrogen bond with LC3B Asp19. The hydrogen bond is indicated as a black dash. The surrounding structure of LC3B is represented in grey. FUNDC1 Tyr18 is colored yellow and labeled in red. The residues of LC3B involved in the interaction with FUNDC1 Tyr18 (Asp19, Ile23, Lys51 and Leu53) are colored green and labeled in black. (B) The ITC fitting results of the FUNDC1 pS_{17} peptide with LC3B mutated at the above-mentioned positions. (C) The hypothetical model of PTR18 (phosphorylated Tyr18, phosphorylated by PyTMs) in the complex structure. The steric clash is highlighted with red ovals.

(K_D = 120.48 ± 5.72 μmol/L) due to the electrostatic repulsion between two electronegative side chains (Fig. 5B). In contrast, the mutation of Lys49 to arginine (K49R) enhances the binding affinity of LC3B for the FUNDC1 pS_{17} peptide with a K_D value of 0.33 ± 0.07 μmol/L (Fig. 5B). This may be attributed to the additional hydrogen bond(s) formed between the arginine residue and the phosphate group of pS_{17} (Fig. 5C).

Additionally, we compared the conformations of Lys49 in our LC3B-FUNDC1 complex structure and apo LC3B (PDB ID: 3VTU) (Fig. 5D). Interestingly, the side chain of LC3B Lys49 undergoes a large structural rearrangement in two structures. In apo LC3B, the Lys49 side chain forms hydrophobic interactions with the aromatic ring of Phe52, occupying the space for the side chain of the FUNDC1 Glu19 in the complex structure. However, in the complex structure, a largely shift (8.2 Å) of the Lys49 side chain is induced by its interaction with the pS_{17} of FUNDC1, providing enough space to accommodate the FUNDC1 LIR, especially the side chain of Glu19 (Fig. 5D).

Taken together, our mutational and structural analyses of LC3B Lys49 reveal that Lys49 is essential in sensing the phosphorylation state of FUNDC1 Ser17 for selective mitophagy (Table 2).

HADDOCK models the interface between FUNDC1 Ser13 and LC3B

Unlike Ser17, FUNDC1 Ser13 is dephosphorylated by PGAM5 phosphatase when cells are treated with hypoxia or mitochondrial uncouplers to enhance the interaction with LC3B (Chen et al., 2014), which was confirmed by our ITC assays *in vitro* (Fig. 1B). In our complex structure, the N-terminus of the FUNDC1 pS_{17} peptide[10–15], including

Table 2. The thermodynamic parameters of the ITC experiments

Peptide	LC3B	ΔH kcal/mol	ΔS cal/mol/K	K_D μmol/L	N
FUNDC1	Wild type	−3.15	15.70	1.78 ± 0.16	1.02
	R10A	−3.56	10.70	11.44 ± 1.14	1.01
	R11A	−4.51	8.90	5.62 ± 0.25	1.12
	F7A	−3.08	14.60	3.52 ± 0.23	1.17
	V46A	−4.86	10.20	1.59 ± 0.19	1.02
	L47A	−2.06	18.3	3.07 ± 0.64	1.19
	D48A	−3.03	17.00	1.17 ± 0.15	0.98
	T50A	−2.65	16.50	2.81 ± 0.31	1.02
FUNDC1 pS_{13}	Wild type	−3.81	11.60	4.63 ± 0.13	1.13
	R10A	−3.67	8.47	28.49 ± 2.97	1.02
	R10E	−2.83	11.00	34.01 ± 3.46	1.03
	R11A	−4.10	8.91	11.17 ± 0.73	1.11
	R11E	−2.39	13.80	17.12 ± 1.05	1.04
FUNDC1 pS_{17}	Wild type	−3.51	16.7	0.59 ± 0.06	1.00
	K49A	−4.90	8.5	3.56 ± 0.21	1.08
	K49E	−4.49	2.87	120.48 ± 5.72	0.93
	K49R	−5.78	10.2	0.33 ± 0.07	1.17
	R70A	−1.80	15.9	15.75 ± 1.50	1.04
	R70E	−2.54	10.8	60.98 ± 3.44	1.03
	D19A	−2.78	16.4	2.39 ± 0.30	1.16
	I23A	−1.53	19.2	4.85 ± 0.25	1.04
	K51A	−2.66	15.0	5.92 ± 0.75	1.03
	L53A	−2.61	18.7	0.97 ± 0.14	1.03
FUNDC1 pY_{18}	Wild type	−0.93	18.7	16.83 ± 2.10	1.16

Ser13, was not visible in the electron density map. Therefore, we generated a LC3B-FUNDC1$^{10-23}$ complex model using HADDOCK to study the interface between FUNDC1 Ser13 and LC3B (Fig. 6). Two hundred refined structures were generated by the HADDOCK run, and the N-terminal end of the FUNDC1 peptide was not converged among these structures (Fig. 6A). Hydrogen bonding interactions between residues in the N-terminus of the FUNDC1 peptide and those in LC3B were further analyzed. The structures suggest that the LC3B forms hydrogen bonding interactions with residues in the N-terminus of FUNDC1 in only 53 structures among the 200 structures. The N-terminus of the FUNDC1 peptide orients to the N-terminal end (Region 1) or the linker of β1β2 (Region 2) of LC3B in the 53 structures (Fig. 6B). Among the 53 structures, 43 structures use LC3B Arg10 and FUNDC1 Ser13 to form hydrogen bonding interactions (Fig. 6C).

To confirm the residues of LC3B that interact with FUNDC1 Ser13, we introduced alanine mutations into region 1 and region 2 of LC3B and performed ITC assays to determine the binding affinities between the unphosphorylated FUNDC1 peptide and LC3B mutants. All thermodynamic parameters of the ITC experiments are listed in Table 2. The ITC results show that the alanine substitution of both LC3B Arg10 and Arg11 (R10A and R11A) notably decreased the binding affinities of LC3B for the FUNDC1 peptide by ~6-fold (K_D = 11.44 ± 1.14 μmol/L) and ~3-fold (K_D = 5.62 ± 0.25 μmol/L), respectively (Fig. 6D). This indicates that the two residues play significant roles in the specific recognition of FUNDC1 Ser13, which is consistent with the interaction information suggested by HADDOCK models. In addition, it has been mentioned above that the phosphorylation of FUNDC1 Ser13 caused a ~3-fold reduction in the binding affinity of LC3B for FUNDC1 peptide (Fig. 1B), while the influence of the LC3B R10A mutation on the binding affinity is much stronger than that of FUNDC1 Ser13 phosphorylation. We speculate that the phosphorylation of FUNDC1 Ser13 may generate steric hindrance for

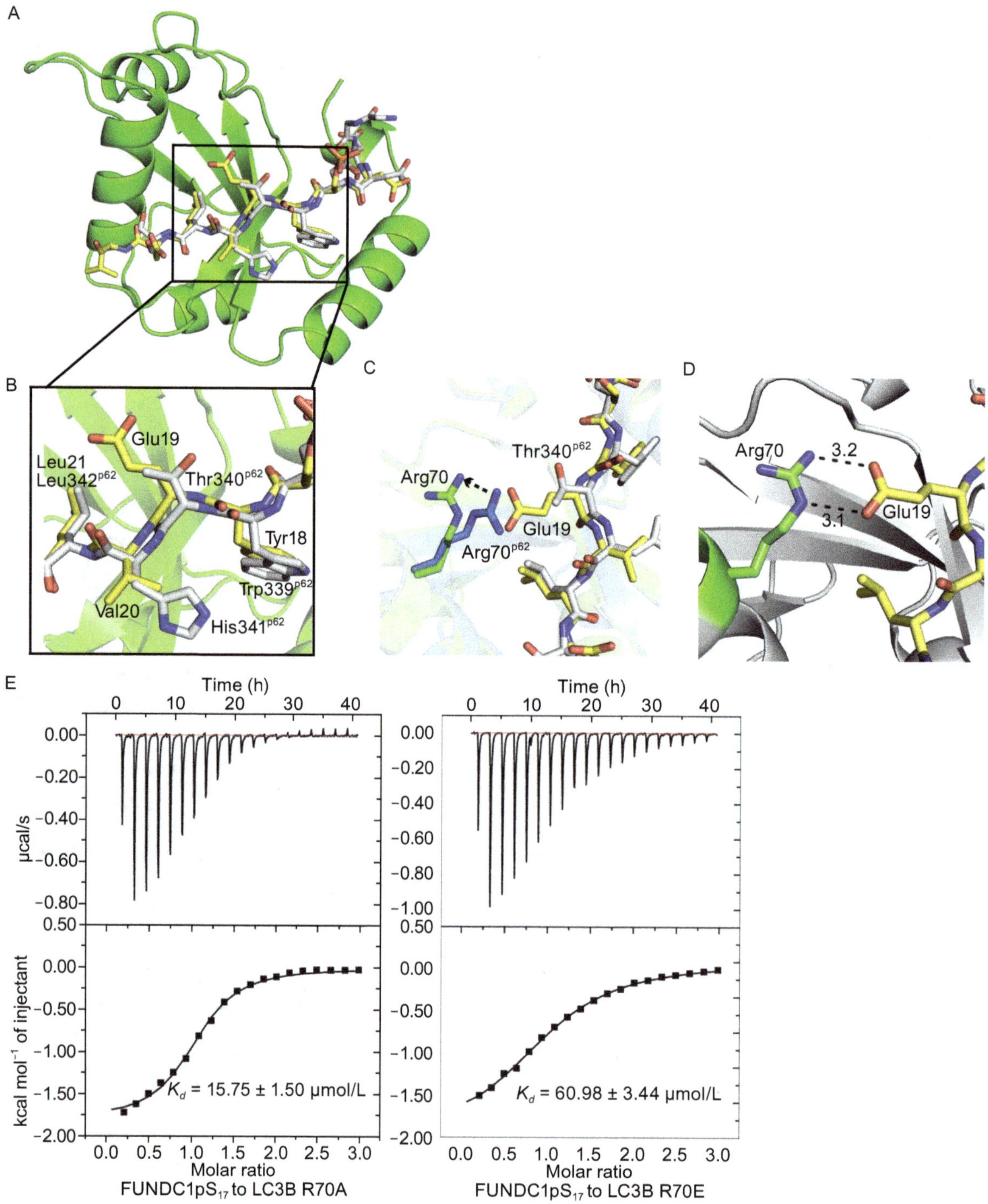

Fig. 4. Comparison of the LC3B-FUNDC1 complex and LC3B-p62 complex. (A) The superimposition of the LC3B-FUNDC1 complex on the LC3B-p62 complex. The superimposition only shows LC3B (green) in the LC3B-FUNDC1 complex (in two complexes, the two LC3B molecules only display a r.m.s. deviation for Cα atoms of 0.489 Å). The FUNDC1 peptide is shown in yellow while the p62 peptide is colored white. (B) Close-up view of the structural comparison of LIRs in FUNDC1 (yellow) and p62 (white). Residues of LIRs in FUNDC1 and p62 are labeled in red and blue, respectively. (C) The difference of LC3B Arg70 (green and blue, respectively) recognized by FUNDC1 Glu19 (yellow) and p62 Thr340 (white). (D) Interactions between LC3B Arg70 and FUNDC1 Glu19. The surrounding structure of LC3B is represented in grey. Salt bonds are indicated as black dashes. (E) The ITC fitting results of FUNDC1 pS_{17} peptide with LC3B mutants at the Arg70 position.

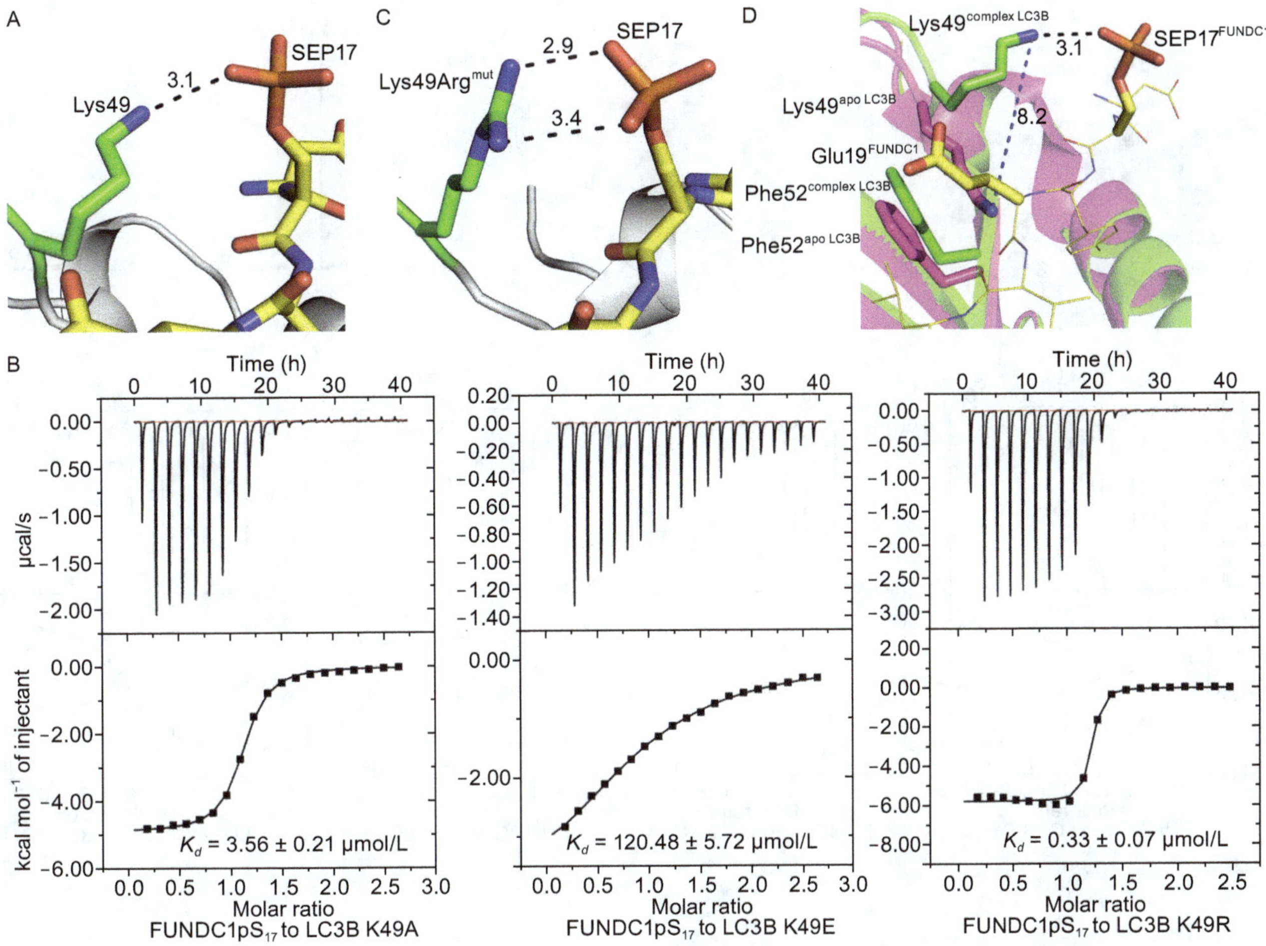

Fig. 5. Molecular basis for the specific recognition between LC3B Lys49 and phosphorylated Ser17 in FUNDC1. (A) Interaction of LC3B Lys49 (green) with the phosphate group of FUNDC1 SEP17 (orange). The surrounding structure of LC3B is represented in grey. Hydrogen bonds are indicated as black dashes. (B) The ITC fitting results of FUNDC1 pS17 peptide with LC3B mutant at the Lys49 position. (C) The hypothetical model of the interaction between Lys49Arg^{mut} (K49R^{mut} mutated in PyMOL) and the phosphate group of FUNDC1 SEP17. (D) The superimposition of the LC3B-FUNDC1 complex and apo LC3B. The LC3B in complex is colored green and labeled in black while the FUNDC1 pS17 peptide is colored yellow and labeled in red. The apo LC3B is colored and labeled in magenta. The shift of LC3B Lys49 is represented by blue dashes.

LC3B binding, while the LC3B R10A mutation directly destroys the interaction between Arg10 of LC3B and Ser13 or nearby Asp14 of FUNDC1 peptide, leading to the decreased binding affinity. Furthermore, we measured the binding affinities of the LC3B mutants R10A and R11A with the FUNDC1 pS$_{13}$ peptide. As expected, the ITC results suggest a ~2-fold reduction compared with the unphosphorylated peptide (K_D = 28.49 ± 2.97 μmol/L and K_D = 11.17 ± 0.73 μmol/L, respectively, Fig. 6E). A stronger decrease occurred when glutamine was introduced into Arg10 (K_D = 34.01 ± 3.46 μmol/L) and Arg11 (K_D = 17.12 ± 1.05 μmol/L), due to the electrostatic repulsion between the two electronegative side chains (Table 2).

Based on these results, a new complex model was generated using HADDOCK for more detailed insights into the interaction between FUNDC1 Ser13 and LC3B (Fig. 6F). In this model, the N-terminus of the FUNDC1pS$_{17}$ peptide lies on the first α-helix (α1) of LC3B with only a few contacts (Fig. 6F). This implies that the N-terminus of this peptide is flexible, which makes sense, as this region is missing in the electron density map. The side chain of LC3B Arg10 forms hydrogen bonds with the side chain and backbone carbonyl group of FUNDC1 Ser13 (Fig. 6G). Alternatively, LC3B Arg11 interacts with the negatively charged side chain of Asp14 in FUNDC1 (Fig. 6G). All HADDOCK results and ITC analyses show that LC3B Arg10 is the key residue that

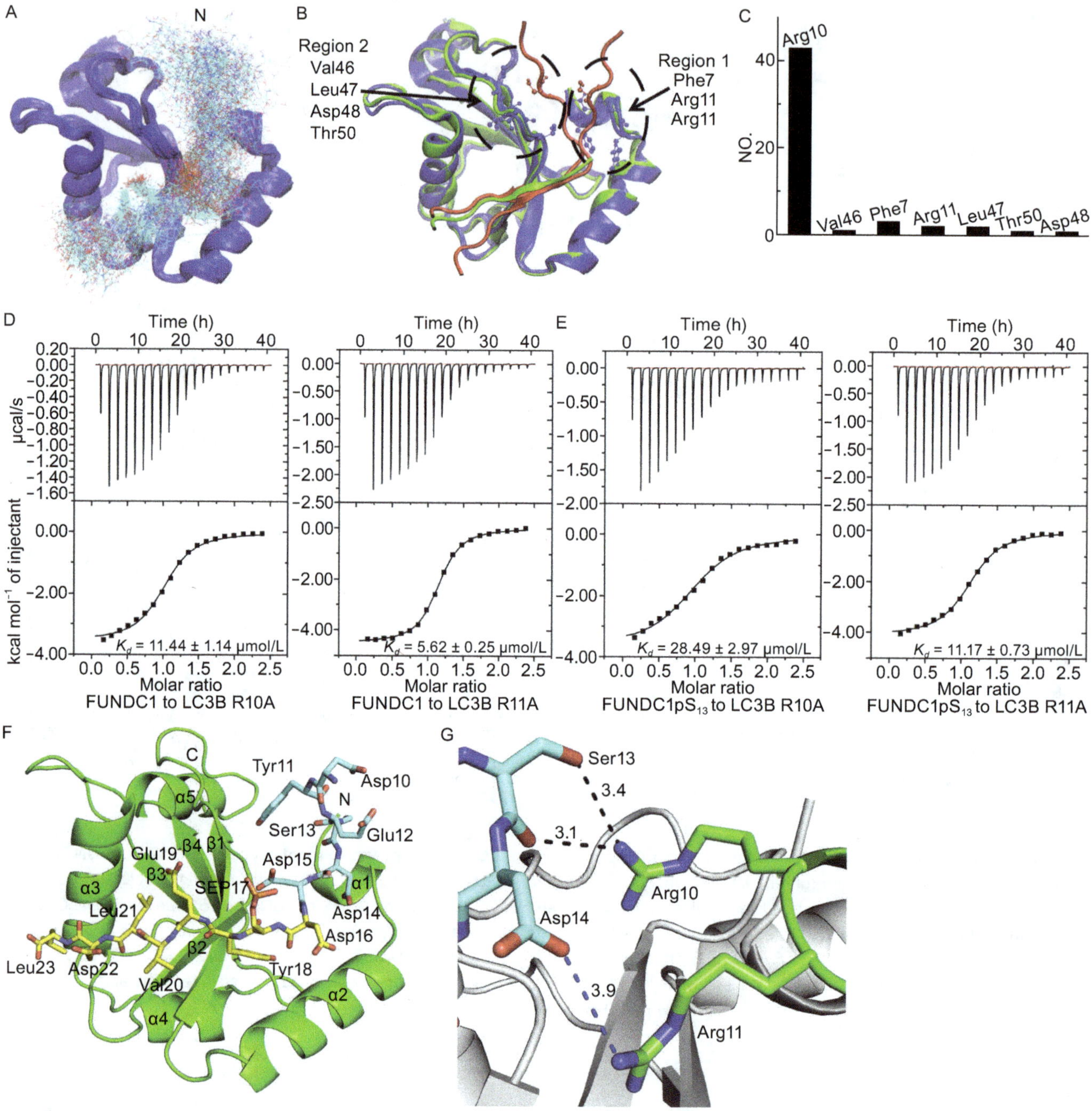

Fig. 6. The HADDOCK model of LC3B-FUNDC1$^{10-23}$ complex. (A) The initial 200 models of the LC3B-FUNDC1$^{10-23}$ complex calculated by HADDOCK. LC3B is colored blue and the FUNDC1 pS$_{17}$ peptide is colored cyan. The N-terminus of the FUNDC1 pS$_{17}$ peptide is marked. (B) The two most possible regions calculated from the statistics of initial HADDOCK models. Our crystal structure is shown in green and the calculated models are colored in blue and red, respectively, and labeled in black. (C) The statistics of the most probable LC3B residues interacting with FUNDC1 Ser13. (D and E) The ITC fitting results of FUNDC1 unphosphorylated peptide and pS$_{13}$ peptide with LC3B mutants at the Arg10 and Arg11 positions. (F) The possible model of the LC3B-FUNDC1$^{10-23}$ complex calculated by HADDOCK. LC3B and FUNDC1 pS$_{17}$ peptide$^{16-23}$ are shown in green and yellow, respectively. The N-terminus of the FUNDC1 pS$_{17}$ peptide$^{10-15}$ is colored and labeled in cyan. (G) The possible interactions of LC3B Arg10 and Arg11 (green) with Ser13 and Asp14 (cyan) of FUNDC1 pS$_{17}$ peptide. Hydrogen bonds and electrostatic interaction are indicated as black and blue dashes, respectively.

interacts with FUNDC1 Ser13 whose phosphorylation may generate steric hindrance for Arg10 binding.

DISCUSSION

FUNDC1, a new receptor for hypoxia-induced mitophagy in mammalian cells, was first reported in 2012 (Liu et al., 2012). Since then, functional studies of the FUNDC1-mediated mitophagy pathway have been extensively reported. Under hypoxia or FCCP stress, FUNDC1 Ser17 is phosphorylated, while Ser13 and Tyr18 are dephosphorylated to enhance the interaction with LC3B and recruit LC3-related autophagosomes to eliminate dysfunctional mitochondria (Liu et al., 2012; Chen et al., 2014; Wu et al., 2014). Thus, post-translational modification, especially phosphorylation, plays an essential role in regulating mitophagy. However, few studies have sufficiently elucidated the phosphorylation impact on the interaction between mitophagy receptors and LC3B from a structural point of view. Moreover, the molecular mechanism of the specific LC3B recognition for FUNDC1 also remains unclear. In this study, we solved the crystal structure of LC3B in complex with the FUNDC1 pS_{17} peptide. The binding interface of the FUNDC1 pS_{17} peptide on LC3B provides the structural elucidation for the specific recognition.

Our structure comparison reveals that, upon the binding with the FUNDC1 pS_{17} peptide, the side chain of LC3B Lys49 undergoes a large structural rearrangement, to accommodate the phosphorylated FUNDC1 peptide. Previous studies have proposed the local conformational change of LC3B when bound with unphosphorylated receptors (Suzuki et al., 2014). However, regarding the interaction with a phosphorylated peptide, the conformational switch of Lys49 in LC3B was revealed for the first time. Suzuki and co-workers presented a similar rearrangement of Lys49 in the LC3A-Atg13 complex (Suzuki et al., 2014). In their structure, Lys49 switches on the hydrophobic interaction surface of LC3A and in turn binds to Val445 of Atg13, with a 6.7 Å shift for Lys49 side chain compared with 8.2 Å in our structure. We speculate that the negative charge of pS_{17} in FUNDC1 induced by phosphorylation may be the reason for the more obvious shift compared with the uncharged side chain of Atg13 Val445. In addition, they demonstrated that the shift of Lys49 side chain is conserved among LC3 homologs in mammals through the structural comparison of several LC3 homologs (Suzuki et al., 2014). This indicates that the LC3 Lys49 not only regulates the binding of LIR as a switch but also takes charge of the specific recognition of post-translational modification of the mitophagy receptors, which is essential for autophagosome formation and the removal of dysfunctional mitochondria.

As mentioned previously, the reversible phosphorylation of FUNDC1 Ser13 and Tyr18 are regulated by different phosphatases and kinases with two different mechanisms, respectively (Liu et al., 2012; Chen et al., 2014). However, Chen et al. confirmed that the reversible phosphorylation of FUNDC1 Ser13 and Tyr18 functionally cooperate to regulate FUNDC1-mediated mitophagy (Chen et al., 2014). Moreover, Wu and co-workers confirmed that Src kinase, which phosphorylates FUNDC1 Tyr18, suppresses phosphorylation of FUNDC1 Ser17 by ULK1 kinase (Wu et al., 2014), providing evidence that ULK1-mediated FUNDC1 pS_{17} may work cooperatively with FUNDC1 Ser13 and Tyr18 during mitophagy. However, how does FUNDC1 respond to different degrees of stress through different states of phosphorylation? Do the conformation and recognition between LC3B and FUNDC1 change under different levels of FUNDC1 phosphorylation? These questions require further research.

Overall, we provide structural insights into the interaction between LC3B and phosphorylated Ser17 and unphosphorylated Ser13 and Tyr18 of FUNDC1, which is the first step to thoroughly explain the molecular mechanism of FUNDC1-dependent mitophagy. Due to the existence of different post-translational modification sites in FUNDC1, further biochemical, structural and cellular experiments should be studied in order to elucidate how cells synergistically regulate these key residues for responding to stress.

MATERIALS AND METHODS

Protein expression and purification

A DNA fragment encoding the full-length (125 amino acids) human LC3B was amplified by PCR from the human brain cDNA library and cloned into the pET-28a(+) expression vector (Novagen), contained an N-terminal 6×His tag and a tobacco etch virus (TEV) protease cleavage site. All mutants of LC3B were generated using a MutanBEST kit (Takara) and confirmed by DNA sequencing. The proteins were expressed in *Escherichia coli* BL21 (DE3) cells (Novagen) cultured in LB medium at 37 °C to OD_{600} = 0.8, then shifted to 16 °C and induced with 0.5 mmol/L IPTG overnight. Bacterial pellets were resuspended in buffer A (20 mmol/L Tris-HCl, 1 mol/L NaCl, pH 8.0) and lysed by sonication on ice. Then, soluble proteins were purified with a Ni^{2+}-chelating column (GE Healthcare), followed by a Superdex 75 column (GE Healthcare). After being cleaved by TEV protease overnight at 16 °C to remove the 6×His tag, the purified protein was concentrated to ~20 mg/mL in buffer B (20 mmol/L Tris-HCl, 200 mmol/L NaCl, 1 mmol/L EDTA, pH 8.0) and stored at −80 °C.

Peptide preparations

Peptides were synthesized by GL Biochem (Shanghai), and stock solutions (5 to 15 mmol/L) were prepared in buffer B. The sequences of the peptides are as follows: (p: phosphorylation; pS: phosphorylated Ser; pY: phosphorylated Tyr)

FUNDC1: DYESDDDSYEVLDLTE
FUNDC1 pS_{17}: DYESDDD-pS_{17}-YEVLDLTE
FUNDC1 pS_{13}: DYE-pS_{13}-DDDSYEVLDLTE
FUNDC1 pY_{18}: DYESDDDS-pY_{18}-EVLDLTE

Isothermal titration calorimetry (ITC)

ITC assays were performed on a MicroCal iTC200 calorimeter (GE Healthcare) at 25 °C. The concentrations of proteins were determined spectrophotometrically. Proteins and peptides were dialyzed against buffer B and adjusted to 0.25 mmol/L and 3 mmol/L, respectively. Curve fitting to a single binding site model was performed by the ITC data analysis module of Origin 7.0 (MicroCal) provided by the manufacturer. The thermodynamic parameters of the ITC experiments are listed in Table 2.

Crystallization, data collection and structure determination

$LC3B^{1-125}$ and $FUNDC1pS_{17}$ $peptide^{10-25}$, mixed at a 1:2 molar ratio, were crystallized in 30% PEG MME 2000, 0.1 mmol/L sodium cacodylate (pH 6.0) at 16 °C by vapor diffusion in sitting drops. Crystals were soaked in cryoprotectant made of mother liquor supplemented with 20% glycerol before being flash-frozen in liquid nitrogen. Data sets were collected on Beamline 17U at Shanghai Synchrotron Radiation Facility (SSRF). The structure of the LC3B-FUNDC1 complex was solved by molecular replacement with the program MOLREP (Vagin and Teplyakov, 2010), using the apo $LC3B^{2-123}$ (PDB ID: 3VTU) as the search model (Rogov et al., 2013). The FUNDC1 pS_{17}^{10-25} peptide was then modeled in COOT (Emsley et al., 2010), and the structure of the LC3B-FUNDC1 complex was refined by the programs REFMAC5 (Murshudov et al., 2011) and PHENIX.refine (Adams et al., 2010). Crystal diffraction data and refinement statistics are displayed in Table 1. Structure analysis was performed using COOT and PyMOL (http://www.pymol.org/).

Coordinates

Coordinates and structure factors for the LC3B-FUNDC1 pS_{17} peptide complex have been deposited in the Protein Data Bank (PDB) under the accession codes 5GMV.

Generation of the LC3B-FUNDC1 pS_{17} $peptide^{10-23}$ complex model by HADDOCK

Our crystal structure containing LC3B and the N-terminus truncated FUNDC1 peptide (Δ10–15) was used in model building, and two rounds of docking were performed using the easy interface of the HADDOCK webserver (de Vries et al., 2010). The first round of the docking procedure was performed as follows. First, coordinates of the missing residues of the FUNDC1 peptide were built using PyMOL and Ser17 in the FUNDC1 peptide was phosphorylated by PyTMs (Warnecke et al., 2014). Then, inputs of the HADDOCK webserver were extracted from the crystal structures. Interface residues involved in inter-chain hydrogen bonding interactions were treated as active residues of their corresponding chains. Passive residues were defined automatically around the active residues. Two hundred refined structures were generated by the HADDOCK run, and the N-terminal of the FUNDC1 peptide was not converged among these structures. In the second round of the docking procedure, Arg10 and Arg11 in LC3B and Ser13 in FUNDC1 peptide were included in active residues. The second docking procedure resulted in another 200 refined structures. The 200 refined structures were divided into 5 clusters by the single linkage cluster method with a distance cut-off of 0.2 Å, and 196 structures were involved in cluster 5, representing 98% of the models HADDOCK generated. Then, we chose the representative conformation of cluster 5 as our model.

ACKNOWLEDGMENTS

We thank Prof. J. H. Wu, Prof. Z. Y. Zhang, Dr L. Xu, Y. Y. Jiang, H. Y. Bao and L. N. Yang for helpful discussions, and H. C. Ou for help with the ITC experiments. We thank the staff of the Beamline BL17U at SSRF for assistance with data collection.

This work was supported by National Natural Science Foundation (Grant No. 31400629); the Strategic Priority Research Program of the Chinese Academy of Science (No. XDB08010101); Ministry Of Science And Technology of China (No. 2016YFA0500700); China Postdoctoral Science Foundation (No. 2015M582009 and 2016T90579) and National Natural Science Foundation (Grant No. 31330018).

COMPLIANCE WITH ETHICS GUIDELINES

Mengqi Lv, Chongyuan Wang, Fudong Li, Junhui Peng, Bin Wen, Qingguo Gong, Yajun Tang and Yunyu Shi declare that they have no conflicts of interest.

This article does not contain any studies with human or animal subjects performed by any of the authors.

REFERENCES

Adams PD, Afonine PV, Bunkoczi G, Chen VB, Davis IW, Echols N, Headd JJ, Hung LW, Kapral GJ, Grosse-Kunstleve RW et al (2010) PHENIX: a comprehensive Python-based system for macromolecular structure solution. Acta Crystallogr Sect D 66:213–221

Bampton ET, Goemans CG, Niranjan D, Mizushima N, Tolkovsky AM (2005) The dynamics of autophagy visualized in live cells: from autophagosome formation to fusion with endo/lysosomes. Autophagy 1:23–36

Batlevi Y, La Spada AR (2011) Mitochondrial autophagy in neural function, neurodegenerative disease, neuron cell death, and aging. Neurobiol Dis 43:46–51

Chen G, Han Z, Feng D, Chen YF, Chen LB, Wu H, Huang L, Zhou CQ, Cai XY, Fu CY et al (2014) A regulatory signaling loop comprising the PGAM5 phosphatase and CK2 controls receptor-mediated mitophagy. Mol Cell 54:362–377

Chu CT, Zhu JH, Dagda R (2007) Beclin 1-independent pathway of damage-induced mitophagy and autophagic stress. Autophagy 3:663–666

de Vries SJ, van Dijk M, Bonvin AM (2010) The HADDOCK web server for data-driven biomolecular docking. Nat Protoc 5:883–897

Deas E, Wood NW, Plun-Favreau H (2011) Mitophagy and Parkinson's disease: the PINK1-parkin link. Biochim Biophys Acta 1813:623–633

Dikic I, Johansen T, Kirkin V (2010) Selective autophagy in cancer development and therapy. Cancer Res 70:3431–3434

Ding WX, Li M, Yin XM (2011) Selective taste of ethanol-induced autophagy for mitochondria and lipid droplets. Autophagy 7:248–249

Egan DF, Shackelford DB, Mihaylova MM, Gelino S, Kohnz RA, Mair W, Vasquez DS, Joshi A, Gwinn DM, Taylor R et al (2011) Phosphorylation of ULK1 (hATG1) by AMP-activated protein kinase connects energy sensing to mitophagy. Science 331:456–461

Emsley P, Lohkamp B, Scott WG, Cowtan K (2010) Features and development of Coot. Acta Crystallogr Sect D 66:486–501

Farre JC, Burkenroad A, Burnett SF, Subramani S (2013) Phosphorylation of mitophagy and pexophagy receptors coordinates their interaction with Atg8 and Atg11. EMBO Rep 14:441–449

Galluzzi L, Kepp O, Trojel-Hansen C, Kroemer G (2012) Mitochondrial control of cellular life, stress, and death. Circ Res 111:1198–1207

Hanna RA, Quinsay MN, Orogo AM, Giang K, Rikka S, Gustafsson AB (2012) Microtubule-associated protein 1 light chain 3 (LC3) interacts with Bnip3 protein to selectively remove endoplasmic reticulum and mitochondria via autophagy. J Biol Chem 287:19094–19104

Ichimura Y, Kumanomidou T, Sou YS, Mizushima T, Ezaki J, Ueno T, Kominami E, Yamane T, Tanaka K, Komatsu M (2008) Structural basis for sorting mechanism of p62 in selective autophagy. J Biol Chem 283:22847–22857

Javadov S, Kuznetsov AV (2013) Mitochondria: the cell powerhouse and nexus of stress. Front Physiol 4:207

Kanki T, Wang K, Cao Y, Baba M, Klionsky DJ (2009) Atg32 is a mitochondrial protein that confers selectivity during mitophagy. Dev Cell 17:98–109

Kanki T, Klionsky DJ, Okamoto K (2011) Mitochondria autophagy in yeast. Antioxid Redox Signal 14:1989–2001

Kim I, Rodriguez-Enriquez S, Lemasters JJ (2007) Selective degradation of mitochondria by mitophagy. Arch Biochem Biophys 462:245–253

Lemasters JJ (2005) Perspective: selective mitochondrial autophagy, or mitophagy, as a targeted defense against oxidative stress, mitochondrial dysfunction, and aging. Rejuv Res 8:3–5

Levine B, Kroemer G (2008) Autophagy in the pathogenesis of disease. Cell 132:27–42

Liu L, Feng D, Chen G, Chen M, Zheng QX, Song PP, Ma Q, Zhu CZ, Wang R, Qi WJ et al (2012) Mitochondrial outer-membrane protein FUNDC1 mediates hypoxia-induced mitophagy in mammalian cells. Nat Cell Biol 14:177–185

Liu L, Sakakibara K, Chen Q, Okamoto K (2014) Receptor-mediated mitophagy in yeast and mammalian systems. Cell Res 24:787–795

McEwan DG, Popovic D, Gubas A, Terawaki S, Suzuki H, Stadel D, Coxon FP, de Stegmann DM, Bhogaraju S, Maddi K et al (2015) PLEKHM1 regulates autophagosome-lysosome fusion through HOPS complex and LC3/GABARAP proteins. Mol Cell 57:39–54

Mizushima N, Levine B, Cuervo AM, Klionsky DJ (2008) Autophagy fights disease through cellular self-digestion. Nature 451:1069–1075

Murphy MP (2013) Mitochondrial dysfunction indirectly elevates ROS production by the endoplasmic reticulum. Cell Metab 18:145–146

Murshudov GN, Skubak P, Lebedev AA, Pannu NS, Steiner RA, Nicholls RA, Winn MD, Long F, Vagin AA (2011) REFMAC5 for the refinement of macromolecular crystal structures. Acta Crystallogr Sect D 67:355–367

Narendra D, Tanaka A, Suen DF, Youle RJ (2008) Parkin is recruited selectively to impaired mitochondria and promotes their autophagy. J Cell Biol 183:795–803

Noda NN, Kumeta H, Nakatogawa H, Satoo K, Adachi W, Ishii J, Fujioka Y, Ohsumi Y, Inagaki F (2008) Structural basis of target recognition by Atg8/LC3 during selective autophagy. Genes Cells 13:1211–1218

Novak I, Kirkin V, McEwan DG, Zhang J, Wild P, Rozenknop A, Rogov V, Lohr F, Popovic D, Occhipinti A et al (2010) Nix is a selective autophagy receptor for mitochondrial clearance. EMBO Rep 11:45–51

Okamoto K, Kondo-Okamoto N, Ohsumi Y (2009) Mitochondria-anchored receptor Atg32 mediates degradation of mitochondria via selective autophagy. Dev Cell 17:87–97

Pankiv S, Clausen TH, Lamark T, Brech A, Bruun JA, Outzen H, Overvatn A, Bjorkoy G, Johansen T (2007) p62/SQSTM1 binds directly to Atg8/LC3 to facilitate degradation of ubiquitinated protein aggregates by autophagy. J Biol Chem 282:24131–24145

Rogov VV, Suzuki H, Fiskin E, Wild P, Kniss A, Rozenknop A, Kato R, Kawasaki M, McEwan DG, Lohr F et al (2013) Structural basis for phosphorylation-triggered autophagic clearance of Salmonella. Biochem J 454:459–466

Rugarli EI, Langer T (2012) Mitochondrial quality control: a matter of life and death for neurons. EMBO J 31:1336–1349

Suzuki H, Tabata K, Morita E, Kawasaki M, Kato R, Dobson RCJ, Yoshimori T, Wakatsuki S (2014) Structural basis of the autophagy-related LC3/Atg13 LIR complex: recognition and interaction mechanism. Structure 22:47–58

Vagin A, Teplyakov A (2010) Molecular replacement with MOLREP. Acta Crystallogr Sect D 66:22–25

Wallace DC (2005) A mitochondrial paradigm of metabolic and degenerative diseases, aging, and cancer: a dawn for evolutionary medicine. Annu Rev Genet 39:359–407

Warnecke A, Sandalova T, Achour A, Harris RA (2014) PyTMs: a useful PyMOL plugin for modeling common post-translational modifications. BMC Bioinform 15:370

Wu WX, Tian WL, Hu Z, Chen G, Huang L, Li W, Zhang XL, Xue P, Zhou CQ, Liu L et al (2014) ULK1 translocates to mitochondria and phosphorylates FUNDC1 to regulate mitophagy. EMBO Rep 15:566–575

Zhu YY, Massen S, Terenzio M, Lang V, Chen-Lindner S, Eils R, Novak I, Dikic I, Hamacher-Brady A, Brady NR (2013) Modulation of serines 17 and 24 in the LC3-interacting region of Bnip3 determines pro-survival mitophagy versus apoptosis. J Biol Chem 288:1099–1113

10

The immunostimulatory effects of retinoblastoma cell supernatant on dendritic cells

Juan Ma[1]✉, Huamin Han[1,2], Li Ma[3], Changzhen Liu[1], Xin Xue[4], Pan Ma[1], Xiaomei Li[1], Hua Tao[1]

[1] CAS Key Laboratory of Pathogenic Microbiology and Immunology, Institute of Microbiology, Chinese Academy of Sciences, Beijing 100101, China
[2] Key Laboratory of Infection and Immunity, Institute of Biophysics, Chinese Academy of Sciences, Beijing 100101, China
[3] Department of Obstetrics and Gynecology, China-Japan Friendship Hospital, Beijing 100029, China
[4] Department of Immunology, Basic Medical Theory of Chinese Medicine, China Academy of Chinese Medical Sciences, Beijing 100700, China
✉ Correspondence: majuan96@sina.com (J. Ma)

ABSTRACT

Dendritic cells (DCs) are crucial for the induction and maintenance of tumor-specific immune responses. Studies have shown that tumor-associated DCs are immunosuppressed in some human tumors. However, phenotype and function of DCs in retinoblastoma (RB) remain unclear. RB cell supernatant (RBcs) was used to treat DCs *in vitro* to explore the effect of RB cells on DCs. DCs were generated from peripheral blood mononuclear cells of healthy donors. On day 5 of culture, DCs were treated with RBcs for 24 h, and then purified using magnetic beads. The maturation of DCs was induced by TNF-α or LPS. After treatment with RBcs, expression of co-stimulatory molecules CD80 and CD86 was elevated in DCs, accompanied by increased production of IL-12p70, TNF-α, IL-6, IL-1β, and IL-8 but decreased production of IL-10. RBcs neither inhibited DC maturation nor promoted DC apoptosis. Moreover, RBcs-exposed DCs stimulated allogenetic T cell proliferation and T cell-derived cytokine production. These results indicate that RBcs can improve DCs' antigen presenting function and capability to activate T cells, suggesting that RB cells may have an immunostimulatory effect on DCs, and DC-based immunotherapy may be adopted in the treatment of RB.

Juan Ma and Huamin Han contributed equally to this work.

KEYWORDS retinoblastoma, dendritic cell, anti-tumor immunity, immunotherapy

INTRODUCTION

Retinoblastoma (RB) is the most common primary intraocular malignant tumor in childhood, and the morbidity of RB is about 11 per million children below age 5 worldwide (Houston et al., 2011). The prognosis of RB patients has been dramatically improved by systematic enucleation (Khetan et al., 2013), external cryotherapy, local thermotherapy (Schueler et al., 2003), and brachytherapy (Merchant et al., 2004). Although these methods are successful at controlling the growth of the primary tumor, they cannot prevent the development of metastasis, which remains universally fatal. Moreover, there are some severe side effects related to radiotherapy or chemotherapy. Cancer immunotherapies have been generally in steady progress in this field over the past decade, particularly in the treatment of metastatic skin melanoma.

Dendritic cells (DCs) are crucial for the induction and maintenance of antitumor immune responses. Tumor-specific antigens bound to molecules of the major histocompatibility complex (MHC) on the surface of DCs are processed, then presented to and recognized by T cells. In addition, DCs provide some critical molecules, such as co-stimulatory signals and cytokines, to the T cells for their full activation. Actually, tumor-infiltrating DCs (TIDCs) are associated with prolonged survival and reduced incidence in some metastatic human tumors (Dieu-Nosjean et al., 2008;

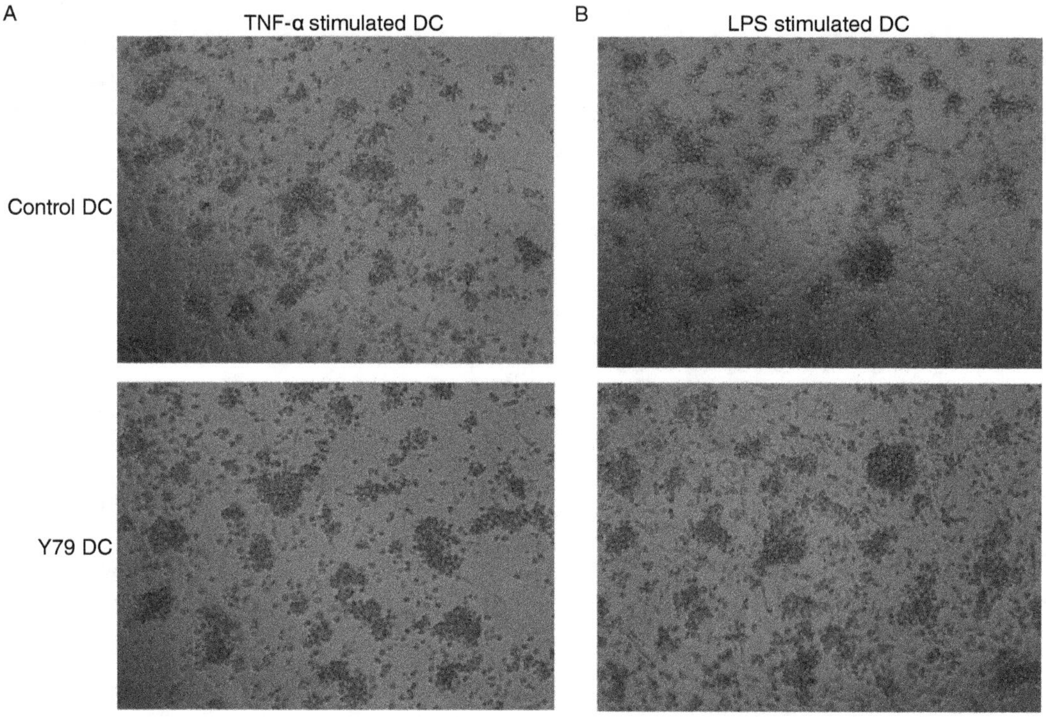

Figure 1. The photomicrograph of DC cultures (200×). Control DCs or RBcs-exposed DCs were treated with 20 ng/mL TNF-α (A) or 1 μg/mL LPS (B) for 24 h. Y79 DC: RBcs-exposed DCs.

Iwamoto et al., 2003; Ladanyi et al., 2007; Nakakubo et al., 2003). However, in some other conditions, TIDCs are functionally compromised. TIDCs are phenotypically and functionally defective in colorectal cancer (Chaux et al., 1997) and melanoma (Ataera et al., 2011; Stoitzner et al., 2008), and a positive correlation of TIDCs with the poor prognosis was found in colorectal cancer (Sandel et al., 2005) and breast cancer (Treilleux et al., 2004). In hepatocellular carcinoma, circulating DCs also exhibit an immature phenotype (Beckebaum et al., 2004).

Until now, the effect of RB on human DCs has not been explored. In the present study, we used RB cell supernatant (RBcs) to mimic the tumor milieu, and performed a detailed study on the phenotype of DCs treated with RBcs. Subsequently, we investigated the effect of RBcs-exposed DCs on allogenetic T cell proliferation and cytokine production. Our study demonstrates that RBcs improves DCs' antigen presenting function and capability to activate T cells, and DC-based immunotherapy may be adopted in the treatment of RB.

RESULTS

Induction of co-stimulatory molecules CD80 and CD86 in DCs by RBcs

Five-day old DCs were treated with or without RBcs for 24 h. On day 6, maturation of DCs was induced by adding 20 ng/mL TNF-α or 1 μg/mL LPS. After 24 h, all DCs appeared as big loosely adherent clumps or isolated floating cells with the typical dendritic morphology (Fig. 1). The expression of DC markers (CD1a and CD83), MHC class molecules (HLA-ABC and HLA-DR) and co-stimulatory molecules (CD40, CD80, and CD86) was determined by flow cytometry (Fig. 2). Compared with control DCs, RBcs-exposed DCs expressed higher levels of CD80 and CD86, but similar levels of CD1a, CD83, HLA-ABC, HLA-DR, and CD40. These data suggest that RB cells may enhance DCs' capacity in priming T cell responses, whereas have no effect on the maturation of DCs.

Induction of IL-12p70, TNF-α, IL-6, IL-1β, IL-8 and inhibition of IL-10 in DCs by RBcs

In addition to co-stimulatory molecules, DC-derived cytokines also play an important role in priming T cell response. Cytokine production in DCs was assayed using CBA Human Inflammation Kit. Compared with control DCs, RBcs-exposed DCs secreted more IL-12p70, TNF-α, IL-6, IL-1β, and IL-8, but less IL-10, a potent immunosuppressive cytokine (Fig. 3). Moreover, alterations in cytokine production induced by RBcs in LPS-matured DCs were more significant than those in TNF-α-matured DCs (Fig. 3B). These results further indicate that RB cells can increase DCs' capacity to activate T cells.

No effect of RBcs on DC apoptosis

As tumor cells can escape immune destruction by inducing DC apoptosis (Kanto et al., 2001; Kiertscher et al., 2000; Ma

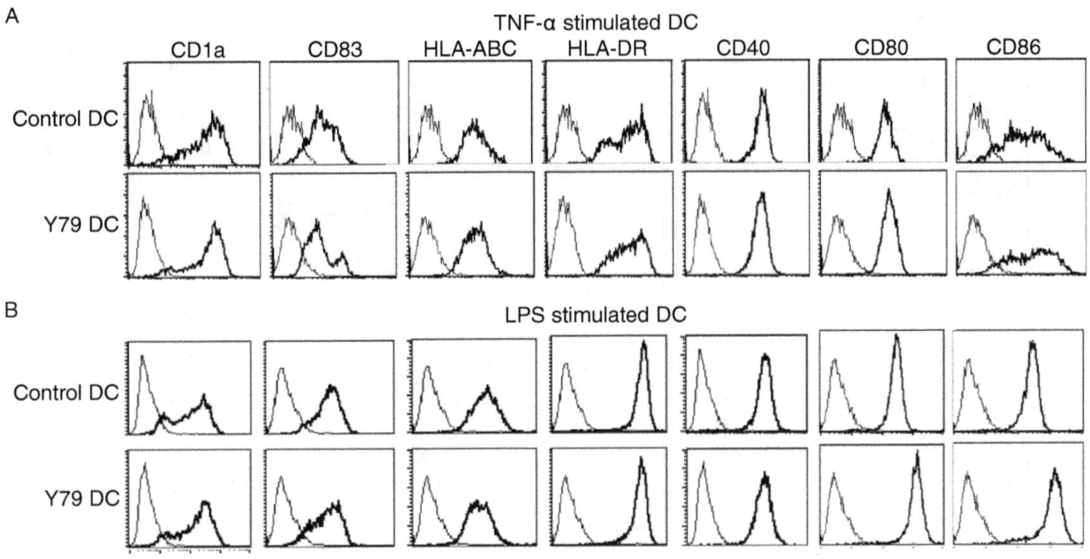

Figure 2. Expression of DC markers, MHC and co-stimulatory molecules in RBcs-exposed DC. Control DCs or RBcs-exposed DCs were treated with 20 ng/mL TNF-α (A) or 1 μg/mL LPS (B) for 24 h. The cells were then harvested for immunofluorescence staining and flow cytometry. Bold lines denote fluorescence when stained with fluorochrome-conjugated antibody to the indicated antigen, and fine lines denote fluorescence when stained with isotype control mAb. Data shown are a representative experiment of five. Y79 DC: RBcs-exposed DCs.

et al., 2010), we evaluated whether RB cells could affect DC apoptosis (Fig. 4). FITC-Annexin-V and propidium iodide (PI) were used to stain DCs, and the proportion of apoptotic cells ($Annexin^+/PI^-$) was determined by flow cytometry. No significant difference in apoptotic proportion was observed between control DCs and RBcs-exposed DCs. DC apoptosis was further assessed by a FACS-based TUNEL assay utilizing FITC-dUTP. It was found that the apoptotic rate of RBcs-exposed DCs was similar to that of control DCs. These data indicate that RB cells have no effect on DC apoptosis.

Stimulation of T cell proliferation by RBcs-exposed DCs

To verify the immunostimulatory effect of RBcs-exposed DCs, DCs treated with or without RBcs were irradiated and then used to stimulate purified allogeneic T cells (2×10^5 per well). Proliferation of T cells was analyzed by measuring the incorporation of [^{3}H]-thymidine. As shown in Fig. 5, a significant increase in T cell proliferation was observed when they were co-cultured with RBcs-exposed DCs at T/DC ratio of 500:1 and 100:1, indicating that DCs' capacity to activate T cells was enhanced after treatment with RBcs.

Induction of T cell-derived cytokines by RBcs-exposed DCs

To further confirm the immunostimulatory effect of RBcs-exposed DCs, levels of T cell-derived cytokines were measured in above T-DC co-culture system using the CBA human Th1/Th2/Th17 cytokine kit. It was observed that RBcs-exposed DCs stimulated T cells to secrete much more cytokines than control DCs at different T/DC ratio, including IFN-γ, IL-2, TNF-α, IL-6, IL-10, and IL-17 (Fig. 6). On the contrary, no IL-4 production was found in T cells treated with either control DCs or RBcs-exposed DCs. These data indicate that RBcs-exposed DCs induce T cells to produce Th1 and Th17 cytokines predominantly.

DISCUSSION

DCs play a critical role in tumor immune surveillance by initiating tumor-specific immune responses. Clinically, infiltration of DCs is correlated with a better prognosis in various human tumors (Dieu-Nosjean et al., 2008; Iwamoto et al., 2003; Ladanyi et al., 2007; Nakakubo et al., 2003), which is attributed to inhibition of tumor growth and metastasis by DCs (Lim et al., 2007; Movassagh et al., 2004; Preynat-Seauve et al., 2007). However, the immune function of DCs is sometimes suppressed due to some factors produced by tumors, such as IL-10, transforming growth factor-β (TGF-β), vascular endothelial growth factor (VEGF), IL-6, macrophage colony-stimulating factor (M-CSF), and prostaglandin E_2 (PGE_2) (Preynat-Seauve et al., 2006; Zou, 2005). The molecular mechanisms of DC dysfunction may be related to activation of the signal transducers and activators of transcription 3 (STAT3) and inhibition of nuclear factor kappa-B (NF-κB) (Gottfried et al., 2008; Wang et al., 2004).

Until now, the role of DCs in RB has not been explored, and the effect of RB on DCs remains unknown. In the

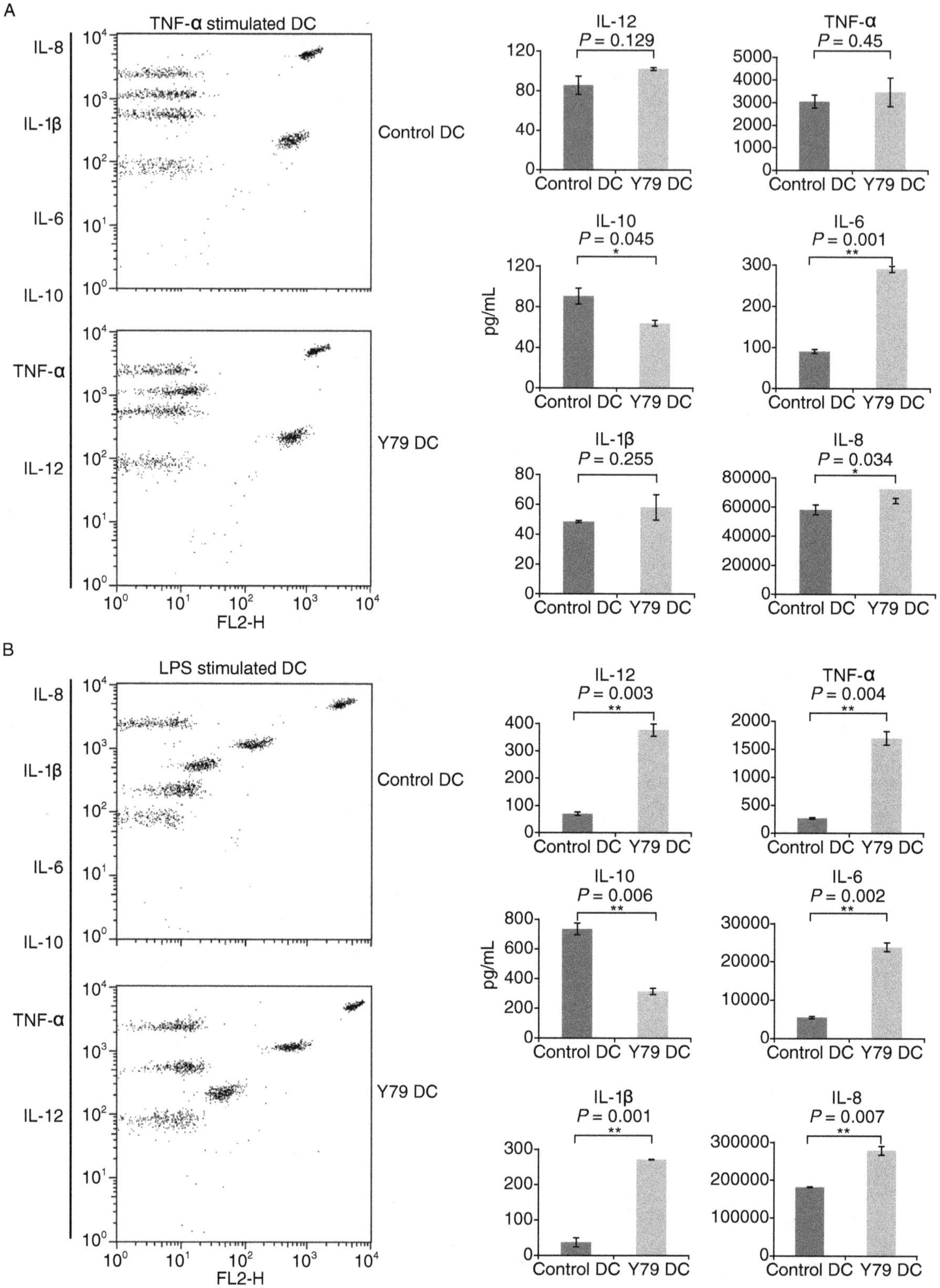

Figure 3. Production of cytokines IL-12, IL-10, TNF-α, IL-1β, IL-6, and IL-8 by RBcs-exposed DCs. Control DCs or RBcs-exposed DCs (4×10^4 cells/well) were treated with 20 ng/mL TNF-α (A) or 1 μg/mL LPS (B) for 24 h. Concentrations of cytokines in cell-free supernatants were analyzed by CBA Human Inflammation Kit that could identify all six kinds of cytokines in a single sample. Quantitative production of a particular kind of cytokine is indicated by PE-labeled-specific Ab staining. The data are mean values ± SD of triplicate determinations. Data shown are a representative experiment of three. *P < 0.05, vs. control DCs. **P < 0.01, vs. control DCs. Y79 DC: RBcs-exposed DCs.

present study, we first identified the alterations in the phenotype of DCs after treatment with RBcs. It was found that RBcs did not inhibit the maturation of DCs, based on the finding that RBcs had no effect on the expression of DC markers CD1a and CD83. Our observation is inconsistent with some other studies in which maturation of DCs is suppressed by tumor cells or tumor condition media (Bharadwaj et al., 2007; Gabrilovich et al., 1996; Ma et al., 2010; Michielsen et al., 2011; Sombroek et al., 2002), suggesting that inhibition of DCs' maturation may be tumor-specific. Because some tumor cells can induce DC apoptosis to escape immune surveillance (Kanto et al., 2001; Kiertscher et al., 2000; Ma et al., 2010), we evaluated the effect of RBcs on DC apoptosis. As shown in the present study, RBcs failed to induce DC apoptosis. There is one report showing that mature DCs can upregulate Bcl-X to resist Fas/CD95-mediated DC apoptosis (Lundqvist et al., 2002).

Our study revealed that RBcs upregulated co-stimulatory molecules CD80 and CD86 in DCs. It has been well established that these molecules are required for optimal T cell activation via binding their receptors CD28 and CD152 (CTLA4) on T cells (Carreno and Collins, 2002; Collins et al., 2005). Therefore, upregulation of CD80 and CD86 by RBcs suggests that RB cells may enhance T cell immune response through DCs. Consistent with this finding, an increase in T cell proliferation was observed when T cells were co-cultured with RBcs-exposed DCs in the present study. In addition, these T cells were found to produce more cytokines than control RBcs-untreated DCs. These results indicate that RB cells improve DCs' capacity to activate T cell, even though phenotype and function of DCs in RB tumor microenvironment still need to be further identified.

In addition to co-stimulatory molecules CD80 and CD86, more cytokines were produced by RBcs-exposed DCs, such as IL-12, an essential cytokine stimulating Th1 responses. Consistent with IL-12 overproduction, RBcs-exposed DCs induced T cells to secrete more IFN-γ. Interestingly, it was found that CD80/86 was synergistic with IL-12 for inducing T cell proliferation and IFN-γ production (Kubin et al., 1994). Some cancer models have demonstrated the anti-tumor activities of IL-12 (Colombo and Trinchieri, 2002). Endogenous IL-12 is required for resistance to transplantable tumors or chemically induced tumors, and recombinant IL-12 treatment inhibits tumor establishment or induces tumor regression (Colombo and Trinchieri, 2002). The anti-tumor mechanisms of IL-12 may also be related to other cytotoxic lymphocytes, such as NK cells, whose maturation and activation are also dependent on IL-12 (Loza and Perussia, 2001).

Contrary to upregulation of IL-12, IL-10 was downregulated in RBcs-exposed DCs. As an immunosuppressor, IL-10 inhibits the expression of MHC class molecules and co-stimulatory molecules, and prevents the production of Th1 cytokines IL-2 and IFN-γ by antigen-presenting cells (APCs). Therefore, IL-10 favors tumor evasion from immune surveillance via suppressing antigen presentation and T cell activation (Lippitz, 2013). IL-10 is overexpressed in a series of human cancers, which is associated with advanced stage and bad prognosis (Lippitz, 2013). Our study shows that RB cells may trigger immune responses by reducing IL-10 production, in addition to increasing IL-12 secretion by DCs.

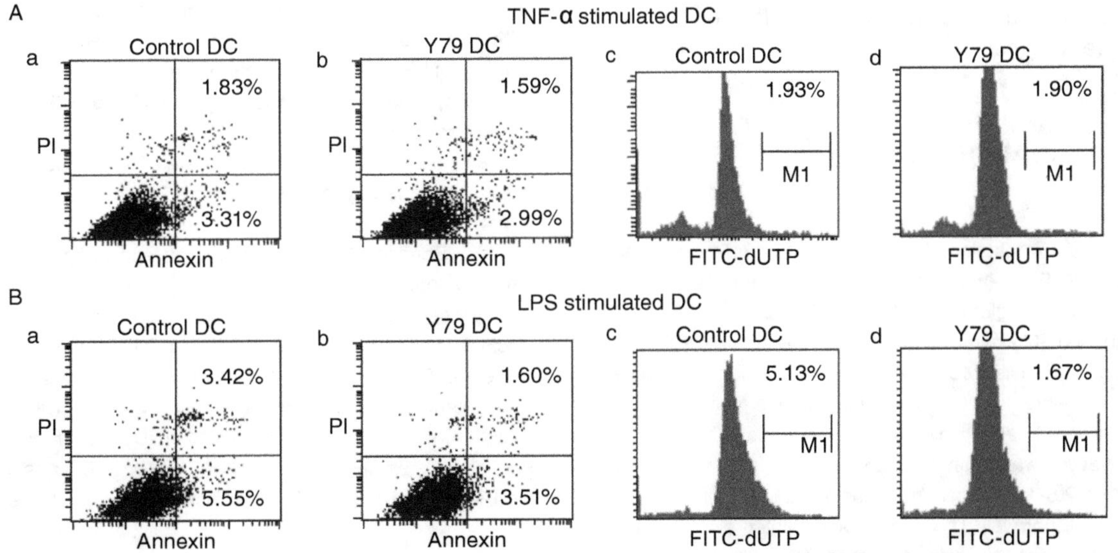

Figure 4. Apoptosis of RBcs-exposed DCs. Control DCs or RBcs-exposed DCs were treated with 20 ng/mL TNF-α (A) or 1 μg/mL LPS (B) for 24 h. DCs were stained with FITC-Annexin-V and propidium iodide (PI), and the proportion of apoptotic cells (Annexin$^+$/PI$^-$) was determined by flow cytometry (a and b). Apoptosis was also assessed by a FACS-based TUNEL assay utilizing FITC-dUTP. The proportion of apoptotic cells (dUTP$^+$) was determined by flow cytometry (c and d). Data shown are one representative experiment of three. Y79 DC: RBcs-exposed DCs.

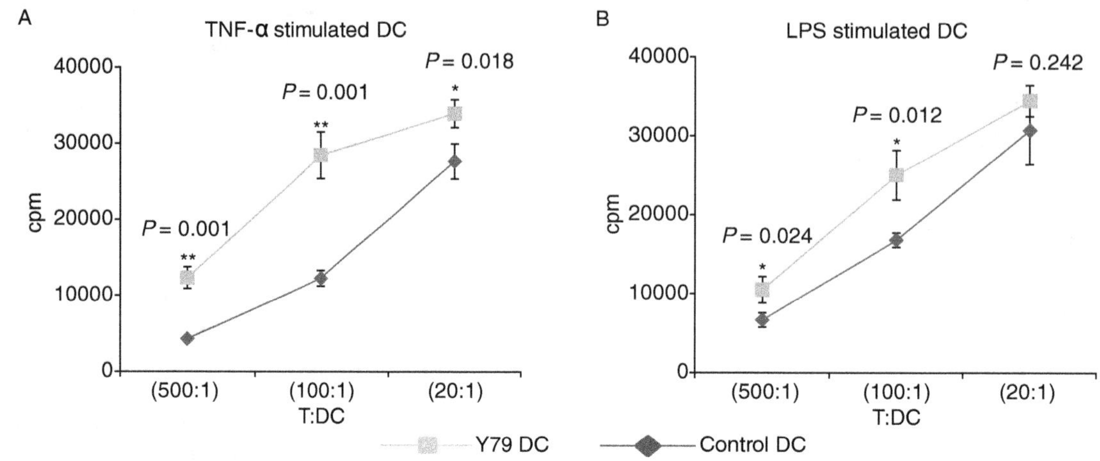

Figure 5. Enhanced allogenetic T cell proliferation stimulated by RBcs-exposed DCs. After treatment with TNF-α (A) or LPS (B) for 24 h, varying numbers (400, 2000, or 10000) of irradiated (30 Gy) DCs were added to triplicate wells containing 2 × 10^5 purified allogeneic T lymphocytes and incubated for 4 days. Cultures were pulsed during the final 8 h of incubation, and incorporation of [^{3}H]-thymidine was measured. The data are mean values ± SD of triplicate determinations. Data shown are a representative experiment of three. *P < 0.05, vs control DCs. **P < 0.01, vs control DCs. Y79 DC: RBcs-exposed DCs.

In summary, we for the first time reveal that RBcs can enhance DCs' capacity to activate T cells based on the following findings: First, RBcs upregulated the expression of co-stimulatory molecules CD80 and CD86 in DCs. Second, RBcs induced the production of IL-12p70, TNF-α, IL-6, IL-1β, and IL-8, whereas reduced the secretion of IL-10 by DCs. Third, RBcs-exposed DCs stimulated T cell proliferation. Finally, RBcs-exposed DCs induced T cells to release Th1/Th17 cytokines predominantly. These results suggest that RB cells may have an immunostimulatory effect on DCs, and immunotherapy aimed at DCs may be a potential way to treat RB.

MATERIALS AND METHODS

RB cell line and DCs

RB cell line Y79 was obtained from ATCC (HTB-18; Rockville, MD), and maintained in RPMI-1640 medium supplemented with 10 mmol/L HEPES, 2 mmol/L L-glutamine, 100 U/mL penicillin, 100 µg/mL streptomycin, and 10% heat-inactivated fetal calf serum (FCS, Sigma Chemical Co., St. Louis, MO, USA). Peripheral blood mononuclear cells (PBMCs) were isolated by Ficoll-Hypaque density gradient centrifugation from healthy individuals. To generate DCs, the mononuclear cell fraction was washed twice with RPMI-1640, suspended in RPMI-1640 at 2.5 × 10^6 cells/mL, and seeded in a 6-well plate (Becton) at 2 mL per well. The plate was incubated at 37°C for 2 h, and the non-adherent cells were discarded. The adherent cells were cultured for 6 days in 2.5 mL RPMI supplemented with 10 mmol/L HEPES, 2 mmol/L L-glutamine, 100 U/mL penicillin, 100 µg/mL streptomycin, 10% heat-inactivated FCS, 50 ng/mL rhGM-CSF, and 20 ng/mL rhIL-4 (R&D System, Minneapolis, MN). Half-volume medium exchange was performed every 3 days with medium containing fresh cytokines.

Treatment of DCs with RBcs

RBcs was prepared by seeding 10-cm dish (Falcon; BD Bioscience, Franklin Lakes, NJ, USA) with 1 × 10^7 RB cell in 10 mL of completed medium for 24 h and centrifuged to remove cell debris. On day 5 of DC culture, RBcs was added to test DCs, while the same volume of culture medium was added to control DCs. On day 6, maturation of DCs was induced by adding 20 ng/mL of TNF-α (R&D System) or 1 µg/mL of LPS (Sigma), and the phenotype of DCs was determined by flow cytometry after incubating with TNF-α or LPS for 24 h. In some experiments, before adding of TNF-α or LPS, DCs were purified on day 6. Briefly, RBcs-treated DCs or control DCs were washed extensively and then purified with microbeads on auto-MACS columns using a Blood Dendritic Cell Isolation kit (Miltenyi Biotech, BergischGladbach, Germany) according to the manufacturer's instructions. In short, isolation of DCs was performed in a two-step procedure. First, cells labeled with the Non-DC Depletion Cocktail comprising with CD14 and CD19 magnetic beads were depleted. Then DCs were positively selected by labeled with DC Enrichment Cocktail comprising with CD1c (BDCA-1), CD304 (BDCA-4/Neuropilin-1), and CD141 (BDCA-3) magnetic beads. Purified DCs samples contained >95% CD1c$^+$ DCs as evaluated by the Blood Dendritic Cell Enumeration Kit (Miltenyi Biotech).

Flow cytometry and Reagents

The following monoclonal antibodies (mAb) conjugated with either fluorescein isothiocyanate (FITC) or phycoerythrin (PE) were used: CD1a, CD83, CD40, CD80, CD86, HLA-ABC, and HLA-DR (BD

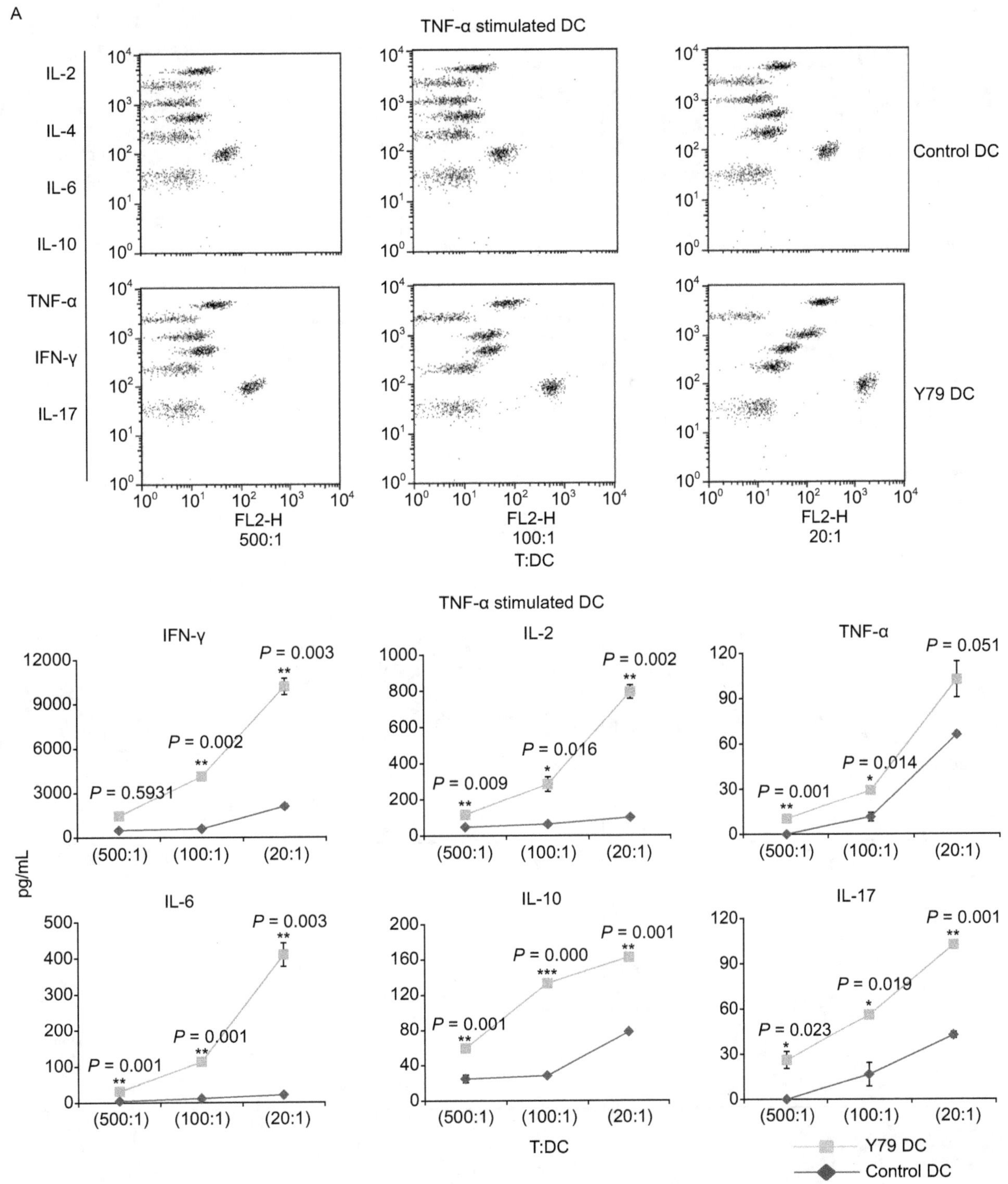

Figure 6. Increased cytokine production of by allogenetic T cell stimulated with RBcs-exposed DCs. After treatment with TNF-α (A) or LPS (B) for 24 h, varying numbers (400, 2000, or 10000) of irradiated (30 Gy) DCs were added to triplicate wells containing 2 × 10^5 purified allogeneic T lymphocytes and incubated for 3 days. The concentrations of cytokines producted in cell-free supernatants were analyzed by CBA Human Th1/Th2/Th17 Cytokine Kit that could identify all seven kinds of cytokines in a single sample. Quantitative production of a particular kind of cytokine is indicated by PE-labeled-specific Ab staining. The data are mean values ± SD of triplicate determinations. Data shown are a representative experiment of three. *P < 0.05, vs. control DCs. **P < 0.01, vs. control DCs. ***P < 0.001, vs. control DCs. Y79 DC: RBcs-exposed DCs.

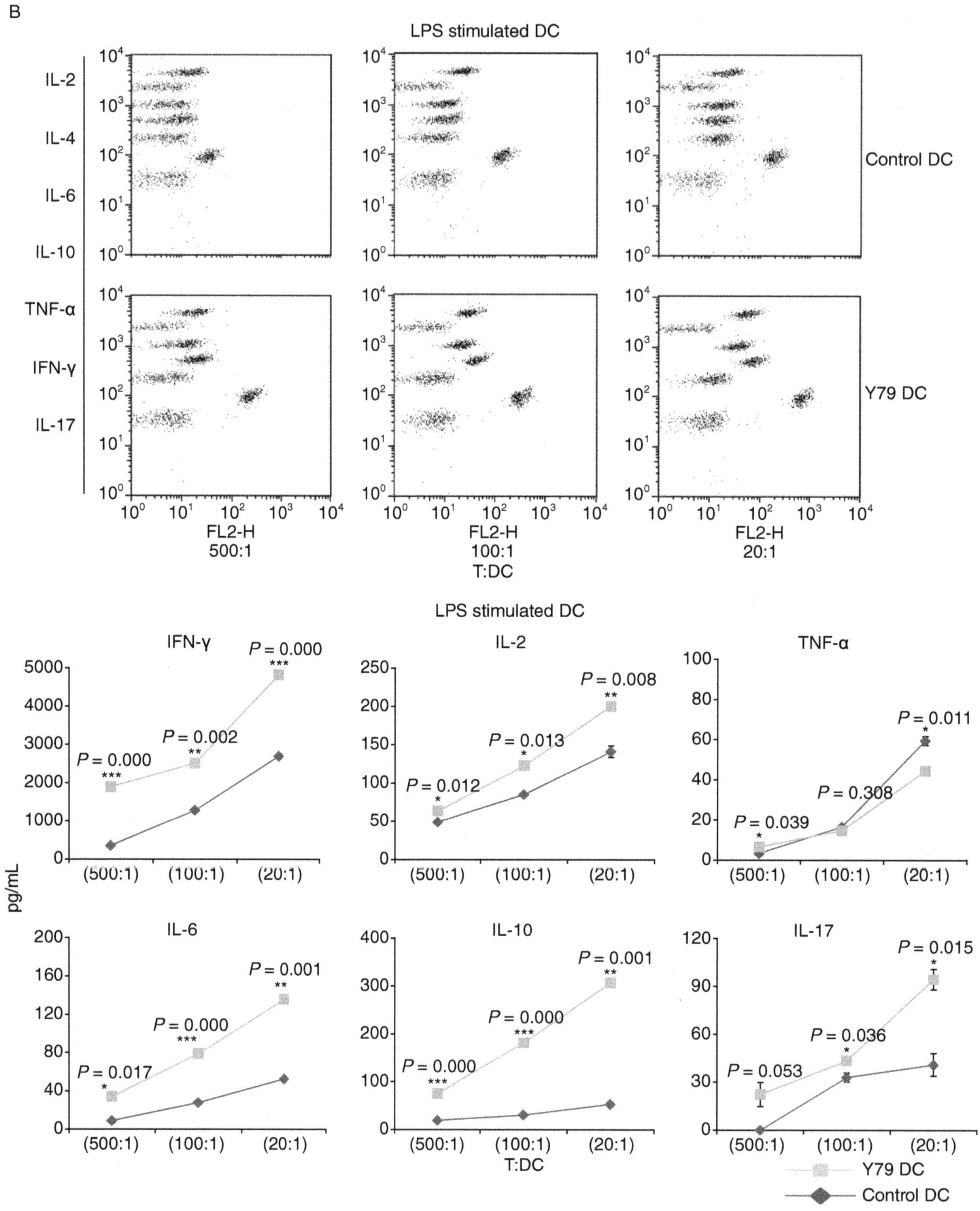

Figure 6. continued.

Pharmingen, San Diego, CA). Mouse IgG isotype control mAbs were purchased from eBioscience (San Diego, CA). To examine apoptosis, DCs were stained by FITC-Annexin-V and propidium iodide (PI) or assessed by a FACS-based TUNEL assay (APO-DIRECTTM Kit; BD Pharmingen). Samples were analyzed using a flow cytometer (FACSCalibur; BD Bioscience) and data were processed using the accompanying software (CellQuest; BD Bioscience).

DC-derived cytokine assays

Purified DCs were cultured in a flat-bottom 96-well micro-culture plate at a density of 4×10^4 cells/well in 0.2 mL of culture medium, in the presence of the TNF-α or LPS. Cell free supernatant was collected 24 h later and levels of IL-12p70, TNF-α, IL-10, IL-6, IL-1β, and IL-8 were measured by Cytometric Bead Array (CBA) assay with the Human Inflammation Kit (BD Bioscience) according to the protocols recommended by the manufacturer.

T cell proliferation assay

Heparinized blood samples were obtained from healthy individuals and PBMCs were isolated by Ficoll-Hypaque centrifugation. $CD3^+$ T cells were purified with microbeads on auto-MACS columns using a pan T cell Isolation kit (Miltenyi Biotech) according to the manufacturer's instructions. Purified $CD3^+$ T cell samples contained >98% $CD3^+$ T cells as determined by flow cytometry. DCs were irradiated with an X-irradiator (Gammacell 40 Exactor; MDS Nordion International, Inc., Ottawa, Ontario, CA) at 30G. Purified allogeneic $CD3^+$ T cell seeded into a round-bottom 96-well micro-culture plate at 2.0×10^5 per well were co-cultured with purified, stimulated, irradiated DCs at T:DC ratio of 20:1, 100:1, or 500:1. All experiments were done in triplicate, and T cells alone were used as the background control. The cultures were incubated for 4 days at 37°C in 5% CO_2 in air, pulsed with $[^3H]$-thymidine (1.0 μCi/10 μL/well) during the last 8 h of incubation, and then harvested onto glass filters with an automated cell harvester. Radioactivity was assessed by liquid scintillation spectrometry (Tomtec, Orange, CT), and expressed as counts per minute.

T cell-derived cytokine assays

Cell free supernatant was collected from T cell-DC co-culture system 72 h later and levels of IFN-γ, TNF-α, IL-10, IL-6, IL-4, IL-2, and IL-17 were measured by CBA assay with the Human Th1/Th2/Th17 Kit (BD Bioscience) according to the protocols recommended by the manufacturer.

Statistical analyses and reproducibility

Experiments were repeated at least twice and usually three times. Results were expressed as mean ± SD. Statistical analyses were performed by independent *t*-test using the computer software PEMS 3.1 for Windows software (Package for Encyclopaedia of Medical Statistics, Chengdu, China) or PASW Statistics 18. $P < 0.05$ was considered significant. Significance was denoted by an asterisk in the figures.

ACKNOWLEDGEMENTS

This work is funded by the grants from the Ministry of Science and Technology of China (S & T major Program: No. 2012ZX1004701-001-002), the National Basic Research Program (973 Program) (No. 2013CB531502), the National Nature Science Foundation of China (Grant Nos. 31170829, 31070783, and 81021003), and Beijing Natural Science Foundation (5112022).

ABBREVIATIONS

APCs, antigen-presenting cells; DCs, Dendritic cells; mAb, monoclonal antibodies; M-CSF, macrophage colony-stimulating factor; MHC, major histocompatibility complex; NF-κB, nuclear factor kappa-B; PBMCs, peripheral blood mononuclear cells; RB, retinoblastoma; RBcs, RB cell supernatant; STAT3, signal transducers and activators of transcription 3; TGF-β, transforming growth factor-β; VEGF, vascular endothelial growth factor.

COMPLIANCE WITH ETHICS GUIDELINES

Juan Ma, Huamin Han, Li Ma, Changzhen Liu, Xin Xue, Pan Ma, Xiaomei Li, and Hua Tao declare that they have no conflict of interest.

All procedures followed were in accordance with the ethical standards of the responsible committee on human experimentation (institutional and national) and with the Helsinki Declaration of 1975, as revised in 2000(5).

REFERENCES

Ataera H, Hyde E, Price KM, Stoitzner P, Ronchese F (2011) Murine melanoma-infiltrating dendritic cells are defective in antigen presenting function regardless of the presence of CD4CD25 regulatory T cells. PLoS One 6:e17515

Beckebaum S, Zhang X, Chen X, Yu Z, Frilling A, Dworacki G, Grosse-Wilde H, Broelsch CE, Gerken G, Cicinnati VR (2004) Increased levels of interleukin-10 in serum from patients with hepatocellular carcinoma correlate with profound numerical deficiencies and immature phenotype of circulating dendritic cell subsets. Clin Cancer Res 10:7260–7269

Bharadwaj U, Li M, Zhang R, Chen C, Yao Q (2007) Elevated interleukin-6 and G-CSF in human pancreatic cancer cell conditioned medium suppress dendritic cell differentiation and activation. Cancer Res 67:5479–5488

Carreno BM, Collins M (2002) The B7 family of ligands and its receptors: new pathways for costimulation and inhibition of immune responses. Annu Rev Immunol 20:29–53

Chaux P, Favre N, Martin M, Martin F (1997) Tumor-infiltrating dendritic cells are defective in their antigen-presenting function and inducible B7 expression in rats. Int J Cancer 72:619–624

Collins M, Ling V, Carreno BM (2005) The B7 family of immune-regulatory ligands. Genome Biol 6:223

Colombo MP, Trinchieri G (2002) Interleukin-12 in anti-tumor immunity and immunotherapy. Cytokine Growth Factor Rev 13:155–168

Dieu-Nosjean MC, Antoine M, Danel C, Heudes D, Wislez M, Poulot V, Rabbe N, Laurans L, Tartour E, de Chaisemartin L et al (2008) Long-term survival for patients with non-small-cell lung cancer with intratumoral lymphoid structures. J Clin Oncol 26:4410–4417

Gabrilovich DI, Chen HL, Girgis KR, Cunningham HT, Meny GM, Nadaf S, Kavanaugh D, Carbone DP (1996) Production of vascular endothelial growth factor by human tumors inhibits the functional maturation of dendritic cells. Nat Med 2:1096–1103

Gottfried E, Kreutz M, Mackensen A (2008) Tumor-induced modulation of dendritic cell function. Cytokine Growth Factor Rev 19:65–77

Houston SK, Murray TG, Wolfe SQ, Fernandes CE (2011) Current update on retinoblastoma. Int Ophthalmol Clin 51:77–91

Iwamoto M, Shinohara H, Miyamoto A, Okuzawa M, Mabuchi H, Nohara T, Gon G, Toyoda M, Tanigawa N (2003) Prognostic value of tumor-infiltrating dendritic cells expressing CD83 in human breast carcinomas. Int J Cancer 104:92–97

Kanto T, Kalinski P, Hunter OC, Lotze MT, Amoscato AA (2001) Ceramide mediates tumor-induced dendritic cell apoptosis. J Immunol 167:3773–3784

Khetan V, Mathur G, Kumar SK, Gopal L (2013) Late recurrence of tumor necessitating enucleation in an adult onset retinoblastoma. Ophthalmic Genet 34:87–89

Kiertscher SM, Luo J, Dubinett SM, Roth MD (2000) Tumors promote altered maturation and early apoptosis of monocyte-derived dendritic cells. J Immunol 164:1269–1276

Kubin M, Kamoun M, Trinchieri G (1994) Interleukin 12 synergizes with B7/CD28 interaction in inducing efficient proliferation and cytokine production of human T cells. J Exp Med 180:211–222

Ladanyi A, Kiss J, Somlai B, Gilde K, Fejos Z, Mohos A, Gaudi I, Timar J (2007) Density of DC-LAMP(+) mature dendritic cells in combination with activated T lymphocytes infiltrating primary cutaneous melanoma is a strong independent prognostic factor. Cancer Immunol Immunother 56:1459–1469

Lim DS, Kim JH, Lee DS, Yoon CH, Bae YS (2007) DC immunotherapy is highly effective for the inhibition of tumor metastasis or recurrence, although it is not efficient for the eradication of established solid tumors. Cancer Immunol Immunother 56:1817–1829

Lippitz BE (2013) Cytokine patterns in patients with cancer: a systematic review. Lancet Oncol 14:e218–e228

Loza MJ, Perussia B (2001) Final steps of natural killer cell maturation: a model for type 1 type 2 differentiation? Nat Immunol 2:917–924

Lundqvist A, Nagata T, Kiessling R, Pisa P (2002) Mature dendritic cells are protected from Fas/CD95-mediated apoptosis by upregulation of Bcl-X(L). Cancer Immunol Immunother 51:139–144

Ma J, Usui Y, Takeuchi M, Okunuki Y, Kezuka T, Zhang L, Mizota A, Goto H (2010) Human uveal melanoma cells inhibit the immunostimulatory function of dendritic cells. Exp Eye Res 91:491–499

Merchant TE, Gould CJ, Wilson MW, Hilton NE, Rodriguez-Galindo C, Haik BG (2004) Episcleral plaque brachytherapy for retinoblastoma. Pediatr Blood Cancer 43:134–139

Michielsen AJ, Hogan AE, Marry J, Tosetto M, Cox F, Hyland JM, Sheahan KD, O'Donoghue DP, Mulcahy HE, Ryan EJ et al (2011) Tumour tissue microenvironment can inhibit dendritic cell maturation in colorectal cancer. PLoS One 6:e27944

Movassagh M, Spatz A, Davoust J, Lebecque S, Romero P, Pittet M, Rimoldi D, Lienard D, Gugerli O, Ferradini L et al (2004) Selective accumulation of mature DC-Lamp+ dendritic cells in tumor sites is associated with efficient T-cell-mediated antitumor response and control of metastatic dissemination in melanoma. Cancer Res 64:2192–2198

Nakakubo Y, Miyamoto M, Cho Y, Hida Y, Oshikiri T, Suzuoki M, Hiraoka K, Itoh T, Kondo S, Katoh H (2003) Clinical significance of immune cell infiltration within gallbladder cancer. Br J Cancer 89:1736–1742

Preynat-Seauve O, Schuler P, Contassot E, Beermann F, Huard B, French LE (2006) Tumor-infiltrating dendritic cells are potent antigen-presenting cells able to activate T cells and mediate tumor rejection. J Immunol 176:61–67

Preynat-Seauve O, Contassot E, Schuler P, French LE, Huard B (2007) Melanoma-infiltrating dendritic cells induce protective antitumor responses mediated by T cells. Melanoma Res 17:169–176

Sandel MH, Dadabayev AR, Menon AG, Morreau H, Melief CJ, Offringa R, van der Burg SH, Janssen-van Rhijn CM, Ensink NG, Tollenaar RA et al (2005) Prognostic value of tumor-infiltrating dendritic cells in colorectal cancer: role of maturation status and intratumoral localization. Clin Cancer Res 11:2576–2582

Schueler AO, Jurklies C, Heimann H, Wieland R, Havers W, Bornfeld N (2003) Thermochemotherapy in hereditary retinoblastoma. Br J Ophthalmol 87:90–95

Sombroek CC, Stam AG, Masterson AJ, Lougheed SM, Schakel MJ, Meijer CJ, Pinedo HM, van den Eertwegh AJ, Scheper RJ, de Gruijl TD (2002) Prostanoids play a major role in the primary tumor-induced inhibition of dendritic cell differentiation. J Immunol 168:4333–4343

Stoitzner P, Green LK, Jung JY, Price KM, Atarea H, Kivell B, Ronchese F (2008) Inefficient presentation of tumor-derived antigen by tumor-infiltrating dendritic cells. Cancer Immunol Immunother 57:1665–1673

Treilleux I, Blay JY, Bendriss-Vermare N, Ray-Coquard I, Bachelot T, Guastalla JP, Bremond A, Goddard S, Pin JJ, Barthelemy-Dubois C et al (2004) Dendritic cell infiltration and prognosis of early stage breast cancer. Clin Cancer Res 10:7466–7474

Wang T, Niu G, Kortylewski M, Burdelya L, Shain K, Zhang S, Bhattacharya R, Gabrilovich D, Heller R, Coppola D et al (2004) Regulation of the innate and adaptive immune responses by Stat-3 signaling in tumor cells. Nat Med 10:48–54

Zou W (2005) Immunosuppressive networks in the tumour environment and their therapeutic relevance. Nat Rev Cancer 5:263–274

11

Advancing chimeric antigen receptor T cell therapy with CRISPR/Cas9

Jiangtao Ren, Yangbing Zhao

Center for Cellular Immunotherapies, Perelman School of Medicine, University of Pennsylvania, Philadelphia, PA 19104-5156, USA
Correspondence: Yangbing@upenn.edu (Y. Zhao)

ABSTRACT

The clustered regularly interspaced short palindromic repeats (CRISPR)/CRISPR-associated 9 (CRISPR/Cas9) system, an RNA-guided DNA targeting technology, is triggering a revolution in the field of biology. CRISPR/Cas9 has demonstrated great potential for genetic manipulation. In this review, we discuss the current development of CRISPR/Cas9 technologies for therapeutic applications, especially chimeric antigen receptor (CAR) T cell-based adoptive immunotherapy. Different methods used to facilitate efficient CRISPR delivery and gene editing in T cells are compared. The potential of genetic manipulation using CRISPR/Cas9 system to generate universal CAR T cells and potent T cells that are resistant to exhaustion and inhibition is explored. We also address the safety concerns associated with the use of CRISPR/Cas9 gene editing and provide potential solutions and future directions of CRISPR application in the field of CAR T cell immunotherapy. As an integration-free gene insertion method, CRISPR/Cas9 holds great promise as an efficient gene knock-in platform. Given the tremendous progress that has been made in the past few years, we believe that the CRISPR/Cas9 technology holds immense promise for advancing immunotherapy.

KEYWORDS CRISPR/Cas9, chimeric antigen receptor, T lymphocytes, adoptive immunotherapy, gene therapy

INTRODUCTION

The clustered regularly interspaced short palindromic repeats (CRISPR)/CRISPR-associated 9 (Cas9) system, a versatile RNA-guided DNA targeting technology, is triggering a revolution in the field of biology. CRISPR/Cas9 has demonstrated great potential for genetic manipulation, even in previously difficult contexts. Here, we review the current development of CRISPR/Cas9 technologies for therapeutic applications, especially chimeric antigen receptor (CAR) T cell-based adoptive immunotherapy. We compare the different methods used to facilitate efficient CRISPR delivery and gene editing in T cells. We explore the potential for genetic manipulation using the CRISPR/Cas9 system to generate universal CAR T cells and potent T cells resistant to exhaustion and inhibition. We discuss safety concerns regarding the specificity and future directions of CRISPR in the field of CAR T cell immunotherapy.

THE CRISPR/CAS9 SYSTEM

The field of genome editing is evolving rapidly. Until a decade ago, zinc-finger nucleases (ZFNs) were the only practical option available for targeted genome editing (Bibikova et al., 2002, 2003; Porteus and Baltimore, 2003; Urnov et al., 2005; Morton et al., 2006; Doyon et al., 2008; Kim et al., 2009; Townsend et al., 2009). Zinc finger proteins recognize target DNA in a modular fashion: each protein consists of at least three zinc finger domains, and a single zinc finger domain interacts with a 3-bp sequence, making them ideal programmable sequence-specific DNA-binding proteins (Pavletich and Pabo, 1991). In 2011, transcription activator-like effector nucleases (TALENs) emerged as a competitive alternative to ZFNs (Boch et al., 2009; Moscou and Bogdanove, 2009; Cermak et al., 2011; Miller et al., 2011; Briggs et al., 2012). Unlike zinc fingers, each repeat domain in TALE proteins recognizes a single base. Four different repeat domains can be mixed and matched to create new DNA-binding proteins, which can be linked to the FokI domain to create a new class of programmable target DNA nucleases (Miller et al., 2011). These molecules enable precise targeting and cutting at a specific genomic locus to generate double-strand breaks (DSBs) followed by nonhomologous end joining (NHEJ) or homology-directed repair (HDR)-mediated repair, thereby enabling precise genome editing. Studies using these two classes of nucleases have led to important scientific discoveries and therapeutic development. In fact, a ZFN-based treatment of HIV that disables the HIV co-receptor C-C chemokine receptor type 5 (CCR5) in human primary T cells is currently in clinical trials and has shown great promise (Perez et al., 2008; Tebas et al., 2014). However, the recognition of the target DNA sequence by these protein-based genome engineering systems is determined by protein sequences. Tedious and complex protein engineering and optimization are therefore required for each specific target DNA sequence, and delivering many of these proteins into cells for simultaneous multiplexed genetic manipulation is challenging. Given these difficulties, their use for large-scale genomic manipulation or genetic screens has been limited.

The CRISPR/Cas9 technology originates from type II CRISPR/Cas9 systems, which provide bacteria with adaptive immunity to viruses, plasmids, and other foreign nucleic acids (Barrangou et al., 2007; Horvath and Barrangou, 2010; Wiedenheft et al., 2012). Type II CRISPR systems incorporate sequences from invading DNA between CRISPR repeat sequences that are encoded as arrays within the bacterial host genome. Transcripts from the CRISPR repeat arrays are processed into CRISPR RNAs (crRNAs) (Deltcheva et al., 2011), each containing a variable sequence transcribed from the invading DNA, which is known as the "protospacer" sequence, and part of the CRISPR repeat. Each crRNA hybridizes with a second RNA, which is known as the transactivating CRISPR RNA (tracrRNA) (Deltcheva et al., 2011), and these two RNAs form a complex with the Cas9 DNA endonuclease (Jinek et al., 2012). The protospacer-encoded portion of the crRNA guides Cas9 to complementary target DNA sequences and cleaves the DNA if they are adjacent to short sequences known as protospacer adjacent motifs (PAMs). The type II CRISPR system from Streptococcus pyogenes has been adapted for inducing sequence-specific DSBs and targeted genome editing. In 2012, Jinek et al. first demonstrated that the Cas9 protein from Streptococcus pyogenes (SpCas9) can bind with a tracrRNA-crRNA RNA complex to induce DSBs *in vitro* at a target DNA sequence by Watson-Crick base pairing of crRNA and target DNA (Jinek et al., 2012). This study also showed that directing Cas9 to bind and cleave a specific DNA sequence did not require an RNA complex. The process can be simply achieved by using a designed, chimeric single guide RNA (sgRNA). In 2013, two groups from MIT and Harvard demonstrated the feasibility of genome editing of human cells using the CRISPR/Cas9 system (Cong et al., 2013; Mali et al., 2013b). These discoveries paved the way and opened the era for the use of CRISPR/Cas9 in genome engineering, including gene editing and gene expression regulation, epigenetic modification, and genome imaging (Cheng et al., 2013; DiCarlo et al., 2013; Gilbert et al., 2013; Hwang et al., 2013; Li et al., 2013; Maeder et al., 2013; Nekrasov et al., 2013; Perez-Pinera et al., 2013; Qi et al., 2013; Shen et al., 2013; Wang et al., 2013; Tanenbaum et al., 2014; Chavez et al., 2015; Hilton et al., 2015; Kearns et al., 2015; Konermann et al., 2015).

GENE EDITING AND THERAPEUTIC APPLICATION OF CRISPR/CAS9 IN HUMAN T CELLS

In addition to generating powerful research tools, genome editing with CRISPR/Cas9 technology holds great promise as a means to produce therapeutic agents or as a therapeutic itself. Although we focus on SpCas9, particularly its use in therapeutic applications and the development of next-generation transformational drugs in T cells, the general outline described here applies to the larger ensemble of CRISPR/Cas9 tools.

The therapeutic potential of CRISPR/Cas9 has already been demonstrated in many aspects. CRISPR/Cas9 has been applied as an antimicrobial agent and has been developed to specifically target antibiotic resistance in highly virulent strains of bacteria (Makarova et al., 2006). Gene therapy applications have also been tested for monogenic diseases. A CFTR gene defect was repaired in cells from human patients with cystic fibrosis *in vitro* in cultured intestinal stem cell organoids using CRISPR-Cas (Schwank et al., 2013). Correction of the defective gene causing hereditary tyrosinaemia was performed in mice after the hydrodynamic injection of CRISPR components. This application led to an expansion of mutation-corrected hepatocytes *in vivo* and resulted in a rescued phenotype in adult mice (Yin et al., 2014). Advancing from the described therapeutic treatment to preventative techniques, muscular dystrophy was prevented via germ line gene editing (Long et al., 2014). The use of CRISPR/Cas9 to treat viral infections, such as HIV and hepatitis B, has also been demonstrated (Zhen et al., 2015). IPSC resistant to HIV-1 was also generated through genome editing (Hu et al., 2014; Ye et al., 2014).

The application of genome editing for therapeutic purpose has begun to overlap with the rapidly evolving field of cancer immunotherapy, particularly for the production of next-generation chimeric antigen receptor (CAR) T cells. These modified T cells armed with tumour-targeting receptors have demonstrated great promise in clinical trials treating various leukaemias and lymphomas and may eventually be used to treat solid cancers (Maus et al., 2014). CARs comprise an extracellular single-chain variable fragment (ScFv) specific to an antigen on tumour cells and an intracellular chimeric signalling domain that drives T cell activation and the killing of tumour cells (Gross et al., 1989; Irving and Weiss, 1991; Maher et al., 2002; Brentjens et al., 2003; Carpenito et al., 2009). To date, the best CAR T cell therapy involves targeting CD19, an antigen expressed by B cells and B cell malignancies. Several other CAR T therapies targeting solid tumours antigens, such as Her2/neu, Mesothelin cMet, GD2, interleukin-13 receptor alpha 2 (IL13Rα2), CEA, and EGFR, are currently under evaluation in different phases of clinical trials.

Currently, most CAR T clinical trials utilize autologous T cells and might therefore be hampered by the poor quality and quantity of T cells and by the time and expense of manufacturing autologous T cell products. CAR T cell therapy could substantially benefit from allogeneic universal donor T cells, as "off-the-shelf" cells could greatly increase the number of patients who could be treated by a single CAR T cell product. However, endogenous TCR on allogeneic T cells may recognize the alloantigens of the recipient, leading to graft-versus-host disease (GVHD); furthermore, the expression of HLA on the surface of allogeneic T cells causes rapid rejection by the host immune system. In this context, ZFNs and TALENs have been used to knock out endogenous T cell receptor genes in T cells, which could prevent unwanted graft-versus-host reactivity (Provasi et al., 2012; Torikai et al., 2012; Poirot et al., 2015). Genome-editing strategies could also be used to prevent or delay the rejection of CAR T cells by the recipient's immune system by eliminating or decreasing the expression of histocompatibility antigens on the donor T cells. Future CAR T cell therapies could benefit from combined modification of endogenous TCR genes, histocompatibility genes, and components of signalling pathways. In a previous study, we reported the use of the CRISPR/Cas9 system to simultaneously disrupt multiple genomic loci. CAR T cells deficient in the expression of endogenous T cell receptor (TCR) and HLA class I (HLA-I) were generated that can be used as universal CAR T cells (Ren et al., 2016).

In addition to enabling the generation of universal CAR T cells, genome editing could be used to enhance CAR T cell function by ablating the genes encoding T cell inhibitory receptors or signalling molecules, such as programmed cell death protein 1 (PD1) or cytotoxic T lymphocyte-associated protein 4 (CTLA4) (Lloyd et al., 2013; Hoos, 2016; Su et al., 2016). Indeed, a clinical trial has recently been approved by the US National Institutes of Health (NIH) Recombinant DNA Advisory Committee (RAC) that will be conducted at the University of Pennsylvania. In this clinical trial, PD1 and the endogenous TCR will be knocked out by CRISPR/Cas9 in NY-ESO-1 TCR transduced T cells. The first clinical trial of CRISPR/Cas9 has been initiated. The trial uses CRISPR/Cas9 to knock out PD1 in T cells of patients with lung cancer; however, CAR or TCR will not be introduced into T cells in this trial (Cyranoski, 2016). Similar trials with PD1-knockout autologous T cells for prostate (NCT02867345), bladder cancer (NCT02863913), and renal cell carcinoma (NCT02867332) are also being initiated. Scientists are seeking to introduce CAR via HDR to eliminate the need to randomly integrate viral delivery systems and control where CAR integrates (Sadelain et al., 2011; Kalos and June, 2013). Notably, simply ablating inhibitory molecules can be a double-edged sword. It is important to investigate whether the removal of some inhibitory signals from the T cells leads to the uncontrolled proliferation of cells or to severe autoimmunity.

DELIVERY OF CRISPR/CAS9 IN T CELLS

The CRISPR/Cas9 system can be directly applied to human cells by transfection with a plasmid that encodes Cas9 and sgRNA (Cong et al., 2013). The viral delivery of CRISPR components has been extensively demonstrated using lentivirus and retrovirus (Shalem et al., 2014; Williams et al., 2016). Gene editing with CRISPR encoded by non-integrating virus, such as adenovirus and adenovirus-associated virus (AAV), has also been reported (Ran et al., 2015; Swiech et al., 2015). Recent discoveries of smaller Cas proteins have enabled and enhanced the combination of this technology with vectors that have gained increasing success for their safety profile and efficiency, such as AAV vectors. Due to their relatively low immunogenicity, AAVs are commonly chosen for *in vivo* gene delivery in somatic gene therapy (Friedland et al., 2015; Ran et al., 2015). CRISPR delivery via Cas9 ribonucleoproteins (RNP) also exhibited efficient gene editing in human cells (Kim et al., 2014).

Gene disruption in T cells had been achieved by lentiviral and adenoviral delivery of CRISPR components into primary T cells. However, these methods cannot site-specifically insert and occasionally disrupt essential genetic elements, and the gene disruption efficiency was not very high (Wang et al., 2014; Li et al., 2015). Recently, a Jurkat T cell-based lentiviral CRISPR toolbox was developed to facilitate the research on T cell function. Given the flexible and easy-to-handle features of Jurkat T cells, and programmability with different Cas9 variants, the toolbox might serve as a useful platform for the study of T cell signal transductions (Chi et al., 2016). Although gene ablation in T cells with DNA nucleofection of CRISPR reagents was also achieved, DNA nucleofection is associated with high toxicity to T cells, which represents a major difficulty for its application (Mandal et al., 2014; Su et al., 2016). Schumann et al. reported the site-specific genome editing of primary human T cells using Cas9 RNPs (Schumann et al., 2015). Cas9 RNPs are recombinant Cas9 proteins complexed with *in vitro*-transcribed sgRNAs. Cas9 RNPs delivered via electroporation efficiently ablate CXCR4 and PD-1 expression in CD4 T cells by introducing insertions or deletions (indels) in the targeted region. Furthermore, the inclusion of a HDR template successfully introduced exogenous DNA into the genome at the Cas9 cleavage site. Deep sequencing results indicated that up to 55% of the treated cells contained indels in the targeted region, with 20% of the cells incorporating the exogenous DNA sequence introduced through the HDR template.

The work by Schumann et al. is joined by other recent reports focused on primary human T cell gene editing using CRISPR/Cas9. Hendel et al. reported the disruption of CCR5 locus in T cells by co-delivering Cas9 mRNA or protein with chemically modified sgRNAs via electroporation, achieving up to 49% target mutagenesis in activated primary human T cells calculated by tracking of indels by decomposition (TIDE) analysis (Hendel et al., 2015).

A major challenge in primary T-cell engineering lies in the limited time frame in which genetic manipulation can be accomplished with high efficiency. Unstimulated primary or naïve T cells are significantly less receptive to exogenous nucleic acid or protein uptake compared with stimulated T cells (Hendel et al., 2015). By contrast, repeated stimulation will lead to T-cell exhaustion and decrease its anti-tumour efficacy. Therefore, protocol optimization will be required to enable the effective application of multiple genetic manipulation techniques on one T-cell product. To this end, we recently reported a method to incorporate disrupting endogenous genes into standard clinical CAR T cell manufacturing processes (Ren et al., 2016). Freshly isolated human T cells were stimulated via anti- CD3/CD28 beads and lentivirally transduced at 1 day post-stimulation to stably express a CAR transgene, and T cells were electroporated at days 3 and 4 with RNA-encoding Cas9 and sgRNA to disrupt TCR, HLA-I, and PD1 simultaneously. The editing efficiency using this combined protocol was donor dependent, with the results indicating >70% CAR transduction efficiency and >60% double-knockout efficiency in most production runs. This manufacturing procedure yielded CAR T cells that were specific to CD19 targets, resistant to host rejection, and incapable of triggering GVHD, thus highlighting the ability to generate multi-functional universal CAR T cells with CRISPR/Cas9 techniques. Similar results were also reported using another approach of CRISPR/Cas9 RNPs targeting the same 3 genes: TCR, B2m, and PD1 (Liu et al., 2017). Compared to multiple deliveries of sgRNAs, multiplex genome editing with Cas9 RNP in T cells reduced the toxicity associated with RNA electroporations at the cost of decreased gene targeting efficiency. The authors used multiple sgRNAs targeting the same gene to improve gene disruption efficiency, which may potentially increase the off-target effects. To further improve the application of multiplex genome editing and to reduce the toxicity associated with multiple electroporation, we developed a One-shot CRISPR system, by incorporation of multiple gRNAs in a CAR lentiviral vector. Efficient multiple gene modification can be achieved by a single electroporation of various Cas9 mRNAs (Ren et al., 2017). A brief summary of various delivery methods of CRISPR/Cas9 components into T cells is schematically presented in Fig. 1. A comparison between different methods for CRISPR gene editing in T cell is presented in Table 1.

These Cas9-based gene-editing techniques will enable the disruption of a wide variety of target genes, including endogenous TCR, the checkpoint receptors PD1 or CTLA-4 in tumour-targeting T cells or the virally targeted chemokine receptors CCR5 and CXCR4 in T cells of HIV patients. The insertion of exogenous DNA sequences using HDR templates demonstrated by Schumann et al. further opens the possibility of precisely integrating transgenic elements, thereby reducing the risk of gene integration in oncogenic sites caused by viruses. However, the current efficiency of site-specific gene knock-in by nuclease-mediated homologous recombination is not comparable to that of standard viral transduction. Thus, the next step for gene editing in T cells will be to develop protocols that enable the combinatorial application of various gene-editing techniques in primary T cells.

SAFETY ISSUES AND CONCERNS

The off-target activities of Cas9 can be measured by directly assessing the potential off-target genomic DNA sites defined by the sequences that have 1–6 nucleotide (nt) differences to the intended target sequence. A given 20-nt target sequence might have hundreds of such potential off-targets within the human genome. The T7 Endonuclease I (T7E1) mutation mismatch assay is commonly used to detect high indel frequencies (>2%–5%). As a result, more sensitive deep sequencing assays are needed to identify lower frequency off-target mutations (Cho et al., 2014; Fu et al., 2014). Exome sequencing was also used for off-target analysis, but the high false negative result rate associated with exome sequencing analysis limits its interpretation of the results. However, accurately predicting the off-target cleavage sites remains a challenge because it is typically biased, given the inability of most available algorithms for off-target prediction to cover all potential off-target sites. The genome-wide detection of DSBs provides a non-biased method to assess the specificity of Cas9-mediated DNA cleavage, and several methods have been developed to meet this purpose. In one method called genome-wide unbiased identification of DSBs enabled by sequencing (GUIDE-Seq), the Cas9-sgRNA induced DSBs are tagged in the genomes of living cells by introducing a blunt, double-stranded oligodeoxynucleotide during the end-joining process following a DSB. The double-stranded oligodeoxynucleotide integration sites are then amplified and deep sequenced (Tsai et al., 2015). A modified high-throughput, genome-wide translocation sequencing (HTGTS) was developed based on linear-amplification-mediated PCR (LAM-PCR HTGTS). LAM-PCR HTGTS enables the detection of DSBs based on translocation to other endogenous or ectopic DSBs using the target DSB as "bait" to capture the "prey" sequences translocated to the target DSB (Frock et al., 2015). A third method, called *in situ* breaks labelling, enrichments on streptavidin, and next generation sequencing (BLESS), captures biotinylated oligonucleotides labelled DSBs in fixed cells using streptavidin. Enriched DSB-containing DNA fragments are PCR amplified and analysed by deep sequencing (Crosetto et al., 2013; Ran et al., 2015). A fourth method called digested genome sequencing (Digenome-Seq) uses isolated genomic DNA for *in vitro* Cas9-mediated digestion followed by whole-genome sequencing to evaluate genome-wide Cas9 off-target effects (Kim et al., 2015).

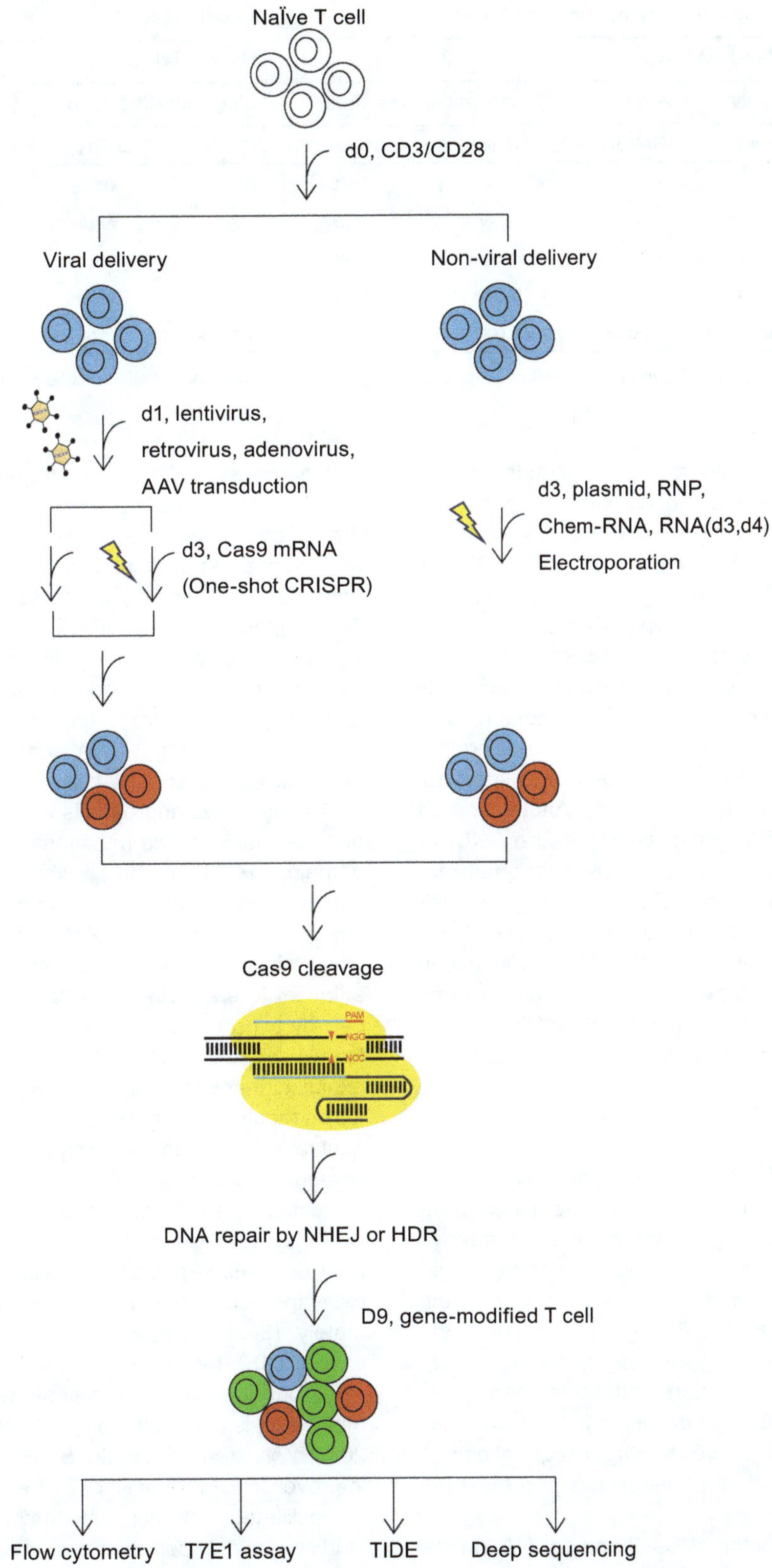

Figure 1. A brief summary of various delivery methods of CRISPR/Cas9 components into T cells.

Table 1. Comparison of various delivery methods of CRISPR/Cas for T cell gene editing

	Viral-delivery				Non-viral-delivery			
	Integrating virus		Integrating-free virus		Single electroporation		Double electroporation	
	Lentivirus	Retrovirus	Adenovirus	AAV	Plasmid	RNP	Chem-RNA	RNA
Integration	Yes	Yes	No	No	Rear	No	No	No
Efficiency	Low	Low	Low	Low	Low	Medium	Medium	High
Electroporation	No	No	No	No	Yes	Yes	Yes	Yes
Toxicity	Medium	Medium	Low	Low	High	Low	Low	Medium
Off-target effects	Medium	Medium	Low	Low	Medium	Rear	Low	Rear

RNP: Cas9 protein-sgRNA complex; Chem-RNA: Cas9 protein or mRNA complexed with chemically modified sgRNA; RNA: Cas9 mRNA and sgRNA.

CRISPR/Cas9 generally disrupts their intended target sites reliably; however, an important question to consider is to what extent these nucleases induce off-target cleavage events, especially in therapeutic application. CRISPR/Cas9 gene editing generates off-target mutations depending upon the experimental setting and cell type (Cho et al., 2014). Pankaj et al. reported an extremely low incidence of off-target mutagenesis of CRISPR in hematopoietic stem cells (Mandal et al., 2014). Recent studies also demonstrated a low incidence of off-target mutagenesis in T cells using lentivirus and adenovirus-delivered CRISPR/Cas9 to knockout CCR5 (Wang et al., 2014; Li et al., 2015). Another report showed no detectable off-target mutations in the CXCR4-knockout CD4 T cells (Hou et al., 2015). We reported rare off-target mutagenesis targeting TRAC or TRBC with Cas9 (Ren et al., 2016). Although these studies all suggest that T cells might be minimally tolerable to CRISPR/Cas9-induced off-target mutagenesis, non-biased strategies combined with deep sequencing for off-target detections should be applied to the selected target gRNA used in the clinical trials.

FUTURE DIRECTIONS

Reducing the off-target effects for safe therapeutic application of CRISPR/Cas9 in immunotherapy remains unresolved. Various approaches have been explored to improve the specificity of CRISPR/Cas9. The choice of proper target sequence is the first and most effective option to improve the specificity. All published studies have suggested that the CRISPR/Cas9 mediated off-target mutagenesis could vary depending on the sgRNA design and target sequence. Predictive algorithms have been developed to facilitate this process by computationally searching target sequences that bear the least similarities to other sequences to reduce the off-target effects.

Precisely tuning the amount of Cas9 and sgRNA in cells is also used to improve the specificity; some studies have demonstrated that a decrease in the amount of CRISPR reagents in cells could reduce off-target effects (Fu et al., 2013; Hsu et al., 2013). Timely controlled Cas9 expression is also demonstrated through a tet-on system (Gonzalez et al., 2014). Furthermore, gene editing with the Cas9 mRNA and protein causes fewer off-target effects compared to plasmids and viruses, likely because the mRNA or RNPs were rapidly degraded after immediate on-target cleavage (Kim et al., 2014). Modifying the sgRNA sequence also improved the specificity. For example, sgRNA with a truncated base-pairing sequence (17 nt instead of 20 nt) enhanced the targeting specificity because truncated sgRNAs have reduced binding affinity with the target DNA and thus are more sensitive to mismatches (Fu et al., 2014).

An alternative approach is to take advantage of the Cas9 nickase that contains mutations in one of the two nuclease domains, HNH or RuvC, which cleave the DNA strand complementary and noncomplementary (respectively) to the sgRNA (Gasiunas et al., 2012; Jinek et al., 2012). A pair of Cas9 nickases could generate two single-strand breaks adjacent to each other on opposite DNA strands when guided by two properly designed sgRNAs (Mali et al., 2013a; Ran et al., 2013; Cho et al., 2014). The paired nickases exhibit higher specificity in editing because the generation of DSBs requires two independent binding events, whereas the nuclease Cas9 requires only one binding event. A similar strategy is to fuse DNA-endonuclease-dead Cas9 (dCas9) to the dimerizing FokI nuclease. The dCas9-FokI fusion is an RNA-guided nuclease that cleaves DNA only when a pair of FokI domains is sufficiently close to form a dimer. Efficient cleavage occurs when two target sites are spaced approximately 13–25 bp apart (Guilinger et al., 2014; Wyvekens et al., 2015). Moreover, because the FokI nuclease activity relies on dimerization, this strategy also reduced unwanted mutagenesis compared to the Cas9 nickase (Guilinger et al., 2014; Wyvekens et al., 2015). However, these approaches improve CRISPR specificity at the cost of reduced efficiency.

Facilitated by the crystal structure of SpCas9, two recent studies have reported more precise genomic edits with rationally engineered CRISPR/Cas9 systems. Slaymaker et al. created systematic single or multiple mutations in the positively charged residues that are predicted to be involved in the interaction with the non-target strand of the target DNA and identified Cas9 mutants that decrease off-target effects without impairing on-target activity (Slaymaker et al., 2016). Using a similar approach, Kleinstiver et al. made a quadruple amino acid substituted SpCas9 that retains high on-target activity with minimal off-target activity. Application and further exploration of high-fidelity Cas9 variants will increase the reliability of CRISPR/Cas9 as both a research tool and a therapeutic approach.

As an integration-free gene insertion method, CRISPR/Cas9 holds great promise as an efficient gene knock-in platform. Conventional CAR expression in T cells requires randomly integrating viral delivery vectors, including lentivirus and retrovirus. However, uncontrolled virus integration in host cell genomes has the potential risk of causing insertional mutagenesis. CRISPR/Cas9 mediates efficient gene knock-in in human cells, embryos, and plants. Schumann et al. reported successful introduction of exogenous DNA into T cells genome at the Cas9 cleavage site. Although the knock-in of large fragments, such as a CAR in T cells, remains challenging, doing so is desirable for therapeutic application and needs further exploration. Sather et al. reported the feasibility of knock-in of a CAR transgene in T cells with MegaTal and an AAV HDR template (Sather et al., 2015). A more recent study showed that targeting a CAR to the TRAC locus greatly enhanced the antitumor activity by reducing tonic activation (Eyquem et al., 2017). Microhomology Mediated End Joining (MMEJ) has been used for gene knock-in in human cells and animals wherein large homology arm is not required, thus facilitating gene delivery with small inserts (Nakade et al., 2014; Sakuma et al., 2015). Previously, some studies have highlighted the possibility of knock-in of large gene cassettes using homology-independent targeted integration strategy, which enables robust DNA knock-in in both dividing and non-dividing cells *in vitro* and, more importantly, *in vivo* (Auer et al., 2014; He et al., 2016; Suzuki et al., 2016).

Given the tremendous progress that has been made in the past several years, we believe that the CRISPR/Cas9 technology holds immense promise for advancing immunotherapy.

ACKNOWLEDGMENTS

This work was supported by a US National Institutes of Health (NIH) Grants YZ (2R01CA120409).

ABBREVIATIONS

AAV, adenovirus-associated virus; CAR, chimeric antigen receptor; CRISPR, clustered regularly interspaced short palindromic repeats; Cas9, CRISPR-associated 9; CTLA4, cytotoxic T lymphocyte-associated protein 4; DSBs, double-strand breaks; GVHD, graft-versus-host disease; HDR, homology-directed repair; HTGTS, high-throughput, genome-wide translocation sequencing; IL13Rα2, interleukin-13 receptor alpha 2; NHEJ, nonhomologous end joining; nt, nucleotide; PAMs, protospacer adjacent motifs; PD1, programmed cell death protein 1; RNP, ribonucleoproteins; ScFv, single-chain variable fragment; sgRNA, single guide RNA; TALENs, transcription activator-like effector nucleases; ZFNs, zinc-finger nucleases

COMPLIANCE WITH ETHICS GUIDELINES

Yangbing Zhao has financial interests due to intellectual property and patents in the field of cell and gene therapy. Conflicts of interest are managed in accordance with University of Pennsylvania policy and oversight. Jiangtao Ren declares that he has no conflicts of interest.

This article does not contain any studies with human or animal subjects performed by the any of the authors.

REFERENCES

Auer TO, Duroure K, De Cian A, Concordet JP, Del Bene F (2014) Highly efficient CRISPR/Cas9-mediated knock-in in zebrafish by homology-independent DNA repair. Genome Res 24:142–153

Barrangou R, Fremaux C, Deveau H, Richards M, Boyaval P, Moineau S, Romero DA, Horvath P (2007) CRISPR provides acquired resistance against viruses in prokaryotes. Science 315:1709–1712

Bibikova M, Golic M, Golic KG, Carroll D (2002) Targeted chromosomal cleavage and mutagenesis in *Drosophila* using zinc-finger nucleases. Genetics 161:1169–1175

Bibikova M, Beumer K, Trautman JK, Carroll D (2003) Enhancing gene targeting with designed zinc finger nucleases. Science 300:764

Boch J, Scholze H, Schornack S, Landgraf A, Hahn S, Kay S, Lahaye T, Nickstadt A, Bonas U (2009) Breaking the code of DNA binding specificity of TAL-type III effectors. Science 326:1509–1512

Brentjens RJ, Latouche JB, Santos E, Marti F, Gong MC, Lyddane C, King PD, Larson S, Weiss M, Riviere I et al (2003) Eradication of systemic B-cell tumors by genetically targeted human T lymphocytes co-stimulated by CD80 and interleukin-15. Nat Med 9:279–286

Briggs AW, Rios X, Chari R, Yang L, Zhang F, Mali P, Church GM (2012) Iterative capped assembly: rapid and scalable synthesis of repeat-module DNA such as TAL effectors from individual monomers. Nucleic Acids Res 40:e117

Carpenito C, Milone MC, Hassan R, Simonet JC, Lakhal M, Suhoski MM, Varela-Rohena A, Haines KM, Heitjan DF, Albelda SM et al (2009) Control of large, established tumor xenografts with genetically retargeted human T cells containing CD28 and CD137 domains. Proc Natl Acad Sci USA 106:3360–3365

Cermak T, Doyle EL, Christian M, Wang L, Zhang Y, Schmidt C, Baller JA, Somia NV, Bogdanove AJ, Voytas DF (2011) Efficient design and assembly of custom TALEN and other TAL effector-based constructs for DNA targeting. Nucleic Acids Res 39:e82

Chavez A, Scheiman J, Vora S, Pruitt BW, Tuttle M, Iyer PRE, Lin S, Kiani S, Guzman CD, Wiegand DJ et al (2015) Highly efficient Cas9-mediated transcriptional programming. Nat Methods 12:326–328

Cheng AW, Wang H, Yang H, Shi L, Katz Y, Theunissen TW, Rangarajan S, Shivalila CS, Dadon DB, Jaenisch R (2013) Multiplexed activation of endogenous genes by CRISPR-on, an RNA-guided transcriptional activator system. Cell Res 23:1163–1171

Chi S, Weiss A, Wang H (2016) A CRISPR-based toolbox for studying T cell signal transduction. Biomed Res Int 2016:5052369

Cho SW, Kim S, Kim Y, Kweon J, Kim HS, Bae S, Kim JS (2014) Analysis of off-target effects of CRISPR/Cas-derived RNA-guided endonucleases and nickases. Genome Res 24:132–141

Cong L, Ran FA, Cox D, Lin S, Barretto R, Habib N, Hsu PD, Wu X, Jiang W, Marraffini LA et al (2013) Multiplex genome engineering using CRISPR/Cas systems. Science 339:819–823

Crosetto N, Mitra A, Silva MJ, Bienko M, Dojer N, Wang Q, Karaca E, Chiarle R, Skrzypczak M, Ginalski K et al (2013) Nucleotide-resolution DNA double-strand break mapping by next-generation sequencing. Nat Methods 10:361–365

Cyranoski D (2016) CRISPR gene-editing tested in a person for the first time. Nature 539:479

Deltcheva E, Chylinski K, Sharma CM, Gonzales K, Chao Y, Pirzada ZA, Eckert MR, Vogel J, Charpentier E (2011) CRISPR RNA maturation by *trans*-encoded small RNA and host factor RNase III. Nature 471:602–607

DiCarlo JE, Norville JE, Mali P, Rios X, Aach J, Church GM (2013) Genome engineering in *Saccharomyces cerevisiae* using CRISPR-Cas systems. Nucleic Acids Res 41:4336–4343

Doyon Y, McCammon JM, Miller JC, Faraji F, Ngo C, Katibah GE, Amora R, Hocking TD, Zhang L, Rebar EJ et al (2008) Heritable targeted gene disruption in zebrafish using designed zinc-finger nucleases. Nat Biotechnol 26:702–708

Eyquem J, Mansilla-Soto J, Giavridis T, van der Stegen SJ, Hamieh M, Cunanan KM, Odak A, Gonen M, Sadelain M (2017) Targeting a CAR to the TRAC locus with CRISPR/Cas9 enhances tumour rejection. Nature 543:113–117

Friedland AE, Baral R, Singhal P, Loveluck K, Shen S, Sanchez M, Marco E, Gotta GM, Maeder ML, Kennedy EM et al (2015) Characterization of *Staphylococcus aureus* Cas9: a smaller Cas9 for all-in-one adeno-associated virus delivery and paired nickase applications. Genome Biol 16:257

Frock RL, Hu J, Meyers RM, Ho YJ, Kii E, Alt FW (2015) Genome-wide detection of DNA double-stranded breaks induced by engineered nucleases. Nat Biotechnol 33:179–186

Fu Y, Foden JA, Khayter C, Maeder ML, Reyon D, Joung JK, Sander JD (2013) High-frequency off-target mutagenesis induced by CRISPR-Cas nucleases in human cells. Nat Biotechnol 31:822–826

Fu Y, Sander JD, Reyon D, Cascio VM, Joung JK (2014) Improving CRISPR-Cas nuclease specificity using truncated guide RNAs. Nat Biotechnol 32:279–284

Gasiunas G, Barrangou R, Horvath P, Siksnys V (2012) Cas9-crRNA ribonucleoprotein complex mediates specific DNA cleavage for adaptive immunity in bacteria. Proc Natl Acad Sci USA 109: E2579–2586

Gilbert LA, Larson MH, Morsut L, Liu Z, Brar GA, Torres SE, Stern-Ginossar N, Brandman O, Whitehead EH, Doudna JA et al (2013) CRISPR-mediated modular RNA-guided regulation of transcription in eukaryotes. Cell 154:442–451

Gonzalez F, Zhu Z, Shi ZD, Lelli K, Verma N, Li QV, Huangfu D (2014) An iCRISPR platform for rapid, multiplexable, and inducible genome editing in human pluripotent stem cells. Cell Stem Cell 15:215–226

Gross G, Waks T, Eshhar Z (1989) Expression of immunoglobulin-T-cell receptor chimeric molecules as functional receptors with antibody-type specificity. Proc Natl Acad Sci USA 86:10024–10028

Guilinger JP, Thompson DB, Liu DR (2014) Fusion of catalytically inactive Cas9 to FokI nuclease improves the specificity of genome modification. Nat Biotechnol 32:577–582

He X, Tan C, Wang F, Wang Y, Zhou R, Cui D, You W, Zhao H, Ren J, Feng B (2016) Knock-in of large reporter genes in human cells via CRISPR/Cas9-induced homology-dependent and independent DNA repair. Nucleic Acids Res 44:e85

Hendel A, Bak RO, Clark JT, Kennedy AB, Ryan DE, Roy S, Steinfeld I, Lunstad BD, Kaiser RJ, Wilkens AB et al (2015) Chemically modified guide RNAs enhance CRISPR-Cas genome editing in human primary cells. Nat Biotechnol 33:985–989

Hilton IB, D'Ippolito AM, Vockley CM, Thakore PI, Crawford GE, Reddy TE, Gersbach CA (2015) Epigenome editing by a CRISPR-Cas9-based acetyltransferase activates genes from promoters and enhancers. Nat Biotechnol 33:510–517

Hoos A (2016) Development of immuno-oncology drugs—from CTLA4 to PD1 to the next generations. Nat Rev Drug Discov 15:235–247

Horvath P, Barrangou R (2010) CRISPR/Cas, the immune system of bacteria and archaea. Science 327:167–170

Hou P, Chen S, Wang S, Yu X, Chen Y, Jiang M, Zhuang K, Ho W, Hou W, Huang J et al (2015) Genome editing of CXCR4 by CRISPR/cas9 confers cells resistant to HIV-1 infection. Sci Rep 5:15577

Hsu PD, Scott DA, Weinstein JA, Ran FA, Konermann S, Agarwala V, Li Y, Fine EJ, Wu X, Shalem O et al (2013) DNA targeting specificity of RNA-guided Cas9 nucleases. Nat Biotechnol 31:827–832

Hu W, Kaminski R, Yang F, Zhang Y, Cosentino L, Li F, Luo B, Alvarez-Carbonell D, Garcia-Mesa Y, Karn J et al (2014) RNA-directed gene editing specifically eradicates latent and prevents new HIV-1 infection. Proc Natl Acad Sci USA 111:11461–11466

Hwang WY, Fu Y, Reyon D, Maeder ML, Tsai SQ, Sander JD, Peterson RT, Yeh JR, Joung JK (2013) Efficient genome editing in zebrafish using a CRISPR-Cas system. Nat Biotechnol 31:227–229

Irving BA, Weiss A (1991) The cytoplasmic domain of the T cell receptor zeta chain is sufficient to couple to receptor-associated signal transduction pathways. Cell 64:891–901

Jinek M, Chylinski K, Fonfara I, Hauer M, Doudna JA, Charpentier E (2012) A programmable dual-RNA-guided DNA endonuclease in adaptive bacterial immunity. Science 337:816–821

Kalos M, June CH (2013) Adoptive T cell transfer for cancer immunotherapy in the era of synthetic biology. Immunity 39:49–60

Kearns NA, Pham H, Tabak B, Genga RM, Silverstein NJ, Garber M, Maehr R (2015) Functional annotation of native enhancers with a Cas9-histone demethylase fusion. Nat Methods 12:401–403

Kim HJ, Lee HJ, Kim H, Cho SW, Kim JS (2009) Targeted genome editing in human cells with zinc finger nucleases constructed via modular assembly. Genome Res 19:1279–1288

Kim S, Kim D, Cho SW, Kim J, Kim JS (2014) Highly efficient RNA-guided genome editing in human cells via delivery of purified Cas9 ribonucleoproteins. Genome Res 24:1012–1019

Kim D, Bae S, Park J, Kim E, Kim S, Yu HR, Hwang J, Kim JI, Kim JS (2015). Digenome-seq: genome-wide profiling of CRISPR-Cas9 off-target effects in human cells. Nat Methods 12:237–243, 231 p following 243

Konermann S, Brigham MD, Trevino AE, Joung J, Abudayyeh OO, Barcena C, Hsu PD, Habib N, Gootenberg JS, Nishimasu H et al (2015) Genome-scale transcriptional activation by an engineered CRISPR-Cas9 complex. Nature 517:583–588

Li JF, Norville JE, Aach J, McCormack M, Zhang D, Bush J, Church GM, Sheen J (2013) Multiplex and homologous recombination-mediated genome editing in *Arabidopsis* and *Nicotiana benthamiana* using guide RNA and Cas9. Nat Biotechnol 31:688–691

Li C, Guan X, Du T, Jin W, Wu B, Liu Y, Wang P, Hu B, Griffin GE, Shattock RJ et al (2015) Inhibition of HIV-1 infection of primary $CD4^+$ T-cells by gene editing of CCR5 using adenovirus-delivered CRISPR/Cas9. J Gen Virol 96:2381–2393

Liu X, Zhang Y, Cheng C, Cheng AW, Zhang X, Li N, Xia C, Wei X, Liu X, Wang H (2017) CRISPR-Cas9-mediated multiplex gene editing in CAR-T cells. Cell Res 27:154–157

Lloyd A, Vickery ON, Laugel B (2013) Beyond the antigen receptor: editing the genome of T-cells for cancer adoptive cellular therapies. Front Immunol 4:221

Long C, McAnally JR, Shelton JM, Mireault AA, Bassel-Duby R, Olson EN (2014) Prevention of muscular dystrophy in mice by CRISPR/Cas9-mediated editing of germline DNA. Science 345:1184–1188

Maeder ML, Linder SJ, Cascio VM, Fu Y, Ho QH, Joung JK (2013) CRISPR RNA-guided activation of endogenous human genes. Nat Methods 10:977–979

Maher J, Brentjens RJ, Gunset G, Riviere I, Sadelain M (2002) Human T-lymphocyte cytotoxicity and proliferation directed by a single chimeric TCRzeta/CD28 receptor. Nat Biotechnol 20:70–75

Makarova KS, Grishin NV, Shabalina SA, Wolf YI, Koonin EV (2006) A putative RNA-interference-based immune system in prokaryotes: computational analysis of the predicted enzymatic machinery, functional analogies with eukaryotic RNAi, and hypothetical mechanisms of action. Biol Direct 1:7

Mali P, Aach J, Stranges PB, Esvelt KM, Moosburner M, Kosuri S, Yang L, Church GM (2013a) CAS9 transcriptional activators for target specificity screening and paired nickases for cooperative genome engineering. Nat Biotechnol 31:833–838

Mali P, Yang L, Esvelt KM, Aach J, Guell M, DiCarlo JE, Norville JE, Church GM (2013b) RNA-guided human genome engineering via Cas9. Science 339:823–826

Mandal PK, Ferreira LM, Collins R, Meissner TB, Boutwell CL, Friesen M, Vrbanac V, Garrison BS, Stortchevoi A, Bryder D et al (2014) Efficient ablation of genes in human hematopoietic stem and effector cells using CRISPR/Cas9. Cell Stem Cell 15:643–652

Maus MV, Grupp SA, Porter DL, June CH (2014) Antibody-modified T cells: CARs take the front seat for hematologic malignancies. Blood 123:2625–2635

Miller JC, Tan S, Qiao G, Barlow KA, Wang J, Xia DF, Meng X, Paschon DE, Leung E, Hinkley SJ et al (2011) A TALE nuclease architecture for efficient genome editing. Nat Biotechnol 29:143–148

Morton J, Davis MW, Jorgensen EM, Carroll D (2006) Induction and repair of zinc-finger nuclease-targeted double-strand breaks in *Caenorhabditis elegans* somatic cells. Proc Natl Acad Sci USA 103:16370–16375

Moscou MJ, Bogdanove AJ (2009) A simple cipher governs DNA recognition by TAL effectors. Science 326:1501

Nakade S, Tsubota T, Sakane Y, Kume S, Sakamoto N, Obara M, Daimon T, Sezutsu H, Yamamoto T, Sakuma T et al (2014) Microhomology-mediated end-joining-dependent integration of donor DNA in cells and animals using TALENs and CRISPR/Cas9. Nat Commun 5:5560

Nekrasov V, Staskawicz B, Weigel D, Jones JD, Kamoun S (2013) Targeted mutagenesis in the model plant *Nicotiana benthamiana* using Cas9 RNA-guided endonuclease. Nat Biotechnol 31:691–693

Pavletich NP, Pabo CO (1991) Zinc finger-DNA recognition: crystal structure of a Zif268-DNA complex at 2.1 A. Science 252:809–817

Perez EE, Wang J, Miller JC, Jouvenot Y, Kim KA, Liu O, Wang N, Lee G, Bartsevich VV, Lee YL et al (2008) Establishment of HIV-1 resistance in $CD4^+$ T cells by genome editing using zinc-finger nucleases. Nat Biotechnol 26:808–816

Perez-Pinera P, Kocak DD, Vockley CM, Adler AF, Kabadi AM, Polstein LR, Thakore PI, Glass KA, Ousterout DG, Leong KW et al (2013) RNA-guided gene activation by CRISPR-Cas9-based transcription factors. Nat Methods 10:973–976

Poirot L, Philip B, Schiffer-Mannioui C, Le Clerre D, Chion-Sotinel I, Derniame S, Potrel P, Bas C, Lemaire L, Galetto R et al (2015) Multiplex genome-edited T-cell manufacturing platform for "Off-the-Shelf" adoptive T-cell immunotherapies. Cancer Res 75:3853–3864

Porteus MH, Baltimore D (2003) Chimeric nucleases stimulate gene targeting in human cells. Science 300:763

Provasi E, Genovese P, Lombardo A, Magnani Z, Liu PQ, Reik A, Chu V, Paschon DE, Zhang L, Kuball J et al (2012) Editing T cell specificity towards leukemia by zinc finger nucleases and lentiviral gene transfer. Nat Med 18:807–815

Qi LS, Larson MH, Gilbert LA, Doudna JA, Weissman JS, Arkin AP, Lim WA (2013) Repurposing CRISPR as an RNA-guided platform for sequence-specific control of gene expression. Cell 152:1173–1183

Ran FA, Hsu PD, Lin CY, Gootenberg JS, Konermann S, Trevino AE, Scott DA, Inoue A, Matoba S, Zhang Y et al (2013) Double nicking by RNA-guided CRISPR Cas9 for enhanced genome editing specificity. Cell 154:1380–1389

Ran FA, Cong L, Yan WX, Scott DA, Gootenberg JS, Kriz AJ, Zetsche B, Shalem O, Wu X, Makarova KS et al (2015) In vivo genome editing using *Staphylococcus aureus* Cas9. Nature 520:186–191

Ren J, Liu X, Fang C, Jiang S, June CH, Zhao Y (2016) Multiplex genome editing to generate universal CAR T cells resistant to PD1 inhibition. Clin Cancer Res 16:1300

Ren J, Zhang X, Liu X, Fang C, Jiang S, June CH, Zhao Y (2017) A versatile system for rapid multiplex genome-edited CAR T cell generation. Oncotarget 8:17002–17011

Sadelain M, Papapetrou EP, Bushman FD (2011) Safe harbours for the integration of new DNA in the human genome. Nat Rev Cancer 12:51–58

Sakuma T, Takenaga M, Kawabe Y, Nakamura T, Kamihira M, Yamamoto T (2015) Homologous recombination-independent large gene cassette knock-in in CHO cells using TALEN and MMEJ-directed donor plasmids. Int J Mol Sci 16:23849–23866

Sather BD, Romano Ibarra GS, Sommer K, Curinga G, Hale M, Khan IF, Singh S, Song Y, Gwiazda K, Sahni J et al (2015) Efficient modification of CCR5 in primary human hematopoietic cells using a megaTAL nuclease and AAV donor template. Sci Transl Med 7:307ra156

Schumann K, Lin S, Boyer E, Simeonov DR, Subramaniam M, Gate RE, Haliburton GE, Ye CJ, Bluestone JA, Doudna JA et al (2015) Generation of knock-in primary human T cells using Cas9 ribonucleoproteins. Proc Natl Acad Sci USA 112:10437–10442

Schwank G, Koo BK, Sasselli V, Dekkers JF, Heo I, Demircan T, Sasaki N, Boymans S, Cuppen E, van der Ent CK et al (2013) Functional repair of CFTR by CRISPR/Cas9 in intestinal stem cell organoids of cystic fibrosis patients. Cell Stem Cell 13:653–658

Shalem O, Sanjana NE, Hartenian E, Shi X, Scott DA, Mikkelsen TS, Heckl D, Ebert BL, Root DE, Doench JG et al (2014) Genome-scale CRISPR-Cas9 knockout screening in human cells. Science 343:84–87

Shen B, Zhang J, Wu H, Wang J, Ma K, Li Z, Zhang X, Zhang P, Huang X (2013) Generation of gene-modified mice via Cas9/RNA-mediated gene targeting. Cell Res 23:720–723

Slaymaker IM, Gao L, Zetsche B, Scott DA, Yan WX, Zhang F (2016) Rationally engineered Cas9 nucleases with improved specificity. Science 351:84–88

Su S, Hu B, Shao J, Shen B, Du J, Du Y, Zhou J, Yu L, Zhang L, Chen F et al (2016) CRISPR-Cas9 mediated efficient PD-1 disruption on human primary T cells from cancer patients. Sci Rep 6:20070

Suzuki K, Tsunekawa Y, Hernandez-Benitez R, Wu J, Zhu J, Kim EJ, Hatanaka F, Yamamoto M, Araoka T, Li Z et al (2016) In vivo genome editing via CRISPR/Cas9 mediated homology-independent targeted integration. Nature 540:144–149

Swiech L, Heidenreich M, Banerjee A, Habib N, Li Y, Trombetta J, Sur M, Zhang F (2015) In vivo interrogation of gene function in the mammalian brain using CRISPR-Cas9. Nat Biotechnol 33:102–106

Tanenbaum ME, Gilbert LA, Qi LS, Weissman JS, Vale RD (2014) A protein-tagging system for signal amplification in gene expression and fluorescence imaging. Cell 159:635–646

Tebas P, Stein D, Tang WW, Frank I, Wang SQ, Lee G, Spratt SK, Surosky RT, Giedlin MA, Nichol G et al (2014) Gene editing of CCR5 in autologous CD4 T cells of persons infected with HIV. N Engl J Med 370:901–910

Torikai H, Reik A, Liu PQ, Zhou Y, Zhang L, Maiti S, Huls H, Miller JC, Kebriaei P, Rabinovich B et al (2012) A foundation for universal T-cell based immunotherapy: T cells engineered to express a CD19-specific chimeric-antigen-receptor and eliminate expression of endogenous TCR. Blood 119:5697–5705

Townsend JA, Wright DA, Winfrey RJ, Fu F, Maeder ML, Joung JK, Voytas DF (2009) High-frequency modification of plant genes using engineered zinc-finger nucleases. Nature 459:442–445

Tsai SQ, Zheng Z, Nguyen NT, Liebers M, Topkar VV, Thapar V, Wyvekens N, Khayter C, Iafrate AJ, Le LP et al (2015) GUIDE-seq enables genome-wide profiling of off-target cleavage by CRISPR-Cas nucleases. Nat Biotechnol 33:187–197

Urnov FD, Miller JC, Lee YL, Beausejour CM, Rock JM, Augustus S, Jamieson AC, Porteus MH, Gregory PD, Holmes MC (2005) Highly efficient endogenous human gene correction using designed zinc-finger nucleases. Nature 435:646–651

Wang H, Yang H, Shivalila CS, Dawlaty MM, Cheng AW, Zhang F, Jaenisch R (2013) One-step generation of mice carrying mutations in multiple genes by CRISPR/Cas-mediated genome engineering. Cell 153:910–918

Wang W, Ye C, Liu J, Zhang D, Kimata JT, Zhou P (2014) CCR5 gene disruption via lentiviral vectors expressing Cas9 and single guided RNA renders cells resistant to HIV-1 infection. PLoS ONE 9:e115987

Wiedenheft B, Sternberg SH, Doudna JA (2012) RNA-guided genetic silencing systems in bacteria and archaea. Nature 482:331–338

Williams MR, Fricano-Kugler CJ, Getz SA, Skelton PD, Lee J, Rizzuto CP, Geller JS, Li M, Luikart BW (2016) A retroviral CRISPR-Cas9 system for cellular autism-associated phenotype discovery in developing neurons. Sci Rep 6:25611

Wyvekens N, Topkar VV, Khayter C, Joung JK, Tsai SQ (2015) Dimeric CRISPR RNA-guided FokI-dCas9 nucleases directed by truncated gRNAs for highly specific genome editing. Hum Gene Ther 26:425–431

Ye L, Wang J, Beyer AI, Teque F, Cradick TJ, Qi Z, Chang JC, Bao G, Muench MO, Yu J et al (2014) Seamless modification of wild-type induced pluripotent stem cells to the natural CCR5Delta32 mutation confers resistance to HIV infection. Proc Natl Acad Sci USA 111:9591–9596

Yin H, Xue W, Chen S, Bogorad RL, Benedetti E, Grompe M, Koteliansky V, Sharp PA, Jacks T, Anderson DG (2014) Genome editing with Cas9 in adult mice corrects a disease mutation and phenotype. Nat Biotechnol 32:551–553

Zhen S, Hua L, Liu YH, Gao LC, Fu J, Wan DY, Dong LH, Song HF, Gao X (2015) Harnessing the clustered regularly interspaced short palindromic repeat (CRISPR)/CRISPR-associated Cas9 system to disrupt the hepatitis B virus. Gene Ther 22:404–412

12

New insights into the regulation of Axin function in canonical Wnt signaling pathway

Xiaomin Song, Sheng Wang, Lin Li✉

State Key Laboratory of Molecular Biology, Institute of Biochemistry and Cell Biology, Shanghai Institutes for Biological Sciences, Chinese Academy of Sciences, Shanghai 200031, China
✉ Correspondence: lli@sibs.ac.cn (L. Li)

ABSTRACT

The Wnt signaling pathway plays crucial roles during embryonic development, whose aberration is implicated in a variety of human cancers. Axin, a key component of canonical Wnt pathway, plays dual roles in modulating Wnt signaling: on one hand, Axin scaffolds the "β-catenin destruction complex" to promote β-catenin degradation and therefore inhibits the Wnt signal transduction; on the other hand, Axin interacts with LRP5/6 and facilitates the recruitment of GSK3 to the plasma membrane to promote LRP5/6 phosphorylation and Wnt signaling. The differential assemblies of Axin with these two distinct complexes have to be tightly controlled for appropriate transduction of the "on" or "off" Wnt signal. So far, there are multiple mechanisms revealed in the regulation of Axin activity, such as post-transcriptional modulation, homo/hetero-polymerization and auto-inhibition. These mechanisms may work cooperatively to modulate the function of Axin, thereby playing an important role in controlling the canonical Wnt signaling. In this review, we will focus on the recent progresses regarding the regulation of Axin function in canonical Wnt signaling.

KEYWORDS Wnt signaling, Axin, post-transcriptional modification, polymerization, auto-inhibition

INTRODUCTION

Wnt signaling is one of the most conserved pathways during evolution, playing essential roles in multiple biological processes like embryonic development and adult tissue homeostasis. Dysregulation of the Wnt pathway underlies the pathogenesis of a variety of human cancers. So far, two types of Wnt signaling pathways have been characterized: canonical Wnt pathway (also known as Wnt/β-catenin pathway) and noncanonical Wnt pathway (including planar cell polarity pathway and Wnt/calcium pathway). The Wnt/β-catenin pathway is distinguished from the others in the way of its dependency on β-catenin, which is nevertheless not required for the two noncanonical Wnt pathways. As a central player in canonical Wnt signaling, β-catenin dictates the signaling intensity through its cellular levels and subcellular localization. In the absence of Wnt signaling, β-catenin is sequestered in the cytoplasm by the "destruction complex" composed of Axin, glycogen synthase kinase 3 (GSK3) and adenomatous polyposis coli (APC), etc. β-catenin is phosphorylated by GSK3 and subsequently subjected to degradation through the ubiquitin-proteasome pathway. Upon Wnt stimulation, low-density lipoprotein receptor-related proteins 5/6 (LRP5/6) is co-clustered with Frizzled (Fz) and Dishevelled (Dvl) and recruits the active Axin and GSK3 complex; these proteins thus form a high-molecular-weight assemblage close to the plasma membrane that is referred to as "LRP signalosome" (Bilic et al., 2007). Following the recruitment of Axin to LRP5/6, the "destruction complex" is disassembled; and β-catenin is released and translocates into the nucleus, where it forms a transcriptional complex with LEF-1/TCF and activates the transcription of Wnt-responsive genes. Recently, Li et al., based on their findings, proposed an alternative model that Wnt stimulation, rather than cause the disassembly of "destruction complex", suppresses β-catenin degradation and leads to saturation of the complex by the accumulated β-catenin, thus sparing the newly synthesized β-catenin, which accumulates in a free cytosolic form and consequently enters the nucleus to bind LEF-1/TCF (Li et al., 2012).

Vertebrates have two Axin isoforms: Axin1 (also referred to as Axin) is constitutively expressed, while Axin2 (also called Conductin) is subjected to the regulation of Wnt signaling (Jho et al., 2002; Leung et al., 2002; Yan et al., 2001). Albeit different in their expression patterns, Axin and Axin2 appear to be functionally, at least partly, interchangeable in canonical Wnt pathway (Chia and Costantini, 2005). Considering its essential role in both "destruction complex" and "LRP signalosome", the activity of Axin has to be tightly controlled. Post-translational modification is the most common mechanism for regulating protein activity. Several ways of modifications, such as phosphorylation, SUMOylation and ubiquitination, have been reported to play a role in regulating Axin stability, sub-cellular localization or association with other proteins (Callow et al., 2011; Fei et al., 2013; Kim et al., 2008; Kim and Jho, 2010; Zhang et al., 2011). Polymerization of Axin also regulates its activity in a way that currently isn't clearly defined. Axin is considered as a protein that is largely unstructured except the N-terminal RGS domain responsible for binding APC and the C-terminal DIX domain for mediating oligomerization. Besides Axin, Dvl and Ccd1 also contain a DIX domain, which is responsible for their homo- and hetero- polymerization (Liu et al., 2011; Schwarz-Romond et al., 2007a; Shiomi et al., 2003). It is reported that DIX-mediated heterotypic interaction between Axin and Dvl may disrupt Axin's association with the "destruction complex" and facilitate its recruitment to the "LRP signalosome" (Mendoza-Topaz et al., 2011; Schwarz-Romond et al., 2007b). Recently, two independent researches carried out by Sung-Eun Kim et al. and our group revealed a new mechanism of "auto-inhibition" behind Axin function, pointing out that conformational changes of Axin may play an important role in regulating its activity (Kim et al., 2013b; Wang et al., 2013). Notably, these mechanisms may be relevant and work cooperatively to fine-tune and regulate the function of Axin.

UBIQUITINATION/SUMOYLATION OF AXIN

Ubiquitination is an important regulatory process for protein, playing diverse roles in modulating protein function. Ubiquitin is covalently attached to the targeted molecules through the sequential activities of three enzymes: ubiquitin-activating enzyme (E1) for ATP-dependent activation of ubiquitin (Ub), ubiquitin-conjugating enzyme (E2) for delivery of Ub, and ubiquitin ligase (E3) for ligation of Ub to the substrates. In this process, the E3 ligases are responsible for substrate selection and the specificity of Ub-linkage. There are seven lysines (K6, K11, K27, K29, K33, K48 and K63) in Ub molecule. Generally, proteins marked by K48-linked poly-Ub chains are destined to proteasome-dependent degradation, while other types of Ub chains could regulate a variety of cellular processes by both proteolytic and non-proteolytic functions (Kulathu and Komander, 2012). Multiple key components in canonical Wnt pathway are regulated by ubiquitination, such as β-catenin, APC and Dvl (Chan et al., 2006; Choi et al., 2004; Ding et al., 2013; Hu et al., 2010; Liu et al., 1999; Liu et al., 2001; Tran et al., 2013; Tran and Polakis, 2012; Wei et al., 2012; Winston et al., 1999). Despite limited reports so far on ubiquitination of Axin, it's increasingly clear that Axin is also subjected to ubiquitin modification. Previously, Stegmeier lab revealed that Axin stability is regulated by Tankyrase and Tankyrase-mediated poly-ADP-ribose modification (PARsylation) of Axin is linked to Axin polyubiquitylation and subsequent degradation by the proteasome (Huang et al., 2009). Later, RNF146 was uncovered to be the E3 ligase for mediating Tankyrase-dependent degradation of Axin, thus playing a positive role in Wnt signaling (Callow et al., 2011; Zhang et al., 2011). Ubiquitin-specific protease 34 (USP34) is also reported to associate with Axin and control its levels, whereby modulating Wnt signaling (Lui et al., 2011). Smurf2 (Smad ubiquitination regulatory factor 2), a C2-WW-HECT type ubiquitin ligase, is suggested to be another E3 ligase for Axin, which also plays a role in regulating its stability (Kim and Jho, 2010). Interestingly, we recently uncovered that Smurf1, an isoform of Smurf2, also targets Axin, but for a non-proteolytic purpose. Our data suggested that Smurf1 ubiquitinates Axin through Lys (K) 29-linked poly-ubiquitin chains, which disrupts its interaction with LRP5/6 rather than leads to its degradation (Fei et al., 2013). Similar to Ubiquitination, SUMO modification of proteins also plays various roles in regulating protein function. However, researches regarding Axin SUMOylation are, so far, very limited. It was reported earlier that SUMOylation of Axin may regulate its function in noncanonical Wnt pathway (Rui et al., 2002). Recently, Kim et al. suggested that SUMOylation may protect Axin from ubiquitination, thus playing a role in regulating Axin stability in canonical Wnt pathway (Kim et al., 2008).

PHOSPHORYLATION/DEPHOSPHORYLATION OF AXIN

Phosphorylation/Dephosphorylation is the most common post-translational modification for regulating protein function. Multiple components of Wnt pathway, such as LRP5/6, Dvl, β-catenin, TCF, are subject to reversible phosphorylation by kinases and phosphatases networks. Axin, as a crucial component of canonical Wnt pathway, is also under the regulation of phosphorylation/dephosphorylation. Without Wnt stimulation, Axin is phosphorylated, which enhances its binding affinity with β-catenin, leading to stabilization of Axin (Jho et al., 1999; Yamamoto et al., 1999); upon Wnt stimulation, Axin is dephosphorylated, leading to a less effective binding with β-catenin and consequently Axin degradation (Strovel et al., 2000; Willert et al., 1999). GSK3 is most likely the key kinase that phosphorylates Axin and facilitates its function in "destruction complex". Besides GSK3, CKI was also observed to phosphorylate Axin both *in vitro* and *in vivo* (Gao et al., 2002). On the other hand, it is found that PP2A and PP2C may be the two phosphatases targeting Axin

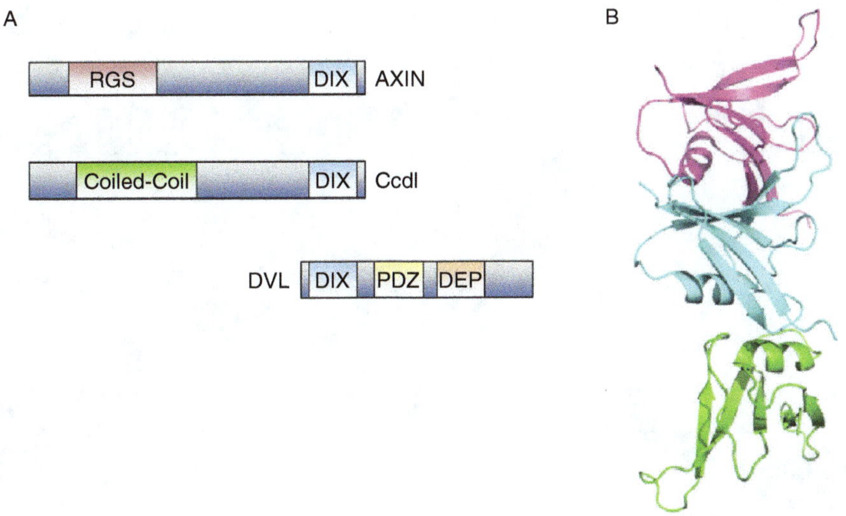

Figure 1. The C-terminal DIX domain of Axin mediates its homo/hetero- polyermerization. Schematic illustration of the domain organization for Axin, Dvl and Ccd1. These three proteins all contain a DIX domain, which mediates their homo- and hetero-interaction. The self-assembly of Axin-DIX or heterotypic interaction with Dvl-DIX may regulate Axin function in canonical Wnt signaling. (B) The three-dimensional structure of Axin DIX domain (PDB code: 1wsp) showed that it forms filaments in the crystal through head-to-tail self-interaction.

(Strovel et al., 2000; Willert et al., 1999). Recently, it was reported that phosphorylation of Axin by CKI may also improve its association with GSK3 and leads to a more active "destruction complex", while protein phosphatase 1 (PP1) interacts with, and dephosphorylates Axin, reversing the effect conferred by CKI and hence contributing to the activation of Wnt signaling (Luo et al., 2007). Later, Jiang et al. reported that Dab2 could block the interaction between Axin and PP1, thus inhibiting Axin dephosphorylation and ultimately leading to inhibition of the Wnt signaling (Jiang et al., 2009). Previously, phosphorylation of Axin by Cyclin A/CDK2 was reported to increase its association with β-catenin (Kim et al., 2004). Recently, Axin was found to be phosphorylated by CDK5, and this phosphorylation facilitates its interaction with GSK3, which plays an essential role for axon development (Fang et al., 2011).

HOMO- AND HETERO-POLYMERIZATION OF AXIN

It is not unusual that proteins modulate their functions through altering their oligomeric states or forming hetero-oligomer with other proteins. Self- or hetero-assembly of proteins also plays important roles in regulating Wnt signaling. For example, aggregation of LRP5/6 is essential for Wnt signaling activation, which may also require the oligomerization of Dvl (Metcalfe et al., 2010). Recently, we identified Caprin-2 as a new component of canonical Wnt signaling through facilitating LRP5/6 phosphorylation (Ding et al., 2008), and we further found that the oligomerization of Caprin-2 is required for its function in Wnt signaling (unpublished data). Axin contains a DIX domain of 80 amino acids located at its C-terminus for mediating homo- or hetero-interaction, which seems to be essential for its function (Choi et al., 2010; Fiedler et al., 2011; Sakanaka and Williams, 1999; Yokoyama et al., 2012). The three-dimensional structure of Axin DIX domain showed that it forms filaments in the crystal through head-to-tail self-interaction (Schwarz-Romond et al., 2007a; Shibata et al., 2007) (Fig. 1B). Besides Axin, the homologous DIX domain is also present at the N-terminus of Dvl and the C-terminus of Ccd1 (also called DIXdc1) (Fig. 1A). It is argued that Dvl may release Axin from the "destruction complex" and deliver Axin to the "LRP signalosome" through DIX-mediated heterotypic interactions between Dvl and Axin (Schwarz-Romond et al., 2007b). Fiedler et al.'s recent work also indicated that Dvl may behave as a dominant-negative modulator of Axin to regulate its function via heterotypic interactions between their DIX domains (Choi et al., 2010; Fiedler et al., 2011). Ccd1 was initially identified as a positive regulator in Wnt signaling (Shiomi et al., 2003). Ccd1-DIX interacts with Dvl-DIX directly, converting latent polymeric Dvl to a biologically active oligomer(s) (Liu et al., 2011; Shiomi et al., 2003). However, the interaction between Ccd1 and Axin does not seem to be dominated by their DIX domains and other domains may also play an important role for their interaction (Wong et al., 2004).

AUTO-INHIBITION OF AXIN

Conformational change is an important way for proteins to regulate their functions. In many cases, when a protein adopts an "auto-inhibited" conformation, its active site for

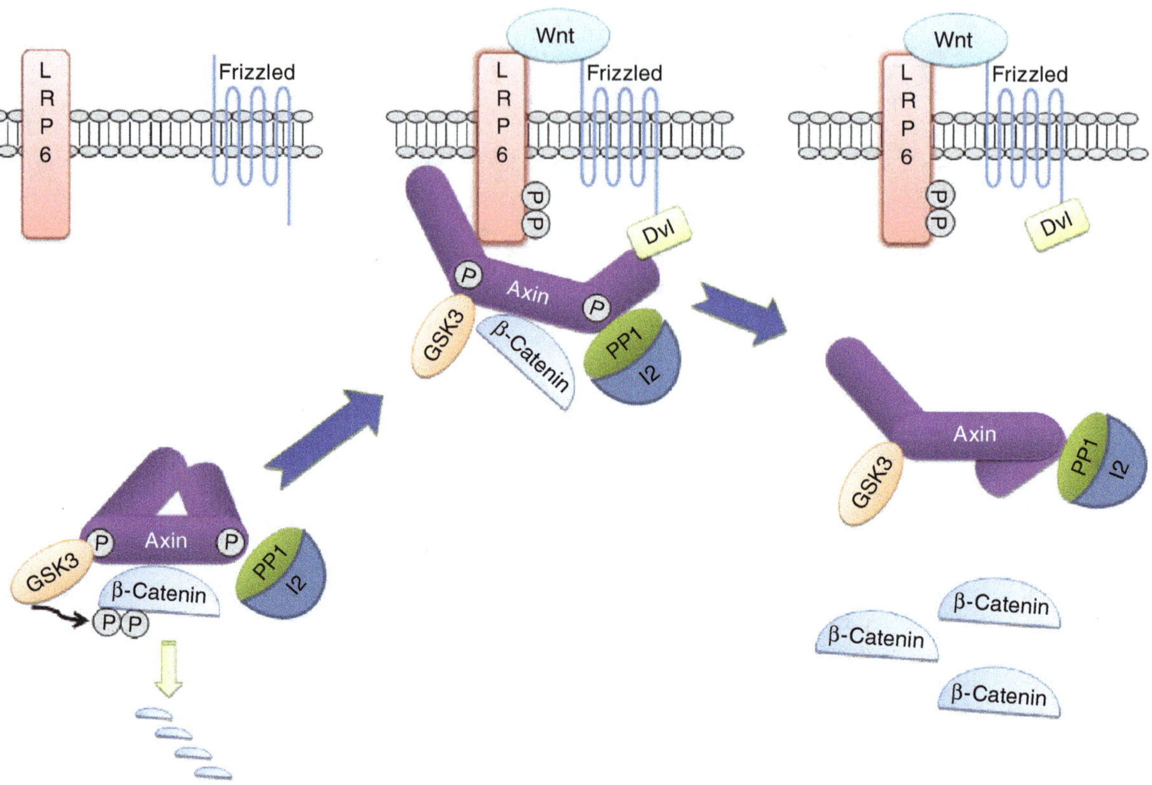

Figure 2. The model of Axin autoinhibition. Based on the findings of Kim et al. and us, we proposed a possible model for Axin "autoinhibiton": without Wnt stimulation, the N-terminus of Axin interacts with its C-terminus, forming "type I" closed conformation, which blocks its binding with LRP5/6 but spares the binding site for β-catenin; upon Wnt stimulation, this type I auto-inhibited state of Axin is released for binding to LRP5/6 and the latter was subsequently phosphorylated, which inhibits the activity of GSK3 and thus moves Axin towards to the dephosphorylated state conferred by PP1. Dephosphorylated Axin forms a "type II" closed conformation through an intramolecular interaction between the C-terminus and the β-catenin binding region, making it unable to bind both of LRP5/6 and β-catenin.

substrates, or binding sites for other partners may be blocked and thus this state is also referred to as an "inactive" state; while, in an "active" state, the autoinhibitory conformation is released, making the active or binding sites available for interaction with substrates or regulatory partners. Normally, this reversible process of conformational change is induced and regulated by post-translational modifications or interactions with other proteins. Previous studies have suggested that Axin N-terminus (including the RGS domain and the linker region between the RGS domain and the GSK binding domain) plays an inhibitory role in Axin's binding with its partners (Chen et al., 2009; Mao et al., 2001). However, the mechanism underlying such inhibition remains elusive. In our recent work, we revealed that the N-terminus of Axin is able to associate with its C-terminus, forming a "closed" conformation, which could be disrupted effectively by a small molecular compound, named as HLY78. HLY78 is an activator of Wnt/β-catenin signaling pathway identified by our group through high-throughput screening, which acts in a Wnt ligand-dependent manner (Wang et al., 2013). Our findings suggested that depending on distinct partners and post-translational modifications, Axin may undergo conformational change during its differential assemblies with "destruction complex" or "Wnt-LRP5/6 signalosome". We further found that HLY78 targets Axin-DIX and triggers the conformational switch of Axin from a "closed" auto-inhibitory state to an "open" active state, leading to an enhanced association of Axin with LRP6 and subsequent activation of LRP6. Meanwhile, an independent work carried out by another group also proposed an "auto-inhibition" model of Axin (Kim et al., 2013b). In their model, phosphorylation by GSK3 makes Axin adopt an "open" conformation for binding to β-catenin or phospho-LRP6, while dephosphorylation by PP1 leads to a "closed" conformation (auto-inhibition) of Axin, making Axin incompetent for interacting with either β-catenin or phospho-LRP6. Notably, according to their findings, the C-terminus of Axin forms a "closed" conformation through an intramolecular interaction with the middle β-catenin binding region rather than the N-terminal region as we found. In our view, these two models are not in conflict with each other and may function alternatively or collaboratively in different cellular contexts. A possible model that accommodates both sets of findings is proposed in Fig. 2: without Wnt stimulation, the N-terminus of Axin interacts with its C-terminus, forming "type I" closed conformation, which blocks its binding with LRP5/6 but spares the one with β-catenin; upon Wnt stimulation, this type I auto-inhibited state of Axin is released for binding to LRP5/6 and the latter was subsequently phosphorylated, which inhibits the activity of GSK3 and thus moves Axin towards to the dephosphorylated state conferred by PP1. Dephosphorylated Axin forms a "type II" closed conformation through an intramolecular interaction between the C-terminus and the β-catenin binding region, making it unable to bind both of LRP5/6 and β-catenin.

MICRORNAS MEDIATED REGULATION OF AXIN

MicroRNA (miRNA) is a small noncoding RNA molecule normally containing 19–25 nucleotides in length, which regulates gene expression both at transcriptional and post-transcriptional level. So far, thousands of transcripts have been reported to subject to microRNA regulation, playing extensive and important roles in eukaryotic development and physiology. Moreover, increasing evidences have implicated the role of miRNA deregulation in human disease including cancer. Recently, there are several researches addressing potential roles of miRNAs in Axin regulation. In a screen for miRNAs affecting Wnt pathway, Silver et al. found that ectopic miR-315 directly inhibits Axin as well as Notum (Silver et al., 2007), thus acting as an activator of Wnt signaling. Later, another miRNA—hsa-miR-34a—reportedly targets both the 5′-UTR and 3′-UTR of *Axin2,* whereby repressing *Axin2* expression (Lee et al., 2009). Recently, Egea et al. revealed that *lethal-7 (let-7)* also targets and represses the translation of *Axin2*. *lethal-7 (let-7)* is one of the first two known miRNAs (the other one is *lin-4*) (Reinhart et al., 2000), which is highly conserved across species and implicated in the regulation of a variety of important genes, such as *RAS* and *HMGA2* (Johnson et al., 2005; Mayr et al., 2007). They found miRNA let-7f promotes β-catenin activity in hMSCs by repressing the translation of *Axin2* (Egea et al., 2012). One the other hand, in a study lately published, Kim et al. reported that upregulation of miR-205 in KB oral cancer cells could trigger cell apoptosis. Further mechanistic study identified *Axin2* as the direct target of miR-205 (Lee et al., 2013). They showed that over-expressed miR-205 in KB oral cancer cells suppresses *Axin2* expression through an interaction with its own binding site at *Axin2* 3′-UTR (64–92). However, it is unclear how reduced *Axin2* expression conferred by miR-205, rather than facilitates tumorigenesis as it usually does, contributes to the suppression of KB oral cancer cells.

EPIGENETIC MODULATIONS OF AXIN IN CANCER

In additional to the post-transcriptional regulations mentioned above in affecting Axin levels, epigenetic modulations could also affect the expression of Axin and the change thereof may relate to tumor development. It is well accepted that genetic changes in components of Wnt signaling are closely related with human cancer. One leading example is colorectal cancer: alterations in Wnt-related genes, mainly APC, have been observed in 90% of colorectal cancer (Miyaki et al., 1994; Najdi et al., 2011). In addition to genetic mutations, epigenetic modification is another common cause for altered gene expression, among which promoter methylation and histone modification are the most frequently observed. Axin is also frequently found mutated in cancer, contributing to the over-activation of Wnt signaling in these cancers, while studies regarding epigenetic alterations of Axin and its relationship to cancer remain relatively scant. As a negative regulator of Wnt signaling, tumor-specific promoter methylation or histone deacetylation will cause decreased Axin expression, which may prompt over-activation of Wnt signaling in cancer. Previously, epigenetic silencing of *Axin2* was reported in colorectal carcinoma with microsatellite instability (MSI+ CRC), and forced expression of *Axin2* results in rapid cell death in an MSI+ CRC cell line, indicating a role of epigenetic silencing of *Axin2* in carcinogenesis of MSI+ CRC (Koinuma et al., 2006). Lately, CpG islands of *Axin2* were found methylated in human neuroendocrine tumors (NETs) and Axin2 expression is correspondingly downregulated, which may contributes to the pathogenesis and growth of NETs (Kim et al., 2013a). A recent work carried out by Yang et al., also indicated that hypermethylated Axin gene may significantly correlate with the progression of lung cancer (Yang et al., 2013). On the other hand, Chen G. et al. reported that overexpression of Menin, a tumor suppressor protein mutated in patients with multiple endocrine neoplasia type 1, is associated with increased H3K4 trimethylation of the Axin2 gene promoter; while, inhibition of Menin expression by siRNA abrogates H3K4 trimethylation and gene expression of Axin2, indicating that Menin functions, at least partly, through affecting histone trimethylation of *Axin2* promoter (Chen et al., 2008). Recently, Histone deacetylase (HDAC) 1 and 2 are implicated in the regulation of Axin expression in non-small cell lung cancer (NSCLC) (Han et al., 2012). X-radiation and siRNA could inhibit expression of HDAC1 and HDAC2 and weaken its inhibitory effect on Axin, which upregulates Axin expression and induces apoptosis of lung cancer cells (Han et al., 2012).

CONCLUDING REMARKS AND PERSPECTIVES

Besides those post-transcriptional modulations mentioned above, another modification—methylation—is also observed for Axin (Cha et al., 2011; Huang et al., 2009). A recent work carried out by Cha et al. revealed that PRMT1 directly interacts with and methylates Axin, thus enhancing Axin stability (Cha et al., 2011; Huang et al., 2009). So far, our knowledge about post-transcriptional modifications of Axin is still very limited. More kinases, phosphatases, E3 ubiquitin ligases and probably other types of post-transcriptional regulations are expected to be identified in the near future to fit into the network that modulates Axin activity in canonical Wnt signaling. In light of current findings, Axin seems to be a flexible and dynamic molecule, which could undergo structural rearrangements for interacting with distinct partners. The three-dimensional structure of the full-length Axin will provide invaluable insights into the details regarding Axin auto-inhibition. On the other hand, increasing evidence indicates a role of Axin in nucleus (Nikuseva Martic et al., 2010; Pecina-Slaus et al., 2011; Schmitz et al., 2013; Willert and Jones, 2006). However, the precise role of nuclear Axin in Wnt signaling remains largely unknown, which warrants further investigation to facilitate our understanding of the function of nuclear Axin as well as its linkage to the cytoplasmic one. As an attractive therapeutic target, much effort has been made to identify chemical compounds modulating Axin activity. So far, most studies focus on small-molecules in improving Axin stability and thus attenuating Wnt signaling, among which inhibitors against Tankyrase are the best explored (Bao et al., 2012; Huang et al., 2009; James et al., 2012; Lu et al., 2009; Shultz et al., 2012). Tankyrase-mediated PARsylation of Axin stimulates Axin degradation via the ubiquitin-proteasome pathway (Huang et al., 2009). Thus, by inhibiting Tankyrase activity, Axin is spared from the consequent ubiquitination and degradation, which may antagonize Wnt signaling and inhibit tumor growth. Wnt/β-catenin signaling normally needs to be blocked for therapeutic purpose in cancer; however, it also plays an important role in hematopoietic stem cell (HSC) formation and self-renewal, and is required for HSC recovery after injury and in transplantation. Hence, our recent identification of HLY78 as a novel Wnt activator, may provide new chances for the circumstances where activation of Wnt signaling is desired, such as for the patients with bone marrow failure or undergoing post-HSC transplantation (Wang et al., 2013).

ACKNOWLEDGEMENTS

L.L and X.S would like to thank the National Basic Research Program (973 Program) (Nos. 2010CB912100 to L.L, and 2011CD966300 to X.S) and the National Natural Science Foundation of China (Grant Nos. 31230044, 30930052 and 91213304 to L.L, and 31100532 to X.S). X.S. gratefully acknowledges the support of SA-SIBS scholarship program.

ABBREVIATIONS

APC, adenomatous polyposis coli; GSK3, glycogen synthase kinase 3; HDAC, histone deacetylase; HSC, hematopoietic stem cell; NETs, neuroendocrine tumors; NSCLC, non-small cell lung cancer; Ub, ubiquitin.

COMPLIANCE WITH ETHICS GUIDELINES

Xiaomin Song, Sheng Wang and Lin Li declare that they have no conflict of interest.

This article does not contain any studies with human or animal subjects performed by the any of the authors.

REFERENCES

Bao R, Christova T, Song S, Angers S, Yan X, Attisano L (2012) Inhibition of tankyrases induces Axin stabilization and blocks Wnt signalling in breast cancer cells. PLoS ONE 7:e48670

Bilic J, Huang YL, Davidson G, Zimmermann T, Cruciat CM, Bienz M, Niehrs C (2007) Wnt induces LRP6 signalosomes and promotes dishevelled-dependent LRP6 phosphorylation. Science 316:1619–1622

Callow MG, Tran H, Phu L, Lau T, Lee J, Sandoval WN, Liu PS, Bheddah S, Tao J, Lill JR et al (2011) Ubiquitin ligase RNF146 regulates tankyrase and Axin to promote Wnt signaling. PLoS ONE 6:e22595

Cha B, Kim W, Kim YK, Hwang BN, Park SY, Yoon JW, Park WS, Cho JW, Bedford MT, Jho EH (2011) Methylation by protein arginine methyltransferase 1 increases stability of Axin, a negative regulator of Wnt signaling. Oncogene 30:2379–2389

Chan DW, Chan CY, Yam JW, Ching YP, Ng IO (2006) Prickle-1 negatively regulates Wnt/beta-catenin pathway by promoting Dishevelled ubiquitination/degradation in liver cancer. Gastroenterology 131:1218–1227

Chen G, AJ, Wang M, Farley S, Lee LY, Lee LC, Sawicki MP (2008) Menin promotes the Wnt signaling pathway in pancreatic endocrine cells. Mol Cancer Res 6:1894–1907

Chen T, Li M, Ding Y, Zhang LS, Xi Y, Pan WJ, Tao DL, Wang JY, Li L (2009) Identification of zinc-finger BED domain-containing 3 (Zbed3) as a novel Axin-interacting protein that activates Wnt/beta-catenin signaling. J Biol Chem 284:6683–6689

Chia IV, Costantini F (2005) Mouse axin and axin2/conductin proteins are functionally equivalent in vivo. Mol Cell Biol 25:4371–4376

Choi J, Park SY, Costantini F, Jho EH, Joo CK (2004) Adenomatous polyposis coli is down-regulated by the ubiquitin-proteasome pathway in a process facilitated by Axin. J Biol Chem 279:49188–49198

Choi SH, Choi KM, Ahn HJ (2010) Coexpression and protein-protein complexing of DIX domains of human Dvl1 and Axin1 protein. BMB Rep 43:609–613

Ding Y, Xi Y, Chen T, Wang JY, Tao DL, Wu ZL, Li YP, Li C, Zeng R, Li L (2008) Caprin-2 enhances canonical Wnt signaling through regulating LRP5/6 phosphorylation. J Cell Biol 182:865–872

Ding Y, Zhang Y, Xu C, Tao QH, Chen YG (2013) HECT domain-containing E3 ubiquitin ligase NEDD4L negatively regulates Wnt signaling by targeting dishevelled for proteasomal degradation. J Biol Chem 288:8289–8298

Egea V, Zahler S, Rieth N, Neth P, Popp T, Kehe K, Jochum M, Ries C (2012) Tissue inhibitor of metalloproteinase-1 (TIMP-1) regulates mesenchymal stem cells through let-7f microRNA and Wnt/beta-catenin signaling. Proc Natl Acad Sci USA 109:E309–E316

Fang WQ, Ip JP, Li R, Ng YP, Lin SC, Chen Y, Fu AK, Ip NY (2011) Cdk5-mediated phosphorylation of Axin directs axon formation during cerebral cortex development. J Neurosci 31:13613–13624

Fei C, Li Z, Li C, Chen Y, Chen Z, He X, Mao L, Wang X, Zeng R, Li L (2013) Smurf1-mediated Lys29-linked non-proteolytic poly-ubiquitination of Axin negatively regulates Wnt/beta-catenin signaling. Mol Cell Biol 33:4095–4105

Fiedler M, Mendoza-Topaz C, Rutherford TJ, Mieszczanek J, Bienz M (2011) Dishevelled interacts with the DIX domain polymerization interface of Axin to interfere with its function in down-regulating beta-catenin. Proc Natl Acad Sci USA 108:1937–1942

Gao ZH, Seeling JM, Hill V, Yochum A, Virshup DM (2002) Casein kinase I phosphorylates and destabilizes the beta-catenin degradation complex. Proc Natl Acad Sci USA 99:1182–1187

Han Y, Zhang Y, Yang LH, Mi XY, Dai SD, Li QC, Xu HT, Yu JH, Li G, Zhao J et al (2012) X-radiation inhibits histone deacetylase 1 and 2, upregulates Axin expression and induces apoptosis in non-small cell lung cancer. Radiat Oncol 7:183

Hu T, Li C, Cao Z, Van Raay TJ, Smith JG, Willert K, Solnica-Krezel L, Coffey RJ (2010) Myristoylated Naked2 antagonizes Wnt-beta-catenin activity by degrading Dishevelled-1 at the plasma membrane. J Biol Chem 285:13561–13568

Huang SM, Mishina YM, Liu S, Cheung A, Stegmeier F, Michaud GA, Charlat O, Wiellette E, Zhang Y, Wiessner S et al (2009) Tankyrase inhibition stabilizes axin and antagonizes Wnt signalling. Nature 461:614–620

James RG, Davidson KC, Bosch KA, Biechele TL, Robin NC, Taylor RJ, Major MB, Camp ND, Fowler K, Martins TJ et al (2012) WIKI4, a novel inhibitor of tankyrase and Wnt/ss-catenin signaling. PLoS ONE 7:e50457

Jho E, Lomvardas S, Costantini F (1999) A GSK3beta phosphorylation site in axin modulates interaction with beta-catenin and Tcf-mediated gene expression. Biochem Biophys Res Commun 266:28–35

Jho EH, Zhang T, Domon C, Joo CK, Freund JN, Costantini F (2002) Wnt/beta-catenin/Tcf signaling induces the transcription of Axin2, a negative regulator of the signaling pathway. Mol Cell Biol 22:1172–1183

Jiang Y, Luo W, Howe PH (2009) Dab2 stabilizes Axin and attenuates Wnt/beta-catenin signaling by preventing protein phosphatase 1 (PP1)–Axin interactions. Oncogene 28:2999–3007

Johnson SM, Grosshans H, Shingara J, Byrom M, Jarvis R, Cheng A, Labourier E, Reinert KL, Brown D, Slack FJ (2005) RAS is regulated by the let-7 microRNA family. Cell 120:635–647

Kim S, Jho EH (2010) The protein stability of Axin, a negative regulator of Wnt signaling, is regulated by Smad ubiquitination regulatory factor 2 (Smurf2). J Biol Chem 285:36420–36426

Kim SI, Park CS, Lee MS, Kwon MS, Jho EH, Song WK (2004) Cyclin-dependent kinase 2 regulates the interaction of Axin with beta-catenin. Biochem Biophys Res Commun 317:478–483

Kim MJ, Chia IV, Costantini F (2008) SUMOylation target sites at the C terminus protect Axin from ubiquitination and confer protein stability. FASEB J 22:3785–3794

Kim JT, Li J, Jang ER, Gulhati P, Rychahou PG, Napier DL, Wang C, Weiss HL, Lee EY, Anthony L et al (2013a) Deregulation of Wnt/beta-catenin signaling through genetic or epigenetic alterations in human neuroendocrine tumors. Carcinogenesis 34:953–961

Kim SE, Huang H, Zhao M, Zhang X, Zhang A, Semonov MV, MacDonald BT, Zhang X, Garcia Abreu J, Peng L et al (2013b) Wnt stabilization of beta-catenin reveals principles for morphogen receptor-scaffold assemblies. Science 340:867–870

Koinuma K, Yamashita Y, Liu W, Hatanaka H, Kurashina K, Wada T, Takada S, Kaneda R, Choi YL, Fujiwara SI et al (2006) Epigenetic silencing of AXIN2 in colorectal carcinoma with microsatellite instability. Oncogene 25:139–146

Kulathu Y, Komander D (2012) Atypical ubiquitylation - the unexplored world of polyubiquitin beyond Lys48 and Lys63 linkages. Nat Rev Mol Cell Biol 13:508–523

Lee I, Ajay SS, Yook JI, Kim HS, Hong SH, Kim NH, Dhanasekaran SM, Chinnaiyan AM, Athey BD (2009) New class of microRNA targets containing simultaneous 5′-UTR and 3′-UTR interaction sites. Genome Res 19:1175–1183

Lee MA, Kim WK, Park HJ, Kang SS, Lee SK (2013) Anti-proliferative activity of hydnocarpin, a natural lignan, is associated with the suppression of Wnt/beta-catenin signaling pathway in colon cancer cells. Bioorg Med Chem Lett 23:5511–5514

Leung JY, Kolligs FT, Wu R, Zhai Y, Kuick R, Hanash S, Cho KR, Fearon ER (2002) Activation of AXIN2 expression by beta-catenin-T cell factor. A feedback repressor pathway regulating Wnt signaling. J Biol Chem 277:21657–21665

Li VS, Ng SS, Boersema PJ, Low TY, Karthaus WR, Gerlach JP, Mohammed S, Heck AJ, Maurice MM, Mahmoudi T et al (2012) Wnt signaling through inhibition of beta-catenin degradation in an intact Axin1 complex. Cell 149:1245–1256

Liu C, Kato Y, Zhang Z, Do VM, Yankner BA, He X (1999) beta-Trcp couples beta-catenin phosphorylation-degradation and regulates Xenopus axis formation. Proc Natl Acad Sci USA 96:6273–6278

Liu J, Stevens J, Rote CA, Yost HJ, Hu Y, Neufeld KL, White RL, Matsunami N (2001) Siah-1 mediates a novel beta-catenin degradation pathway linking p53 to the adenomatous polyposis coli protein. Mol Cell 7:927–936

Liu YT, Dan QJ, Wang J, Feng Y, Chen L, Liang J, Li Q, Lin SC, Wang ZX, Wu JW (2011) Molecular basis of Wnt activation via the DIX domain protein Ccd1. J Biol Chem 286:8597–8608

Lu J, Ma Z, Hsieh JC, Fan CW, Chen B, Longgood JC, Williams NS, Amatruda JF, Lum L, Chen C (2009) Structure-activity relationship studies of small-molecule inhibitors of Wnt response. Bioorg Med Chem Lett 19:3825–3827

Lui TT, Lacroix C, Ahmed SM, Goldenberg SJ, Leach CA, Daulat AM, Angers S (2011) The ubiquitin-specific protease USP34 regulates axin stability and Wnt/beta-catenin signaling. Mol Cell Biol 31:2053–2065

Luo W, Peterson A, Garcia BA, Coombs G, Kofahl B, Heinrich R, Shabanowitz J, Hunt DF, Yost HJ, Virshup DM (2007) Protein phosphatase 1 regulates assembly and function of the beta-catenin degradation complex. EMBO J 26:1511–1521

Mao J, Wang J, Liu B, Pan W, Farr GH 3rd, Flynn C, Yuan H, Takada S, Kimelman D, Li L et al (2001) Low-density lipoprotein receptor-related protein-5 binds to Axin and regulates the canonical Wnt signaling pathway. Mol Cell 7:801–809

Mayr C, Hemann MT, Bartel DP (2007) Disrupting the pairing between let-7 and Hmga2 enhances oncogenic transformation. Science 315:1576–1579

Mendoza-Topaz C, Mieszczanek J, Bienz M (2011) The Adenomatous polyposis coli tumour suppressor is essential for Axin complex assembly and function and opposes Axin's interaction with Dishevelled. Open Biol 1:110013

Metcalfe C, Mendoza-Topaz C, Mieszczanek J, Bienz M (2010) Stability elements in the LRP6 cytoplasmic tail confer efficient signalling upon DIX-dependent polymerization. J Cell Sci 123:1588–1599

Miyaki M, Konishi M, Kikuchi-Yanoshita R, Enomoto M, Igari T, Tanaka K, Muraoka M, Takahashi H, Amada Y, Fukayama M et al (1994) Characteristics of somatic mutation of the adenomatous polyposis coli gene in colorectal tumors. Cancer Res 54:3011–3020

Najdi R, Holcombe RF, Waterman ML (2011) Wnt signaling and colon carcinogenesis: beyond APC. J Carcinog 10:5

Nikuseva Martic T, Pecina-Slaus N, Kusec V, Kokotovic T, Musinovic H, Tomas D, Zeljko M (2010) Changes of AXIN-1 and beta-catenin in neuroepithelial brain tumors. Pathol Oncol Res 16:75–79

Pecina-Slaus N, Martic TN, Kokotovic T, Kusec V, Tomas D, Hrascan R (2011) AXIN-1 protein expression and localization in glioblastoma. Collegium Antropologicum 35(Suppl 1):101–106

Reinhart BJ, Slack FJ, Basson M, Pasquinelli AE, Bettinger JC, Rougvie AE, Horvitz HR, Ruvkun G (2000) The 21-nucleotide let-7 RNA regulates developmental timing in *Caenorhabditis elegans*. Nature 403:901–906

Rui HL, Fan E, Zhou HM, Xu Z, Zhang Y, Lin SC (2002) SUMO-1 modification of the C-terminal KVEKVD of Axin is required for JNK activation but has no effect on Wnt signaling. J Biol Chem 277:42981–42986

Sakanaka C, Williams LT (1999) Functional domains of axin. Importance of the C terminus as an oligomerization domain. J Biol Chem 274:14090–14093

Schmitz Y, Rateitschak K, Wolkenhauer O (2013) Analysing the impact of nucleo-cytoplasmic shuttling of beta-catenin and its antagonists APC, Axin and GSK3 on Wnt/beta-catenin signalling. Cell Signal 25:2210–2221

Schwarz-Romond T, Fiedler M, Shibata N, Butler PJ, Kikuchi A, Higuchi Y, Bienz M (2007a) The DIX domain of Dishevelled confers Wnt signaling by dynamic polymerization. Nat Struct Mol Biol 14:484–492

Schwarz-Romond T, Metcalfe C, Bienz M (2007b) Dynamic recruitment of axin by Dishevelled protein assemblies. J Cell Sci 120:2402–2412

Shibata N, Tomimoto Y, Hanamura T, Yamamoto R, Ueda M, Ueda Y, Mizuno N, Ogata H, Komori H, Shomura Y et al (2007) Crystallization and preliminary X-ray crystallographic studies of the axin DIX domain. Acta Crystallogr Sect F Struct Biol Cryst Commun 63:529–531

Shiomi K, Uchida H, Keino-Masu K, Masu M (2003) Ccd1, a novel protein with a DIX domain, is a positive regulator in the Wnt signaling during zebrafish neural patterning. Curr Biol 13:73–77

Shultz MD, Kirby CA, Stams T, Chin DN, Blank J, Charlat O, Cheng H, Cheung A, Cong F, Feng Y et al (2012) [1,2,4]triazol-3-ylsulfanylmethyl)-3-phenyl-[1,2,4]oxadiazoles: antagonists of the Wnt pathway that inhibit tankyrases 1 and 2 via novel adenosine pocket binding. J Med Chem 55:1127–1136

Silver SJ, Hagen JW, Okamura K, Perrimon N, Lai EC (2007) Functional screening identifies miR-315 as a potent activator of Wingless signaling. Proc Natl Acad Sci USA 104:18151–18156

Strovel ET, Wu D, Sussman DJ (2000) Protein phosphatase 2Calpha dephosphorylates axin and activates LEF-1-dependent transcription. J Biol Chem 275:2399–2403

Tran H, Polakis P (2012) Reversible modification of adenomatous polyposis coli (APC) with K63-linked polyubiquitin regulates the assembly and activity of the beta-catenin destruction complex. J Biol Chem 287:28552–28563

Tran H, Bustos D, Yeh R, Rubinfeld B, Lam C, Shriver S, Zilberleyb I, Lee MW, Phu L, Sarkar AA et al (2013) HectD1 E3 ligase modifies adenomatous polyposis coli (APC) with polyubiquitin to promote the APC-axin interaction. J Biol Chem 288:3753–3767

Wang S, Yin J, Chen D, Nie F, Song X, Fei C, Miao H, Jing C, Ma W, Wang L et al (2013) Small-molecule modulation of Wnt signaling via modulating the Axin-LRP5/6 interaction. Nat Chem Biol 9:579–585

Wei W, Li M, Wang J, Nie F, Li L (2012) The E3 ubiquitin ligase ITCH negatively regulates canonical Wnt signaling by targeting dishevelled protein. Mol Cell Biol 32:3903–3912

Willert K, Jones KA (2006) Wnt signaling: is the party in the nucleus? Genes Dev 20:1394–1404

Willert K, Shibamoto S, Nusse R (1999) Wnt-induced dephosphorylation of axin releases beta-catenin from the axin complex. Genes Dev 13:1768–1773

Winston JT, Strack P, Beer-Romero P, Chu CY, Elledge SJ, Harper JW (1999) The SCFbeta-TRCP-ubiquitin ligase complex associates specifically with phosphorylated destruction motifs in IkappaBalpha and beta-catenin and stimulates IkappaBalpha ubiquitination in vitro. Genes Dev 13:270–283

Wong CK, Luo W, Deng Y, Zou H, Ye Z, Lin SC (2004) The DIX domain protein coiled-coil-DIX1 inhibits c-Jun N-terminal kinase activation by Axin and dishevelled through distinct mechanisms. J Biol Chem 279:39366–39373

Yamamoto H, Kishida S, Kishida M, Ikeda S, Takada S, Kikuchi A (1999) Phosphorylation of axin, a Wnt signal negative regulator, by glycogen synthase kinase-3beta regulates its stability. J Biol Chem 274:10681–10684

Yan D, Wiesmann M, Rohan M, Chan V, Jefferson AB, Guo L, Sakamoto D, Caothien RH, Fuller JH, Reinhard C et al (2001) Elevated expression of axin2 and hnkd mRNA provides evidence that Wnt/beta -catenin signaling is activated in human colon tumors. Proc Natl Acad Sci USA 98:14973–14978

Yang LH, Xu HT, Li QC, Jiang GY, Zhang XP, Zhao HY, Xu K, Wang EH (2013) Abnormal hypermethylation and clinicopathological significance of Axin gene in lung cancer. Tumour Biol 34:749–757

Yokoyama N, Markova NG, Wang HY, Malbon CC (2012) Assembly of Dishevelled 3-based supermolecular complexes via phosphorylation and Axin. J Mol Signal 7:8

Zhang Y, Liu S, Mickanin C, Feng Y, Charlat O, Michaud GA, Schirle M, Shi X, Hild M, Bauer A et al (2011) RNF146 is a poly(ADP-ribose)-directed E3 ligase that regulates axin degradation and Wnt signalling. Nat Cell Biol 13:623–629

13

Establishment and function of tissue-resident innate lymphoid cells in the skin

Jie Yang[1,2], Luming Zhao[1], Ming Xu[1], Na Xiong[1✉]

[1] Centre for Molecular Immunology and Infectious Diseases and Department of Veterinary and Biomedical Sciences, The Pennsylvania State University, University Park, PA 16802, USA
[2] Department of Immunology, The University of Texas MD Anderson Cancer Center, 7455 Fannin Street, Houston, TX 77054, USA
✉ Correspondence: nux1@psu.edu (N. Xiong)

ABSTRACT

Innate lymphoid cells (ILCs) are a newly classified family of immune cells of the lymphoid lineage. While they could be found in both lymphoid organs and non-lymphoid tissues, ILCs are preferentially enriched in barrier tissues such as the skin, intestine, and lung where they could play important roles in maintenance of tissue integrity and function and protection against assaults of foreign agents. On the other hand, dysregulated activation of ILCs could contribute to tissue inflammatory diseases. In spite of recent progress towards understanding roles of ILCs in the health and disease, mechanisms regulating specific establishment, activation, and function of ILCs in barrier tissues are still poorly understood. We herein review the up-to-date understanding of tissue-specific relevance of ILCs. Particularly we will focus on resident ILCs of the skin, the outmost barrier tissue critical in protection against various foreign hazardous agents and maintenance of thermal and water balance. In addition, we will discuss remaining outstanding questions yet to be addressed.

KEYWORDS innate lymphoid cells, skin, migration, chemokine receptor, homeostasis, inflammation

INTRODUCTION

Innate lymphoid cells (ILCs) are the newly identified members of the innate immune system. ILCs have the property of classic lymphoid cells but lack rearranged antigen-specific receptors (Spits et al., 2013; Walker et al., 2013). ILCs encompass conventional natural killer (NK) cells and lymphoid tissue inducer (LTi) cells, which were first discovered in 1975 and 1997, respectively (Kiessling et al., 1975a; Kiessling et al., 1975b; Mebius et al., 1997), and other recently described subsets. ILCs can be generally identified by their expression of certain subunits of cytokine receptors such as CD25 (interleukin (IL)-2 receptor α) and CD127 (IL-7 receptor α) but no cell-surface molecules that identify the other types of immune cells (lineage marker-negative or Lin^-) (Artis and Spits, 2015). While ILCs could be found in nearly every tissue and organ, they are preferentially enriched in barrier tissues such as the skin, intestine and lung, where they participate in local tissue homeostasis and inflammation (Artis and Spits, 2015; McKenzie et al., 2014).

SUBSETS OF INNATE LYMPHOID CELLS

In analog to helper T (Th) cells, ILCs are conventionally divided into three groups based on their differential expression of effector cytokines and developmental requirements for transcription factors (Spits et al., 2013). The group 1 ILCs are composed of ILCs capable of producing interferon γ (IFNγ), which include NK cells and non-cytotoxic helper-like ILC1s. NK cells are innate effector lymphocytes that are capable of inducing granule-mediated cytotoxicity by expressing perforin and granzymes, as well as producing IFNγ and tumor necrosis factor α (TNFα) (Herberman et al., 1975a; Herberman et al., 1975b; Kiessling et al., 1975a; Kiessling et al., 1975b; Vivier et al., 2011). On the other hand, non-cytotoxic ILC1s are innate helper lymphocytes that are capable of producing IFNγ, TNFα and other Th1-associated cytokines, but lack cytotoxic abilities (Artis and Spits, 2015; Bernink et al., 2013; Vonarbourg et al., 2010). Recent studies identified additional developmental and phenotypic differences between NK and ILC1s (Constantinides et al., 2014;

Klose et al., 2014). While eomesodermin (Eomes) is required for development of NK cells, T-bet and GATA3 are critical for ILC1s (Daussy et al., 2014; Gordon et al., 2012; Yagi et al., 2014). In addition, NK cells re-circulate in the blood while ILC1s predominantly reside in tissues (Sojka et al., 2014). Group 2 ILCs produce Th2-associated cytokines such as IL-4, IL-5, and IL-13 when activated, and depend on GATA3 and RORα for their development (Halim et al., 2012; Hoyler et al., 2012). Group 3 ILCs are highly heterogeneous, although they preferentially produce Th17-assciated cytokines including IL-17A and IL-22 and require RORγt for the development. Based on their expression of the chemokine receptor CCR6, group 3 ILCs can be divided into two subsets (Klose et al., 2013; Sawa et al., 2011). $CCR6^+$ ILC3s consist of $CD4^+$ and $CD4^-$ LTi cells, while $CCR6^-$ ILC3s are composed of two subpopulations that have distinct expression of the natural cytotoxicity receptor (NCR) NKp46 (Klose et al., 2013; Rankin et al., 2013).

ILCs of the different groups can show certain functional plasticity. For example, in response to IL-12 and IL-18, $CCR6^-NKp46^-$ ILC3s increase expression of T-bet and differentiate into $CCR6^-NKp46^+$ ILC3s, which can further increase T-bet and decrease RORγt expression and differentiate into ex-$ROR\gamma t^+$ ILC3s that display ILC1-like functions with increased production of IFNγ and decreased IL-17A and IL-22 (Artis and Spits, 2015; Bernink et al., 2013; Cella et al., 2010; Klose et al., 2013; Vonarbourg et al., 2010). Using the single cell transcription analysis, a recent study found that there are significant heterogeneities within ILCs and that some of them display transcription profiles associated with more than one group of ILCs, consistent with the notion that they are of transition stages (Gury-BenAri et al., 2016).

DEVELOPMENT OF INNATE LYMPHOID CELLS

Common lymphoid progenitor cells of the bone marrow give rise to all lymphocytes, including ILCs. Several intermediate progenitors have been identified that have the potential to give rise to different subsets of ILCs. Committed ILC precursors can differentiate into all ILC subsets, including NK cells, but not T and B cells (Seillet et al., 2014; Yu et al., 2014). They are termed α-lymphoid precursor (αLP) cells because they express the integrin α4β7, but also known as common innate lymphoid progenitor (CILP). αLP cells express the chemokine receptor CXCR6 and require the transcription factor NFIL3 for development (Seillet et al., 2014; Yu et al., 2014). Downstream of αLP, two populations of ILC precursor cells have been identified based on the expression of inhibitor of DNA binding 2 (Id2) and promyeloid leukemia zinc finger (PLZF) (Constantinides et al., 2014; Klose et al., 2014). Common helper-like innate lymphoid progenitor (CHILP) cells are $Id2^+Lin^-IL\text{-}7R\alpha^+\alpha4\beta7^+CD25^-$ and dependent on GATA3 for development. CHILP can give rise to non-cytotoxic ILC1s, ILC2s and ILC3s, but no NK cells (Klose et al., 2014). Innate lymphoid progenitor (ILCP) cells are $PLZF^{high}Lin^-IL\text{-}7R\alpha^+cKit^+\alpha4\beta7^{high}CXCR6^-$. ILCP differentiates efficiently into the majority of ILC subsets except NK cells and LTi cells (Constantinides et al., 2014). ILCP is thus placed downstream of CHILP because ILCP has a more restricted differentiation potential than CHILP. More recently, a population of the transcription factor TCF-1^+ early ILC progenitor (EILP) cells were also described (Yang et al., 2015). The TCF-1^+ EILP cells are $IL\text{-}7R\alpha^{neg/low\text{-}}$ $Thy1^-Lin^-CD122^-CXCR6^-$ and express low levels of PLZF. EILP could develop into all subsets of ILCs, including conventional NK cells, but not T/B cells (Yang et al., 2015). TCF-1 knockout mice have reduced EILP as well as NKP and CHILP, suggesting that TCF-1 is critical for development of EILP and that NKP and CHILP are downstream of EILP (Yang et al., 2015). The relationship between EILP and αLP is not fully characterized even though EILP seems to functionally overlap with αLP in their ability to differentiate into ILCs but not T/B cells. Detailed developmental pathway and regulation of the ILC differentiation have been extensively reviewed (Ishizuka et al., 2016; Serafini et al., 2015; Yang and Bhandoola, 2016; Zook and Kee, 2016).

FUNCTION OF INNATE LYMPHOID CELLS IN MAINTENANCE OF TISSUE HOMEOSTASIS AND HOST PROTECTION

ILCs are enriched in barrier tissues where they play important roles in maintenance of tissue homeostasis, regulation of equilibrium between the host and microbiota, and immunity against pathogens (Artis and Spits, 2015; McKenzie et al., 2014; Walker et al., 2013). In addition, ILCs participate in the development of lymphoid tissues, tissue remodeling, and metabolic homeostasis (Artis and Spits, 2015; McKenzie et al., 2014; Walker et al., 2013).

Functions of NK cells of the group 1 ILC family have been well established, which include killing virus-infected cells through cytolysis mediated by granzymes and perforin, and targeting tumor cells that have lost expression of the class I major histocompatibility complex (MHCI) (Biron et al., 1999; Vivier et al., 2012). The other non-cytotoxic ILC1s can produce high amounts of IFNγ and TNFα, contributing to the protection of hosts against bacteria and intracellular parasites (Klose et al., 2014; Klose et al., 2013). More recently, it was also reported that ILC1s play an important role in the early protection against tumor genesis (Dadi et al., 2016).

ILC2s are a critical innate source of type 2 cytokines, such as IL-4, IL-5, IL-9 and IL-13, which can induce eosinophilia, mastocytosis, activation of alternatively activated macrophages, goblet cell hyperplasia and muscle contractility, and tissue repair (Pulendran and Artis, 2012). ILC2s can be activated by cytokines such as IL-25, IL-33, and thymic stromal lymphopoietin (TSLP) (Hams et al., 2014; Kim et al., 2013; McHedlidze et al., 2013). During parasitic helminth infection, ILC2s are identified as an important source of IL-13 critical for the parasite expulsion (Moro et al., 2016; Neill et al., 2010; Price et al., 2010). ILC2s also express amphiregulin and are involved in tissue repairing after virus infection (Monticelli et al., 2011). In addition, growing evidence demonstrates roles of ILC2s in metabolic homeostasis (Hams et al., 2013; Molofsky et al., 2013; Stanya et al., 2013). ILC2s in visceral adipose tissue produce IL-5 and IL-13 and maintain responses of eosinophil and alternatively activated macrophages, lack of which results in increased adiposity and insulin resistance (Molofsky et al., 2013). ILCs can respond to changes in nutrients such as Vitamin A, deficiency of which leads to reduced ILC3s but increased ILC2s in intestines and is associated with defective antibacterial immunity but enhanced anti-helminth immunity (Spencer et al., 2014). Circadian clock also regulates cytokine production of intestinal ILC2s through vasoactive intestinal peptides (Nussbaum et al., 2013).

The role of LTi cells of the group 3 ILCs family in the development of lymphoid organs of lymph nodes (LNs) and Peyer's patches has long been recognized (De Togni et al., 1994; Eberl et al., 2004). Lymphotoxin $\alpha 1\beta 2$ produced by LTi cells binds to lymphotoxin β receptor on stromal cells to induce the production of chemokines and upregulation of adhesion molecules, which recruit leukocytes to form lymphoid structures (van de Pavert et al., 2009). ILC3s have also been implicated in tissue repair following inflammation or damage in multiple tissues, including intestines, thymi, and lungs (Dudakov et al., 2012; Sawa et al., 2011; Taube et al., 2011). ILC3s are an important source of IL-22, which promotes epithelial cells to express antimicrobial peptides to protect against pathogens (Sonnenberg et al., 2011; Zheng et al., 2008). In addition, ILC3s interact with stromal cells and other immune cells to maintain balance between the host and microbiota by limiting dissemination of commensal bacteria and inappropriate immune responses to them (Sonnenberg et al., 2012).

PARTICIPATION OF INNATE LYMPHOID CELLS IN PATHOGENESIS OF DISEASES

Although ILCs normally play host-protective roles, dysregulated activation of ILCs could lead to inflammatory diseases (Artis and Spits, 2015; McKenzie et al., 2014; Walker et al., 2013). IFNγ-producing ILCs, including ILC1s and ex-RORγt$^+$ ILC3s, can induce inflammation in the intestines of both mice and humans (Bernink et al., 2013; Buonocore et al., 2010; Fuchs et al., 2013). In the adipose tissue, high fat diet could activate ILC1s to produce IFNγ and promote inflammatory macrophage polarization and development of insulin resistance (O'Sullivan et al., 2016).

Chronic ILC2 activation was reported to contribute to a large variety of tissue inflammatory disorders such as asthma in the lung and atopic dermatitis in the skin, which are generally associated with over-production of the type 2 inflammatory cytokines such as IL-4, IL-5, and IL-13. ILC2s expand and produce the type 2 cytokines in mice with allergic lung inflammation (Chang et al., 2011; Halim et al., 2014), consistent with the increase of ILC2s in peripheral blood of patients with asthma (Bartemes et al., 2014; Moffatt et al., 2010). Increased frequency of ILC2s was found in lesional skin of mouse models and human patients with atopic dermatitis (Imai et al., 2013; Kim et al., 2013; Salimi et al., 2013).

ILC3s could contribute to inflammatory diseases and tumorigenesis in various tissues. IL-17- and IL-22-producing ILC3s have been linked with the inflammatory skin disease psoriasis, inflammatory bowl disease, obesity-associated asthma, and multiple sclerosis (Buonocore et al., 2010; Kim et al., 2014b; Pantelyushin et al., 2012; Perry et al., 2012; Powell et al., 2012; Teunissen et al., 2014; Villanova et al., 2014). IL-22 produced by ILC3s can contribute to development of colon cancer in a mouse model, consistent with the presence of IL-22-producing CD3$^-$ cells within human colorectal carcinoma tumors (Kirchberger et al., 2013).

INNATE LYMPHOID CELLS ARE PREDOMINANTLY TISSUE-RESIDENT CELLS

Specific functions of ILCs in homeostasis and inflammation are associated with their unique localization in peripheral tissues. Recently, several groups reported that ILCs are predominantly tissue-resident cells that do not circulate in the body. In parabiotic studies, helper-like ILCs in both non-lymphoid tissues and lymphoid organs such as the small intestine, salivary gland, lung, mensenteric lymph node, and adipose tissue of two joined adult mice with shared blood circulation systems remain locally and do not move from one mouse to another (Gasteiger et al., 2015; O'Sullivan et al., 2016). However, conventional NK cells are mostly circulating cells (Peng et al., 2013; Sojka et al., 2014). Furthermore, systemic inflammation or infection did not mobilize tissue-resident ILCs into the circulation either (Gasteiger et al., 2015). These suggest that ILCs or their progenitor cells might settle in specific peripheral tissues at early ontogenic stages and are maintained and expanded locally. Consistent with this notion, progenitor cells of ILCs have been identified in lymph nodes of humans that could give rise to all innate lymphoid cells in the *in vitro* development analysis (Scoville et al., 2016). In addition, the bone marrow reconstitution in human patients with severe combined immunodeficiency (SCID) due to mutation of the gene encoding the common γ chain cytokine receptor subunit IL-2Rγ or the tyrosine kinase JAK3 did not restore presence of ILCs in various peripheral tissues (Vely et al., 2016), supporting the notion that tissue-resident ILCs are established at early ontogenic stages. Mechanisms regulating specific localization and maintenance of ILCs in various peripheral tissues are still poorly understood.

MIGRATION AND ESTABLISHMENT OF INNATE LYMPHOID CELLS IN THE SKIN

Skin is the outmost barrier tissue constantly exposed to assaults of various foreign agents. Skin is enriched with ILCs and ILCs of all the three groups could be found in the skin (Yang et al., 2016). Based on expression of the transcription factor GATA3 and the cytokines such as IL-4 and IL-5, ILC2s account for a major fraction of total skin ILCs in adult mice (Roediger et al., 2013; Yang et al., 2016). Although few ILCs of the skin in mice express the transcription factor RORγt, a significant fraction of skin ILCs are capable of producing IL-17A, suggesting that they are ILC3s or ILC3-like cells independent of RORγt (Yang et al., 2016). However, their lineage relationship with other RORγt-expressing ILC3s is not known. The group 1 ILCs, including NK cells, are present in the skin of mice (Luci et al., 2009; Yang et al., 2016), and they account for a smaller fraction of total skin ILCs than ILC2s or ILC3s in adult mice (Yang et al., 2016). On the other hand, in the skin of fetal and newborn mice, there are abundant NK1.1$^+$ ILC1-type cells although their origin and lineage association are not known (Almeida et al., 2015). As in mice, ILCs of all the three groups are found in the healthy skin of humans (Dyring-Andersen et al., 2014; Ebert et al., 2006; Salimi et al., 2013; Teunissen et al., 2014; Villanova et al., 2014). Based on their expression of the prostaglandin D2 receptor CRTH2, ILC2s account for 25%–40% of skin ILCs in humans (Dyring-Andersen et al., 2014; Salimi et al., 2013; Teunissen et al., 2014; Villanova et al., 2014). ILC3s, including both NCR$^+$ and NCR$^-$ subsets, account for about 50% of total skin ILCs in humans (Dyring-Andersen et al., 2014; Teunissen et al., 2014; Villanova et al., 2014). The rest are ILC1s and others uncharacterized.

Since ILCs share many common regulatory pathways with helper T cells for their development and function, we investigated whether skin-specific ILCs are programmed in skin-draining lymph nodes to acquire their skin-homing property for establishment of their skin residency, paralleling the process by which the skin-homing property of conventional αβ T cells is imprinted in skin-draining lymph nodes (Yang et al., 2016). Among the most skin-specific homing molecules expressed on skin-homing T cells and ILCs is the chemokine receptor CCR10 (Sigmundsdottir and Butcher, 2008; Sigmundsdottir et al., 2007; Xiong et al., 2012; Yang et al., 2016). A ligand for CCR10, CCL27, is highly and specifically expressed by keratinocytes of the healthy skin in both humans and mice (Homey et al., 2000; Morales et al., 1999). The adhesion molecules E- and P-selectin ligands and chemokine receptors CCR4, CCR6, and CCR8 are also involved in the skin-homing process in the healthy skin (Austrup et al., 1997; Campbell and Butcher, 2002; Campbell et al., 1999; Picker et al., 1993; Reiss et al., 2001; Weninger et al., 2000). In addition, different sets of homing molecules are involved in migration of immune cells to the inflamed skin under specific inflammatory conditions than those under homeostatic conditions (Lonsdorf et al., 2009; Masopust and Schenkel, 2013; Mora and von Andrian, 2006). For example, high induction of the chemokines CCL1 and CCL18 is found in the lesional skin of patients with atopic dermatitis (Gombert et al., 2005; Pivarcsi et al., 2004). The chemokine receptor CXCR3 is involved in migration of T cells to the inflamed skin, where its ligands CXCL9, CXCL10, and CXCL11 are produced (Flier et al., 2001). While CCL27 could be upregulated in the skin under inflammatory conditions of allergic contact dermatitis, atopic dermatitis, and psoriasis (Homey et al., 2002; Moed et al., 2004), it is profoundly suppressed in severe psoriatic skin lesions (Gudjonsson et al., 2010; Quaranta et al., 2014; Riis et al., 2011a). In the same patient, CCL27 expression is much lower in psoriatic lesional skin than in peri-lesional skin (Karakawa et al., 2014; Riis et al., 2011a), suggesting that its expression is associated with severity of the disease. Indeed, CCL27 is upregulated in the skin by the inflammatory cytokines TNFα and IL-1β at the early phase of inflammation (Riis et al., 2011b; Vestergaard et al., 2004) but is suppressed by pathognomonic cytokines such as IL-17A and IFNγ at the late, severe stage of skin inflammation (Kanda et al., 2005; Karakawa et al., 2014). We recently found that CCR10 is critical for migration of T cells into the skin under steady-state conditions but is dispensable for their infiltration into inflamed skin (Fu et al., 2016; Xia et al., 2014). Like T cells, ILCs are also programmed in skin-draining lymph nodes to acquire the CCR10 expression for their migration to the skin under steady-state conditions (Yang et al., 2016). In CCR10-knockout mice, ILCs generated in skin-draining lymph nodes is defective in migration into the skin and increasingly diverted into spleens and other sites where CCR10 ligands are not expressed (Yang et al., 2016). Moreover, skin-specific CD207$^+$ dendritic cells are involved in the programming of skin-homing CCR10$^+$ ILCs in skin-draining lymph nodes (Yang et al., 2016) (Fig. 1).

Dependent on environmental cues of lymphoid organs in which T cells and ILCs are activated, they could be also imprinted to acquire homing properties to other distinct peripheral tissues (Campbell and Butcher, 2002; Kantele et al., 1999; Masopust and Schenkel, 2013; Mora et al., 2003; Mora et al., 2005; Mora and von Andrian, 2006; Rott et al., 1997; Weninger et al., 2000). A recent study reported that ILCs undergo a homing receptor switch by upregulation of CCR9 and α4β7 and downregulation of CCR7 in intestine-draining mesenteric lymph nodes in a retinoic acid-dependent manner for their migration into the intestine (Kim et al., 2015). Together with the findings in the skin, these results suggest that specific epithelial tissues and their associated lymph nodes form an integral unit for generation and maintenance of tissue-specific ILCs. However, there might be other regulatory mechanisms in establishment of tissue-specific ILCs (Kim et al., 2015; Yang et al., 2016). In our study of the generation of skin-specific CCR10$^+$ ILCs in skin-draining lymph nodes, local inflammatory conditions suppress programming of skin-homing CCR10$^+$ ILCs (Yang et al., 2016) (Fig. 1). Molecular factors regulating the distinctive generation and establishment of skin-specific ILCs under homeostatic and inflammatory conditions are mostly unknown.

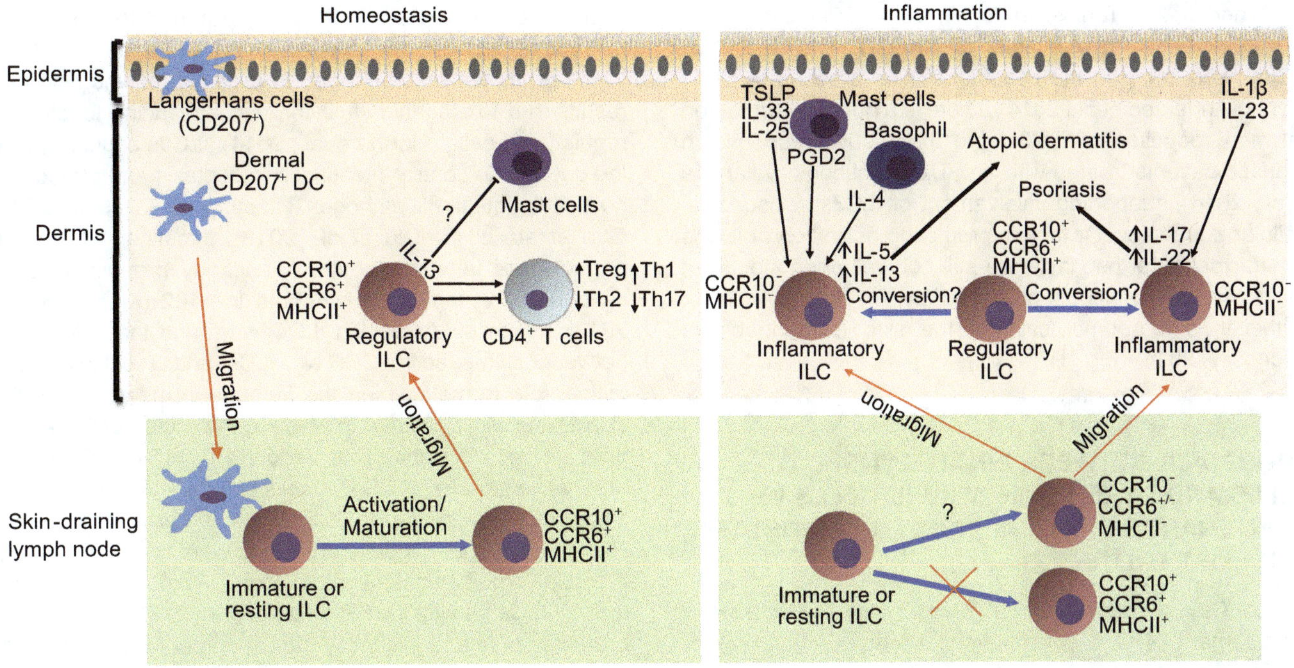

Figure 1. Schematic illustration of establishment and function of ILCs in the skin homeostasis and inflammation. Under homeostatic conditions, CCR10$^+$ ILCs with regulatory properties are preferentially generated in skin-draining lymph nodes and migrate into the skin where they promote the local immune homeostasis by controlling balanced presence and activation of other immune cells. Under inflammatory conditions, there is reduced generation of CCR10$^+$ regulatory ILCs in skin-draining lymph nodes. In addition, there is also increased conversion of regulatory CCR10$^+$ ILCs to inflammatory CCR10$^-$ ILCs, which promote immune activation and inflammatory processes in diseases such as atopic dermatitis and psoriasis, depending cytokines produced by the activated inflammatory ILCs.

INTERACTION BETWEEN INNATE LYMPHOID CELLS AND LOCAL IMMUNE AND NON-IMMUNE CELLS IN PARTICIPATION OF SKIN INFLAMMATION

Compared to the healthy skin, ILC2s are increased in the skin of patients with atopic dermatitis (Kim et al., 2013; Salimi et al., 2013), suggesting that ILC2s are involved in atopic dermatitis, an inflammatory disease associated with over-active type 2 immune responses to environmental allergens and barrier dysfunction (Leung, 2013). In a mouse model of atopic dermatitis induced by topical application of Calcipotriene (a synthetic Vitamin D3 derivative), ILC2-derived IL-5 and IL-13 were reported to play an important role in promoting the disease development (Kim et al., 2013) (Fig. 1). It was also found that skin ILC2s are activated by TSLP produced by skin keratinocytes in response to the topical Calcipotriene-treatment (Kim et al., 2013). Another study identified the role of IL-33 and IL-25 in activating skin ILC2s to promote atopic dermatitis (Salimi et al., 2013). Consistent with this, transgenic overexpression of IL-33 in keratinocytes promotes accumulation of ILC2s in the skin and induces atopic dermatitis-like disease (Imai et al., 2013). Prostaglandin D2 (PGD2), acting through its receptor CRTH2 expressed on human skin ILC2s, induces migration and cytokine production of ILC2s in human atopic dermatitis skin (Xue et al., 2014). Mast cells could be an important source of prostaglandin D2 for activation of ILC2s (Barnig et al., 2013; Roediger et al., 2013). IL-4 derived from basophils was also shown to regulate the proliferation of skin ILC2s, which express the IL-4 receptor IL-4Rα, and induce atopic dermatitis-like disease (Kim et al., 2014a). Given the role of E-cadherin in inhibiting the cytokine production of ILC2s via binding to the inhibitory receptor KLRG1 expressed on skin ILC2s, downregulation of E-cadherin in atopic dermatitis skin might also contribute to the activation of skin ILC2s (Salimi et al., 2013).

Increasing evidence implicates involvement of ILC3s in psoriasis, a skin inflammatory disease largely resulting from overproduction of Th17-associated cytokines such as IL-17A (Dyring-Andersen et al., 2014; Pantelyushin et al., 2012; Teunissen et al., 2014; Villanova et al., 2014). In a mouse model of psoriasis induced by topical application of Imiquimod, it was found that ILC3s are a cellular source of IL-17A and IL-22 that mediate the Imiquimod-induced psoriasis-like disease (Pantelyushin et al., 2012). In patients with psoriasis, NKp44$^+$ ILC3s are increased in both lesional and non-lesional skin and NKp44$^-$ ILC3s have the potential to differentiate to NKp44$^+$ ILC3s in response to stimulation of

IL-1β and IL-23 (Teunissen et al., 2014; Villanova et al., 2014). Increased percentages of $NKG2D^+$ $ROR\gamma t^+CD56^+$ ILC3s were found in the psoriatic skin in a separate study (Dyring-Andersen et al., 2014). There is a positive correlation between reduction of ILC3s and therapeutic response in psoriatic patients treated with anti-TNFα antibody (Villanova et al., 2014), supporting involvement of ILC3s in psoriasis. Cellular and molecular factors promoting activation of ILC3s in participation of psoriasis are still not well understood but could involve cytokines produced by myeloid, lymphoid, epithelial, and stromal cells of the skin (Gasteiger and Rudensky, 2014) (Fig. 1).

CROSSTALK BETWEEN INNATE LYMPHOID CELLS AND OTHER SKIN IMMUNE CELLS IN MAINTENANCE OF LOCAL TISSUE HOMEOSTASIS AND HOST PROTECTION

While ILCs are relatively well studied for their participation in development of skin inflammatory diseases, their physiological role in maintenance of the skin homeostasis and host protection is still poorly understood. Since ILC2s are capable of producing amphiregulin, a member of epithelial growth factor family, they could potentially contribute to wound healing (Salimi et al., 2013). Similarly, IL-22 produced by ILC3s is an important growth factor of keratinocytes. Indeed, both ILC2s and ILC3s were recently reported to promote the wound healing process in mouse models (Li et al., 2016; Rak et al., 2016). The ILC2 response is promoted by the cutaneous injury in an IL-33-dependent manner (Rak et al., 2016), while ILC3s are recruited into wounded dermis by epidermal Notch1, which functions as a damage response signal (Li et al., 2016).

Skin ILCs interact with other skin-resident immune cells to maintain the immune homeostatic condition. In the steady-state skin, the preferential interaction of ILC2s and mast cells has been detected (Barnig et al., 2013; Roediger et al., 2013). IL-13 produced by skin ILC2s suppresses the function of mast cells in homeostatic conditions, suggesting a regulatory role of skin ILC2s (Roediger et al., 2013). ILCs might also interact with T cells to regulate their homeostasis in the skin (Yang et al., 2016), as reported in recent studies on the cross-regulation between ILCs and T cells in several other barrier tissues. ILC2s express MHCII and activate T cells to induce production of IL-2, which in turn promote the proliferation of ILC2s and their expression of type 2 cytokines for worm expulsion in the gut (Oliphant et al., 2014). In a model of allergic lung inflammation, IL-13 produced by ILC2s promotes the migration of dendritic cells to draining lymph nodes where they prime T cells to differentiate into Th2 cells (Halim et al., 2014). The co-stimulatory molecule OX40L and ILC2-derived IL-4 were reportedly involved in crosstalk between ILC2s and T cells (Drake et al., 2014). Intestinal MHCII-expressing ILC3s have the capacity to present antigens to $CD4^+$ T cells and dampen their responses to microbiota (Hepworth et al., 2013). In addition, intestinal microbiota induces the production of GM-CSF by ILC3s, which promotes macrophages to produce retinoic acid and results in a tolerogenic state through enhancing function of regulatory T cells (Mortha et al., 2014). ILC3s could regulate the gut flora to control intestinal homeostasis and protect the host from intestinal pathogen infection (Goto et al., 2014; Guo et al., 2015; Qiu et al., 2013). Intestinal ILC3s also promote the immunoglobulin A antibody production by B cells through membrane lymphotoxin α1β2 (Kruglov et al., 2013). In our study of skin ILCs, we found that interaction between skin-specific $CCR10^+$ ILCs and T cells plays a critical role in maintaining the immune homeostasis of the skin (Yang et al., 2016). In *Rag1*$^{-/-}$ mice that lack T and B cells, there are significantly reduced $CCR10^+$ ILCs in the skin, while transfer of $CD4^+$ T cells could partially restore the homeostatic presence of $CCR10^+$ ILCs (Yang et al., 2016). Reciprocally, ILCs regulate homeostasis of resident T cells in the skin. When transferred into *Il2rg*$^{-/-}$*Rag2*$^{-/-}$ mice, which lack all ILCs as well as T/B cells, donor $CD4^+$ T cells in skin of hosts contain significantly higher percentages of IL-17A-producing Th17 and IL-5-producing Th2 cells but much lower percentages of IFNγ-producing Th1 cells than those transferred into *Rag1*$^{-/-}$ recipients that have ILCs (Yang et al., 2016). In absence of ILCs, there are also significantly reduced presence of regulatory T cells in the skin, indicating an important role of ILCs in regulating the homeostasis of skin T cells (Yang et al., 2016) (Fig. 1).

PERSPECTIVES

The research over past several years has started to reveal roles of ILCs in the skin and unravel mechanisms regulating establishment and function of the skin-resident ILCs. However, several major questions remain to be addressed. First, physiological functions of skin-resident ILCs in host protection and immune homeostatic maintenance are still poorly understood. While ILCs have been generally implicated in protection against infection, control of commensal bacteria and regulation of immune activation in various barrier tissues, the function of resident ILCs specifically in the skin is not well studied. Considering that ILCs are maintained locally in different barrier tissues, functions and regulation of ILCs in the skin and other barrier tissues are likely regulated differently with local environmental cues. For example, the CCR10/CCL27 axis is involved in the regulation of resident ILCs in the skin but not other sites (Yang et al., 2016). Further studies are needed to understand how tissue-specific factors are involved in regulation and function of skin-specific ILCs. Associated with this, molecular mechanisms regulating tissue-specific establishment of ILCs in the skin versus other barrier tissues need further investigation. Furthermore, functions of different groups of ILCs in the skin require delineation at different ontogenic stages. For example, although group 1 ILCs, including conventional NK cells, are abundantly present in the skin of both humans and mice, particularly in the newborn and young (Batista et al., 2013; Ebert et al., 2006; Luci et al., 2009; Teunissen et al., 2014; Villanova et al., 2014), their roles in establishment of skin homeostasis are mostly unknown and need further investigation.

Second, mechanisms regulating differential involvement of ILCs in the skin homeostasis and inflammation are largely unknown. Studies of ILCs of the skin and other barrier tissues have showed that ILCs could be involved in both homeostatic regulation and tissue inflammatory processes. However, it is currently unknown whether and how ILCs involved in homeostatic regulation are different from ILCs involved in immune inflammatory processes in the skin. Our studies have found that $CCR10^+$ ILCs are dominant in the healthy skin where they promote balanced maintenance of resident T cells (Yang et al., 2016). In immune dysregulatory or inflammatory conditions of $Rag1^{-/-}$, $Foxp3^{-/-}$ or topical Calcipotriol-treated wild-type mice, $CCR10^+$ ILCs are decreased while $CCR10^-$ ILCs are increased in the skin (Yang et al., 2016), suggesting that $CCR10^+$ and $CCR10^-$ ILCs might be differently involved in skin immune homeostatic regulation and activation (Fig. 1). Further study on how $CCR10^+$ and $CCR10^-$ skin ILCs are regulated and function should help understand basic biology of the skin-specific ILCs in the health and disease.

Third, the involvement of ILCs in promoting diseases has raised interest in modulating ILC functions for the treatment of diseases. In other tissues, antibodies against ILC2-activating cytokines, such as IL-25 and IL-33, as well as those targeting ILC2-produced cytokines, including IL-5 and IL-13, have been found to suppress ILC2 proliferation and function in tissue inflammatory diseases in the lung (Gauvreau et al., 2014; Hambly and Nair, 2014). Blockage of the binding of prostaglandin D2 to CRTH2 expressed on human ILC2s inhibits the migration and cytokine production of ILC2s (Barnig et al., 2013; Xue et al., 2014). Montelukast, a leukotriene receptor antagonist, inhibits the cytokine production of ILC2s by blocking interaction between leukotriene D4 and cysteinyl leukotriene receptor 1 (CysLT1R) expressed by ILC2s (Doherty et al., 2013). ILC3s have been targeted in the treatment of multiple sclerosis by anti-IL-2R antibody (Perry et al., 2012). Antagonists of RORγt, which can regulate the activity of Th17, might also target ILC3s (Bernink et al., 2015; Solt et al., 2011; Withers et al., 2016). However, while there is a report showing that inhibition of RORγt reduces differentiation of $CD127^+$ ILC1s to ILC3s (Bernink et al., 2015), inhibition of RORγt does not impair function of mature ILC3s (Bernink et al., 2015; Withers et al., 2016). How to target ILCs in the skin to modulate their cytokine production for treatment of skin inflammatory diseases is still understudied and might represent a new area of study for disease treatment.

ACKNOWLEDGEMENTS

Research reported in this publication was partly supported by the National Institute of Allergy and Infectious Diseases and the National Institute of Arthritis and Musculoskeletal and Skin Diseases of the National Institutes of Health under Award Numbers AR064831 and AI071043 (to N.X). The content is solely the responsibility of the authors and does not necessarily represent the official views of the National Institutes of Health. J.Y. is also supported by an institutional match fund of a NIH pre-doctoral training grant in "Animal Models of Inflammation" (T32 AI074551) from The Pennsylvania State University.

COMPLIANCE WITH ETHICS GUIDELINES

Jie Yang, Luming Zhao, Ming Xu, and Na Xiong declare that they have no conflict of interest. All institutional and national guidelines for the care and use of laboratory animals were followed.

ABBREVIATIONS

αLP, α-lymphoid precursor; CHILP, common helper-like innate lymphoid progenitor; CILP, common innate lymphoid progenitor; EILP, early ILC progenitor; Id2, inhibitor of DNA binding 2; IFNγ, interferon γ; IL-2, interleukin-2; ILCP, innate lymphoid progenitor; ILCs, innate lymphoid cells; LNs, lymph nodes; LTi, lymphoid tissue inducer; MHCI, class I major histocompatibility complex; NCR, natural cytotoxicity receptor; NK, natural killer; PGD2, prostaglandin D2; PLZF, promyeloid leukemia zinc finger; SCID, severe combined immunodeficiency; Th, helper T; TNFα, tumor necrosis factor α; TSLP, thymic stromal lymphopoietin.

REFERENCES

Almeida FF, Tenno M, Brzostek J, Li JL, Allies G, Hoeffel G, See P, Ng LG, Fehling HJ, Gascoigne NR et al (2015) Identification of a novel lymphoid population in the murine epidermis. Sci Rep 5:12554

Artis D, Spits H (2015) The biology of innate lymphoid cells. Nature 517:293–301

Austrup F, Vestweber D, Borges E, Lohning M, Brauer R, Herz U, Renz H, Hallmann R, Scheffold A, Radbruch A, Hamann A (1997) P- and E-selectin mediate recruitment of T-helper-1 but not T-helper-2 cells into inflammed tissues. Nature 385:81–83

Barnig C, Cernadas M, Dutile S, Liu X, Perrella MA, Kazani S, Wechsler ME, Israel E, Levy BD (2013) Lipoxin A4 regulates natural killer cell and type 2 innate lymphoid cell activation in asthma. Sci Transl Med 5:174ra126

Bartemes KR, Kephart GM, Fox SJ, Kita H (2014) Enhanced innate type 2 immune response in peripheral blood from patients with asthma. J Allergy Clin Immunol 134(671–678):e674

Batista MD, Ho EL, Kuebler PJ, Milush JM, Lanier LL, Kallas EG, York VA, Chang D, Liao W, Unemori P et al (2013) Skewed distribution of natural killer cells in psoriasis skin lesions. Exp Dermatol 22:64–66

Bernink JH, Peters CP, Munneke M, te Velde AA, Meijer SL, Weijer K, Hreggvidsdottir HS, Heinsbroek SE, Legrand N, Buskens CJ et al (2013) Human type 1 innate lymphoid cells accumulate in inflamed mucosal tissues. Nat Immunol 14:221–229

Bernink JH, Krabbendam L, Germar K, de Jong E, Gronke K, Kofoed-Nielsen M, Munneke JM, Hazenberg MD, Villaudy J, Buskens CJ et al (2015) Interleukin-12 and -23 control plasticity of CD127(+) Group 1 and Group 3 innate lymphoid cells in the intestinal lamina propria. Immunity 43:146–160

Biron CA, Nguyen KB, Pien GC, Cousens LP, Salazar-Mather TP (1999) Natural killer cells in antiviral defense: function and regulation by innate cytokines. Annu Rev Immunol 17:189–220

Buonocore S, Ahern PP, Uhlig HH, Ivanov II, Littman DR, Maloy KJ, Powrie F (2010) Innate lymphoid cells drive interleukin-23-dependent innate intestinal pathology. Nature 464:1371–1375

Campbell DJ, Butcher EC (2002) Rapid acquisition of tissue-specific homing phenotypes by CD4(+) T cells activated in cutaneous or mucosal lymphoid tissues. J Exp Med 195:135–141

Campbell JJ, Haraldsen G, Pan J, Rottman J, Qin S, Ponath P, Andrew DP, Warnke R, Ruffing N, Kassam N et al (1999) The chemokine receptor CCR4 in vascular recognition by cutaneous but not intestinal memory T cells. Nature 400:776–780

Cella M, Otero K, Colonna M (2010) Expansion of human NK-22 cells with IL-7, IL-2, and IL-1beta reveals intrinsic functional plasticity. Proc Natl Acad Sci USA 107:10961–10966

Chang YJ, Kim HY, Albacker LA, Baumgarth N, McKenzie AN, Smith DE, Dekruyff RH, Umetsu DT (2011) Innate lymphoid cells mediate influenza-induced airway hyper-reactivity independently of adaptive immunity. Nat Immunol 12:631–638

Constantinides MG, McDonald BD, Verhoef PA, Bendelac A (2014) A committed precursor to innate lymphoid cells. Nature 508:397–401

Dadi S, Chhangawala S, Whitlock BM, Franklin RA, Luo CT, Oh SA, Toure A, Pritykin Y, Huse M, Leslie CS, Li MO (2016) Cancer immunosurveillance by tissue-resident innate lymphoid cells and innate-like T cells. Cell 164:365–377

Daussy C, Faure F, Mayol K, Viel S, Gasteiger G, Charrier E, Bienvenu J, Henry T, Debien E, Hasan UA et al (2014) T-bet and Eomes instruct the development of two distinct natural killer cell lineages in the liver and in the bone marrow. J Exp Med 211:563–577

De Togni P, Goellner J, Ruddle NH, Streeter PR, Fick A, Mariathasan S, Smith SC, Carlson R, Shornick LP, Strauss-Schoenberger J et al (1994) Abnormal development of peripheral lymphoid organs in mice deficient in lymphotoxin. Science 264:703–707

Doherty TA, Khorram N, Lund S, Mehta AK, Croft M, Broide DH (2013) Lung type 2 innate lymphoid cells express cysteinyl leukotriene receptor 1, which regulates TH2 cytokine production. J Allergy Clin Immunol 132:205–213

Drake LY, Iijima K, Kita H (2014) Group 2 innate lymphoid cells and CD4+ T cells cooperate to mediate type 2 immune response in mice. Allergy 69:1300–1307

Dudakov JA, Hanash AM, Jenq RR, Young LF, Ghosh A, Singer NV, West ML, Smith OM, Holland AM, Tsai JJ et al (2012) Interleukin-22 drives endogenous thymic regeneration in mice. Science 336:91–95

Dyring-Andersen B, Geisler C, Agerbeck C, Lauritsen JP, Gudjonsdottir SD, Skov L, Bonefeld CM (2014) Increased number and frequency of group 3 innate lymphoid cells in nonlesional psoriatic skin. Br J Dermatol 170:609–616

Eberl G, Marmon S, Sunshine MJ, Rennert PD, Choi Y, Littman DR (2004) An essential function for the nuclear receptor RORγ(t) in the generation of fetal lymphoid tissue inducer cells. Nat Immunol 5:64–73

Ebert LM, Meuter S, Moser B (2006) Homing and function of human skin γδ T cells and NK cells: relevance for tumor surveillance. J Immunol 176:4331–4336

Flier J, Boorsma DM, van Beek PJ, Nieboer C, Stoof TJ, Willemze R, Tensen CP (2001) Differential expression of CXCR3 targeting chemokines CXCL10, CXCL9, and CXCL11 in different types of skin inflammation. J Pathol 194:398–405

Fu Y, Yang J, Xiong N (2016) Cutting edge: skin CCR10+ CD8+ T cells support resident regulatory T cells THROUGH the B7.2/receptor axis to regulate local immune homeostasis and response. J Immunol 196:4859–4864

Fuchs A, Vermi W, Lee JS, Lonardi S, Gilfillan S, Newberry RD, Cella M, Colonna M (2013) Intraepithelial type 1 innate lymphoid cells are a unique subset of IL-12- and IL-15-responsive IFN-γ-producing cells. Immunity 38:769–781

Gasteiger G, Rudensky AY (2014) Interactions between innate and adaptive lymphocytes. Nat Rev Immunol 14:631–639

Gasteiger G, Fan X, Dikiy S, Lee SY, Rudensky AY (2015) Tissue residency of innate lymphoid cells in lymphoid and nonlymphoid organs. Science 350:981–985

Gauvreau GM, O'Byrne PM, Boulet LP, Wang Y, Cockcroft D, Bigler J, FitzGerald JM, Boedigheimer M, Davis BE, Dias C et al (2014) Effects of an anti-TSLP antibody on allergen-induced asthmatic responses. N Engl J Med 370:2102–2110

Gombert M, Dieu-Nosjean MC, Winterberg F, Bunemann E, Kubitza RC, Da Cunha L, Haahtela A, Lehtimaki S, Muller A, Rieker J et al (2005) CCL1-CCR8 interactions: an axis mediating the recruitment of T cells and Langerhans-type dendritic cells to sites of atopic skin inflammation. J Immunol 174:5082–5091

Gordon SM, Chaix J, Rupp LJ, Wu J, Madera S, Sun JC, Lindsten T, Reiner SL (2012) The transcription factors T-bet and Eomes control key checkpoints of natural killer cell maturation. Immunity 36:55–67

Goto Y, Obata T, Kunisawa J, Sato S, Ivanov II, Lamichhane A, Takeyama N, Kamioka M, Sakamoto M, Matsuki T et al (2014) Innate lymphoid cells regulate intestinal epithelial cell glycosylation. Science 345:1254009

Gudjonsson JE, Ding J, Johnston A, Tejasvi T, Guzman AM, Nair RP, Voorhees JJ, Abecasis GR, Elder JT (2010) Assessment of the psoriatic transcriptome in a large sample: additional regulated genes and comparisons with in vitro models. J Invest Dermatol 130:1829–1840

Guo X, Liang Y, Zhang Y, Lasorella A, Kee BL, Fu YX (2015) Innate lymphoid cells control early colonization resistance against intestinal pathogens through ID2-dependent regulation of the microbiota. Immunity 42:731–743

Gury-BenAri M, Thaiss CA, Serafini N, Winter DR, Giladi A, Lara-Astiaso D, Levy M, Salame TM, Weiner A, David E et al (2016) The spectrum and regulatory landscape of intestinal innate lymphoid cells are shaped by the microbiome. Cell 166(1231–1246):e1213

Halim TY, MacLaren A, Romanish MT, Gold MJ, McNagny KM, Takei F (2012) Retinoic-acid-receptor-related orphan nuclear receptor alpha is required for natural helper cell development and allergic inflammation. Immunity 37:463–474

Halim TY, Steer CA, Matha L, Gold MJ, Martinez-Gonzalez I, McNagny KM, McKenzie AN, Takei F (2014) Group 2 innate lymphoid cells are critical for the initiation of adaptive T helper 2 cell-mediated allergic lung inflammation. Immunity 40:425–435

Hambly N, Nair P (2014) Monoclonal antibodies for the treatment of refractory asthma. Curr Opin Pulm Med 20:87–94

Hams E, Locksley RM, McKenzie AN, Fallon PG (2013) Cutting edge: IL-25 elicits innate lymphoid type 2 and type II NKT cells that regulate obesity in mice. J Immunol 191:5349–5353

Hams E, Armstrong ME, Barlow JL, Saunders SP, Schwartz C, Cooke G, Fahy RJ, Crotty TB, Hirani N, Flynn RJ et al (2014) IL-25 and type 2 innate lymphoid cells induce pulmonary fibrosis. Proc Natl Acad Sci USA 111:367–372

Hepworth MR, Monticelli LA, Fung TC, Ziegler CG, Grunberg S, Sinha R, Mantegazza AR, Ma HL, Crawford A, Angelosanto JM et al (2013) Innate lymphoid cells regulate CD4+ T-cell responses to intestinal commensal bacteria. Nature 498:113–117

Herberman RB, Nunn ME, Holden HT, Lavrin DH (1975a) Natural cytotoxic reactivity of mouse lymphoid cells against syngeneic and allogeneic tumors. II. Characterization of effector cells. Int J Cancer 16:230–239

Herberman RB, Nunn ME, Lavrin DH (1975b) Natural cytotoxic reactivity of mouse lymphoid cells against syngeneic acid allogeneic tumors. I. Distribution of reactivity and specificity. Int J Cancer 16:216–229

Homey B, Wang W, Soto H, Buchanan ME, Wiesenborn A, Catron D, Muller A, McClanahan TK, Dieu-Nosjean MC, Orozco R et al (2000) Cutting edge: the orphan chemokine receptor G protein-coupled receptor-2 (GPR-2, CCR10) binds the skin-associated chemokine CCL27 (CTACK/ALP/ILC). J Immunol 164:3465–3470

Homey B, Alenius H, Muller A, Soto H, Bowman EP, Yuan W, McEvoy L, Lauerma AI, Assmann T, Bunemann E et al (2002) CCL27-CCR10 interactions regulate T cell-mediated skin inflammation. Nat Med 8:157–165

Hoyler T, Klose CS, Souabni A, Turqueti-Neves A, Pfeifer D, Rawlins EL, Voehringer D, Busslinger M, Diefenbach A (2012) The transcription factor GATA-3 controls cell fate and maintenance of type 2 innate lymphoid cells. Immunity 37:634–648

Imai Y, Yasuda K, Sakaguchi Y, Haneda T, Mizutani H, Yoshimoto T, Nakanishi K, Yamanishi K (2013) Skin-specific expression of IL-33 activates group 2 innate lymphoid cells and elicits atopic dermatitis-like inflammation in mice. Proc Natl Acad Sci USA 110:13921–13926

Ishizuka IE, Constantinides MG, Gudjonson H, Bendelac A (2016) The innate lymphoid cell precursor. Annu Rev Immunol 34:299–316

Kanda N, Koike S, Watanabe S (2005) IL-17 suppresses TNF-alpha-induced CCL27 production through induction of COX-2 in human keratinocytes. J Allergy Clin Immunol 116:1144–1150

Kantele A, Zivny J, Hakkinen M, Elson CO, Mestecky J (1999) Differential homing commitments of antigen-specific T cells after oral or parenteral immunization in humans. J Immunol 162:5173–5177

Karakawa M, Komine M, Hanakawa Y, Tsuda H, Sayama K, Tamaki K, Ohtsuki M (2014) CCL27 is downregulated by interferon γ via epidermal growth factor receptor in normal human epidermal keratinocytes. J Cell Physiol 229(12):1935–1945

Kiessling R, Klein E, Pross H, Wigzell H (1975a) "Natural" killer cells in the mouse. II. Cytotoxic cells with specificity for mouse Moloney leukemia cells. Characteristics of the killer cell. Eur J Immunol 5:117–121

Kiessling R, Klein E, Wigzell H (1975b) "Natural" killer cells in the mouse. I. Cytotoxic cells with specificity for mouse Moloney leukemia cells. Specificity and distribution according to genotype. Eur J Immunol 5:112–117

Kim BS, Siracusa MC, Saenz SA, Noti M, Monticelli LA, Sonnenberg GF, Hepworth MR, Van Voorhees AS, Comeau MR, Artis D (2013) TSLP elicits IL-33-independent innate lymphoid cell responses to promote skin inflammation. Sci Transl Med 5:170ra116

Kim BS, Wang K, Siracusa MC, Saenz SA, Brestoff JR, Monticelli LA, Noti M, Tait Wojno ED, Fung TC, Kubo M, Artis D (2014a) Basophils promote innate lymphoid cell responses in inflamed skin. J Immunol 193:3717–3725

Kim HY, Lee HJ, Chang YJ, Pichavant M, Shore SA, Fitzgerald KA, Iwakura Y, Israel E, Bolger K, Faul J et al (2014b) Interleukin-17-producing innate lymphoid cells and the NLRP3 inflammasome facilitate obesity-associated airway hyperreactivity. Nat Med 20:54–61

Kim MH, Taparowsky EJ, Kim CH (2015) Retinoic acid differentially regulates the migration of innate lymphoid cell subsets to the gut. Immunity 43:107–119

Kirchberger S, Royston DJ, Boulard O, Thornton E, Franchini F, Szabady RL, Harrison O, Powrie F (2013) Innate lymphoid cells sustain colon cancer through production of interleukin-22 in a mouse model. J Exp Med 210:917–931

Klose CS, Kiss EA, Schwierzeck V, Ebert K, Hoyler T, d'Hargues Y, Goppert N, Croxford AL, Waisman A, Tanriver Y, Diefenbach A (2013) A T-bet gradient controls the fate and function of CCR6-RORγt+ innate lymphoid cells. Nature 494:261–265

Klose CS, Flach M, Mohle L, Rogell L, Hoyler T, Ebert K, Fabiunke C, Pfeifer D, Sexl V, Fonseca-Pereira D et al (2014) Differentiation of type 1 ILCs from a common progenitor to all helper-like innate lymphoid cell lineages. Cell 157:340–356

Kruglov AA, Grivennikov SI, Kuprash DV, Winsauer C, Prepens S, Seleznik GM, Eberl G, Littman DR, Heikenwalder M, Tumanov AV, Nedospasov SA (2013) Nonredundant function of soluble LTalpha3 produced by innate lymphoid cells in intestinal homeostasis. Science 342:1243–1246

Leung DY (2013) New insights into atopic dermatitis: role of skin barrier and immune dysregulation. Allergol Int 62:151–161

Li Z, Hodgkinson T, Gothard EJ, Boroumand S, Lamb R, Cummins I, Narang P, Sawtell A, Coles J, Leonov G et al (2016) Epidermal Notch1 recruits RORγ(+) group 3 innate lymphoid cells to orchestrate normal skin repair. Nat Commun 7:11394

Lonsdorf AS, Hwang ST, Enk AH (2009) Chemokine receptors in T-cell-mediated diseases of the skin. J Invest Dermatol 129:2552–2566

Luci C, Reynders A, Ivanov II, Cognet C, Chiche L, Chasson L, Hardwigsen J, Anguiano E, Banchereau J, Chaussabel D et al (2009) Influence of the transcription factor RORγt on the development of NKp46+ cell populations in gut and skin. Nat Immunol 10:75–82

Masopust D, Schenkel JM (2013) The integration of T cell migration, differentiation and function. Nat Rev Immunol 13:309–320

McHedlidze T, Waldner M, Zopf S, Walker J, Rankin AL, Schuchmann M, Voehringer D, McKenzie AN, Neurath MF, Pflanz S, Wirtz S (2013) Interleukin-33-dependent innate lymphoid cells mediate hepatic fibrosis. Immunity 39:357–371

McKenzie AN, Spits H, Eberl G (2014) Innate lymphoid cells in inflammation and immunity. Immunity 41:366–374

Mebius RE, Rennert P, Weissman IL (1997) Developing lymph nodes collect CD4+CD3− LTbeta+ cells that can differentiate to APC, NK cells, and follicular cells but not T or B cells. Immunity 7:493–504

Moed H, Boorsma DM, Tensen CP, Flier J, Jonker MJ, Stoof TJ, von Blomberg BM, Bruynzeel DP, Scheper RJ, Rustemeyer T, Gibbs S (2004) Increased CCL27-CCR10 expression in allergic contact dermatitis: implications for local skin memory. J Pathol 204:39–46

Moffatt MF, Gut IG, Demenais F, Strachan DP, Bouzigon E, Heath S, von Mutius E, Farrall M, Lathrop M, Cookson WO (2010) A large-scale, consortium-based genomewide association study of asthma. N Engl J Med 363:1211–1221

Molofsky AB, Nussbaum JC, Liang HE, Van Dyken SJ, Cheng LE, Mohapatra A, Chawla A, Locksley RM (2013) Innate lymphoid type 2 cells sustain visceral adipose tissue eosinophils and alternatively activated macrophages. J Exp Med 210:535–549

Monticelli LA, Sonnenberg GF, Abt MC, Alenghat T, Ziegler CG, Doering TA, Angelosanto JM, Laidlaw BJ, Yang CY, Sathaliyawala T et al (2011) Innate lymphoid cells promote lung-tissue homeostasis after infection with influenza virus. Nat Immunol 12:1045–1054

Mora JR, von Andrian UH (2006) T-cell homing specificity and plasticity: new concepts and future challenges. Trends Immunol 27:235–243

Mora JR, Bono MR, Manjunath N, Weninger W, Cavanagh LL, Rosemblatt M, Von Andrian UH (2003) Selective imprinting of gut-homing T cells by Peyer's patch dendritic cells. Nature 424:88–93

Mora JR, Cheng G, Picarella D, Briskin M, Buchanan N, von Andrian UH (2005) Reciprocal and dynamic control of CD8 T cell homing by dendritic cells from skin- and gut-associated lymphoid tissues. J Exp Med 201:303–316

Morales J, Homey B, Vicari AP, Hudak S, Oldham E, Hedrick J, Orozco R, Copeland NG, Jenkins NA, McEvoy LM, Zlotnik A (1999) CTACK, a skin-associated chemokine that preferentially attracts skin-homing memory T cells. Proc Natl Acad Sci USA 96:14470–14475

Moro K, Kabata H, Tanabe M, Koga S, Takeno N, Mochizuki M, Fukunaga K, Asano K, Betsuyaku T, Koyasu S (2016) Interferon and IL-27 antagonize the function of group 2 innate lymphoid cells and type 2 innate immune responses. Nat Immunol 17:76–86

Mortha A, Chudnovskiy A, Hashimoto D, Bogunovic M, Spencer SP, Belkaid Y, Merad M (2014) Microbiota-dependent crosstalk between macrophages and ILC3 promotes intestinal homeostasis. Science 343:1249288

Neill DR, Wong SH, Bellosi A, Flynn RJ, Daly M, Langford TK, Bucks C, Kane CM, Fallon PG, Pannell R et al (2010) Nuocytes represent a new innate effector leukocyte that mediates type-2 immunity. Nature 464:1367–1370

Nussbaum JC, Van Dyken SJ, von Moltke J, Cheng LE, Mohapatra A, Molofsky AB, Thornton EE, Krummel MF, Chawla A, Liang HE, Locksley RM (2013) Type 2 innate lymphoid cells control eosinophil homeostasis. Nature 502:245–248

Oliphant CJ, Hwang YY, Walker JA, Salimi M, Wong SH, Brewer JM, Englezakis A, Barlow JL, Hams E, Scanlon ST et al (2014) MHCII-mediated dialog between group 2 innate lymphoid cells and CD4(+) T cells potentiates type 2 immunity and promotes parasitic helminth expulsion. Immunity 41:283–295

O'Sullivan TE, Rapp M, Fan X, Weizman OE, Bhardwaj P, Adams NM, Walzer T, Dannenberg AJ, Sun JC (2016) Adipose-resident group 1 Innate lymphoid cells promote obesity-associated insulin resistance. Immunity 45:428–441

Pantelyushin S, Haak S, Ingold B, Kulig P, Heppner FL, Navarini AA, Becher B (2012) Rorγt+ innate lymphocytes and γδ T cells initiate psoriasiform plaque formation in mice. J Clin Invest 122:2252–2256

Peng H, Jiang X, Chen Y, Sojka DK, Wei H, Gao X, Sun R, Yokoyama WM, Tian Z (2013) Liver-resident NK cells confer adaptive immunity in skin-contact inflammation. J Clin Invest 123:1444–1456

Perry JS, Han S, Xu Q, Herman ML, Kennedy LB, Csako G, Bielekova B (2012) Inhibition of LTi cell development by CD25 blockade is associated with decreased intrathecal inflammation in multiple sclerosis. Sci Transl Med 4:145ra106

Picker LJ, Treer JR, Ferguson-Darnell B, Collins PA, Bergstresser PR, Terstappen LW (1993) Control of lymphocyte recirculation in man. II. Differential regulation of the cutaneous lymphocyte-associated antigen, a tissue-selective homing receptor for skin-homing T cells. J Immunol 150:1122–1136

Pivarcsi A, Gombert M, Dieu-Nosjean MC, Lauerma A, Kubitza R, Meller S, Rieker J, Muller A, Da Cunha L, Haahtela A et al (2004) CC chemokine ligand 18, an atopic dermatitis-associated and dendritic cell-derived chemokine, is regulated by staphylococcal products and allergen exposure. J Immunol 173:5810–5817

Powell N, Walker AW, Stolarczyk E, Canavan JB, Gokmen MR, Marks E, Jackson I, Hashim A, Curtis MA, Jenner RG et al (2012) The transcription factor T-bet regulates intestinal inflammation mediated by interleukin-7 receptor+ innate lymphoid cells. Immunity 37:674–684

Price AE, Liang HE, Sullivan BM, Reinhardt RL, Eisley CJ, Erle DJ, Locksley RM (2010) Systemically dispersed innate IL-13-expressing cells in type 2 immunity. Proc Natl Acad Sci USA 107:11489–11494

Pulendran B, Artis D (2012) New paradigms in type 2 immunity. Science 337:431–435

Qiu J, Guo X, Chen ZM, He L, Sonnenberg GF, Artis D, Fu YX, Zhou L (2013) Group 3 Innate lymphoid cells inhibit T-cell-mediated intestinal inflammation through aryl hydrocarbon receptor signaling and regulation of microflora. Immunity 39:386–399

Quaranta M, Knapp B, Garzorz N, Mattii M, Pullabhatla V, Pennino D, Andres C, Traidl-Hoffmann C, Cavani A, Theis FJ et al (2014). Intraindividual genome expression analysis reveals a specific molecular signature of psoriasis and eczema. Sci Transl Med 6:244ra290

Rak GD, Osborne LC, Siracusa MC, Kim BS, Wang K, Bayat A, Artis D, Volk SW (2016) IL-33-dependent group 2 innate lymphoid cells promote cutaneous wound healing. J Invest Dermatol 136:487–496

Rankin LC, Groom JR, Chopin M, Herold MJ, Walker JA, Mielke LA, McKenzie AN, Carotta S, Nutt SL, Belz GT (2013) The transcription factor T-bet is essential for the development of NKp46+ innate lymphocytes via the Notch pathway. Nat Immunol 14:389–395

Reiss Y, Proudfoot AE, Power CA, Campbell JJ, Butcher EC (2001) CC chemokine receptor (CCR)4 and the CCR10 ligand cutaneous T cell-attracting chemokine (CTACK) in lymphocyte trafficking to inflamed skin. J Exp Med 194:1541–1547

Riis JL, Johansen C, Vestergaard C, Bech R, Kragballe K, Iversen L (2011a) Kinetics and differential expression of the skin-related chemokines CCL27 and CCL17 in psoriasis, atopic dermatitis and allergic contact dermatitis. Exp Dermatol 20:789–794

Riis JL, Johansen C, Vestergaard C, Otkjaer K, Kragballe K, Iversen L (2011b) CCL27 expression is regulated by both p38 MAPK and IKKβ signalling pathways. Cytokine 56:699–707

Roediger B, Kyle R, Yip KH, Sumaria N, Guy TV, Kim BS, Mitchell AJ, Tay SS, Jain R, Forbes-Blom E et al (2013) Cutaneous immunosurveillance and regulation of inflammation by group 2 innate lymphoid cells. Nat Immunol 14:564–573

Rott LS, Rose JR, Bass D, Williams MB, Greenberg HB, Butcher EC (1997) Expression of mucosal homing receptor alpha4beta7 by circulating CD4+ cells with memory for intestinal rotavirus. J Clin Invest 100:1204–1208

Salimi M, Barlow JL, Saunders SP, Xue L, Gutowska-Owsiak D, Wang X, Huang LC, Johnson D, Scanlon ST, McKenzie AN et al (2013) A role for IL-25 and IL-33-driven type-2 innate lymphoid cells in atopic dermatitis. J Exp Med 210:2939–2950

Sawa S, Lochner M, Satoh-Takayama N, Dulauroy S, Berard M, Kleinschek M, Cua D, Di Santo JP, Eberl G (2011) RORγt+ innate lymphoid cells regulate intestinal homeostasis by integrating negative signals from the symbiotic microbiota. Nat Immunol 12:320–326

Scoville SD, Mundy-Bosse BL, Zhang MH, Chen L, Zhang X, Keller KA, Hughes T, Chen L, Cheng S, Bergin SM et al (2016) A progenitor cell expressing transcription factor RORγt generates all human innate lymphoid cell subsets. Immunity 44:1140–1150

Seillet C, Rankin LC, Groom JR, Mielke LA, Tellier J, Chopin M, Huntington ND, Belz GT, Carotta S (2014) Nfil3 is required for the development of all innate lymphoid cell subsets. J Exp Med 211:1733–1740

Serafini N, Vosshenrich CA, Di Santo JP (2015) Transcriptional regulation of innate lymphoid cell fate. Nat Rev Immunol 15:415–428

Sigmundsdottir H, Butcher EC (2008) Environmental cues, dendritic cells and the programming of tissue-selective lymphocyte trafficking. Nat Immunol 9:981–987

Sigmundsdottir H, Pan J, Debes GF, Alt C, Habtezion A, Soler D, Butcher EC (2007) DCs metabolize sunlight-induced vitamin D3 to 'program' T cell attraction to the epidermal chemokine CCL27. Nat Immunol 8:285–293

Sojka DK, Plougastel-Douglas B, Yang L, Pak-Wittel MA, Artyomov MN, Ivanova Y, Zhong C, Chase JM, Rothman PB, Yu J et al (2014) Tissue-resident natural killer (NK) cells are cell lineages distinct from thymic and conventional splenic NK cells. eLife 3: e01659

Solt LA, Kumar N, Nuhant P, Wang Y, Lauer JL, Liu J, Istrate MA, Kamenecka TM, Roush WR, Vidovic D et al (2011) Suppression of TH17 differentiation and autoimmunity by a synthetic ROR ligand. Nature 472:491–494

Sonnenberg GF, Fouser LA, Artis D (2011) Border patrol: regulation of immunity, inflammation and tissue homeostasis at barrier surfaces by IL-22. Nat Immunol 12:383–390

Sonnenberg GF, Monticelli LA, Alenghat T, Fung TC, Hutnick NA, Kunisawa J, Shibata N, Grunberg S, Sinha R, Zahm AM et al (2012) Innate lymphoid cells promote anatomical containment of lymphoid-resident commensal bacteria. Science 336:1321–1325

Spencer SP, Wilhelm C, Yang Q, Hall JA, Bouladoux N, Boyd A, Nutman TB, Urban JF Jr, Wang J, Ramalingam TR et al (2014) Adaptation of innate lymphoid cells to a micronutrient deficiency promotes type 2 barrier immunity. Science 343:432–437

Spits H, Artis D, Colonna M, Diefenbach A, Di Santo JP, Eberl G, Koyasu S, Locksley RM, McKenzie AN, Mebius RE et al (2013) Innate lymphoid cells—a proposal for uniform nomenclature. Nat Rev Immunol 13:145–149

Stanya KJ, Jacobi D, Liu S, Bhargava P, Dai L, Gangl MR, Inouye K, Barlow JL, Ji Y, Mizgerd JP et al (2013) Direct control of hepatic glucose production by interleukin-13 in mice. J Clin Invest 123:261–271

Taube C, Tertilt C, Gyulveszi G, Dehzad N, Kreymborg K, Schneeweiss K, Michel E, Reuter S, Renauld JC, Arnold-Schild D et al (2011) IL-22 is produced by innate lymphoid cells and limits inflammation in allergic airway disease. PLoS ONE 6: e21799

Teunissen MB, Munneke JM, Bernink JH, Spuls PI, Res PC, Te Velde A, Cheuk S, Brouwer MW, Menting SP, Eidsmo L et al (2014) Composition of innate lymphoid cell subsets in the human skin: enrichment of NCR ILC3 in lesional skin and blood of psoriasis patients. J Invest Dermatol 134(9):2351–2360

van de Pavert SA, Olivier BJ, Goverse G, Vondenhoff MF, Greuter M, Beke P, Kusser K, Hopken UE, Lipp M, Niederreither K et al (2009) Chemokine CXCL13 is essential for lymph node initiation and is induced by retinoic acid and neuronal stimulation. Nat Immunol 10:1193–1199

Vely F, Barlogis V, Vallentin B, Neven B, Piperoglou C, Ebbo M, Perchet T, Petit M, Yessaad N, Touzot F et al (2016) Evidence of innate lymphoid cell redundancy in humans. Nat Immunol 17:1291–1299

Vestergaard C, Johansen C, Christensen U, Just H, Hohwy T, Deleuran M (2004) TARC augments TNF-alpha-induced CTACK production in keratinocytes. Exp Dermatol 13:551–557

Villanova F, Flutter B, Tosi I, Grys K, Sreeneebus H, Perera GK, Chapman A, Smith CH, Di Meglio P, Nestle FO (2014) Characterization of innate lymphoid cells in human skin and blood demonstrates increase of NKp44+ ILC3 in psoriasis. J Invest Dermatol 134:984–991

Vivier E, Raulet DH, Moretta A, Caligiuri MA, Zitvogel L, Lanier LL, Yokoyama WM, Ugolini S (2011) Innate or adaptive immunity? The example of natural killer cells. Science 331:44–49

Vivier E, Ugolini S, Blaise D, Chabannon C, Brossay L (2012) Targeting natural killer cells and natural killer T cells in cancer. Nat Rev Immunol 12:239–252

Vonarbourg C, Mortha A, Bui VL, Hernandez PP, Kiss EA, Hoyler T, Flach M, Bengsch B, Thimme R, Holscher C et al (2010) Regulated expression of nuclear receptor RORγt confers distinct functional fates to NK cell receptor-expressing RORγt(+) innate lymphocytes. Immunity 33:736–751

Walker JA, Barlow JL, McKenzie AN (2013) Innate lymphoid cells—how did we miss them? Nat Rev Immunol 13:75–87

Weninger W, Ulfman LH, Cheng G, Souchkova N, Quackenbush EJ, Lowe JB, von Andrian UH (2000) Specialized contributions by alpha(1,3)-fucosyltransferase-IV and FucT-VII during leukocyte rolling in dermal microvessels. Immunity 12:665–676

Withers DR, Hepworth MR, Wang X, Mackley EC, Halford EE, Dutton EE, Marriott CL, Brucklacher-Waldert V, Veldhoen M, Kelsen J et al (2016) Transient inhibition of ROR-γt therapeutically limits intestinal inflammation by reducing TH17 cells and preserving group 3 innate lymphoid cells. Nat Med 22:319–323

Xia M, Hu S, Fu Y, Jin W, Yi Q, Matsui Y, Yang J, McDowell MA, Sarkar S, Kalia V, Xiong N (2014) CCR10 regulates balanced maintenance and function of resident regulatory and effector T cells to promote immune homeostasis in the skin. J Allergy Clin Immunol 134(634–644):e610

Xiong N, Fu Y, Hu S, Xia M, Yang J (2012) CCR10 and its ligands in regulation of epithelial immunity and diseases. Protein Cell 3:571–580

Xue L, Salimi M, Panse I, Mjosberg JM, McKenzie AN, Spits H, Klenerman P, Ogg G (2014) Prostaglandin D2 activates group 2 innate lymphoid cells through chemoattractant receptor-homologous molecule expressed on TH2 cells. J Allergy Clin Immunol 133:1184–1194

Yagi R, Zhong C, Northrup DL, Yu F, Bouladoux N, Spencer S, Hu G, Barron L, Sharma S, Nakayama T et al (2014) The transcription factor GATA3 is critical for the development of all IL-7Ralpha-expressing innate lymphoid cells. Immunity 40:378–388

Yang Q, Bhandoola A (2016) The development of adult innate lymphoid cells. Curr Opin Immunol 39:114–120

Yang Q, Li F, Harly C, Xing S, Ye L, Xia X, Wang H, Wang X, Yu S, Zhou X et al (2015) TCF-1 upregulation identifies early innate lymphoid progenitors in the bone marrow. Nat Immunol 16:1044–1050

Yang J, Hu S, Zhao L, Kaplan DH, Perdew GH, Xiong N (2016) Selective programming of CCR10(+) innate lymphoid cells in skin-draining lymph nodes for cutaneous homeostatic regulation. Nat Immunol 17:48–56

Yu X, Wang Y, Deng M, Li Y, Ruhn KA, Zhang CC, Hooper LV (2014) The basic leucine zipper transcription factor NFIL3 directs the development of a common innate lymphoid cell precursor. eLife. doi:10.7554/eLife.04406

Zheng Y, Valdez PA, Danilenko DM, Hu Y, Sa SM, Gong Q, Abbas AR, Modrusan Z, Ghilardi N, de Sauvage FJ, Ouyang W (2008) Interleukin-22 mediates early host defense against attaching and effacing bacterial pathogens. Nat Med 14:282–289

Zook EC, Kee BL (2016) Development of innate lymphoid cells. Nat Immunol 17:775–782

14

Human BDCA2$^+$CD123$^+$CD56$^+$ dendritic cells (DCs) related to blastic plasmacytoid dendritic cell neoplasm represent a unique myeloid DC subset

Haisheng Yu[1,3], Peng Zhang[2,3], Xiangyun Yin[1,3], Zhao Yin[4], Quanxing Shi[4], Ya Cui[2], Guanyuan Liu[5], Shouli Wang[4], Pier Paolo Piccaluga[6], Taijiao Jiang[2]✉, Liguo Zhang[1]✉

[1] Key Laboratory of Immunity and Infection, Institute of Biophysics, Chinese Academy of Sciences, Beijing 100101, China
[2] Key Laboratory of Protein and Peptide Pharmaceuticals, Institute of Biophysics, Chinese Academy of Sciences, Beijing 100101, China
[3] Graduate School of the Chinese Academy of Sciences, Beijing 100080, China
[4] Department of Cardiology, 306th Hospital of PLA, Beijing 100101, China
[5] Department of Gynecology and Obstetrics, Beijing Chaoyang Hospital, Capital Medical University, Beijing 100020, China
[6] Department of Experimental, Diagnostic, and Specialty Medicine, Hematopathology & Hematology Sections, Molecular Pathology Laboratory, S. Orsola-Malpighi Hospital, Bologna University, Bologna 40126, Italy
✉ Correspondence: taijiao@moon.ibp.ac.cn (T. Jiang), liguozhang@ibp.ac.cn (L. Zhang)

ABSTRACT

Dendritic cells (DCs) comprise two functionally distinct subsets: plasmacytoid DCs (pDCs) and myeloid DCs (mDCs). pDCs are specialized in rapid and massive secretion of type I interferon (IFN-I) in response to nucleic acids through Toll like receptor (TLR)-7 or TLR-9. In this report, we characterized a CD56$^+$ DC population that express typical pDC markers including CD123 and BDCA2 but produce much less IFN-I comparing with pDCs. In addition, CD56$^+$ DCs cluster together with mDCs but not pDCs by genome-wide transcriptional profiling. Accordingly, CD56$^+$ DCs functionally resemble mDCs by producing IL-12 upon TLR4 stimulation and priming naïve T cells without prior activation. These data suggest that the CD56$^+$ DCs represent a novel mDC subset mixed with some pDC features. A CD4$^+$CD56$^+$ hematological malignancy was classified as blastic plasmacytoid dendritic cell neoplasm (BPDCN) due to its expression of characteristic molecules of pDCs. However, we demonstrated that BPDCN is closer to CD56$^+$ DCs than pDCs by global gene-expression profiling. Thus, we propose that the CD4$^+$CD56$^+$ neoplasm may be a tumor counterpart of CD56$^+$ mDCs but not pDCs.

Haisheng Yu and Peng Zhang have contributed equally to this work

KEYWORDS dendritic cells, CD56$^+$ DC, pDC, mDC, BPDCN

INTRODUCTION

Dendritic cells (DCs) are professional antigen-presenting cells found in virtually all tissues. The main function of DCs is to induce T-cell activation, polarization and expansion against invading pathogens while maintaining tolerance to self antigens (Steinman, 2007). Three DC subsets have been identified in the human blood: plasmacytoid dendritic cell (pDC) and two subsets of myeloid DC (mDC) expressing CD1c (BDCA1) and CD141 (BDCA3) respectively (Dzionek et al., 2000; Ziegler-Heitbrock et al., 2010). pDCs are characterized by their specific expression of CD123, BDCA2, BDCA4 and ILT7 (Dzionek et al., 2000; Cao et al., 2006; Rissoan et al., 2002). In addition, pDCs express high levels of interferon regulatory factor 7 (IRF7), Toll like receptor

(TLR)-7 and TLR-9. Accordingly, they produce large amounts of type I interferon (IFN-I) upon stimulation with nucleic-acid ligands of TLR7/TRL9 (Siegal et al., 1999; Cella et al., 1999). Transcription factors E2-2 and Spi B are specifically expressed in pDCs and play important roles in their development and maintenance (Cisse et al., 2008; Schotte et al., 2004).

Blastic plasmacytoid dendritic cell neoplasm (BPDCN) is a rare hematological malignancy characterized by the clonal proliferation of pDC-like cells (Facchetti et al., 2008). Neoplastic cells from BPDCN patients express pDC specific molecules, such as BDCA2, CD123, CD4, TCL1, Bcl11A, CD2AP and Spi B (Petrella and Facchetti, 2010). Additionally, BPDCN cells can produce IFN-I upon TLR7 and TLR9 ligands stimulation although at much lower level comparing with pDCs (Chaperot et al., 2001). Interestingly, BPDCN cells also express CD56, which is not present on the majority of pDCs (Grouard et al., 1997). A small population of DCs in normal or FLT3 mobilized human blood do express CD56, which have been proposed as a pDC subpopulation that gives rise to BPDCN (Petrella et al., 2002; Comeau, 2002).

In this report, we characterized these BDCA2$^+$ CD123$^+$CD56$^+$ DCs from human peripheral blood. Although they express some pDC featured molecules, the transcriptomic and functional characterizations indicate that they belong to mDCs rather than pDCs.

RESULTS

CD56$^+$ DCs are functionally distinct from pDCs

BDCA2 and CD123 are commonly used for the identification of human pDCs in the blood and lymphoid tissues. A subpopulation of BDCA2$^+$CD123$^+$CD56$^+$ (CD56$^+$) DCs have been reported in human blood and proposed as a pDC subpopulation related to BPDCN (Petrella et al., 2002; Comeau, 2002). We first confirmed the presence of CD56$^+$ subset in the HLA-DR$^+$, CD11c$^-$, BDCA2$^+$ and CD123$^+$ DCs in human blood (Fig. 1A and 1B). The CD56$^+$ subset represents about 5% of the BDCA2 and CD123 double positive cells (Fig. 1B, n = 15). It was reported that pDCs can be divided into two populations by their CD2 expression (Matsui et al., 2009). We found that the CD56$^+$ DCs universally express CD2 (Fig. 2A).

Next, we purified the CD56$^+$ and CD56$^-$ cells to compare their IFN-I production by TLR7 or TLR9 stimulation. In order to get higher purity, we added CD2 in our purification protocol, thus CD123$^+$CD2$^+$CD56$^+$ and CD123$^+$CD2$^-$CD56$^-$ DCs were sorted by FACS for further analysis. Usually the purity of CD2$^-$CD56$^-$ cells was over 95% and the purity of CD2$^+$CD56$^+$ cells was over 80% with contamination of some CD2$^+$CD56$^-$ cells (Fig. 2B). High level IFNα production upon TLR7 or TLR9 stimulation is the key characteristic of pDCs. However, purified CD56$^+$ DCs produced much less IFNα upon CpG A stimulation comparing to CD56$^-$ cells from same donors (around 10%) (Fig. 2C, left panel). Similar results were also observed when the purified DCs were stimulated with influenza virus (Fig. 2C, right panel). In order to exclude the effect of CD56 antibody binding during purification, we also compared the IFNα production capacity of CD56$^+$ and CD56$^-$ DCs by intracellular staining without prior purification. When enriched DCs were stimulated with CpG A and stained IFNα intracellularly, above 70% of CD56$^-$ cells but only 10% CD56$^+$ cells expressed IFNα (Fig. 2D, left panel). Similar results were observed by R848, a TLR7 ligand stimulation (Fig. 2D, right panel).

CD56$^+$ DCs clustered together with mDCs but not pDCs by transcriptomic analysis

In order to compare the gene expression of CD56$^+$ DCs and other DC subsets at whole genome level, we analyzed the transcriptome of both CD56$^-$ pDCs and CD56$^+$ DCs by RNA sequencing (RNA-seq). Next, we compared the transcriptional profile of CD56$^+$ DCs with cDNA array data of human DC subsets retrieved from public databases (Robbins et al., 2008). There were 772 genes expressed in CD56$^+$ DCs at higher level (over 2 times) than those of CD56$^-$ DCs and 2398 genes *vice versa*. Among those, 648 genes highly expressed in CD56$^+$ cells (Fig. 3A, left panel) and 1557 lowly expressed ones (Fig. 3B, left panel) were present in the cDNA array data of human blood DC subsets (Robbins et al., 2008). And the names of those differentially expressed genes are provided in Table S1 and S2. Intriguingly, 547 (91%) CD56$^+$ DC highly expressed genes present at higher level in mDCs than pDCs based on cDNA array data (Fig. 3A, right panel; Table S1). Similarly, 1292 (83%) CD56$^+$ DCs lowly expressed genes presented at lower level in mDCs than pDCs (Fig. 3B, right panel; Table S2).

To clarify the relationship of CD56$^+$ DCs and other DC subsets, we normalized our RNA-Seq data with cDNA data, then did hierarchical clustering with complete linkage and principal component analysis (PCA) (11,372 genes). Both analyses showed that CD56$^-$ DCs clustered together with pDCs as expected, while the CD56$^+$ DCs clustered together with BDCA3 and BDCA1 mDCs (Fig. 3C and 3D).

CD56$^+$ DCs are functionally analogous to mDCs

TLR expression patterns of pDCs and mDCs are drastically different. pDCs preferentially express TLR7 and TLR9 while mDCs express higher levels of TLR2 and TLR4 (Liu 2005; Cerboni et al., 2013). We found that the expression level of TLR4 (Fig. 4A) and the amount of TNFα produced upon LPS stimulation (Fig. 4B and 4C) were similar between CD56$^+$ DCs and mDCs. Moreover, CD56$^+$ DCs could produce IL-12 at a comparable level to mDCs at both RNA (Fig. 4D) and protein level (Fig. 4E). While pDCs did not produce IL-12 (Fig. 4D and 4E) as previously reported (Kadowaki et al., 2001).

Different from pDCs, mDCs could prompt T-cell proliferation without prior stimulation or activation. In order to

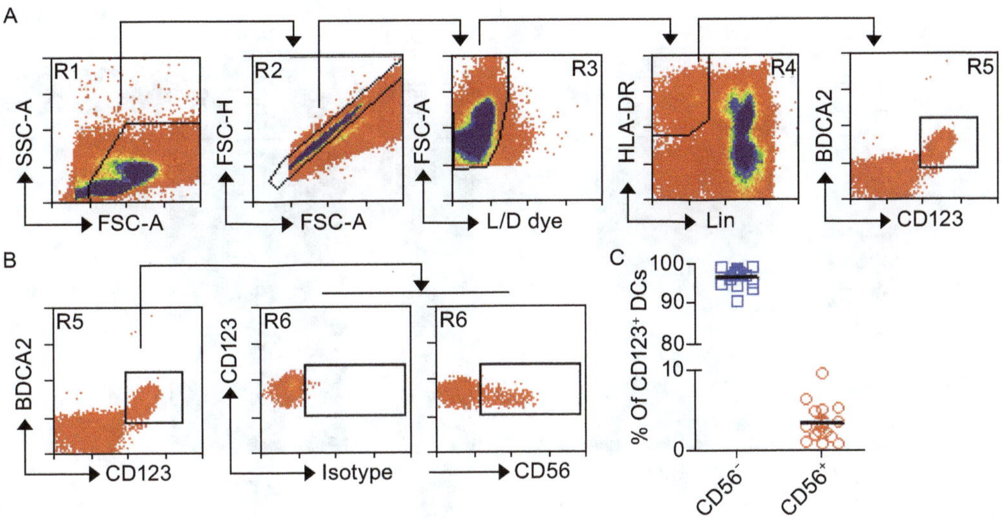

Figure 1. CD56⁺ DCs in human blood. (A) Human PBMC were first gated based on light scatters, and dead cells were excluded by yellow fluorescent reactive dye (L/D dye), then Lin⁻HLA-DR⁺BDCA2⁺CD123⁺ dendritic cells (DCs) were gated for further analysis. (B) Gated Lin⁻HLA-DR⁺BDCA2⁺CD123⁺ blood DCs were divided into two subsets by their CD56 expression. (C) Summarized data show relative abundance of CD56⁻ and CD56⁺ populations in BDCA2⁺CD123⁺ DCs (Mean ± SD, n = 15).

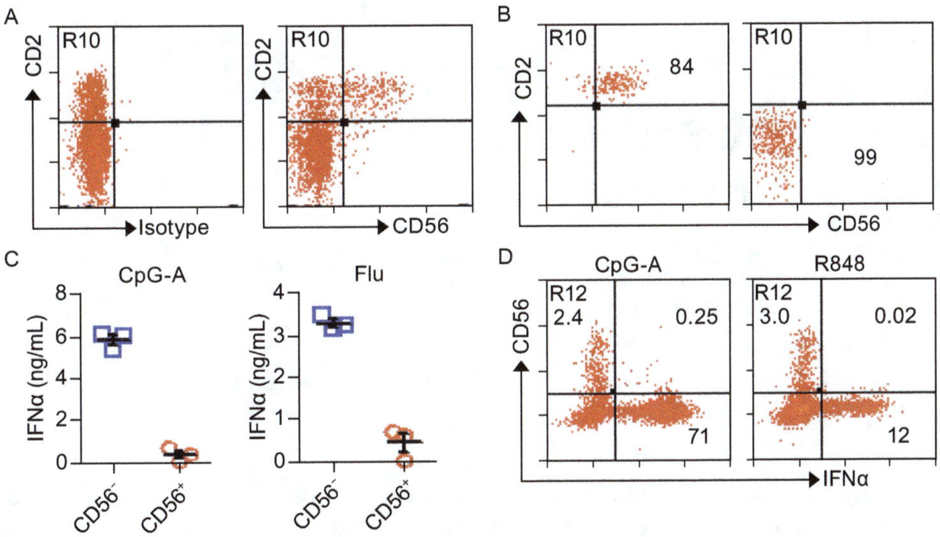

Figure 2. CD56⁺ DCs are functionally distinct from pDCs. (A) Human PBMC were gated on Lin⁻HLA-DR⁺CD11c⁻CD123⁺ DCs, and the gated DCs were analyzed by their CD2 and CD56 expression. (B) Lineage depleted PBMC were gated on Lin⁻HLA-DR⁺CD11c⁻CD123⁺ DCs, then CD2⁺CD56⁺ and CD2⁻CD56⁻ DCs were sorted by FACS. The purities of sorted cells are shown in each plot. (C) 10,000 purified CD123⁺CD56⁻ and CD123⁺CD56⁺ DCs were stimulated with CpG A or influenza virus (Flu) for 24 h, then interferon α (IFNα) in the supernatants were analyzed by ELISA. Data summarized from 3 independent donors. (D) Lineage depleted PBMCs were stimulated with CpG A or R848 then analyzed by FACS for surface markers and IFNα intracellularly. Lin⁻HLADR⁺CD11c⁻CD123⁺ cells were gated and plotted for CD56 and IFNα. The numbers in each quadrant represent percentage of total gated cells. One out of four independent experiments is shown.

further characterize the relationship between CD56⁺ DCs, pDCs and mDCs, we examined more gene expression that related to antigen presentation. MHC II molecules and CD86 were highly expressed on CD56⁺ DCs compared to pDCs at RNA level (Table S3). These were confirmed at protein level by FACS analysis (Fig. 4F). Gamma-interferon-inducible

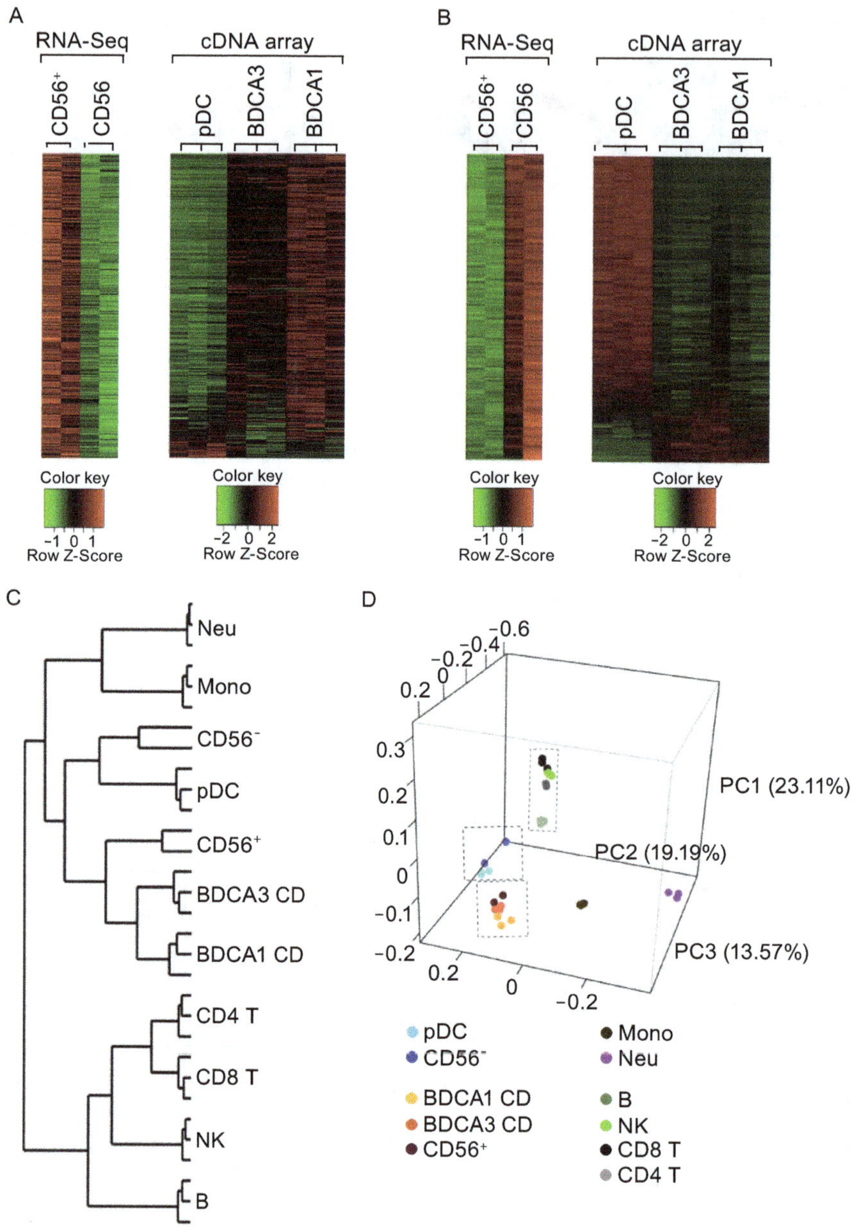

Figure 3. Comparing gene expression of CD56$^+$ DCs and other DC subsets. (A) 648 genes expressed in CD123$^+$CD56$^+$ DCs at least two fold higher than in CD123$^-$ CD56$^-$ DCs obtained by RNA-Seq (left panel) and their expressions in pDCs, BDCA3 mDCs and BDCA1 mDCs extracted from cDNA array data (right panel). (B) 1557 lowly expressed genes in CD123$^+$CD56$^+$ relative to CD123$^-$ CD56$^-$ DCs (left panel) and their expressions in pDCs, BDCA3 mDCs and BDCA1 mDCs extracted from cDNA array data (right panel). (C) Clustering of CD123$^+$CD56$^+$ and CD123$^-$ CD56$^-$ DCs with other immune cell lineages isolated from human peripheral blood based on their gene expression profiles. (D) The principal component analysis (PCA) results. Mono, monocytes; Neu, neutrophils; NK, natural killer cells.

lysosomal thiolreductase (GILT) is an essential enzyme for antigen processing, (Singh and Cresswell 2010; Maric et al., 2001) and we found that GILT and cathepsins were expressed at much higher levels in CD56$^+$ DCs than pDCs (Table S3). We next examined the antigen presentation by CD56$^+$ DCs, pDCs and mDCs. Allogeneic naïve T cells were co-cultured with purified DCs and analyzed by tritium incorporation. The CD56$^+$ DCs stimulated comparable T-cell proliferation to mDCs, while pDCs did not induce significant T-cell expansion (Fig. 4G).

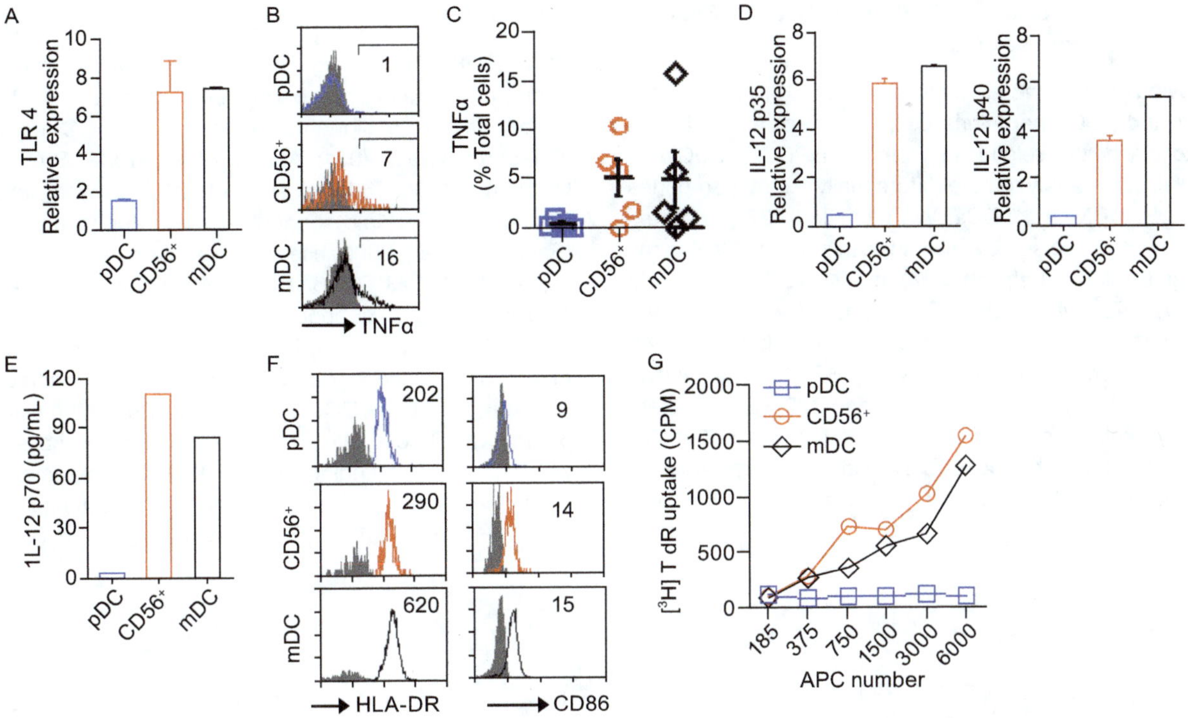

Figure 4. CD56$^+$ DCs functionally resemble mDCs. (A) The expression of TLR4 was quantified by RT-PCR with purified DCs. Error bars represent standard derivation from duplicated PCR samples of the same donor. Data shown are from one representative of 3 independent donors. (B) Lineage depleted PBMC were stimulated with LPS for 6 h, then the stimulated cells were collected and stained with surface markers followed by intracellular staining of TNFα. CD123$^+$CD56$^-$ (pDC)· CD123$^+$CD56$^+$ (CD56$^+$) and CD123$^-$ CD11C$^+$ (mDC) in the Lin$^-$ HLA-DR$^+$ gate were analyzed for TNFα expression. (C) Summarized data of (B) from 5 independent donors. (D) Purified DCs were stimulated with mixture of LPS, CpG and IFNγ for 6 h, the expression of IL-12 p35 and p40 were analyzed by RT-PCR. Data shown are representative of 3 independent experiments. (E) Purified DCs were stimulated mixture of LPS, CpG and IFNγ for 24 h, and the expression of IL-12 p70 were analyzed by ELISA. Data shown are representative of 2 independent experiments. (F) Expression of HLA-DR and CD86 on different blood DC subsets were analyzed by FACS. (G) Purified allogenic naive T cells were cocultured with different numbers of CD123$^+$CD56$^-$ (pDC, circle)· CD123$^+$CD56$^+$ (CD56$^+$, square) and CD123$^-$CD11C$^+$ (mDC, triangle) DCs . After 5 days, the cocultured cells were pulsed with 1 mCi [^{3}H] thymidine for 18 h before harvest. Data shown are one representative of 4 experiments.

CD56$^+$ DCs are unique mDCs mixed with some pDC features

CD56$^+$ DCs expressed BDCA2 and CD123, which were considered as hallmarks of pDCs (Fig. 1B). In order to further characterize the relationship between CD56$^+$ DCs and pDCs, we examined more pDC-specific genes at both RNA and protein levels in other DC subsets. And we found that BDCA4 and ILT7, two additional pDC specific surface markers, were expressed on CD56$^+$ DCs but not mDCs (Fig. 5A). Both E2-2 and SpiB, the two important transcription factors for pDC lineage development, were expressed in CD56$^+$ DCs at slightly lower levels than pDCs in the RNA-seq data (Table S4) and confirmed by RT-PCR (Fig. 5B, left two panels). It is worth to pointing out that the expression levels of E2-2 and SpiB in CD56$^+$ DCs were much higher than in mDCs (Fig. 5B, left two panels). In addition, pDC highly expressed genes IRF7, BCL11A and CD2AP (Marafioti et al., 2008) were also expressed in CD56$^+$ DCs at a slightly lower level compared to pDCs (Table S4). However, other pDC specific genes, such as Granzyme B, TCL1A (Herling et al., 2003), PACSIN1 (Esashi et al., 2012) and BAD-LAMP (Defays et al., 2011), were expressed at least 10 times higher in pDCs than in CD56$^+$ DCs in RNA-seq data (Table S4).

We have already shown that CD56$^+$ DCs produced very low level IFNα upon CpG stimulation. And this is correlated with the lower expression level of TLR9 in CD56$^+$ DCs than pDCs. However, the expression of TLR9 in CD56$^+$ DCs is clearly higher than in mDCs (Fig. 5B, far right panel). Interestingly, CD56$^+$ DCs produced comparable levels of TNFα to those of pDCs upon CpG B stimulation (Fig. 5C–E). These results argue that CD56$^+$ DCs are unique comparing other mDCs by maintaining certain pDC characteristic features.

BPDCN is closely related to CD56$^+$ DCs, but not pDCs by global gene expression profiling

In a previous report, we have found that BPDCN shows both myeloid and pDC characteristics (Cerboni et al., 2013). In order to clarify the relationship among BPDCN, pDC and CD56$^+$ DCs, we analyzed the differentially expressed genes between BPDCN and pDCs. We identified 127 genes express at higher levels in BPDCN than in pDC in cDNA array data (Fig. 6A, left panel; Table S7), but 93 (73%) of them are highly expressed in CD56$^+$ DCs compared to pDCs in our RNA-seq data (Fig. 6A, right panel). Similarly, there are 1143 genes expressed at lower levels in BPDCN cells than in pDCs (Fig. 6B, left panel; Table S8), and 778 (68%) of them are highly expressed in the CD56$^+$ DCs than pDCs (Fig. 6B, right panel). The global gene expression profiling data suggests that BPDCN is more closely related to the CD56$^+$ DCs than pDCs.

DISCUSSION

CD4$^+$CD56$^+$ hematologic malignancy with primary cutaneous presentation has been well documented and is proposed as the malignant counterpart of pDC precursors (Chaperot et al., 2001; Adachi et al., 1994; Brody 1995; Uchiyama 1998; Petrella et al., 1999) and was named as blastic plasmacytoid dendritic cell neoplasm (BPDCN) (Facchetti et al., 2008). Here in this report, we characterized a BDCA2$^+$CD123$^+$CD56$^+$ blood DC subset, which has been considered as a subpopulation of pDCs (Petrella et al., 2002). We demonstrated that the CD56$^+$ DCs clustered with mDCs by transcriptional profiling at whole genome level. Accordingly, CD56$^+$ DCs responded to TLR4 stimulation and prime naïve T cells without prior stimulation. And this agrees with our previous finding that BPDCN are characterized by their mixed pDC and myeloid signature (Sapienza et al., 2014). However, CD56$^+$ DCs are distinguished from other

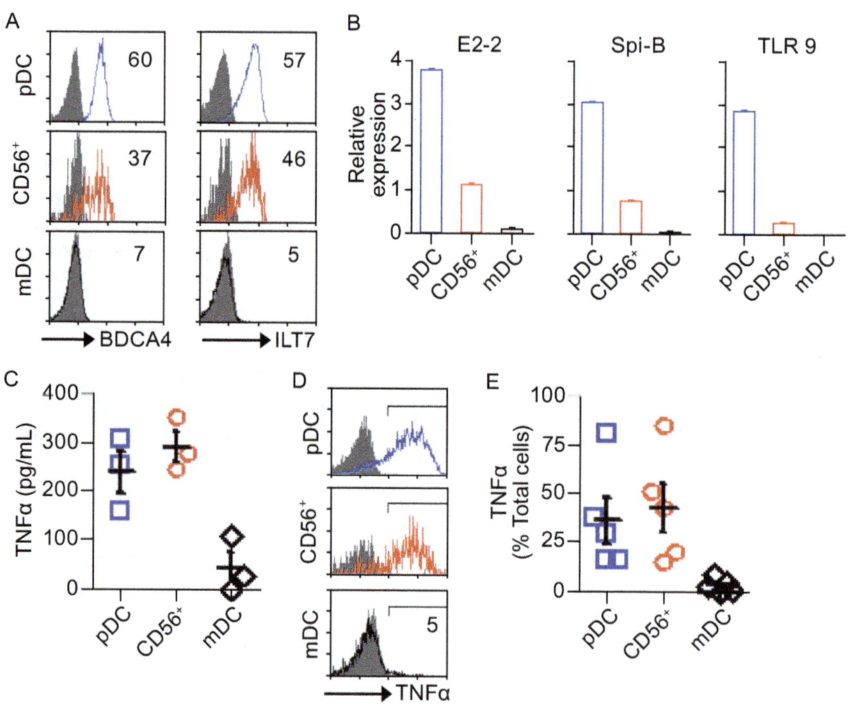

Figure 5. CD56$^+$ DCs adopt certain characteristics attributed to pDCs. (A) The expression of BDCA4 and ILT7 in different DC subpopulations from PBMC were analyzed by FACS. Data shown are one representative of 3 donors. (B) CD123$^+$CD56$^-$ (pDC), CD123$^+$CD56$^+$ (CD56$^+$) and CD123$^-$CD11C$^+$ (mDC) were purified from PBMC, E2-2, Spi-B and TLR9 expression in those three DC populations were analyzed by RT-PCR. Error bars represent standard derivation from duplicated PCR samples of same donor. Data shown are one representative of 3 donors. (C) Purified DCs were stimulated with CpG B for 24 h, and the expression of TNFα were analyzed by ELISA. Data shown are summarized from 3 independent experiments. (D) Lineage depleted PBMCs were stimulated with CpG B for 6 h, then stimulated cells were collected and stained with Lin, HLA-DR, CD123, CD11C and CD56 followed by intracellular staining of TNFα. The Lin$^-$ HLA-DR$^+$ cells were divided into CD123$^+$CD56$^-$ (pDC), CD123$^+$CD56$^+$ (CD56$^+$) and CD123$^-$CD11C$^+$ (mDC) subsets, TNFα expression in each DC subset were analyzed. (E) Summarized data of (D) from 5 independent donors.

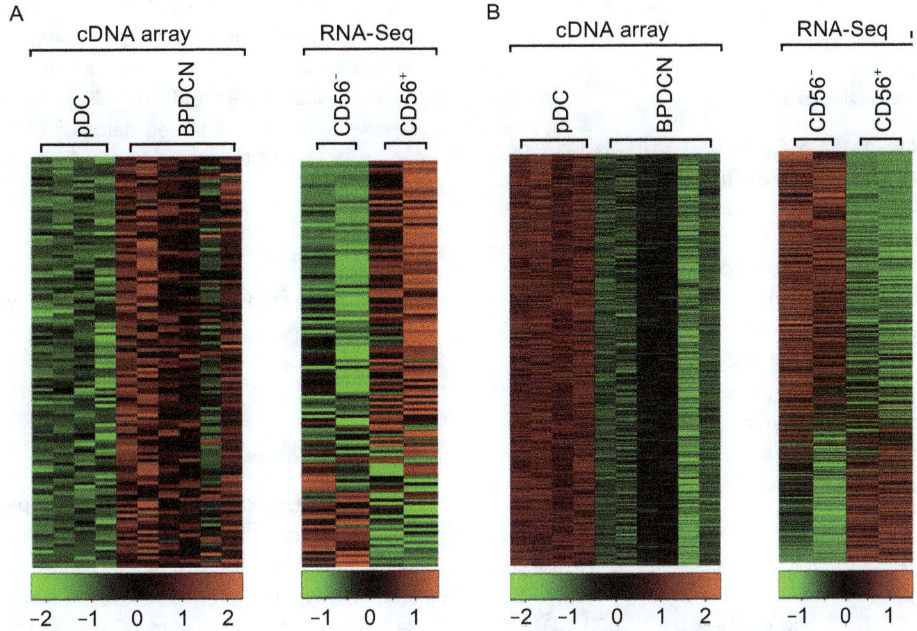

Figure 6. Global gene expression of BPDCN, pDCs and CD56$^+$ DCs. (A) Heat map of 127 genes highly expressed in BPDCN comparing with pDCs by cDNA array (left panel) and their expressions in CD56$^+$ DCs relative to CD56$^-$ pDCs by RNA-Seq (right panel). (B) Heat map of 1143 genes highly expressed in pDCs comparing with BPDCN by cDNA array (left panel) and their expressions in CD56$^+$ DCs relative to CD56$^-$ pDCs by RNA-Seq (right panel).

mDCs by expressing pDC specific molecules. Thus, we propose that CD56$^+$ DCs represent a unique mDC subset and their function in immune response warrant further characterization.

Besides BPDCN, another pDC related tumoral condition, pDC proliferations in patients with myeloid disorders (PPMD), has also been reported (Petrella and Facchetti 2010). The tumor cells of BPDCN and PPDM share the common characteristic phenotype of pDCs: lin$^-$, CD4$^+$, CD68$^+$, CLA$^+$, CD123$^+$, BDCA2$^+$, Bcl11A$^+$ and CD2AP$^+$. However, CD56 expression, which is the hallmark of BPDCN, is not detected in PPDM (Petrella and Facchetti 2010; Vermi et al., 2004). Granzyme B (GrB) expresses at high level in pDCs (Rissoan et al., 2002) and is also positive in PPMD (Vermi et al., 2004). However, most reported BPDCN cases are GrB negative (Chaperot et al., 2001; Petrella et al., 2002; Pilichowska et al., 2007; An et al., 2013). Here in this study, we show CD56$^+$ DCs express much lower level of GrB compared to pDC (Table S4). We speculate that PPMD may be the tumor counterpart of pDC, while BPDCN may represent the counterpart of CD123$^+$CD56$^+$ DCs. This hypothesis needs to be further investigated in the future by transcription profiling at whole genome level for both DC subsets and their tumor counterparts.

It is worth pointing out that the human CD123$^+$CD56$^+$ DCs are similar to a $CX_3CR1^+CD8\alpha^+$ mouse DC subset, which also exhibits mixed characteristics of pDC and mDC (Ghosh et al., 2010). The $CX_3CR1^+CD8\alpha^+$ mouse DCs are developmentally dependent on E2-2 and express pDC specific genes (Spi-B, Siglec H, Bst2). However, CX_3CR1^+ $CD8\alpha^+$ DCs do not produce IFNα upon TLR7 and TLR9 stimulation (Ghosh et al., 2010). The $CX_3CR1^+CD8\alpha^+$ DC is transcriptionally similar to mouse CD8$^-$ mDCs, but not to CD8$^+$ mDCs (Ghosh et al., 2010). The developmental relationship between human CD56$^+$ DCs and mouse $CX_3CR1^+CD8\alpha^+$ DCs needs further investigation.

Taken together, we have demonstrated that human CD56$^+$ DCs were very close to mDC by both transcriptomic analysis and functional characterization. Thus, we propose that CD56$^+$ DC should be classified as mDC, but not pDCs. And we propose that BPDCN should be classified as mDC but not pDC related leukemia.

MATERIALS AND METHODS

FACS analysis of DC subsets in human blood

For analysis of DC subsets in human peripheral blood, single cells were first gated by light scatters and dead cells were excluded by yellow fluorescent reactive live/dead cell dye staining (Life Technologies, Carlsbad, USA), then the lineage positive cells were gated out by specific markers (CD3, CD14, CD16, CD19). HLA-DR$^+$CD123$^+$BDCA2$^+$ cells were further divided by their CD2 and CD56 expression. Antibody stained peripheral blood mononuclear cells (PBMC) were analyzed with LSRfortessa (BD Biosciences, Franklin Lakes, NJ, USA) and data were analyzed with Summit 4.3

(DAKO, Denmark). Detailed information of antibodies used for flow cytometry analysis was shown in Table S5.

DC enrichment and *in vitro* stimulation

DCs were enriched by lineage (CD3, CD14, CD16, CD19) depletion, then lineage negative cells (2×10^5 in 200 μL culture medium) were stimulated with TLR ligands or viruses for 4 h followed by another 2 h in the presence of Golgi Blocker (BD Biosciences, Franklin Lakes, NJ, USA). Cells were stained with surface markers and permeablized and stained with antibodies against various cytokines. TLR ligands were purchased from Invivogen (San Diego, CA, USA) and used at following concentrations: LPS (1 μg/mL), CpG 2216 (2 μg/mL), R848 (2 μg/mL). Heat inactivated influenza virus A/PR8/34 was used at 10 MOI for stimulation.

DC purification and stimulation

The lineage (CD3, CD14, CD16, CD19, CD20) depleted PBMC were stained with HLA-DR (APC-Cy7), CD2 (PE-Cy7), CD11c (APC), CD56 (PerCP-cy5.5) and CD123 (BV421). $Lin^-HLA\text{-}DR^+CD123^+CD11c^-$ $CD2^+CD56^+$ and $Lin^-HLA\text{-}DR^+CD123^+CD11c^-$ $CD2^-CD56^-$ DCs were sorted with BD FACSAriaIII (BD Biosciences, Franklin Lakes, NJ, USA). In some experiments, $Lin^-HLA\text{-}DR^+CD123^-CD11c^+$ mDC were sorted as control. Purified DCs (1×10^4 in 200 μL culture medium) were stimulated with different TLR ligands at the concentration mentioned above. For IL-12 production, purified DCs were stimulated with LPS, CpG 2006 (1 μmol/L) and IFNγ (50 ng/mL). IFNα, IL-12 and TNFα levels in the supernatant were quantified by ELISA.

RNA-seq and data analysis

RNA extraction and sequencing were done by BGI Tech (Shenzhen, Guangdong, China). The gene expression level was measured by the number of uniquely mapped reads per kilobase of exon region per million mappable reads (RPKM). The RNA-seq raw data was aligned to the reference genome (hg19) (Trapnell et al., 2009). The gene expressions of other cell types used in our study were obtained from cDNA array data by Scott H Robbins (Robbins et al., 2008). The target gene regions of microarray probes were retrieved from the annotation file for human U133 Plus 2.0 (http://www.affymetrix.com). To compare gene expressions from both RNA-seq and cDNA array experiments, we only considered the reads that have at least one nucleotide overlap with the gene regions targeted by microarray probes. Heat maps were generated with differentially expressed genes between $CD2^+CD56^+$ and $CD2^-CD56^-$ DCs (larger than 2 times difference). Transcriptome comparison of $CD2^+CD56^+$ DCs and other blood cells lineages were carried out with two different mathematical methods, hierarchical clustering with complete linkage (Eisen et al., 1998) and principal component analysis (PCA) (Alter et al., 2000). Detailed methods of RNA-seq and data analysis were provided in the supplemental file.

BPDCN microarray data of 6 cryopreserved tissue samples (Sapienza et al., 2014) by Affymetrix microarray platform were normalized with the rank invariant method using Lumi (Bioconductor) (Du et al., 2008). Statistics for differential expression were applied by Limma (Bioconductor), (Smyth 2004) and we defined BPDCN significant differential expressed genes on normalized data as adjusted *P*-value less than 0.05. One hundred and twenty seven BPDCN highly expressed genes and 1143 BPDCN lowly expressed genes were matched to our RNA-seq datasets of pDC and $CD56^+$ DCs (genes with RPKM less than 10 were deleted). Heat maps were displayed using the R software.

RT-PCR

RNA was extracted from purified DCs and reverse transcribed with Oligo dT primers. Quantitative real time PCR (RT-PCR) was performed using cDNA with EF1α gene as internal control. Primers for RT-PCR are listed in Table S6.

T-cell proliferation assay

Human $CD4^+CD45RA^+$ naïve T cells were purified from PBMC by negative selection with magnetic beads (Miltenyi Biotech, Germany). Different numbers of FACS purified DCs were cultured with 5×10^4 allogeneic naïve T cells in 96-well round-bottomed plates in 200 μL culture medium. After 5 days of DC and T cell co-culture, cells were pulsed with 1 mCi [^{3}H]-thymidine for 18 h before harvest. Radioactive uptake was measured with a TopCount NXT micro-plate scintillation and luminescence counter (PerkinElmer).

ACKNOWLEDGMENTS

This work was supported by grants from Beijing Municipal Science & Technology Commission (SCW 2014-09), Chinese Academy of Science (KJZD-EW-L10-02) and Ministry of Health (2013ZX10001002) to L.Z. The authors thank Dr. Lishan Su for critical reading this manuscript and Mr. Anthony Curtis for English editing. H.Y. and L.Z. designed the project, analyzed data and wrote the paper; P.Z., Y.C. and T.J. analyzed and interpreted the RNA-seq data and wrote the paper; P. P. participated in the analysis of BPDCN data; X.Y., Z.Y., Q.S., S.W. and G.L. performed experiments and analyzed data. The authors declare no financial conflict of interest in relation to this work.

ABBREVIATIONS

BPDCN, blastic plasmacytoid dendritic cell neoplasm; DCs, dendritic cells; IFN-I, type I interferon; IRF7, interferon regulatory factor 7; mDC, myeloid DC; PBMC, peripheral blood mononuclear cell; PCA, principal component analysis; pDC, plasmacytoid DC; TLR, toll like receptor.

COMPLIANCE WITH ETHICS GUIDELINES

Haisheng Yu, Peng Zhang, Xiangyun Yin, Zhao Yin, Quanxing Shi, Ya Cui, Guanyuan Liu, Shouli Wang, Pier Paolo Piccaluga, Taijiao Jiang and Liguo Zhang declare that they have no conflict of interest. All procedures followed were in accordance with the ethical standards of the responsible committee on human experimentation (institutional and national) and with the Helsinki Declaration of 1975, as revised in 2000 (5). Informed consent was obtained from all patients for being included in the study.

REFERENCES

Adachi M, Maeda K, Takekawa M et al (1994) High expression of CD56 (N-CAM) in a patient with cutaneous CD4-positive lymphoma. Am J Hematol 47(4):278–282

Alter O, Brown PO, Botstein D (2000) Singular value decomposition for genome-wide expression data processing and modeling. Proc Natl Acad Sci USA 97(18):10101–10106

An HJ, Yoon DH, Kim S et al (2013) Blastic plasmacytoid dendritic cell neoplasm: a single-center experience. Ann Hematol 92 (3):351–356

Brody JP, Allen S, Schulman P et al (1995) Acute agranular CD4-positive natural killer cell leukemia. Comprehensive clinicopathologic studies including virologic and in vitro culture with inducing agents. Cancer 75(10):2474–2483

Cao W, Rosen DB, Ito T et al (2006) Plasmacytoid dendritic cell-specific receptor ILT7-Fc epsilonRI gamma inhibits Toll-like receptor-induced interferon production. J Exp Med 203 (6):1399–1405

Cella M, Jarrossay D, Facchetti F et al (1999) Plasmacytoid monocytes migrate to inflamed lymph nodes and produce large amounts of type I interferon. Nat Med 5(8):919–923

Cerboni S, Gentili M, Manel N (2013) Diversity of pathogen sensors in dendritic cells. Adv Immunol 120:211–237

Chaperot L, Bendriss N, Manches O et al (2001) Identification of a leukemic counterpart of the plasmacytoid dendritic cells. Blood 97 (10):3210–3217

Cisse B, Caton ML, Lehner M et al (2008) Transcription factor E2-2 is an essential and specific regulator of plasmacytoid dendritic cell development. Cell 135(1):37–48

Comeau MR (2002) Van der Vuurst de Vries AR, Maliszewski CR, Galibert L. CD123bright plasmacytoid predendritic cells: progenitors undergoing cell fate conversion? J Immunol 169(1):75–83

Defays A, David A, de Gassart A et al (2011) BAD-LAMP is a novel biomarker of nonactivated human plasmacytoid dendritic cells. Blood 118(3):609–617

Du P, Kibbe WA, Lin SM (2008) lumi: a pipeline for processing Illumina microarray. Bioinformatics 24(13):1547–1548

Dzionek A, Fuchs A, Schmidt P et al (2000) BDCA-2, BDCA-3, and BDCA-4: three markers for distinct subsets of dendritic cells in human peripheral blood. J Immunol 165(11):6037–6046

Eisen MB, Spellman PT, Brown PO, Botstein D (1998) Cluster analysis and display of genome-wide expression patterns. Proc Natl Acad Sci USA 95(25):14863–14868

Esashi E, Bao M, Wang YH, Cao W, Liu YJ (2012) PACSIN1 regulates the TLR7/9-mediated type I interferon response in plasmacytoid dendritic cells. Eur J Immunol 42(3):573–579

Facchetti F, Jones D, Petrella T (2008) Blastic plasmacytoid dendritic cell neoplasm. In: Swerdlow S, Campo E, Harris N et al (eds) WHO classification of tumours of haematopoietic and lymphoid tissues, 4th edn. International Agency for Research on Cancer (IARC), Lyon, pp 145–147

Ghosh HS, Cisse B, Bunin A, Lewis KL, Reizis B (2010) Continuous expression of the transcription factor e2-2 maintains the cell fate of mature plasmacytoid dendritic cells. Immunity 33(6):905–916

Grouard G, Rissoan MC, Filgueira L, Durand I, Banchereau J, Liu YJ (1997) The enigmatic plasmacytoid T cells develop into dendritic cells with interleukin (IL)-3 and CD40-ligand. J Exp Med 185 (6):1101–1111

Herling M, Teitell MA, Shen RR, Medeiros LJ, Jones D (2003) TCL1 expression in plasmacytoid dendritic cells (DC2s) and the related $CD4^+$ $CD56^+$ blastic tumors of skin. Blood 101(12):5007–5009

Kadowaki N, Ho S, Antonenko S et al (2001) Subsets of human dendritic cell precursors express different toll-like receptors and respond to different microbial antigens. J Exp Med 194(6):863–869

Liu YJ (2005) IPC: professional type 1 interferon-producing cells and plasmacytoid dendritic cell precursors. Annu Rev Immunol 23:275–306

Marafioti T, Paterson JC, Ballabio E et al (2008) Novel markers of normal and neoplastic human plasmacytoid dendritic cells. Blood 111(7):3778–3792

Maric M, Arunachalam B, Phan UT et al (2001) Defective antigen processing in GILT-free mice. Science 294(5545):1361–1365

Matsui T, Connolly JE, Michnevitz M et al (2009) CD2 distinguishes two subsets of human plasmacytoid dendritic cells with distinct phenotype and functions. J Immunol 182(11):6815–6823

Petrella T, Facchetti F (2010) Tumoral aspects of plasmacytoid dendritic cells: what do we know in 2009? Autoimmunity 43 (3):210–214

Petrella T, Dalac S, Maynadie M et al (1999) $CD4^+$ $CD56^+$ cutaneous neoplasms: a distinct hematological entity? Groupe Francais d'Etude des Lymphomes Cutanes (GFELC). Am J Surg Pathol 23 (2):137–146

Petrella T, Comeau MR, Maynadie M et al (2002) 'Agranular $CD4^+$ $CD56^+$ hematodermic neoplasm' (blastic NK-cell lymphoma) originates from a population of $CD56^+$ precursor cells related to plasmacytoid monocytes. Am J Surg Pathol 26(7):852–862

Pilichowska ME, Fleming MD, Pinkus JL, Pinkus GS (2007) $CD4^+$/$CD56^+$ hematodermic neoplasm ("blastic natural killer cell lymphoma"): neoplastic cells express the immature dendritic cell marker BDCA-2 and produce interferon. Am J Clin Pathol 128 (3):445–453

Rissoan MC, Duhen T, Bridon JM et al (2002) Subtractive hybridization reveals the expression of immunoglobulin-like transcript 7, Eph-B1, granzyme B, and 3 novel transcripts in human plasmacytoid dendritic cells. Blood 100(9):3295–3303

Robbins SH, Walzer T, Dembele D et al (2008) Novel insights into the relationships between dendritic cell subsets in human and mouse revealed by genome-wide expression profiling. Genome Biol 9(1):R17

Sapienza MR, Fuligni F, Agostinelli C et al (2014) Molecular profiling of blastic plasmacytoid dendritic cell neoplasm reveals a unique pattern and suggests selective sensitivity to NF-kB pathway inhibition. Leukemia 28:1606–1616

Schotte R, Nagasawa M, Weijer K, Spits H, Blom B (2004) The ETS transcription factor Spi-B is required for human plasmacytoid dendritic cell development. J Exp Med 200(11):1503–1509

Siegal FP, Kadowaki N, Shodell M et al (1999) The nature of the principal type 1 interferon-producing cells in human blood. Science 284(5421):1835–1837

Singh R, Cresswell P (2010) Defective cross-presentation of viral antigens in GILT-free mice. Science 328(5984):1394–1398

Smyth GK (2004) Linear models and empirical bayes methods for assessing differential expression in microarray experiments. Stat Appl Genet Mol Biol 3:1–25

Steinman RM (2007) Lasker Basic Medical Research Award. Dendritic cells: versatile controllers of the immune system. Nat Med 13(10):1155–1159

Trapnell C, Pachter L, Salzberg SL (2009) TopHat: discovering splice junctions with RNA-Seq. Bioinformatics 25(9):1105–1111

Uchiyama N, Ito K, Kawai K, Sakamoto F, Takaki M, Ito M (1998) $CD2^-$, $CD4^+$, $CD56^+$ agranular natural killer cell lymphoma of the skin. Am J Dermatopathol 20(5):513–517

Vermi W, Facchetti F, Rosati S et al (2004) Nodal and extranodal tumor-forming accumulation of plasmacytoid monocytes/interferon-producing cells associated with myeloid disorders. Am J Surg Pathol 28(5):585–595

Ziegler-Heitbrock L, Ancuta P, Crowe S et al (2010) Nomenclature of monocytes and dendritic cells in blood. Blood 116(16):e74–e80

15

Direct conversion of human fibroblasts into retinal pigment epithelium-like cells by defined factors

Kejing Zhang[1], Guang-Hui Liu[2]✉, Fei Yi[3], Nuria Montserrat[4,5], Tomoaki Hishida[1], Concepcion Rodriguez Esteban[1], Juan Carlos Izpisua Belmonte[1,4]✉

[1] Gene Expression Laboratory, Salk Institute for Biological Studies, 10010 North Torrey Pines Road, La Jolla, CA 92037, USA
[2] National Laboratory of Biomacromolecules, Institute of Biophysics, Chinese Academy of Sciences, Beijing 100101, China
[3] Department of Molecular and Cellular Physiology, Stanford University School of Medicine, 265 Campus Drive, Stanford, CA 94305, USA
[4] Center for Regenerative Medicine in Barcelona, Dr. Aiguader 88, 08003 Barcelona, Spain
[5] Biomedical Research Networking Center in Bioengineering, Biomaterials and Nanomedicine (CIBER-BBN), Barcelona, Spain
✉ Correspondence: ghliu@ibp.ac.cn (G.-H. Liu), belmonte@salk.edu, izpisua@cmrb.eu (J. C. Izpisua Belmonte)

ABSTRACT

The generation of functional retinal pigment epithelium (RPE) is of great therapeutic interest to the field of regenerative medicine and may provide possible cures for retinal degenerative diseases, including age-related macular degeneration (AMD). Although RPE cells can be produced from either embryonic stem cells or induced pluripotent stem cells, direct cell reprogramming driven by lineage-determining transcription factors provides an immediate route to their generation. By monitoring a human RPE specific Best1::GFP reporter, we report the conversion of human fibroblasts into RPE lineage using defined sets of transcription factors. We found that Best1::GFP positive cells formed colonies and exhibited morphological and molecular features of early stage RPE cells. Moreover, they were able to obtain pigmentation upon activation of Retinoic acid (RA) and Sonic Hedgehog (SHH) signaling pathways. Our study not only established an ideal platform to investigate the transcriptional network regulating the RPE cell fate determination, but also provided an alternative strategy to generate functional RPE cells that complement the use of pluripotent stem cells for disease modeling, drug screening, and cell therapy of retinal degeneration.

KEYWORDS retinal pigment epithelium, fibroblasts, direct conversion

INTRODUCTION

The retinal pigment epithelium (RPE) is a pigmented monolayer of epithelium residing outside of the neurosensory retina where it supports metabolic and cellular processes of retinal photoreceptors. Dysfunction and degeneration of RPE lead to photoreceptor loss in many sight-threatening diseases, including the leading causes of blindness in the developed world, age-related macular degeneration (AMD) (Khandhadia et al., 2012). Currently, treatments available for these diseases are limited and do not offer a cure for the cell loss. Therefore, the generation of functional RPE cells is of great therapeutic interest to the field of regenerative medicine and offers possible cures for retina degeneration diseases (Schwartz et al., 2012).

Induced RPE differentiation has been achieved previously using embryonic stem cells (ESCs) and induced pluripotent stem cells (iPSCs) (Carr et al., 2009; Lu et al., 2009; Zhu et al., 2013). Although differentiating RPE cells from hESCs is promising (Zhang et al., 2013), the application of hESCs for therapeutic purposes remains challenging due to un-eased ethical concerns. iPSC technology holds the potential to be used to generate patient specific cells for autologous cell transplantation, which has created enormous expectations and circumvented some ethical debates (Takahashi and Yamanaka, 2006; Takahashi et al., 2007). However, differentiation of RPE cells from iPSCs is inefficient, time-consuming, and variable among different iPSC lines (Buchholz et al., 2009). Moreover, iPSCs present several safety concerns such as the genetic and epigenetic aberrations they are carrying, as well as the potential risk for tumor formation (Panopoulos et al., 2011). Recently, advances in direct cell lineage conversion have suggested a potential solution to these issues. The fact that one somatic cell type can be readily converted into another implicates access to abundant resources of any clinically relevant cell type. Moreover, as direct lineage conversion bypasses the pluripotent state, it could theoretically reduce the risk of tumorigenicity after transplantation (Ben-David and Benvenisty, 2011). Using distinct sets of transcription factors, direct lineage conversion technology has been applied to produce various cell types, including neurons, hepatocytes and neural stem cells (Vierbuchen et al., 2010; Huang et al., 2011; Kim et al., 2011; Pang et al., 2011; Sekiya and Suzuki, 2011; Giorgetti et al., 2012; Liu et al., 2012; Zhang et al., 2012).

Here we report the successful development of a RPE-specific Best1::GFP reporter, which faithfully represented human RPE lineage commitment during hESC differentiation. Using this reporter system, we show that a defined set of transcription factors can reprogram human fibroblasts into Best1::GFP$^+$ colonies. These Best1::GFP$^+$ cells exhibit specific morphological and molecular features of RPE lineage and are capable of pigmentation. Our study not only provided a powerful system to study the nature of cellular identity and plasticity of RPE lineage, but also offered a new path to produce functional RPE cells for regenerative therapy and drug development in the future.

RESULTS

Establishment of a human RPE specific reporter system

A proper reporter system is essential for monitoring efficient cell fate conversion in reprogramming studies. A series of truncated versions of human *Bestrophin1* (*Best1*) gene and *Rpe65* gene promoter, have been shown to mediate robust ocular reporter gene expression that is restricted to the RPE in transgenic mice (Boulanger et al., 2000; Esumi et al., 2004; Acland et al., 2005; Esumi et al., 2009). To test their functionality in human cells, these promoter fragments were cloned into the upstream of EGFP in a lentiviral vector, pGreenZeo (Fig. 1A). We then performed lentivirus infection to test the promoter specificity in different human cell types including human foreskin fibroblasts (HFF), HEK293T cells, H9 hESCs, as well as human primary RPE cells (HRPE). Distinct from the non-specific expression of the Rpe65(0.8k) or Rpe65(1k) reporter in all of the examined cell types, the Best1(0.6 kb) (spanning from -588 to +58) and Best1(1 kb) (spanning from -1020 to +38) reporters exhibited high selectivity in driving strong EGFP expression in human RPE cells (Fig. 1A). In the following studies, data presented were generated using Best1(0.6 kb) reporter as Best1::GFP reporter, if not specifically indicated. In order to further confirm the specificity of the reporter, we tested the reporter in H9 ESC differentiated cells at day 60, using a reported protocol (Meyer et al., 2009), which gave rise to a mixed population of various cell types including RPE cells. We observed that the Best1::GFP reporter was only expressed in RPE-like cells, which displayed a morphologically characteristic hexagonal shape, but not in other neural or fibroblast-like cells (Fig. 1B).

We next evaluated whether the Best1::GFP reporter works properly during the directed RPE differentiation from hESCs. Accordingly, H9 hESCs were transduced with Best1::GFP lentiviral reporter and ten individual colonies were chosen to verify the integration of Best1::GFP expression cassette by genomic DNA PCR (data not shown). Positive colonies were subcultured and expanded for further studies. For RPE differentiation, we developed a rapid and efficient protocol, which is modified from two recent publications (Boucherie et al., 2013; Zhu et al., 2013) (Fig. 2A, group Activin A). With this approach, pigmented cells appeared in a monolayer form 3 weeks after Activin A treatment. Immunostaining showed that these cells expressed typical RPE markers including microphthalmia-associated transcription factor (MITF) and Best1, as well as a tight junction protein, ZO-1, indicating a successful commitment of RPE from hESCs (Fig. 2E). Given the reported roles of Retinoic acid (RA) and Sonic Hedgehog (SHH) signaling in RPE maturation and pigmentation (Okada et al., 2004; Zahabi et al., 2012), we set up an RA plus SHH experimental group. As shown in Fig. 2A and 2B, RA plus SHH treatment enhanced the pigmentation despite resulting in a lower yield of RPE cells compared to the Activin A treatment group.

Monitoring the Best1::GFP expression represents a direct way to visualize the dynamics of RPE differentiation. In both the "Activin A" and "RA plus SHH" treatment groups, Best1::GFP$^+$ cells were initially observed at around day 21 of differentiation, and the percentage of Best1::GFP$^+$ cells increased to approximate 50% after 40 days of differentiation (Fig. 2C). Consistently, time-course studies showed that the levels of *Best1* mRNA were strongly upregulated after 3 weeks of differentiation from H9 hESCs (Fig. 2D). Interestingly, Best1::GFP was highly expressed in non-pigmented and lightly pigmented RPE cells, while when cells were heavily pigmented, GFP fluorescence diminished or even became undetectable (Fig. 2E c–c″, d–d″). Since endogenous Best1 protein was still expressed in highly pigmented cells (Fig. 2E d–d″), the absence of GFP fluorescence in these cells may be attributed to the absorption of fluorescence by melanin. To further characterize the mature status

A

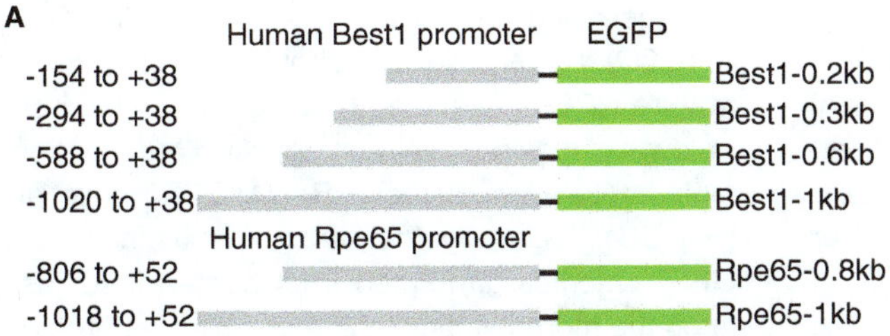

	293T	HFF	H9 hESC	HRPE
Best1-0.2kb	+	+	-	+
Best1-0.3kb	+	+	-	+
Best1-0.6kb	-	-	-	+++
Best1-1kb	-	-	-	+++
Rpe65-0.8kb	+	+	+	+
Rpe65-1kb	+	+	+	+

B

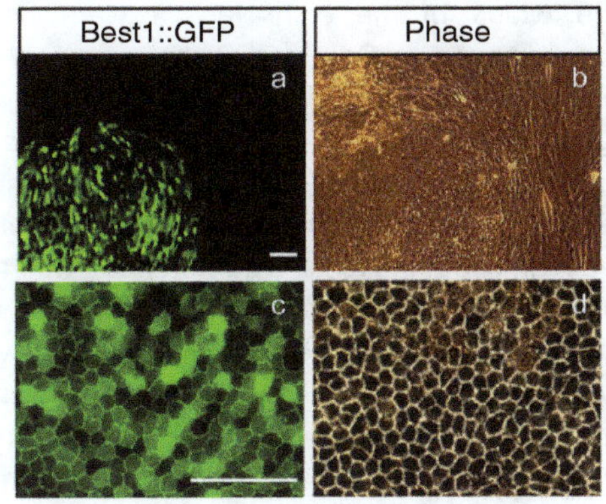

Figure 1. Generating human RPE specific reporters. (A) Schematic representation of the promoter regions of human *Best1* or *Rpe65* gene for reporter constructs. The table shows the relative GFP levels of each reporter observed in different cell lines. HRPE, human primary RPE cells; -, no expression; +, weak expression; +++, strong expression. (B) Best1::GFP specifically expressed in RPE-like cells differentiated from H9 hESC. All scale bars: 100 μm.

of Best1::GFP$^+$ cells that followed the protocol with Activin A treatment, we performed qPCR analysis at day 40. The results demonstrated that Best1::GFP$^+$ cells did express the early eye field genes including *Pax6*, *Lhx2*, *Six6* and *Rax* (Zuber et al., 2003) (Fig. 2F). Two key genes for RPE differentiation, *Mitf* and *Otx2* (Boulanger et al., 2000; Steingrimsson et al., 2004), were markedly upregulated. Other mature RPE hallmark genes were also highly expressed in Best1::GFP$^+$ cells, including *Best1*, pigment epithelium derived factor (*Pedf*), RPE-specific protein 65 kDa (*Rpe65*), cellular retinaldehyde-binding protein (*Cralbp*) involved in vitamin A metabolism, tyrosinase (*Tyr*) and tyrosinase related protein 2 (*Tyrp2*), both involved in pigment synthesis (Martinez-Morales et al., 2004; Strauss, 2005) (Fig. 2F). These gene expression patterns of Best1::GFP$^+$ cells are similar to those of human primary RPE cells (Fig. 2F). Collectively, these data demonstrate the successful establishment of a human RPE specific reporter system, which faithfully represented RPE lineage commitment during ESC differentiation.

Definition of transcription factor pool

We hypothesized that transcription factors known to instruct RPE formation during development might also facilitate the conversion of other somatic cell types into RPEs. To test this idea, we constructed retroviral expression vectors encoding six candidate transcription factors (TFs): *Rax*, *Crx*, *Pax6*, *Mitf*, *Otx2*, and *Nrl* that have been reported to participate in different stages of retina development and RPE specification (Zuber et al., 2003; Martinez-Morales et al., 2004; Steingrimsson et al., 2004) (Fig. 3A). After transduction of HFF with a combination of these 6TFs together with Best1::EGFP reporter lentivirus, cells were cultured with hESC culture medium (CDF12) to prime the lineage conversion for 7 days. The cells were subsequently cultured in a similar condition that was used to induce RPE differentiation from hESCs (Fig. 3B). However, such a strategy resulted in neither obvious morphological change nor Best1::GFP$^+$ cells from parental HFF. This raised a possibility that other factors may be required to increase the epigenetic "plasticity" of the original human fibroblasts. Based on the "clue" that two "Yamanaka factors" (*cMyc* and *Klf4*) are highly expressed in adult RPEs (Salero et al., 2012), we supplemented *cMyc* and *Klf4* into the previous six transcription factor cocktail. qPCR analysis verified the expression of all eight transcription factors (8TFs) in HFF transduced with this new cocktail for three days (Fig. 3C). Using the co-transduced retroviral vector encoding mCherry as an indicator, we found that the retrovirus infection system in HFF achieved a transduction efficiency for individual factors of 77% (Fig. 3D). This experiment suggests that almost 12% of parental HFF cells could have the opportunity to express all of the eight TFs to initiate the reprogramming process.

Best1::GFP$^+$ cells generated by direct reprogramming of human fibroblasts

We next performed the lineage conversion experiments by transduction of HFF with 8TFs. Encouragingly, a number of Best1::GFP$^+$ colonies emerged at 12 days post-transduction. These Best1::GFP$^+$ cell colonies kept actively proliferating and became tightly compacted with a high cytoplasmic to nuclear ratio at day 21. Later, Best1::GFP$^+$ colonies gradually spread out on matrigel and exhibited a cobblestone-like morphology at day 35 (Fig. 4A). These morphological observations suggested a RPE progenitor-like status of the generated Best1::GFP$^+$ cell colonies. To determine which of the eight factors were necessarily required for generating Best1::GFP$^+$ colonies, we removed each TF-encoding vector from the 8TFs cocktail one at a time (Fig. 4B). As shown in Fig. 4B, excluding *cMyc*, *MitfA* or *Otx2* eliminated GFP$^+$ colony formation, and the number of Best1::GFP$^+$ colonies was dramatically reduced when *Rax*, or *Crx* was removed. Interestingly, when *Pax6* was removed, the number of Best1::GFP$^+$ colony slightly reduced, while the number of colonies was not discernibly changed when *Klf4* or *Nrl* was ruled out (Fig. 4B). However, when any combination of two or three factors of *Klf4*, *Pax6* and *Nrl* was removed from the 8TFs pool, we could not detect any Best1::GFP$^+$ colonies (data not shown). These data indicated that *cMyc*, *Mitf*, *Otx2*, *Rax*, and *Crx* are crucial for reprogramming of human fibroblasts into Best1::GFP$^+$ cells, while *Klf4*, *Nrl* and *Pax6* individually are not so important and could be omitted or replaced by other factors.

Characterization of HFF-derived Best1::GFP$^+$ cells

To assess whether Best1::GFP$^+$ colonies hold the potential to achieve known characteristics of cultured RPE cells, we evaluated the molecular signatures of the generated Best1::GFP$^+$ cells induced by 8TFs. Immunofluorescence analysis demonstrated that most of the Best1::GFP$^+$ colonies were positive for MITF staining (Fig. 5A). We observed that only some of the colonies expressed Pax6, which is consistent with the notion that down-regulated expression of *Pax6* occurs upon RPE maturation (Fig. 5A and data not shown). To induce the further maturation of the Best1::GFP$^+$ colonies, we picked the Best1::GFP$^+$ colonies onto matrigel coated wells at day 21, and then the cells were maintained in the medium consisting of Activin A alone, or RA plus SHH (Fig. 3B). Both conditions promoted further growth of the Best1::GFP$^+$ colonies. At day 35 post-transduction, cells derived from Best1::GFP$^+$ colonies exhibited cobblestone-like morphology resembling RPE cells differentiated from hESCs, and expressed RPE-specific markers including MITF, ZO-1 and Best1 (Fig. 5A). More importantly, in the RA plus SHH treatment group, pigmented cells were observed at day 35, indicating that Best1::GFP$^+$ cells have achieved further maturation giving rise to functional RPE cells (Fig. 5B). To further characterize their identity, we purified the Best1::GFP$^+$ cells by flow cytometry analysis (FACS), and performed qPCR to analyze the transcripts of RPE-specific factors. Compared to the control HFF cells, the generated Best1::GFP$^+$ cells expressed not only the transcription factors we introduced, but also other eye field genes such as *Six3* and *Lhx2*. Additionally, a series of mature RPE marker genes, including *Best1*, *Cralbp*, *Pedf*, and *Tyrp2*, were also dramatically induced in Best1::GFP$^+$ cells directly converted from HFF, and notably these genes are involved in different functions of RPE, e.g. vitamin A metabolism, pigment synthesis and growth factor secretion etc. (Strauss, 2005) (Fig. 5C). Nevertheless, it should be noted that some of the mature RPE markers such as *Rpe65* and *Tyr* still remained at low expression levels, suggesting the functionally incomplete maturation of these Best1::GFP$^+$ cells at this stage (Fig. 5C). Taken together, these results indicate that Best1::GFP$^+$ colonies generated from human fibroblasts recapitulated the morphological and molecular features of RPE cells, which, upon further maturation, may represent an invaluable RPE replacement for the field of regenerative medicine.

DISCUSSION

Encouraged by the first report of the positive implication from prospective clinical trials transplanting hESCs derived RPE cells into two patients (Schwartz et al., 2012), RPE cells generated from pluripotent stem cells have attracted increasing attention for the promise of regenerative medicine (Zhang et al., 2013). Current methods differentiating RPE cells from human pluripotent cells include spontaneous differentiation, monolayer differentiation or 3D culture using mouse and human ESCs (Meyer et al., 2009; Osakada et al., 2009; Nakano et al., 2012). Although several of these recent methods have significantly advanced the yield and accelerated differentiation, all methods to date result in a mixture of RPE cells and neural retina cells, and thus require stringent selection before any therapeutic application. However, the only method of RPE selection described so far is the manual picking and expansion of pigmented cells, a time and labor consuming process. In this study, we generated a RPE specific reporter, which faithfully represents RPE cell differentiation and allows for easy cell sorting.

Lineage conversion of one somatic cell type to another is an alternative approach for generating specific cell types bypassing a pluripotent state. This approach has been successful in various cases, from neural lineages to cardiomyocytes and hepatocytes (Sancho-Martinez et al., 2012; Yi et al., 2012). Here we show that a small set of transcription factors can convert adult human fibroblasts into Best1::GFP positive cells, which bear many morphological and molecular features of RPE such as the expression of *Mitf*, *Best1*, *Cralbp*, and *Tyrp2*. Moreover, these cells are able to obtain pigmentation, a typical characteristic of functional RPE cells.

Although we cannot rule out the possibility that other RPE-inducing factors have been overlooked, our study indeed established a platform to study the transcriptional network that regulates the conversion from human fibroblasts to RPE cells. Consistent with the previous reports that Mitf and Otx2 are crucial for RPE development and function (Martinez-Morales et al., 2004; Bharti et al., 2006), our data showed that Mitf and Otx2 are necessary for Best1::GFP$^+$ colony formation. Notably, mature RPE cells are polarized, highly pigmented cells which have very limited ability for proliferation, a feature which is not suitable to be used for cell

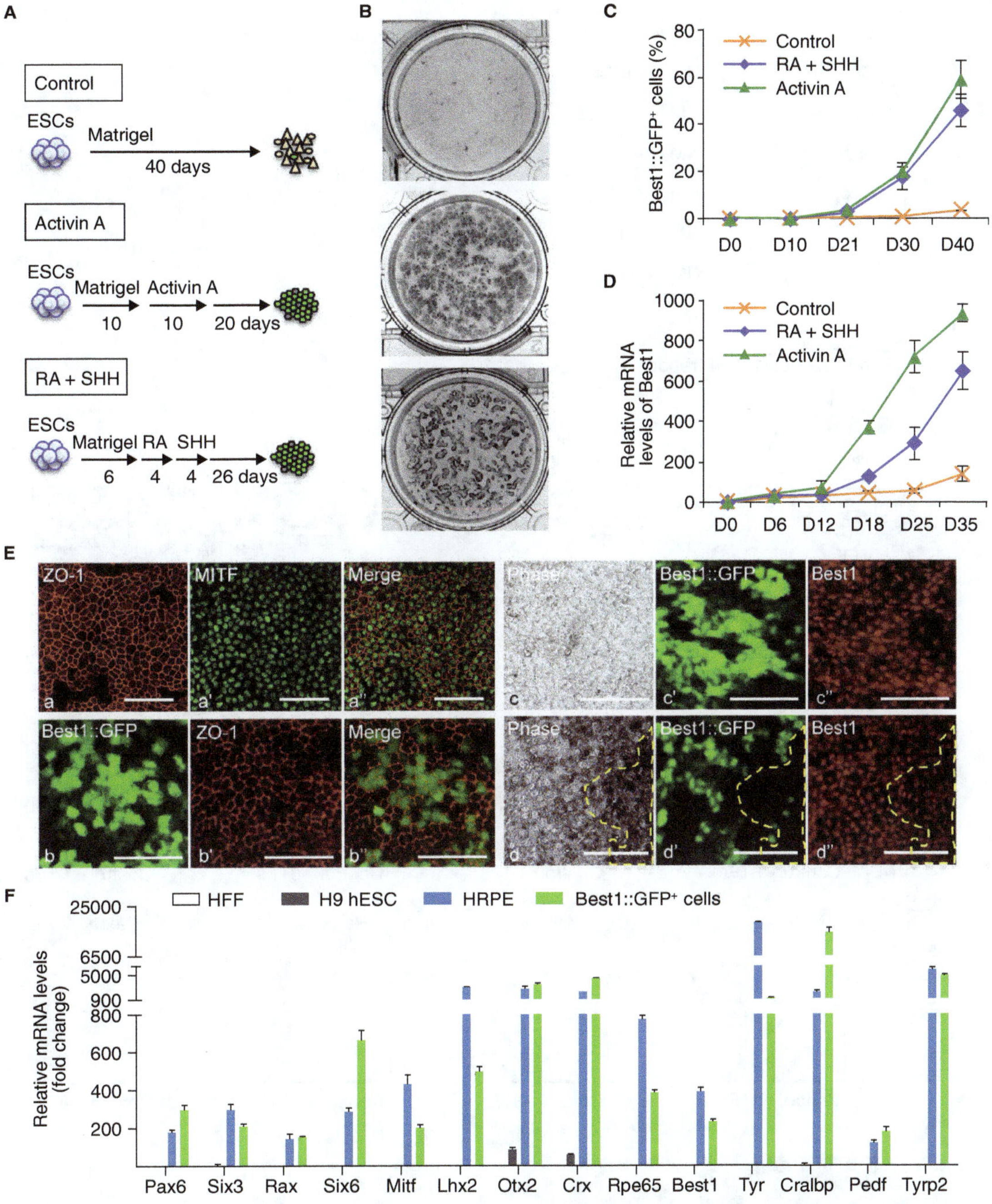

Figure 2. Characterization of Best1::GFP positive cells during hESC differentiation towards RPE. (A) Schematic illustration outlining differentiation of H9 hESCs with Best1::GFP reporter into RPE under different conditions. (B) Pigmentation of RPE cells differentiated from H9 hESCs using three conditions at day 40. (C) Percentages of Best1::GFP$^+$ cells calculated by FACS analysis in indicated days of H9 hESC RPE differentiation, with different treatments as indicated in Fig. 2A. Data are presented as mean ± s.d. of three biological replicates. (D) Q-PCR analysis of *Best1* mRNA expression levels in indicated days of RPE differentiation. Expression was normalized to levels in undifferentiated H9 hESC. (E) Immunofluorescence staining and phase contrast microscopy of RPE differentiation from H9 hESCs. a–a'', Immunofluorescence analysis of ZO-1 and MITF expression at day 45. b–b'', Best1::GFP and ZO-1 expression at day 45. c–c'', Best1::GFP expression and immunostaining of Best1 protein. Best1::GFP is highly expressed in non-pigmented or slightly pigmented cells at day 30. d–d'', Best1::GFP expression and immunostaining of Best1 protein at day 45. Yellow dash line highlighted area indicates diminished Best1::GFP expression in heavily pigmented cells. All scale bars: 100 μm. (F) Relative fold changes of RPE specific genes in HFF, H9 hESCs, HRPE cells and Best1::GFP$^+$ cells which were sorted by FACS at day 40 of hESC RPE differentiation. Relative expressions were normalized to levels in HFF.

A

Gene	Accession Number	Length (bp)
Crx	NM_000554.2	900
Pax6	NM_000280.2	1269
Mitf-A	NM_198159.2	1563
Otx2	NM_021728.2	891
Nrl	NM_006177.2	714
Rax	NM_013435.2	1041

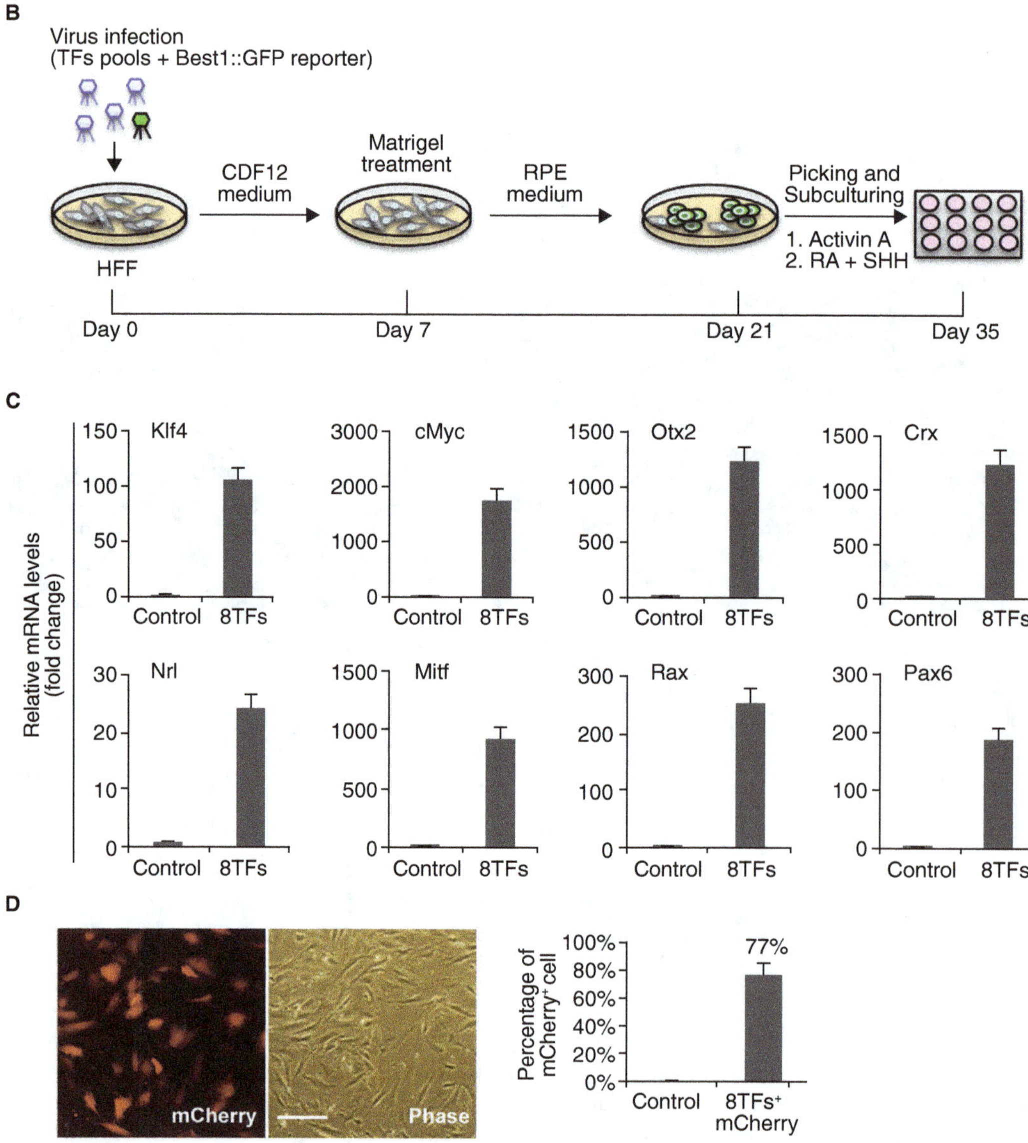

Figure 3. Strategy for transcription factors-mediated human fibroblast to RPE transdifferentiation. (A) Table listing the selected 6 transcription factors (6TFs). (B) Experimental design to induce the conversion of HFF into RPE. (C) qPCR analysis of individual exogenous TF expression 3 days after HFFs were transduced with 8TFs. HFF cells transduced with mCherry were included as the control. Data are presented as mean ± s.d. of three replicates. (D) mCherry as indicator for the transduction efficiency of HFF cells 3 days after retroviral 8TFs plus mCherry were infected. The percentage of mCherry positive cells was indicated. Scale bars: 100 μm.

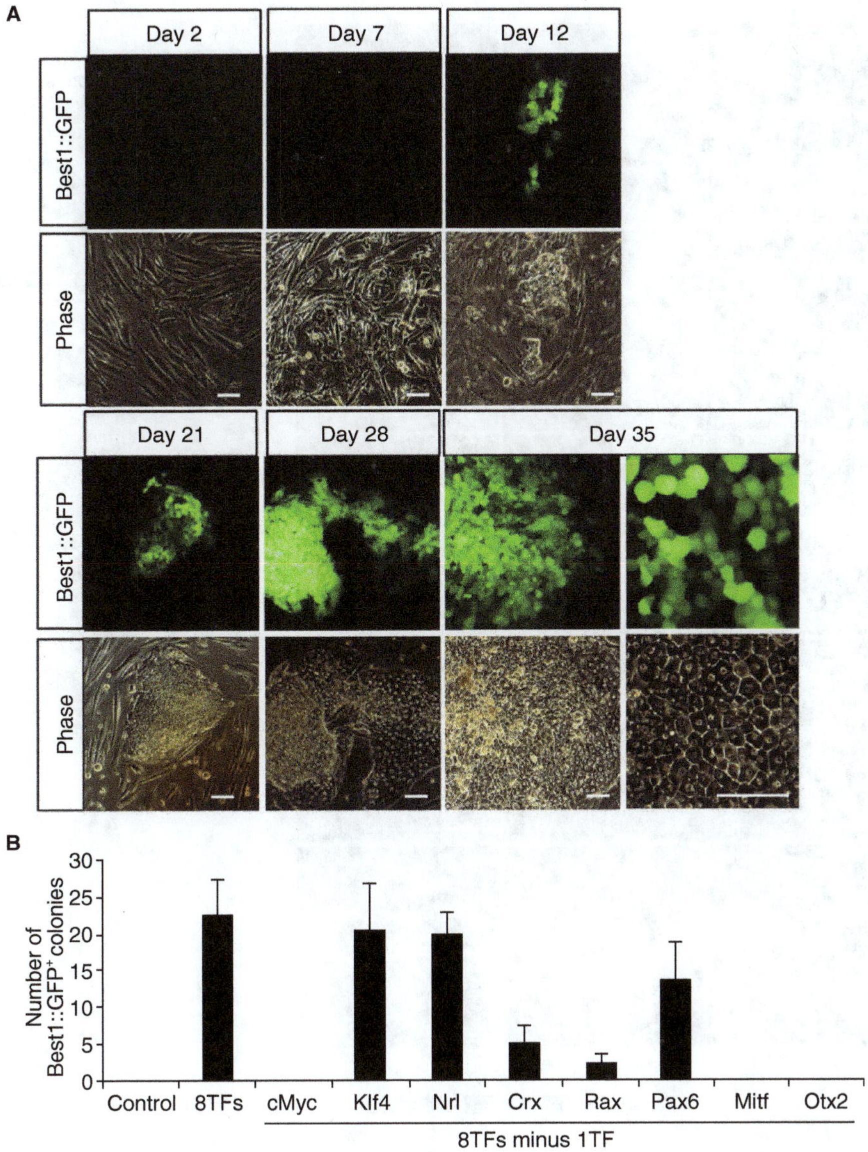

Figure 4. Eight transcription factors induced Best1::GFP$^+$ colonies from human fibroblasts. (A) Morphological changes of HFF during the process of RPE lineage conversion. Best1::GFP$^+$ cells exhibited cobblestone-like morphology at day 35. All scale bars: 100 μm. (B) Effects of individual factor withdrawal from 8 transcription factors (8TFs) on Best1::GFP$^+$ colony formation. HFF cells transduced with mCherry were served as the control. Data are presented as mean ± s.d. of three replicates.

transplantation (Salero et al., 2012). On the other hand, Best1::GFP$^+$ colonies we converted from adult fibroblasts showed a progenitor-like status which is expandable. Developing an inducible expression system capable of shutting down some progenitor genes' expression at certain time points would be crucial for controlling the switch between the proliferation and maturation status of these Best1::GFP$^+$ cells. In addition to transcription factors, culture condition has also been highlighted as a critical factor determining the lineage conversion process (Efe et al., 2011; Kurian et al., 2013). Here in this report, we provided a matrigel based culture condition combined with RA and SHH treatment that supports the generation of pigmented cells from the Best1::GFP$^+$ cells. Although we have not tested whether the RPE cells we obtained are fully functional, e.g. by cell transplantation in animal models, with further methodological optimization, this approach would facilitate the functional RPE cells' generation from human fibroblasts. In summary, our findings provide a powerful system not only for studying the molecular nature of cell identity and plasticity, but also for developing therapeutic strategies for retinal degenerative diseases.

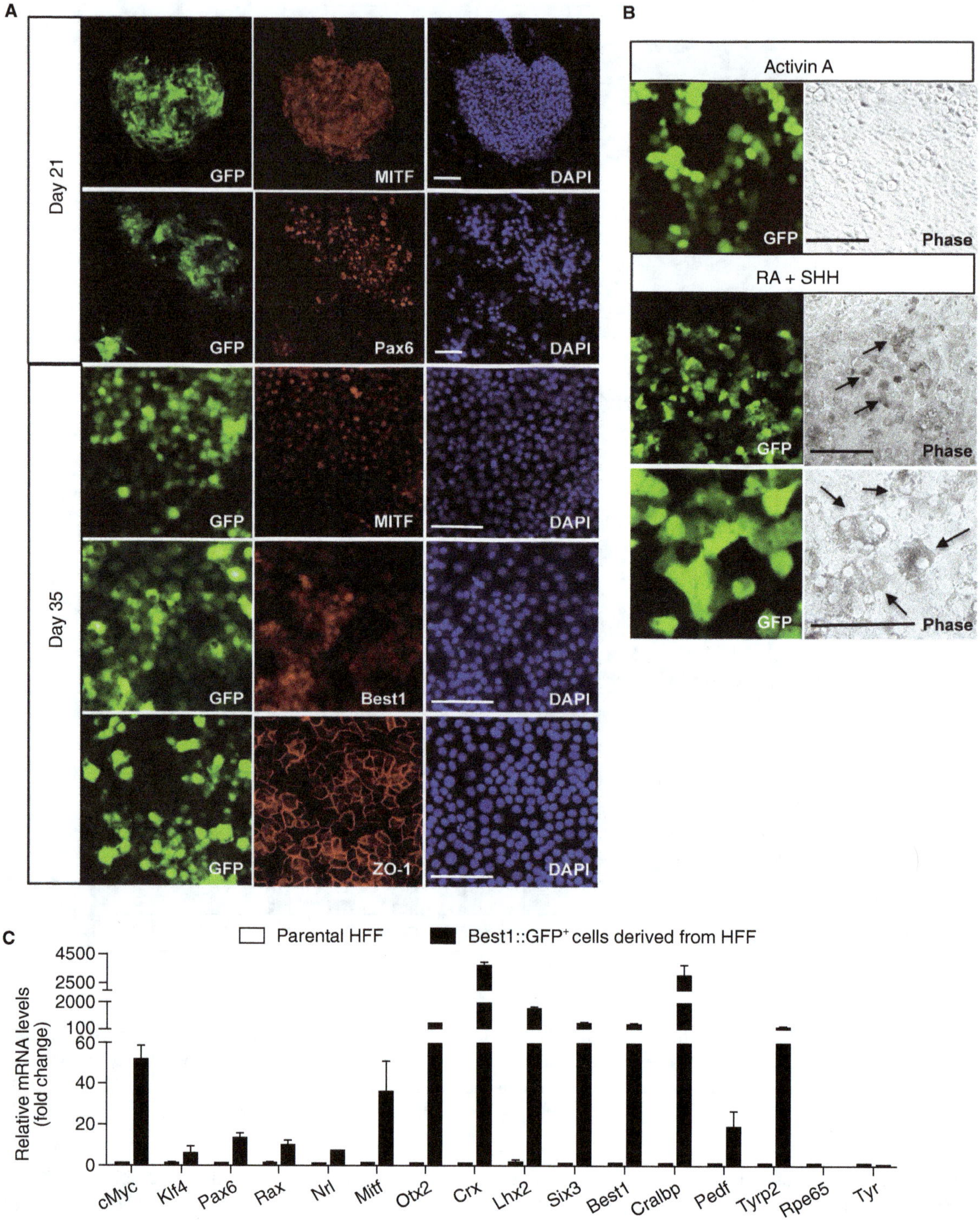

Figure 5. Characterization of Best1::GFP⁺ cells converted from human fibroblasts. (A) Immunostaining of MITF, Pax6, Best1 and ZO-1 in Best1::GFP⁺ colonies. DAPI was used to counterstain the nuclei (blue). All scale bars: 100 μm. (B) Pigmented cells indicated by arrows were observed at day 35 post-transduction in the Best1::GFP⁺ colonies generated by treatment with RA plus SHH, but not by Activin A treatment. All scale bars: 100 μm. (C) qPCR analysis of RPE-specific gene expression. 8TF-induced Best1::GFP⁺ colonies from HFF were picked and subcultured in RA plus SHH condition. Best1::GFP⁺ cells were sorted for qPCR analysis at day 35. Relative expressions were normalized to levels in parental HFF. Data are presented as mean ± s.d. of three replicates.

MATERIAL AND METHODS

Cell culture

H9 and H1 hESCs (WiCell Research) were maintained on a layer of mitotically inactivated mouse embryonic fibroblasts (MEFs) in hESC medium (CDF12 medium): DMEM/F12 (Invitrogen) supplemented with 0.1 mmol/L non-essential amino acids (Invitrogen), 1 mmol/L GlutaMAX (Invitrogen), 20% Knockout Serum Replacement (Invitrogen), 55 μmol/L β-mercaptoethanol (Invitrogen) and 10 ng/mL bFGF (Joint Protein Central). hESCs were also cultured in mTeSR medium (StemCell Technologies). Plates were coated with Matrigel (BD Biosciences), and medium was changed daily.

Human primary RPE cells (HRPE) (Lonza) used as a positive control were cultured in basic RPE medium consisting of DMEM/F-12 supplemented with 0.1 mmol/L non-essential amino acids, 1 mmol/L GlutaMAX, 1% N2 supplement (Invitrogen) and 5% Knockout Serum Replacement (Invitrogen).

RPE differentiation from hESCs

hESCs were dissociated as cell clumps and plated on 1% ESC qualified-Matrigel for 1 h. Attached aggregates of cells were covered with 2% ESC qualified-Matrigel diluted in N2B27 medium consisting of DMEM/F-12 supplemented with 0.1 mmol/L non-essential amino acids, 1 mmol/L GlutaMAX, 1% N2 supplement (Invitrogen) and 2% B27 (Invitrogen). After overnight incubation, fresh medium without Matrigel was added and changed every other day.

For RA plus SHH treatment, Retinoic acid (RA) (500 nmol/L, Sigma) was supplemented into N2B27 medium from day 7, then gradually reduced to 200 nmol/L by day 10. From day 11 to day 14, medium was changed into N2B27 medium supplemented with Sonic Hedgehog (SHH) (25 nmol/L, Prospec). Later, medium was changed into basic RPE medium consisting of DMEM/F-12 supplemented with 0.1 mmol/L non-essential amino acids, 1 mmol/L GlutaMAX, 1% N2 supplement and 5% Knockout Serum Replacement.

For Activin A treatment, the medium was changed to DMEM supplemented with 0.1 mmol/L non-essential amino acids, 1 mmol/L GlutaMAX, 20% Knockout Serum Replacement (Invitrogen), and Activin A (100 nmol/L, Prospec) from day 10 to day 20. Later medium was changed into the basic RPE medium as described above.

Production of lentivirus and retrovirus

For generating lentiviral reporters, DNA fragments as described in Fig. 1A were obtained by PCR amplification from H1 hESC-derived genomic DNA template and cloned into pGreenZeo lentiviral vector (System bioscience). Corresponding packaging plasmids are pMDL, pCMV_VSVG and pRSV_REV. For transcription factors mediated lineage conversion, the human cDNAs listed in Fig. 3A were cloned into the pMX-gateway vector (Clontech). Coding region for mCherry was cloned as control. Corresponding packaging plasmids are pCMV-GAG-Pol and pCMV-VSV-G. HEK293T cells were seeded at a density between 6.0–8.5 × 10^4 cells/cm^2 and transfected by Lipofectamine 2000 (Invitrogen) 16 h later. Individual supernatants containing virus were harvested at 48 h and 72 h post-transfection and filtered with 0.45 μm PVDF membrane (Millipore).

Conversion of human fibroblasts into Best1::GFP$^+$ cells

Human foreskin fibroblasts (HFF-1, HFF-693) were plated on Matrigel-coated six-well plates at 75,000 cells per well. The next day, cells were infected with an equal ratio of a combination of eight retroviruses encoding PAX6, RAX, CRX, MITF-A, OTX2, NRL, KLF4 and c-MYC (8F) as well as pGZ-BEST1-GFP lentivirus. The plates were infected by spinfection of the cells at 1850 r/min for 1 h at room temperature in the presence of polybrene (4 μg/mL) and put back in the incubator without medium change. 24 h later, the medium was switched to CDF12 medium with medium changes every day. Cells were covered with 2% ESC qualified-Matrigel diluted in N2B27 medium at day 7. After overnight incubation, fresh RPE medium without Matrigel was added and changed every other day. At day 21, Best1::GFP$^+$ colonies were picked and cultured in Matrigel coated 12 well plates, followed by 10 days treatment with 100 ng/mL Activin A or 500 nmol/L RA plus 25 ng/mL SHH in base RPE medium.

Immunofluorescence staining

Cells were fixed using 4% paraformaldehyde in PBS at 4°C for 1 h. After fixation, cells were exposed to 0.3% Triton X-100 in PBS for 5 min at RT. Cells were blocked with 5% BSA in PBS for 30 min and incubated with primary antibody for 1 h at RT or overnight at 4°C. Washing was conducted with PBS followed by incubation with a corresponding secondary antibody for 1 h at RT. DAPI, was used to stain nuclei. Primary antibodies were obtained from the following sources: Mouse anti-Pax6 (1:1000, DSHB); Mouse anti-MITF (1:50, Millipore); Rabbit anti-ZO-1 (1:200, Sigma); Mouse anti-Best1 (1:200, Millipore).

Quantitative PCR (qPCR)

Total cellular RNA was isolated using Trizol reagent (Invitrogen), or RNeasy Micro Kit (Qiagen) for a small amount of cell samples according to the manufacturer's recommendations. Total RNA was treated with 2 μg of DNase 1 (Invitrogen) and used for cDNA synthesis using iScript reverse transcription supermix (Bio-Rad). Real-time PCR was performed using the SYBR Green PCR Master Mix (Applied Biosystems). The expression levels of respective genes were normalized to corresponding *GAPDH* values and are shown as fold change relative to the value of the control sample. All the assays were performed in triplicate. Primer sequences are listed in Table S1.

Flow cytometry analysis and cell sorting

H9 hESCs were infected with lentiviral Best1::GFP reporter and then underwent RPE differentiation. Cells were harvested using TrypLE (Invitrogen), washed once with PBS and resuspended with 1× PBS/10% FCS medium. A minimum of 10,000 cells in the living population were analyzed by using a BD LSRII flow cytometry machine equipped with five different lasers and the BD FACSDiva software. Percentages of Best1::GFP$^+$ cells are presented after the subtraction of isotype background and refer to the total living population analyzed. Results are representative of at least three independent experiments with a minimum of two technical replicates per experiment. For cell sorting, cells were treated as described above and sorted with a BDAria II FACS sorter (BD Biosystems).

ACKNOWLEDGEMENTS

G.H.L. is supported by the Strategic Priority Research Program of the Chinese Academy of Sciences (XDA01020312), National Natural Science Foundation of China (Grant Nos. 81271266, 31222039, and 31201111), the Thousand Young Talents program of China, National Laboratory of Biomacromolecules (2013kf05, and 2013kf11), and State Key Laboratory of Drug Research (SIMM1302KF-17). JCIB was supported by CIBER-BBN and grants from MINECO, Fundacion Cellex, G. Harold and Leila Y. Mathers Charitable Foundation, The Leona M. and Harry B. Helmsley Charitable Trust and The Ellison Medical Foundation.

ABBREVIATIONS

AMD, age-related macular degeneration; ESC, embryonic stem cell; FACS, fluorescence activated cell sorting; iPSC, induced pluripotent stem cell; MEF, mouse embryonic fibroblast; RA, retinoic acid; RPE, retinal pigment epithelium; SHH, Sonic Hedgehog; TF, transcription factor.

COMPLIANCE WITH ETHICS GUIDELINES

All procedures followed were in accordance with the ethical standards of the responsible committee on human experimentation (Center of Regenerative Medicine in Barcelona, Spain) and with the Helsinki Declaration of 1975, as revised in 2000 (5). Informed consent was obtained from all patients for being included in the study.

Kejing Zhang, Guang-Hui Liu, Fei Yi, Nuria Montserrat, Tomoaki Hishida, Concepcion Rodriguez Esteban, Juan Carlos Izpisua Belmonte declare that they have no conflict of interest.

REFERENCES

Acland GM, Aguirre GD, Bennett J, Aleman TS, Cideciyan AV, Bennicelli J, Dejneka NS, Pearce-Kelling SE, Maguire AM, Palczewski K et al (2005) Long-term restoration of rod and cone vision by single dose rAAV-mediated gene transfer to the retina in a canine model of childhood blindness. Mol Ther 12:1072–1082

Ben-David U, Benvenisty N (2011) The tumorigenicity of human embryonic and induced pluripotent stem cells. Nat Rev Cancer 11:268–277

Bharti K, Nguyen MT, Skuntz S, Bertuzzi S, Arnheiter H (2006) The other pigment cell: specification and development of the pigmented epithelium of the vertebrate eye. Pigment Cell Res 19:380–394

Boucherie C, Mukherjee S, Henckaerts E, Thrasher AJ, Sowden JC, Ali RR (2013) Brief report: self-organizing neuroepithelium from human pluripotent stem cells facilitates derivation of photoreceptors. Stem Cells 31:408–414

Boulanger A, Liu S, Henningsgaard AA, Yu S, Redmond TM (2000) The upstream region of the Rpe65 gene confers retinal pigment epithelium-specific expression in vivo and in vitro and contains critical octamer and E-box binding sites. J Biol Chem 275:31274–31282

Buchholz DE, Hikita ST, Rowland TJ, Friedrich AM, Hinman CR, Johnson LV, Clegg DO (2009) Derivation of functional retinal pigmented epithelium from induced pluripotent stem cells. Stem Cells 27:2427–2434

Carr AJ, Vugler AA, Hikita ST, Lawrence JM, Gias C, Chen LL, Buchholz DE, Ahmado A, Semo M, Smart MJ et al (2009) Protective effects of human iPS-derived retinal pigment epithelium cell transplantation in the retinal dystrophic rat. PLoS ONE 4: e8152

Efe JA, Hilcove S, Kim J, Zhou H, Ouyang K, Wang G, Chen J, Ding S (2011) Conversion of mouse fibroblasts into cardiomyocytes using a direct reprogramming strategy. Nat Cell Biol 13:215–222

Esumi N, Oshima Y, Li Y, Campochiaro PA, Zack DJ (2004) Analysis of the VMD2 promoter and implication of E-box binding factors in its regulation. J Biol Chem 279:19064–19073

Esumi N, Kachi S, Hackler L Jr, Masuda T, Yang Z, Campochiaro PA, Zack DJ (2009) BEST1 expression in the retinal pigment epithelium is modulated by OTX family members. Hum Mol Genet 18:128–141

Giorgetti A, Marchetto MC, Li M, Yu D, Fazzina R, Mu Y, Adamo A, Paramonov I, Cardoso JC, Monasterio MB et al (2012) Cord blood-derived neuronal cells by ectopic expression of Sox2 and c-Myc. Proc Natl Acad Sci USA 109:12556–12561

Huang P, He Z, Ji S, Sun H, Xiang D, Liu C, Hu Y, Wang X, Hui L (2011) Induction of functional hepatocyte-like cells from mouse fibroblasts by defined factors. Nature 475:386–389

Khandhadia S, Cherry J, Lotery AJ (2012) Age-related macular degeneration. Adv Exp Med Biol 724:15–36

Kim J, Efe JA, Zhu S, Talantova M, Yuan X, Wang S, Lipton SA, Zhang K, Ding S (2011) Direct reprogramming of mouse fibroblasts to neural progenitors. Proc Natl Acad Sci USA 108:7838–7843

Kurian L, Sancho-Martinez I, Nivet E, Aguirre A, Moon K, Pendaries C, Volle-Challier C, Bono F, Herbert JM, Pulecio J et al (2013) Conversion of human fibroblasts to angioblast-like progenitor cells. Nat Methods 10:77–83

Liu GH, Yi F, Suzuki K, Qu J, Izpisua Belmonte JC (2012) Induced neural stem cells: a new tool for studying neural development and neurological disorders. Cell Res 22:1087–1091

Lu B, Malcuit C, Wang S, Girman S, Francis P, Lemieux L, Lanza R, Lund R (2009) Long-term safety and function of RPE from human embryonic stem cells in preclinical models of macular degeneration. Stem Cells 27:2126–2135

Martinez-Morales JR, Rodrigo I, Bovolenta P (2004) Eye development: a view from the retina pigmented epithelium. BioEssays 26:766–777

Meyer JS, Shearer RL, Capowski EE, Wright LS, Wallace KA, McMillan EL, Zhang SC, Gamm DM (2009) Modeling early retinal development with human embryonic and induced pluripotent stem cells. Proc Natl Acad Sci USA 106:16698–16703

Nakano T, Ando S, Takata N, Kawada M, Muguruma K, Sekiguchi K, Saito K, Yonemura S, Eiraku M, Sasai Y (2012) Self-formation of optic cups and storable stratified neural retina from human ESCs. Cell Stem Cell 10:771–785

Okada Y, Shimazaki T, Sobue G, Okano H (2004) Retinoic-acid-concentration-dependent acquisition of neural cell identity during in vitro differentiation of mouse embryonic stem cells. Dev Biol 275:124–142

Osakada F, Ikeda H, Sasai Y, Takahashi M (2009) Stepwise differentiation of pluripotent stem cells into retinal cells. Nat Protoc 4:811–824

Pang ZP, Yang N, Vierbuchen T, Ostermeier A, Fuentes DR, Yang TQ, Citri A, Sebastiano V, Marro S, Sudhof TC et al (2011) Induction of human neuronal cells by defined transcription factors. Nature 476:220–223

Panopoulos AD, Ruiz S, Izpisua Belmonte JC (2011) iPSCs: induced back to controversy. Cell Stem Cell 8:347–348

Salero E, Blenkinsop TA, Corneo B, Harris A, Rabin D, Stern JH, Temple S (2012) Adult human RPE can be activated into a multipotent stem cell that produces mesenchymal derivatives. Cell Stem Cell 10:88–95

Sancho-Martinez I, Baek SH, Izpisua Belmonte JC (2012) Lineage conversion methodologies meet the reprogramming toolbox. Nat Cell Biol 14:892–899

Schwartz SD, Hubschman JP, Heilwell G, Franco-Cardenas V, Pan CK, Ostrick RM, Mickunas E, Gay R, Klimanskaya I, Lanza R (2012) Embryonic stem cell trials for macular degeneration: a preliminary report. Lancet 379:713–720

Sekiya S, Suzuki A (2011) Direct conversion of mouse fibroblasts to hepatocyte-like cells by defined factors. Nature 475:390–393

Steingrimsson E, Copeland NG, Jenkins NA (2004) Melanocytes and the microphthalmia transcription factor network. Annu Rev Genet 38:365–411

Strauss O (2005) The retinal pigment epithelium in visual function. Physiol Rev 85:845–881

Takahashi K, Yamanaka S (2006) Induction of pluripotent stem cells from mouse embryonic and adult fibroblast cultures by defined factors. Cell 126:663–676

Takahashi K, Tanabe K, Ohnuki M, Narita M, Ichisaka T, Tomoda K, Yamanaka S (2007) Induction of pluripotent stem cells from adult human fibroblasts by defined factors. Cell 131:861–872

Vierbuchen T, Ostermeier A, Pang ZP, Kokubu Y, Sudhof TC, Wernig M (2010) Direct conversion of fibroblasts to functional neurons by defined factors. Nature 463:1035–1041

Yi F, Liu GH, Izpisua Belmonte JC (2012) Rejuvenating liver and pancreas through cell transdifferentiation. Cell Res 22:616–619

Zahabi A, Shahbazi E, Ahmadieh H, Hassani SN, Totonchi M, Taei A, Masoudi N, Ebrahimi M, Aghdami N, Seifinejad A et al (2012) A new efficient protocol for directed differentiation of retinal pigmented epithelial cells from normal and retinal disease induced pluripotent stem cells. Stem Cells Dev 21:2262–2272

Zhang W, Duan S, Li Y, Xu X, Qu J, Zhang W, Liu GH (2012) Converted neural cells: induced to a cure? Protein Cell 3:91–97

Zhang K, Yi F, Liu GH, Izpisua Belmonte JC (2013) New march towards the regeneration of sensation and cognition: hear more, see more and learn more. J Mol Cell Biol 5:151–153

Zhu Y, Carido M, Meinhardt A, Kurth T, Karl MO, Ader M, Tanaka EM (2013) Three-dimensional neuroepithelial culture from human embryonic stem cells and its use for quantitative conversion to retinal pigment epithelium. PLoS ONE 8:e54552

Zuber ME, Gestri G, Viczian AS, Barsacchi G, Harris WA (2003) Specification of the vertebrate eye by a network of eye field transcription factors. Development 130:5155–5167

16

Homocysteine activates T cells by enhancing endoplasmic reticulum-mitochondria coupling and increasing mitochondrial respiration

Juan Feng, Silin Lü, Yanhong Ding, Ming Zheng✉, Xian Wang✉

Department of Physiology and Pathophysiology, School of Basic Medical Science, Peking University, Key Laboratory of Molecular Cardiovascular Science, Ministry of Education, Beijing 100191, China

✉ Correspondence: zhengm@bjmu.edu.cn (M. Zheng), xwang@bjmu.edu.cn (X. Wang)

ABSTRACT

Hyperhomocysteinemia (HHcy) accelerates atherosclerosis by increasing proliferation and stimulating cytokine secretion in T cells. However, whether homocysteine (Hcy)-mediated T cell activation is associated with metabolic reprogramming is unclear. Here, our *in vivo* and *in vitro* studies showed that Hcy-stimulated splenic T-cell activation in mice was accompanied by increased levels of mitochondrial reactive oxygen species (ROS) and calcium, mitochondrial mass and respiration. Inhibiting mitochondrial ROS production and calcium signals or blocking mitochondrial respiration largely blunted Hcy-induced T-cell interferon γ (IFN-γ) secretion and proliferation. Hcy also enhanced endoplasmic reticulum (ER) stress in T cells, and inhibition of ER stress with 4-phenylbutyric acid blocked Hcy-induced T-cell activation. Mechanistically, Hcy increased ER-mitochondria coupling, and uncoupling ER-mitochondria by the microtubule inhibitor nocodazole attenuated Hcy-stimulated mitochondrial reprogramming, IFN-γ secretion and proliferation in T cells, suggesting that juxtaposition of ER and mitochondria is required for Hcy-promoted mitochondrial function and T-cell activation. In conclusion, Hcy promotes T-cell activation by increasing ER-mitochondria coupling and regulating metabolic reprogramming.

KEYWORDS homocysteine, T cell, mitochondria, endoplasmic reticulum stress

INTRODUCTION

Homocysteine (Hcy) is a sulfur-containing non-constitutive amino acid derived from the essential amino acid methionine. Numerous clinical studies have established hyperhomocysteinemia (HHcy) as an independent risk factor for cardiovascular diseases in humans. We and others have demonstrated that HHcy accelerates atherosclerosis by affecting the immuno-inflammatory response, stimulating chemokine/cytokine secretion in monocytes and T cells, and repressing regulatory T-cell functions (Feng et al., 2009; Ma et al., 2013; Zeng et al., 2003; Zhang et al., 2002). Although reactive oxygen species (ROS) were identified as mediators in the process, the exact mechanism of Hcy-mediated lymphocyte activation is not well understood. Recent studies have revealed the intercrossing of pathways in the regulation of immune and metabolic systems (MacIver et al., 2013). Alterations in metabolism at both cellular and tissue levels affect specific T-cell functions (MacIver et al., 2013). However, whether the cellular metabolism is changed during Hcy-induced T-cell activation is unclear, and if changed, the underlying mechanism is unknown.

The endoplasmic reticulum (ER) and mitochondria are important cellular organelles in maintaining cell homeostasis, and dysfunction of each part is associated with multiple disorders, especially metabolic diseases. Emerging evidence showed that in addition to their distinct functions in regulating cellular processes, the ER and mitochondria interact with each other physically and functionally. The ER-mitochondria interaction enables the exchange of signals, lipids, and proteins (Szabadkai and Duchen, 2008). Enhancing ER-mitochondria coupling can increase mitochondrial calcium flux from the ER, thereby promoting mitochondrial respiration and adenosine triphosphate (ATP) production (Balaban, 2009; Brown, 1992). ER stress induced in HeLa cells increased ER-mitochondria coupling, thus elevating mitochondrial respiration and bioenergetics (Bravo et al., 2011). More recently, it was reported that chronic enrichment of hepatic ER-mitochondria coupling led to mitochondrial dysfunction in obesity (Arruda et al., 2014). It has been reported that ER and mitochondrial homeostasis play important and emerging functional roles respectively in T-cell development, innate immune responses, dendritic cell and macrophage functions (Bettigole and Glimcher, 2015; Cubillos-Ruiz et al., 2015; Martinon et al., 2010; Martinon and Glimcher, 2011; Staton et al., 2011). However, whether the coupling between the ER and mitochondria is involved in the regulation of Hcy-mediated T cell activation has not yet been studied.

In the present study, we explored the possible role of ER-mitochondria coupling and mitochondrial metabolism in Hcy-activated T cells. Our data showed that Hcy activates T cells through inducing ER stress. This subsequently enhances ER-mitochondria coupling and mitochondrial metabolic reprogramming which includes increased mitochondrial ROS production, calcium signal, membrane potential, and mass.

Figure 1. Reprogramming of mitochondrial metabolism in T cells from HHcy mice. Flow cytometry of splenic T cells from mice fed with or without Hcy and stained with MitoSOX Red (A) or Rhod-2 (B). Traces of OCR of splenic T cells from control or HHcy mice (C) or stimulated with anti-CD3 antibody for additional 24 h (D) as measured by the XF24 metabolic analyzer, with additions of mitochondrial effectors at time points indicated (left), and quantification of the basal OCR, max capacity OCR, and ATP-linked OCR (right). ECAR of splenic T cells from control or HHcy mice (E) or stimulated with anti-CD3 antibody for additional 24 h (F) as measured by the XF24 metabolic analyzer, and quantification of the basal and maximal ECAR. Flow cytometry of T cells stained with rhodamine 123 (G). (H) Confocol images of T cells loaded with MitoTracker Green. Results are mean ± SEM of six mice per group. *, $P < 0.05$ vs. control.

RESULTS

Reprogramming of mitochondrial metabolism in T cells from HHcy mice

Hcy treated mice were established to investigate the *in vivo* role of Hcy in regulating T cell mitochondrial metabolism. Our previous study found that ROS serves as a mediator in concanavalin A-activated T-cell proliferation potentiated by Hcy (Zhang et al., 2002). Here we further determined whether mitochondrial ROS was changed in response to Hcy stimulation. In consistent with our previous study (Zhang et al., 2002), the global ROS levels in T cells isolated from HHcy mice were increased (data not shown). HHcy also greatly increased mitochondrial ROS levels, as determined by the mitochondrial-specific superoxide anion probe MitoSOX Red, from 1.15 ± 0.13 mean fluorescence intensity [MFI] in cells from control mice to 1.39 ± 0.02 in cells from HHcy mice (Fig. 1A), indicating that increased mitochondrial ROS, the main source of ROS, may respond to HHcy-induced T-cell ROS production. This finding is in general agreement with a previous study in mice vaccinated with the lymphocytic choriomeningitis virus, in which authors found that mitochondrial ROS regulated T-cell activation and resulted in increased interleukin 2 production and antigen-specific expansion (Sena et al., 2013). We next examined the mitochondrial content of calcium, another important metabolism-associated mitochondrial signal, by loading T cells with the mitochondrial calcium probe Rhod-2. HHcy significantly increased Rhod-2 positive T cells to a 2.3-fold of control cells, from 3.67% ± 0.7% in control T cells vs. 8.46% ± 0.3% in HHcy treated cells (Fig. 1B). These results revealed that *in vivo* HHcy regulates both mitochondrial ROS and calcium signals in T cells.

Because mitochondrial ROS levels and calcium signals are closely linked with mitochondrial metabolism, we then determined mitochondrial oxidative phosphorylation in response to HHcy. T cells were sequentially treated with the ATP synthase inhibitor oligomycin, the uncoupler carbonylcyanltiep-trifluoro-methoxyphenylhydrazone (FCCP), and the combination of the electron-transport-chain inhibitor rotenone with antimycin A, and oxygen consumption rate (OCR), attributed to basal, ATP coupled, and maximal respiration, was monitored. HHcy increased both basal and maximal OCR, from 49 ± 8 pmol/min in control cells to 90 ± 12 pmol/min in HHcy treated cell for basal OCR and 70 ± 10 pmol/min to 120 ± 9 pmol/min for maximal OCR, while the ATP-coupled OCR was not altered, although with a slight trend of increase (Fig. 1C). To further confirm the effect of HHcy on T cell mitochondrial respiration, we also examined OCR in T cells isolated from HHcy mice in the presence of anti-CD3 antibody *in vitro* for additional 24 h. Likewise, the overall OCRs were increased, with basal OCR from 21 ± 11 pmol/min in control cells to 50 ± 10 pmol/min in HHcy treated cell, maximal OCR from 17 ± 12 pmol/min to 47 ± 14 pmol/min, and ATP-coupled OCR from 18 ± 13 pmol/min to 40 ± 11 pmol/min (Fig. 1D). Accompanying with the upregulated OCR, basal and maximal extracellular acidification rates (ECAR) were also increased by HHcy with (Fig. 1F) or without (Fig. 1E) anti-CD3 antibody *in vitro* for additional 24 h. Taken together, our results suggest that HHcy enhances mitochondrial respiration, probably through regulating mitochondrial ROS or calcium signals.

Increased mitochondrial respiration and metabolism are usually associated with increased mitochondrial membrane potential (Bravo et al., 2011), so we next detected the mitochondrial membrane potential in T cells from HHcy mice. Flow cytometry data showed that the mitochondrial membrane potential, as indicated by rhodamine 123 positive cells, were largely increased in T cells from HHcy mice as compared with the cells from control mice, from 3.74 ± 1.08 in control cells to 8.28 ± 0.03 in HHcy cells (Fig. 1G).

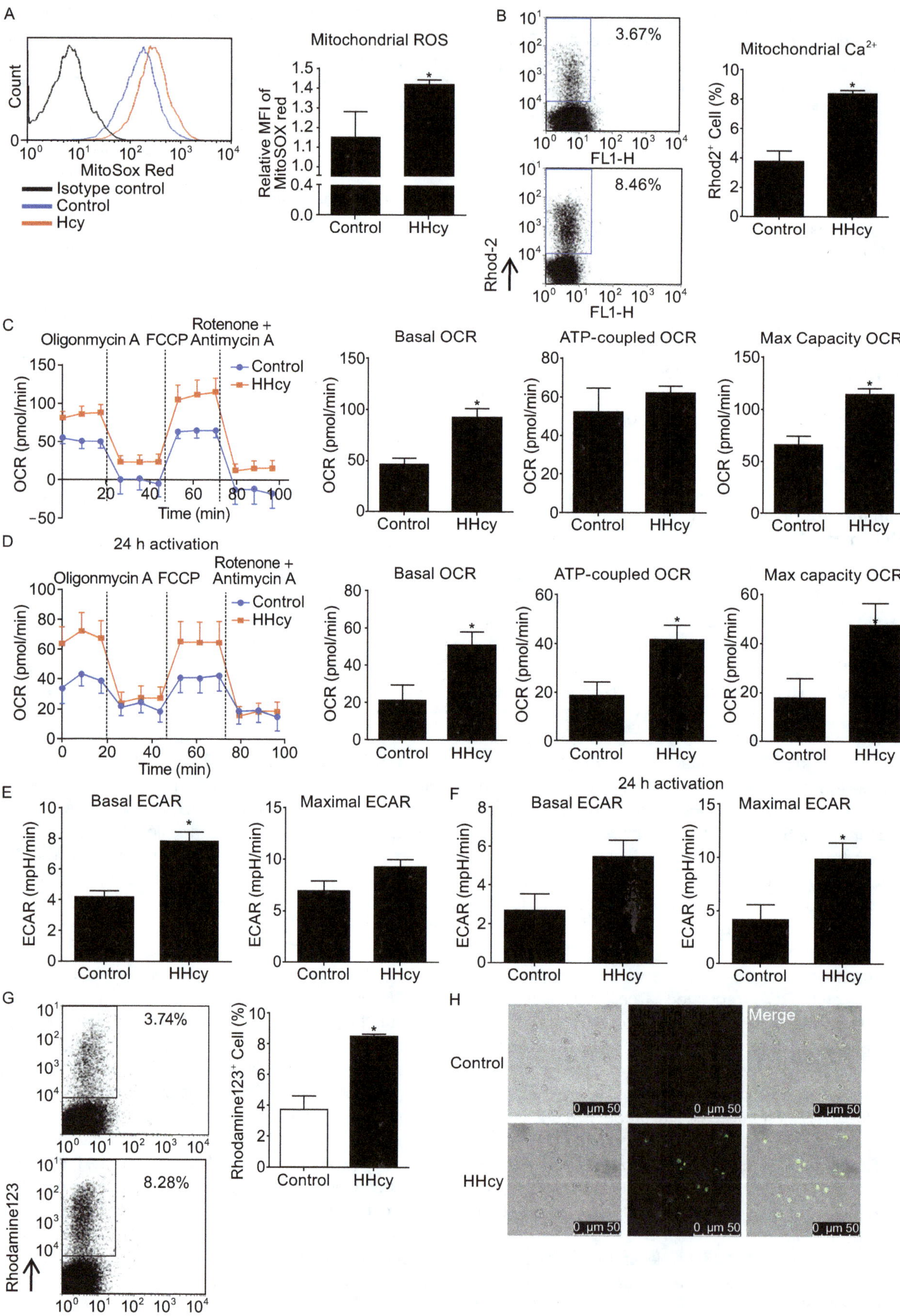
A
Count
MitoSox Red
Isotype control
Control
Hcy
Mitochondrial ROS
Relative MFI of MitoSOX red
Control
HHcy
B
3.67%
8.46%
FL1-H
Rhod-2
Mitochondrial Ca2+
Rhod2+ Cell (%)
Control
HHcy
C
Oligonmycin A
FCCP
Rotenone + Antimycin A
Control
HHcy
OCR (pmol/min)
Time (min)
Basal OCR
ATP-coupled OCR
Max Capacity OCR
D
24 h activation
Oligonmycin A
FCCP
Rotenone + Antimycin A
Basal OCR
ATP-coupled OCR
Max capacity OCR
E
Basal ECAR
Maximal ECAR
ECAR (mpH/min)
F
24 h activation
Basal ECAR
Maximal ECAR
G
3.74%
8.28%
Rhodamine123
Rhodamine123+ Cell (%)
Control
HHcy
H
Merge
Control
HHcy
0 μm 50

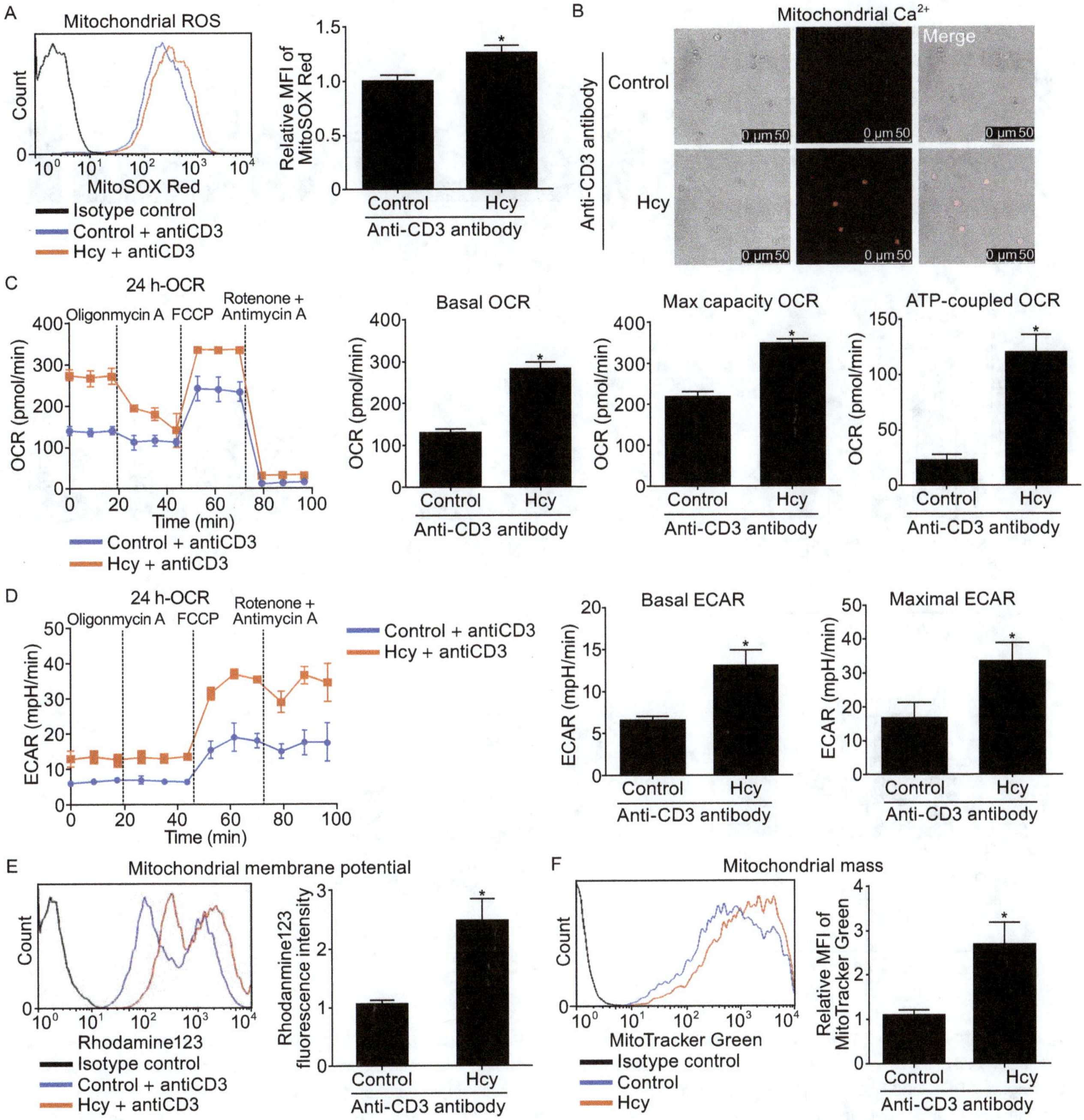

Figure 2. Altered T-cell mitochondrial metabolism in response to Hcy stimulation. T cells were incubated with or without Hcy (50 μmol/L) in the presence of anti-CD3 antibody for 24 h. (A) Flow cytometry of T cells stained with MitoSOX Red (left), and quantification (right). (B) Confocol images of T cells loaded with Rho-2. (C) Traces of OCR of T cells as measured by the XF24 metabolic analyzer, with additions of mitochondrial effectors at time points indicated (left), and quantification of the basal OCR, max capacity OCR, and ATP-linked OCR (right). (D) Traces of ECAR of T cells as measured by the XF24 metabolic analyzer, and quantification of the basal and maximal ECAR (right). Flow cytometry of T cells stained with rhodamine 123 (E) and MitoTracker Green (F). Data are mean ± SEM from 3 independent experiments. *, $P < 0.05$ vs. control.

Furthermore, staining of mitochondria probe MitoTracker Green showed that mitochondrial mass was significantly increased in HHcy-treated cells (Fig. 1H).

Collectively, our data from *in vivo* experiments showed that HHcy regulates mitochondrial metabolism through increasing mitochondrial ROS and calcium signals, and

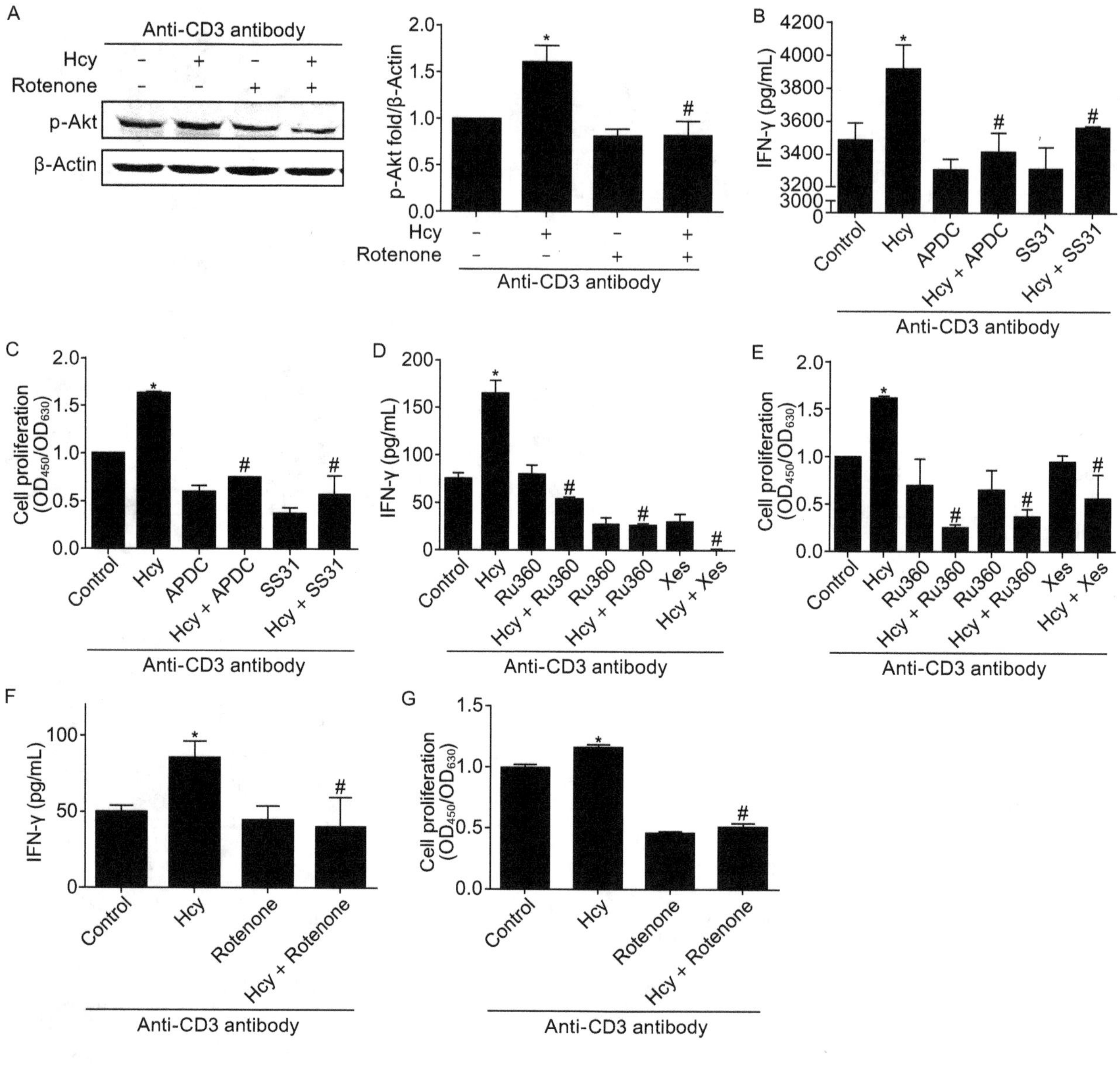

Figure 3. Mitochondrial metabolism mediates Hcy-induced T cell activation. (A) Western blot of phosphorylated Akt in T cells. (B) ELISA analysis of T-cell IFN-γ secretion in response to Hcy (50 μmol/L, 24 h) stimulation in the presence of mitochondrial ROS scavengers APDC and SS31; (C) T cell proliferation by CCK8 assay. T-cell IFN-γ secretion (D) or proliferation (E) in response to Hcy stimulation with or without the mitochondrial Ca^{2+} uniporter inhibitors Ru360 (10 μmol/L), RuRed (10 μmol/L) or IP3 receptor inhibitor Xestospongin C (Xes, 2 μmol/L). IFN-γ secretion (F) or cell proliferation (G) in response to Hcy with or without mitochondrial OXPHOS inhibitor rotenone (1 μmol/L). Data are mean ± SEM from 3 independent experiments. *, $P < 0.05$ vs. control. #, $P < 0.05$ vs. Hcy.

enhancing mitochondrial respiration, membrane potential, and mass.

Hcy regulates mitochondrial metabolism in T cells *in vitro*

We further investigated the effect of Hcy in cultured T cells. Similarly, Hcy (50 μmol/L) stimulation for 24 h increased mitochondrial ROS levels, as indicated by the MitoSOX Red MFI, from 0.99 ± 0.02 in control to 1.26 ± 0.03 in Hcy-treated cells (Fig. 2A). Mitochondrial calcium content in cultured T cells was also increased by Hcy stimulation for 24 h, as showed by the increased fluorescent signals of Rhod-2 (Fig. 2B). Moreover, in consistent with our results from HHcy mice, Hcy increased OCR in cultured T cells, with the basal OCR from 120 ± 10 pmol/min in control cells to 274 ± 22 pmol/min in Hcy treated cells, maximal OCR from 223 ± 11 pmol/min to 351 ± 9 pmol/min, and ATP-coupled OCR from

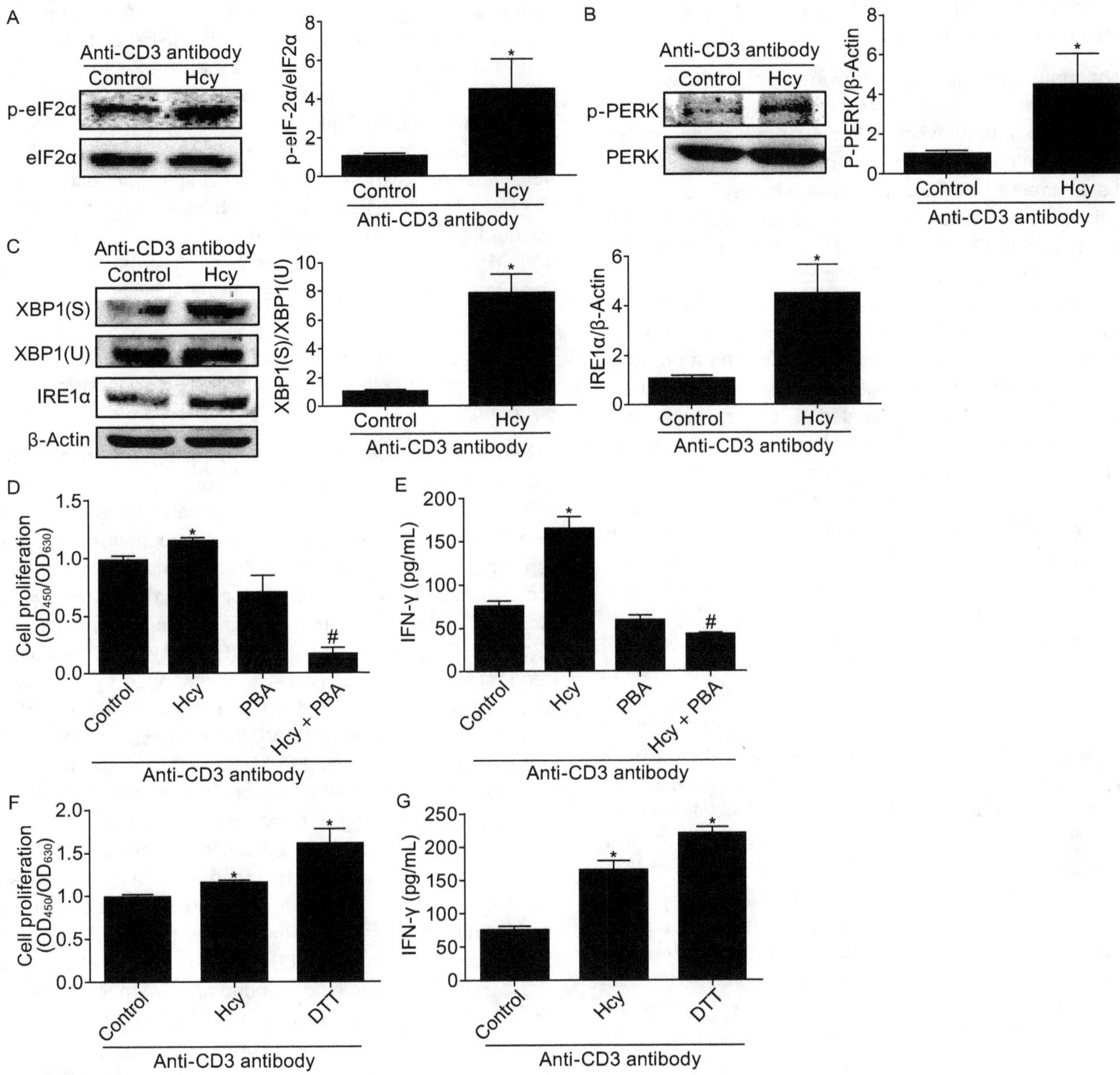

Figure 4. Hcy enhances T cell ER stress. T cells were incubated with or without Hcy (50 μmol/L) for 4 h with anti-CD3 antibody. Western blot analysis of phosphorylated eIF2α (A), phosphorylated PERK (B), and protein levels of IRE1α and spliced (S) and unspliced (U) XBP1 (C) in T cells with or without Hcy stimulation. β-Actin as a protein loading control. (D) CCK8 assay of cell proliferation with Hcy stimulation with or without ER stress inhibitor PBA or (F) ER stress stimulator DTT. ELISA of IFN-γ with Hcy stimulation with or without PBA (E) or DTT (G). Data are mean ± SEM from 3 independent experiments. *: $P < 0.05$ vs. Control. #, $P < 0.05$ vs. Hcy.

25 ± 6 pmol/min to 123 ± 21 pmol/min (Fig. 2C). At the same time, basal and maximal ECAR were also upregulated by Hcy (Fig. 2D). Likewise, Hcy enhanced mitochondrial membrane potential, as indicated by rhodamine 123 fluorescence intensity, from 1.04 ± 0.08 in control T cells to 1.31 ± 0.07 in Hcy-treated cells (Fig. 2E), and increased mitochondrial mass, indicated by the MFI of MitoTracker Green, from 1.09 ± 0.04 in control to 1.30 ± 0.05 in Hcy-treated cells (Fig. 2F).

Therefore, both *in vivo* and *in vitro* data confirmed that Hcy regulates T cell mitochondrial metabolism, through increasing mitochondrial ROS production and calcium content, mitochondrial respiration, membrane potential, and mass.

Mitochondrial metabolism reprogramming mediates Hcy-induced T cell activation

Mitochondrial metabolism is closely involved in multiple cell processes in immune system (MacIver et al., 2013), so we sought to understand whether Hcy-altered mitochondrial metabolism contributed to Hcy-induced T-cell activation. We at first examined the Akt phosphorylation in response to Hcy stimulation, which indicates the activation of TCR/CD28 signal pathway. While Hcy (50 μmol/L, 24 h) increased the phosphorylation level of Akt in T cells to about 1.5-fold of the control cells, mitochondrial respiration chain inhibitor rotenone (20 nmol/L) totally abolished this Hcy-increased Akt phosphorlation to basal level (Fig. 3A), indicating a possible involvement of mitochondrial metabolism in T cell activation. We next explored if Hcy-mediated T cell mitochondrial metabolism was related to IFN-γ secretion and cell proliferation, two markers of T-cell activation (Feng et al., 2009). IFN-γ secretion by Hcy stimulation was increased to 112.65% ± 5.59% of that in control cells, and the Hcy-increased T-cell IFN-γ secretion was totally inhibited by ammonium 1-pyrrolidinedithiocarbamate (APDC; 100 μmol/L) or SS31 (50 μmol/L), two mitochondrial-specific ROS scavengers, to a level comparable to that in control cells (Fig. 3B). Similar results were obtained from cell proliferation experiments. Both APDC and SS31 completely inhibited Hcy-increased T-cell proliferation: fold of the control (OD_{450}/OD_{630} absorbance) was decreased from 1.64 ± 0.00 in Hcy-treated cells to 0.75 ± 0.00 or 0.57 ± 0.14 in cells with APDC or SS31, respectively (Fig. 3C). Likewise, inhibition of mitochondrial calcium overload by the mitochondrial calcium uniporter inhibitors Ru360 (10 μmol/L) or RuRed (10 μmol/L), or by IP3 receptor inhibitor Xestospongin C (2 μmol/L), all abolished Hcy-increased IFN-γ secretion and proliferation (Fig. 3D and 3E). These results suggest that the increased mitochondrial ROS production and calcium signal participate in Hcy-mediated T-cell activation. Furthermore, inhibition of mitochondrial respiration by rotenone (20 nmol/L) greatly abolished Hcy-increased IFN-γ secretion from 80.71 ± 15.10 pg/mL in control cells to 40.81 ± 18.16 pg/mL in rotenone-treated cells (Fig. 3F), and cell proliferation from 1.15 ± 0.03 to 0.51 ± 0.03 fold of control (OD_{450}/OD_{630} absorbance) in rotenone-treated cells (Fig. 3G). Therefore, our data here suggest that Hcy may activate T cells by reprogramming mitochondrial metabolism.

Hcy enhances ER stress in T cells

The ER is physically and functionally linked with mitochondria (Bravo et al., 2011). Cellular stress stimuli including ER stress could lead to an oxidative stress response, which subsequently activates immune responses (Katika et al., 2011). We next investigated the possibility of whether Hcy-induced mitochondrial metabolic changes and T-cell activation are related to ER stress. We found that in cultured mouse splenic T cells, treatment with Hcy (50 μmol/L, 4 h) increased the protein levels of ER stress markers including p-eIF2a (Fig. 4A), p-PERK (Fig. 4B), IRE-1α, and spliced XBP-1 (Fig. 4C) to 4.0-, 4.4-, 7.2-, and 4.2-fold comparing with control cells. More importantly, inhibition of ER stress with PBA (5 mmol/L) almost completely abolished Hcy-induced T-cell proliferation and IFN-γ secretion (Fig. 4D and 4E), whereas stimulation of ER stress with dithiothreitol (DTT) (0.1 mmol/L) *per se* was sufficient to mimic Hcy-induced T-cell proliferation and IFN-γ secretion (Fig. 4F and 4G). Thus, our results suggest that ER stress might be an important player in Hcy-induced T-cell activation.

Hcy reprograms mitochondrial metabolism and activates T cells by enhancing ER-mitochondria coupling

ER interacts with mitochondria, forming dynamic networks to induce metabolic responses (Bravo et al., 2011), so we wondered whether ER-mitochondria coupling is associated with Hcy-induced reprogramming of mitochondrial metabolism and T-cell activation. We first examined the levels of proteins at the mitochondria-associated membrane (MAM), including mitofusin 2 (MFN2), glutathione peroxidase 7 (Gpx7), ER resident protein 44 (ERP44), and voltage-dependent anion channel (VDAC), in response to Hcy stimulation. Hcy (50 μmol/L, 4 h) increased levels of MFN2 (Fig. 5A), Gpx7 (Fig. 5B), VDAC (Fig. 5D), and slightly increased ERP44 level (Fig. 5C) to 3.2-, 7.4-, 4.8-, and 2.2-fold of that in control cells. This indicates that Hcy enhances the coupling of mitochondria and ER.

To further visualize the interactions of mitochondria and ER, we used structured illumination microscopy (SIM) and found that the colocalization of ER and mitochondria was significantly increased with Hcy treatment for 4 h, from 0.41 ± 0.03 in control T cells to 0.60 ± 0.01 in Hcy-treated T cells (Fig. 5E). The ER and mitochondria are dynamic organelles, moving along the microtubule and interacting with each other (Bravo et al., 2011). In the present study, inhibition of microtubules by nocodazole (10 μmol/L) for one hour decreased the Hcy-induced ER and mitochondria colocalization, from 0.60 ± 0.01 in Hcy-treated T cells to 0.11 ± 0.02 in nocodazole-pretreated T cells (Fig. 5E). Taken together, our data suggest that Hcy increases ER-mitochondria interaction which requires microtubule-directed movement.

To determine whether the increased ER-mitochondria interaction contributes to the reprogramming of mitochondrial metabolism, we examined the effects of nocodazole on Hcy-mediated mitochondrial ROS level and calcium signal as well as mitochondrial membrane potential. Nocodazole greatly attenuated the Hcy-induced mitochondrial membrane potential, as indicated by relative rhodamine 123 MFI, from 2.60 ± 0.33 to 0.95 ± 0.10, comparable to the 1.26 ± 0.22 MFI in untreated control cells (Fig. 5F). Likewise, nocodazole (10 μmol/L) completely abolished Hcy-increased mitochondrial calcium signal to the level of control cells, as shown by Rhod 2 MFI (Fig. 5G). Moreover, nocodazole largely

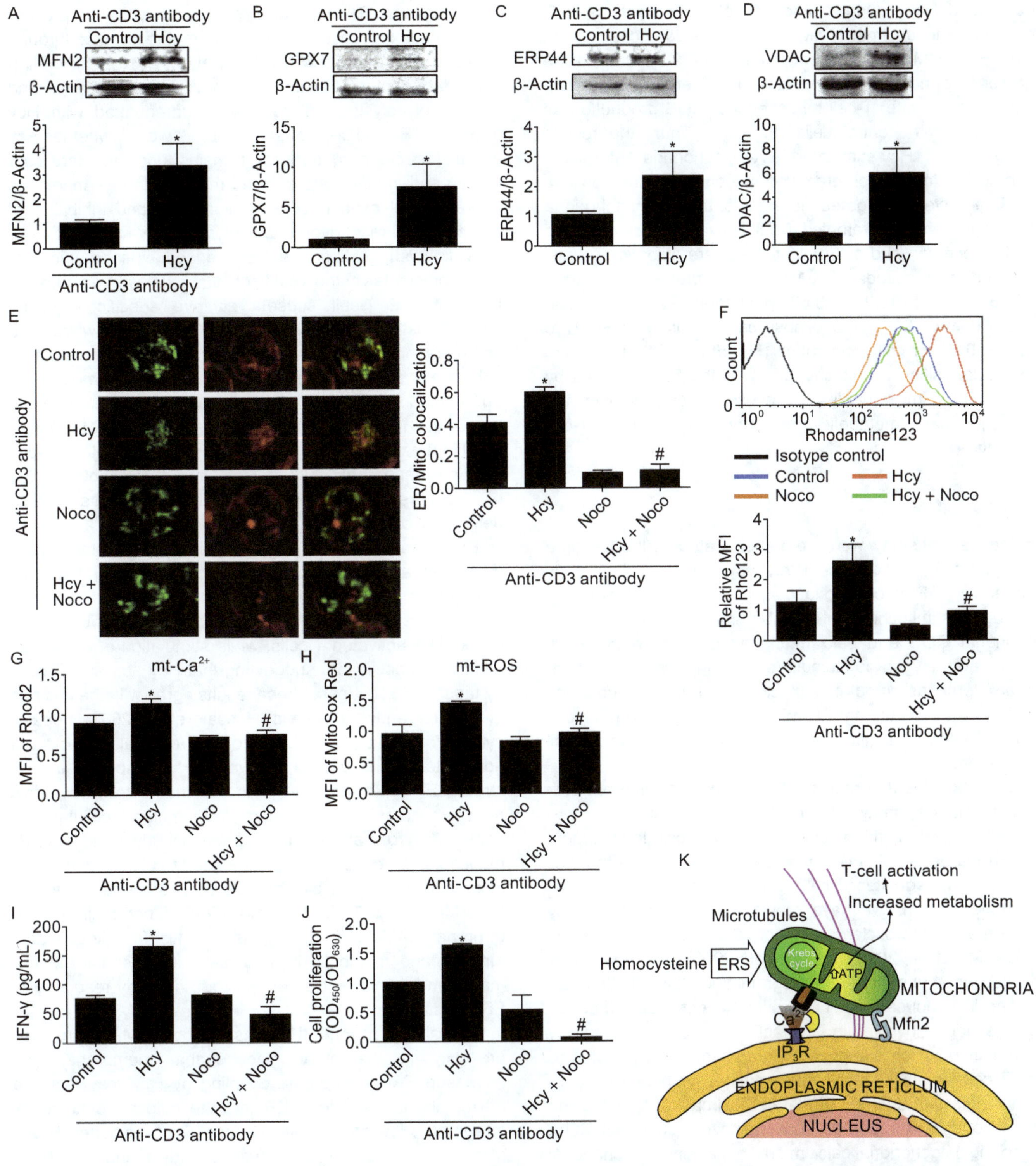

Figure 5. Hcy reprograms mitochondrial metabolism and activates T cells through extending ER-mitochondria interaction. Western blot analysis of MFN2 (A), GPX7 (B), ERP44 (C), and VDAC (D). (E) Structured illumination microscopy of ER-mitochondria colocalization (left) and quantification of the Manders' coefficiency (right) of T cells with or without Hcy treatment for 4 h and nocodazole pretreated for one hour. (F), (G) and (H) MFI in T cells loaded with rhodamine 123, Rhod-2, and MitoSOX Red respectively, and quantified by FACS after Hcy stimulation for 24 h with or without nocodazole pretreatment. (I) IFN-γ secretion and (J) cell proliferation in response to Hcy with or without nocodazole pretreatment. (K) Proposed model of T-cell activation in response to Hcy stimulation. Data are mean ± SEM from 3 independent experiments. *, $P < 0.05$ vs. Control. #, $P < 0.05$ vs. Hcy.

decreased Hcy-induced mitochondrial ROS level, as indicated by the changes of MitoSox Red fluorescence from 1.45 ± 0.01 MFI in Hcy-treated cells to 0.98 ± 0.04 in the presence of nocodazole pretreatment, whereas nocodazole alone did not alter T-cell mitochondrial ROS production as compared with control cells (Fig. 5H). Thus, the results suggest that Hcy-stimulated ER-mitochondria interaction contributes to the reprogrammed mitochondrial metabolism.

Finally, we investigated the effect of the ER-mitochondria interaction on Hcy-promoted T-cell activation. Hcy-activated IFN-γ secretion and cell proliferation were also significantly blocked with nocodazole (10 μmol/L) pretreatment, from 164.8 ± 14.36 to 49.27 ± 13.68 pg/mL for IFN-γ secretion and fold of control (OD_{450}/OD_{630} absorbance) from 1.64 ± 0.01 to 0.07 ± 0.05 for cell proliferation (Fig. 5I and 5J). Therefore, the juxtaposition of mitochondria with the ER mediated by cytoskeleton microtubules is necessary for Hcy-promoted mitochondrial metabolic reprogramming and T-cell activation.

DISCUSSION

In the present study, we have demonstrated for the first time that, both *in vivo* and *in vitro*, Hcy activates T cells by enhancing ER-mitochondria coupling, which subsequently reprograms mitochondrial metabolism by increasing mitochondrial ROS and calcium signal, membrane potential, mass, and oxygen consumption. Inhibiting these mitochondrial alterations completely blocks Hcy-induced T-cell activation. And disturbing ER-mitochondria coupling inhibits both Hcy-induced mitochondrial reprogramming and T-cell activation.

Lymphocytes and memory $CD8^+$ T cells at different activation stages show distinct metabolic signatures, and metabolic programming is necessary for T-cell development (O'Sullivan et al., 2014). These evidences suggest that cell metabolism regulates T-cell function and cell fate in the immune system. We have previously reported that HHcy accelerates atherosclerosis by activating T cells (Feng et al., 2009; Ma et al., 2013; Zhang et al., 2002). Here we investigated whether metabolic reprogramming plays a critical role during Hcy-induced T-cell activation. Mitochondrial metabolism is indeed involved in T cell activation in response to Hcy stimulation, which is in general agreement with other studies of T cells in mice vaccinated with the lymphocytic choriomeningitis virus (Sena et al., 2013). Importantly, inhibiting mitochondrial changes by scavenging mitochondrial ROS, blocking mitochondrial calcium flux, or ceasing mitochondrial respiration, all abolished Hcy-induced T-cell proliferation and IFN-γ secretion, further supporting our hypothesis that Hcy regulates T-cell activation by reprogramming mitochondrial metabolism.

Emerging evidence highlights that T-cell activation leads to marked shifts in cell metabolism to protect against pathogens and to orchestrate the action of other immune cells (Macl et al., 2013). Quiescent T cells require predominantly ATP-generating processes, whereas proliferating effector T cells require high metabolic flux through growth-promoting pathways (Macl et al., 2013). Our preliminary data showed that both oxidative phosphorylation and aerobic glycolysis in T cells were upregulated with Hcy treatment (Figs. 1C–F, 2C, and 2D). However, whether Hcy modulates bioenergetic shift remains unclear. Here, our *in vivo* and *in vitro* data showed that Hcy largely increases overall T cell oxygen consumption rate, and inhibition of mitochondrial respiration totally blocked Hcy-mediated T cell activation, suggesting that Hcy-induced bioenergenic process participates in the T-cell activation. It has been reported that different T-cell subsets achieve specific functions through distinct energetic and biosynthetic pathways, and cell metabolic changes at celluar and systemic levels may enhance or suppress specific T-cell functions (Gerriets et al., 2015; Macl et al., 2013). We have previously reported the acceleration of atherosclerosis by HHcy via activating proinflammatory T cells and repressing regulatory T-cell function (Feng et al., 2009). How different metabolic modes of Hcy treatment on T-cell subtypes merits further determination.

ER is physically and functionally linked with mitochondria (Bravo et al., 2011). In addition to the finding that Hcy reprograms mitochondrial metabolism, we found that Hcy also triggers ER stress in T cells, and inhibition of ER stress blocks Hcy-activated T cells, while stimulation of ER stress mimics this activation, suggesting that Hcy activates T cells by triggering ER stress. These results agree with the findings in vascular endothelial cells (Hossain et al., 2003), hepatocytes (Yang et al., 2014), neurons (Wei et al., 2014), and adipocytes (Li et al., 2013), where Hcy also causes ER stress. ER interacts with mitochondria forming dynamic networks across all cells to regulate multiple cellular processes (Bravo et al., 2011). Our present study found that ER-mitochondria coupling in response to Hcy stimulation increased, as shown by upregulated ER-mitochondria proteins MFN2, Gpx7, ERP44, and VDAC. Since maintenance of proper MAMs and Ca^{2+} flux is essential for mitochondrial function and metabolic homeostasis, increased MAM formation may represent a generalized and early response to ER stress.

ER calcium channel IP3R resides on MAM site, in close proximity to the outer mitochondrial membrane. The increased ER-mitochondria coupling by Hcy may facilitate the uptake of Ca^{2+} from ER into the mitochondrial matrix through the low-affinity mitochondrial Ca^{2+} uniporter (MCU) (Arruda et al., 2014). Our present study found that T-cell activation was mediated by abnormal Ca^{2+} transport from the ER to mitochondria. Inhibition of mitochondrial Ca^{2+} overload by the mitochondrial MCU inhibitors Ru360 and RuRed or IP3 receptor inhibitor Xestospongin C all abolished Hcy-increased IFN-γ secretion and T-cell proliferation. Moreover, we found that inhibition of the Hcy-enhanced ER-mitochondria interaction by dissembling microtubules blocks mitochondrial calcium flux from the ER and decreases mitochondrial ROS levels and membrane potential. Functionally, inhibiting ER-mitochondria coupling by nocodazole diminished Hcy-induced T-cell activation. These findings suggest a working diagram that, the microtubule-directed increase of ER-mitochondria coupling as an early adaptive response to Hcy-mediated ERS, drives Ca^{2+} accumulation in the mitochondria that, in turn, leads to mitochondrial reprogramming, increased ROS generation, metabolism, and T-cell activation (Fig. 5K). Thus, it is reasonable to speculate that tightening of the ER-mitochondria interface, as an earlier event rather than a consequence of mitochondrial metabolic changes, may be responsible for promoting adaptive mitochondrial responses induced by Hcy.

Although the immune system has a critical role in protecting against diseases, energy demands must be finely regulated. We are beginning to understand how pathways that regulate early T-cell activation induced by Hcy are linked to those regulating metabolism. Because of the fundamental importance of cellular bioenergetics control in normal cellular physiology and pathophysiology, as well as the role of T-cell activation in diverse human pathologies, identification of this early process has broad implications. Manipulation of metabolic programs at an early stage may offer therapeutic opportunities for altering the immune responses in diseases. We have reported that HHcy promotes insulin resistance by inducing ER stress and inflammation in the adipocytes (Li et al., 2013). And a recent study by other groups revealed that obesity in mice led to tightened ER-mitochondria contact and increased mitochondrial oxidative stress and impaired metabolic homeostasis in the liver (Arruda et al., 2014). Therefore, our present study may have implications beyond Hcy-induced T-cell activation.

Taken together, our findings have revealed for the first time that the metabolic axis, consisting of ER stress, ER-mitochondria coupling, and mitochondrial metabolism, is involved in the Hcy-promoted T-cell activation. This work provides supporting evidence for an important role of mitochondrial metabolism in regulating early activation of T cells, and sheds new light on understanding the mechanisms underlying the pathogenesis of HHcy-accelerated metabolic syndrome.

MATERIALS AND METHODS

Animal and treatment

Six-week-old female C57BL/6J mice were provided by the Animal Center of Peking University Health Science Center (Beijing, China). Mice were fed normal mouse chow supplemented with or without 1.8 g/L D.L-Hcy (Sigma Chemical Co., St. Louis, MO) in drinking water (n = 6 for each) for 2 weeks, as we previously reported (Dai et al., 2006) with some modifications. The protocol was approved by the Committee on the Ethics of Animal Experiments of the Health Science Center of Peking University. All animal studies were performed in compliance with the U.S. Department of Health and Human Services Guide for the Care and Use of Laboratory Animals.

T-cell proliferation and IFN-γ production

T cells were isolated and cultured as previously reported (Feng et al., 2009). The anti-CD3 antibody (1 μg/mL) was pre-coated onto all microplates overnight. Then T cells at 3 × 10^5 cells/well in the anti-CD3-antibody-precoated microplates, with or without Hcy (50 μmol/L) for 4 h were used for evaluating ERS, or for 24 h were used for proliferation assay. Inhibitors were added one hour before the addition of Hcy. CCK8 (20 μL/well, DOJINDO, Kyushu Island, Japan) was added 6 h before analysis, and experiments were performed in triplicate. The absorbance was determined at 450-nm wavelength with a reference wavelength of 630 nm by use of a microplate reader (Bio-Rad Laboratories, Hercules, CA, USA). IFN-γ in the cell supernatant was measured by ELISA (NeoBioscience Technology, Shenzhen, China).

Flow cytometry

T cells were stained with MitoTracker green (100 nmol/L), Fluo-3 (1 mmol/L), Rhod-2 (5 mmol/L) for 30 min or with Brefeldin A (5 μg/mL) for 5 h, then analyzed by FACS flow cytometry with Cell QuestPro software (BD Biosciences, USA).

Western blot analysis

The protocol of Western blot analysis was as previously reported (Feng et al., 2009) with some modifications of indicated primary antibodies.

Metabolism assays

Oxygen consumption rates (OCR) and extracellular acidification rates (ECAR) were measured using the XF-24 Extracellular Flux Analyzer (Seahorse Bioscience). The protocol was as reported (van der Windt et al., 2012) with some modifications. In total, 6 × 10^5 cells with or without Hcy treatment for 24 h were seeded per well in 24-well XF microplates coated with Poly-L-lysine hydrobromide. Cell culture medium was replaced with XF Assay Medium supplemented with 11 mmol/L glucose and 1 mmol/L pyruvate and incubated in a CO_2-free incubator at 37°C for 1 h for temperature and pH equilibration. The baseline oxygen consumption rate (OCR) was measured, then wells were injected sequentially with oligomycin (1 μmol/L) to measure the ATP-link OCR, oxidative phosphorylation uncoupler carbonylcyanltiep-trifluoro-methoxyphenylhydrazone (FCCP, 1 μmol/L) to determine maximal respiration, and rotenone (1 μmol/L) and antimycin A (1 μmol/L) to determine non-mitochondrial respiration. Experiments were performed with 3 or 4 wells of each plate used as technical replicates, and each experiment had at least three replicates. OCR was normalized by the amount of cells in each well.

For cell glycolysis measurement, T cells seeded in XF24 microplates as described above were cultured in XF base medium supplemented with 2 mmol/L L-glutamine in CO_2-free incubator for 1 h. After three baseline ECAR measurements, cells were added sequentially with glucose (10 mmol/L) to measure the glucose metabolism, oligomycin (1 μmol/L) to measure the glycolytic capacity, and 2-deoxy-D-glucose (2-DG) (100 mmol/L) to measure the non-glycolytic acidification. Experimental treatments were performed on 3 or 4 wells of each plate as technical replicates and each experiment had at least 3 biological replicates. ECAR was normalized by the amount of protein in each well.

Confocal microscopy

T cells were loaded with Rhod-2 in HBSS buffer supplemented with 0.02% pluronic acid F-127 for 30 min at 37°C, washed with phosphate buffered saline (PBS) buffer 3 times, and then incubated in the PBS buffer for 2 h at room temperature to allow for complete deesterification. Fluorescence was visualized by confocal microscopy with a 488-nm laser line. For mitochondrial network and ER colocalization, T cells were loaded with MitoTrack Green and ER-Tracker™ Red. One focal plane was analyzed. The images obtained were deconvolved and the background was subtracted by use of ImageJ. Colocalization between organelles was quantified by use of the Manders' algorithm as described (Costes et al., 2004).

Reagents and antibodies

D.L-Hcy, nocodazole, ammonium 1-pyrrolidinedithiocarbamate (APDC), and 4-phenylbutyric acid (PBA), and anti-voltage-dependent anion channel (VDAC) antibody were from Sigma Chemical Co. Anti-X-box binding protein 1 (anti-XBP-1) antibody was from Santa Cruz Biotechnology. Anti-inositol-requiring enzyme 1α (anti-IRE1α), anti-p-PKR-like endoplasmic reticulum kinase (anti-p-PERK), and anti-p-eukaryotic initiation factor 2 (anti-p-eIF2α) antibodies were from Cell Signal Technology. MitoTracker Green, Rhod-2, ER-Tracker™ Red, and MitoSOX Red were from Invitrogen. Anti-mitofusin 2 (anti-MFN2) antibody was from ABGENT. APC-IFN-γ was from BD Pharmagen. Brefeldin A (BFA) was from Biolegend. Anti-endoplasmic reticulum resident protein 44 (anti-ERP44) and anti-glutathione peroxidase 7 (anti-Gpx7) antibodies were gifts from Prof. Chih-chen Wang (Institute of Biophysics, Chinese Academy of Sciences, Beijing).

Statistical analysis

All data are reported as mean ± SEM. Data were analyzed by use of the GraphPad Prism software. Statistical analysis involved one-way ANOVA followed by the Student-Newman-Keuls test for multiple comparisons and Student's unpaired *t*-test for comparisons between two groups. $P < 0.05$ was considered statistically significant.

ACKNOWLEDGEMENTS

This work was supported by the National Natural Science Foundation of China (Grant Nos. 91439206, 31230035 to Xian Wang; No. 81370006 to Juan Feng), and the National Basic Research Program (973 Program) (No. 2013CB531203 to Ming Zheng). We would like to thank the Center for Biological Imaging (CBI), Institute of Biophysics, Chinese Academy of Science for our Structured Illumination Microscopy work and we are grateful to Ms. Shuoguo Li for her help of in analyzing images.

ABBREVIATIONS

APDC, ammonium 1-pyrrolidinedithiocarbamate; ATP, adenosine triphosphate; BFA, Brefeldin A; DTT, dithiothreitol; eIF2a, eukaryotic initiation factor 2; ER, endoplasmic reticulum; ERP44, ER resident protein 44; FCCP, carbonylcyanltiep-trifluoro-methoxyphenylhydrazone; Gpx7, glutathione peroxidase 7; HHcy, hyperhomocysteinemia; Hcy, homocysteine; IFN-γ, interferon γ; IP3R, inositol 1,4,5-trisphosphate receptor; IRE-1α, inositol-requiring enzyme 1α; MAM, mitochondria-associated membrane; MCU, mitochondrial calcium uniporter; MFI, mean fluorescence intensity; MFN2, mitofusin 2; OCR, oxygen consumption rate; PBA, 4-phenylbutyric acid; PBS, phosphate buffered saline; p-PERK, p-PKR-like endoplasmic reticulum kinase; ROS, reactive oxygen species; SIM, structured illumination microscopy; TCR, T cell receptor; VDAC, voltage-dependent anion channel; XBP-1, X-box binding protein 1.

COMPLIANCE WITH ETHICS GUIDELINES

Juan Feng, Silin Lü, Yanhong Ding, Ming Zheng, and Xian Wang declare that they have no conflict of interest. All institutional and national guidelines for the care and use of laboratory animals were followed.

REFERENCES

Arruda AP, Pers BM, Parlakgül G, Güney E, Inouye K, Hotamisligil GS (2014) Chronic enrichment of hepatic endoplasmic reticulum-mitochondria contact leads to mitochondrial dysfunction in obesity. Nat Med 20:1427–1435

Balaban RS (2009) The role of Ca(2+) signaling in the coordination of mitochondrial ATP production with cardiac work. Biochim Biophys Acta 1787:1334–1341

Bettigole SE, Glimcher LH (2015) Endoplasmic reticulum stress in immunity. Annu Rev Immunol 33:107–138

Bravo R, Vicencio JM, Parra V, Troncoso R, Munoz JP, Bui M, Quiroga C, Rodriguez AE, Verdejo HE, Ferreira J, Iglewski M, Chiong M, Simmen T, Zorzano A, Hill JA, Rothermel BA, Szabadkai G, Lavandero S (2011) Increased ER-mitochondrial coupling promotes mitochondrial respiration and bioenergetics during early phases of ER stress. J Cell Sci 124:2143–2152

Brown GC (1992) Control of respiration and ATP synthesis in mammalian mitochondria and cells. Biochem J 284:1–13

Costes SV, Daelemans D, Cho EH, Dobbin Z, Pavlakis G, Lockett S (2004) Automatic and quantitative measurement of protein-protein colocalization in live cells. Biophys J 86:3993–4003

Cubillos-Ruiz JR, Silberman PC, Rutkowski MR, Chopra S, Perales-Puchalt A, Song M, Zhang S, Bettigole SE, Gupta D, Holcomb K, Ellenson LH, Caputo T, Lee AH, Conejo-Garcia JR, Glimcher LH (2015) ER stress sensor XBP1 controls anti-tumor immunity by disrupting dendritic cell homeostasis. Cell 161:1527–1538

Dai J, Li W, Chang L, Zhang Z, Tang C, Wang N, Zhu Y, Wang X (2006) Role of redox factor-1 in hyperhomocysteinemia-accelerated atherosclerosis. Free Radic Biol Med 41:1566–1577

Feng J, Zhang Z, Kong W, Liu B, Xu Q, Wang X (2009) Regulatory T cells ameliorate hyperhomocysteinaemia-accelerated atherosclerosis in apoE-/- mice. Cardiovasc Res 84:155–163

Gerriets VA, Kishton RJ, Nichols AG, Macintyre AN, Inoue M, Ilkayeva O, Winter PS, Liu X, Priyadharshini B, Slawinska ME, Haeberli L, Huck C, Turka LA, Wood KC, Hale LP, Smith PA, Schneider MA, MacIver NJ, Locasale JW, Newgard CB, Shinohara ML, Rathmell JC (2015) Metabolic programming and PDHK1 control CD4+ T cell subsets and inflammation. J Clin Invest 125:194–207

Hossain GS, van Thienen JV, Werstuck GH, Zhou J, Sood SK, Dickhout JG, de Koning AB, Tang D, Wu D, Falk E, Poddar R, Jacobsen DW, Zhang K, Kaufman RJ, Austin RC (2003) TDAG51 is induced by homocysteine, promotes detachment-mediated programmed cell death, and contributes to the cevelopment of atherosclerosis in hyperhomocysteinemia. J Biol Chem 278:30317–30327

Katika MR, Hendriksen PJ, van Loveren H, Peijnenburg A (2011) Exposure of Jurkat cells to bis (tri-n-butyltin) oxide (TBTO) induces transcriptomics changes indicative for ER- and oxidative stress, T cell activation and apoptosis. Toxicol Appl Pharmacol 254:311–322

Li Y, Zhang H, Jiang C, Xu M, Pang Y, Feng J, Xiang X, Kong W, Xu G, Li Y, Wang X (2013) Hyperhomocysteinemia Promotes Insulin Resistance by Inducing Endoplasmic Reticulum Stress in Adipose Tissue. J Biol Chem 288:9583–9592

Ma K, Lv S, Liu B, Liu Z, Luo Y, Kong W, Xu Q, Feng J, Wang X (2013) CTLA4-IgG ameliorates homocysteine-accelerated atherosclerosis by inhibiting T-cell overactivation in apoE(-/-) mice. Cardiovasc Res 97:349–359

MacIver NJ, Michalek RD, Rathmell JC (2013) Metabolic regulation of T lymphocytes. Annu Rev Immunol 31:259–283

Martinon F, Glimcher LH (2011) Regulation of innate immunity by signaling pathways emerging from the endoplasmic reticulum. Curr Opin Immunol 23:35–40

Martinon F, Chen X, Lee AH, Glimcher LH (2010) TLR activation of the transcription factor XBP1 regulates innate immune responses in macrophages. Nat Immunol 11:411–418

O'Sullivan D, van der Windt GJ, Huang SC, Curtis JD, Chang CH, Buck MD, Qiu J, Smith AM, Lam WY, DiPlato LM, Hsu FF, Birnbaum MJ, Pearce EJ, Pearce EL (2014) Memory CD8(+) T cells use cell-intrinsic lipolysis to support the metabolic programming necessary for development. Immunity 41:75–88

Sena LA, Li S, Jairaman A, Prakriya M, Ezponda T, Hildeman DA, Wang CR, Schumacker PT, Licht JD, Perlman H, Bryce PJ, Chandel NS (2013) Mitochondria are required for antigen-specific T cell activation through reactive oxygen species signaling. Immunity 38:225–236

Staton TL, Lazarevic V, Jones DC, Lanser AJ, Takagi T, Ishii S, Glimcher LH (2011) Dampening of death pathways by schnurri-2 is essential for T-cell development. Nature 472:105–109

Szabadkai G, Duchen MR (2008) Mitochondria: the hub of cellular Ca2+ signaling. Physiology (Bethesda) 23:84–94

van der Windt GJ, Everts B, Chang CH, Curtis JD, Freitas TC, Amiel E, Pearce EJ, Pearce EL (2012) Mitochondrial respiratory capacity is a critical regulator of CD8+ T cell memory development. Immunity 36:68–78

Wei HJ, Xu JH, Li MH, Tang JP, Zou W, Zhang P, Wang L, Wang CY, Tang XQ (2014) Hydrogen sulfide inhibits homocysteine-induced endoplasmic reticulum stress and neuronal apoptosis in rat hippocampus via upregulation of the BDNF-TrkB pathway. Acta Pharmacol Sin 35:707–715

Yang X, Xu H, Hao Y, Zhao L, Cai X, Tian J, Zhang M, Han X, Ma S, Cao J, Jiang Y (2014) Endoplasmic reticulum oxidoreductin 1α mediates hepatic endoplasmic reticulum stress in homocysteine-induced atherosclerosis. Acta Biochim Biophys Sin (Shanghai) 46:902–910

cells use cell-intrinsic lipolysis to support the metabolic programming necessary for development. Immunity 41:75–88

Sena LA, Li S, Jairaman A, Prakriya M, Ezponda T, Hildeman DA, Wang CR, Schumacker PT, Licht JD, Perlman H, Bryce PJ, Chandel NS (2013) Mitochondria are required for antigen-specific T cell activation through reactive oxygen species signaling. Immunity 38:225–236

Staton TL, Lazarevic V, Jones DC, Lanser AJ, Takagi T, Ishii S, Glimcher LH (2011) Dampening of death pathways by schnurri-2 is essential for T-cell development. Nature 472:105–109

Szabadkai G, Duchen MR (2008) Mitochondria: the hub of cellular Ca2+ signaling. Physiology (Bethesda) 23:84–94

van der Windt GJ, Everts B, Chang CH, Curtis JD, Freitas TC, Amiel E, Pearce EJ, Pearce EL (2012) Mitochondrial respiratory capacity is a critical regulator of CD8+ T cell memory development. Immunity 36:68–78

Wei HJ, Xu JH, Li MH, Tang JP, Zou W, Zhang P, Wang L, Wang CY, Tang XQ (2014) Hydrogen sulfide inhibits homocysteine-induced endoplasmic reticulum stress and neuronal apoptosis in rat hippocampus via upregulation of the BDNF-TrkB pathway. Acta Pharmacol Sin 35:707–715

Yang X, Xu H, Hao Y, Zhao L, Cai X, Tian J, Zhang M, Han X, Ma S, Cao J, Jiang Y (2014) Endoplasmic reticulum oxidoreductin 1α mediates hepatic endoplasmic reticulum stress in homocysteine-induced atherosclerosis. Acta Biochim Biophys Sin (Shanghai) 46:902–910

17

miR-10a inhibits cell proliferation and promotes cell apoptosis by targeting BCL6 in diffuse large B-cell lymphoma

Qian Fan[1], Xiangrui Meng[1], Hongwei Liang[2], Huilai Zhang[1], Xianming Liu[1], Lanfang Li[1], Wei Li[1], Wu Sun[3], Haiyang Zhang[3], Ke Zen[2], Chen-Yu Zhang[2], Zhen Zhou[2]✉, Xi Chen[2]✉, Yi Ba[3]✉

[1] Department of Lymphoma, Sino-US Center for Lymphoma and Leukemia, Tianjin Medical University Cancer Institute and Hospital, National Clinical Research Center of Cancer, Key Laboratory of Cancer Prevention and Therapy, Tianjin 300060, China

[2] State Key Laboratory of Pharmaceutical Biotechnology, NJU Advanced Institute of Life Sciences, Jiangsu Engineering Research Center for MicroRNA Biology and Biotechnology, School of Life Sciences, Nanjing University, Nanjing 210093, China

[3] Department of Digestion, Tianjin Medical University Cancer Institute and Hospital, National Clinical Research Center of Cancer, Key Laboratory of Cancer Prevention and Therapy, Tianjin 300060, China

✉ Correspondence: zhenzhou@nju.edu.cn (Z. Zhou), xichen@nju.edu.cn (X. Chen), bayi@tjmuch.com (Y. Ba)

ABSTRACT

The BCL6 (B-Cell Lymphoma 6) gene is a proto-oncogene that is often expressed in diffuse large B-cell lymphomas (DLBCLs). BCL6 loss of function can kill DLBCL cells, demonstrating that BCL6 is necessary for the survival of DLBCL cells and could be a therapeutic target. In this study, we found that BCL6 protein levels were consistently upregulated in DLBCL tissues, whereas its mRNA levels varied randomly in tissues, suggesting that a post-transcriptional mechanism was involved in BCL6 regulation. We used bioinformatics analysis to search for miRNAs, which potentially target BCL6, and identified specific targeting sites for miR-10a in the 3′-untranslated region (3′-UTR) of BCL6. We further identified an inverse correlation between miR-10a levels and BCL6 protein levels, but not mRNA levels, in DLBCL tumor tissue samples. By overexpressing or knocking down miR-10a in DLBCL cells, we experimentally validated that miR-10a directly recognizes the 3′-UTR of the BCL6 transcript and regulated BCL6 expression. Furthermore, we demonstrated that negatively regulating BCL6 by miR-10a suppressed the proliferation and promoted apoptosis of DLBCL cells.

Qian Fan, Xiangrui Meng, Hongwei Liang, and Huilai Zhang have contributed equally to this work.

KEYWORDS microRNA, miR-10a, BCL6, DLBCL, proliferation, apoptosis

INTRODUCTION

Worldwide, diffuse large B-cell lymphoma (DLBCL) is the most common type of lymphoma, accounting for 30%–40% of newly diagnosed non-Hodgkin lymphoma (NHL) cases (Yatomi, 2012). Enormous progress has been made in therapy, and the average five-year overall survival is approximately 50% (Gerrard et al., 2013). Despite improvements in the treatments, DLBCL is still associated with a high mortality rate (Westin and Fayad, 2009); approximately one-third of patients with DLBCL will be refractory to therapy or relapse (Van Den Neste et al., 2015).

The BCL6 (B-Cell Lymphoma 6) gene is a member of the BTB-POZ family and is the most frequently involved oncogene in DLBCL (Parekh et al., 2007). Recent studies have demonstrated that BCL6 plays an important role in the formation of germinal center (GC) B cells, which are the cells of origin of DLBCLs (Basso and Dalla-Favera, 2012; Ding et al., 2015). BCL6 can impact DLBCL through modulating B-cell activation, differentiation, cell cycle arrest and apoptosis (Polo et al., 2004; Shaffer et al., 2000). In addition, BCL6 is involved in the development of CD4+ T-follicular helper cells that play a critical role during the generation of germinal centers (Yu et al., 2009; Hollister et al., 2013). Thus, most B cell lymphomas arise from GC B cells need continued expression of BCL6 to maintain their survival (Hatzi et al., 2013; Bertolo et al., 2013). Targeted inhibition of these BCL6 functions has emerged as the basis for the rational design of lymphoma therapies and combinatorial regimens.

MicroRNAs (miRNAs) are a class of 19-24-nucleotide-long short noncoding RNAs, which regulate genes in a sequence-specific manner (Krol et al., 2010; Koscianska and Krzyzosiak, 2014). They play key roles in regulating the translation and degradation of mRNAs by antisense complementarity to specific mRNA, resulting in either direct RNA degradation or inhibition of protein translation (Koscianska and Krzyzosiak, 2014; Pillai et al., 2007). Although the biological functions of most miRNAs are not yet fully understood, it has been suggested that they are involved in various biological processes, including cell proliferation, cell death, stress resistance, and fat metabolism, through regulation of gene expression (Ivey and Srivastava, 2015). Increasing evidence has indicated that miRNAs, in fact, may be key regulators of various fundamental biological processes. In the past years, many research groups have focused on the potential clinical application of microRNAs as diagnostic or therapeutic tools for patients with DLBCL (Jung and Aguiar, 2009; Wang et al., 2014). Some papers have reported the deregulation of miR-10 family members in several human cancers (Khan et al., 2015; Zeng and Li, 2014), such as gastric, bladder, cervical and thyroid cancer. *Anja* and colleagues found that miR-10a is downregulated in hematological tumor cell lines (Agirre et al., 2008), and miR-10a was reported to be downregulated in DLBCL (Roehle et al., 2008). Early studies indicated that miR-10a could regulate the development and activation of immunocytes by targeting BCL6 and its co-repressor Ncor2, which impacts the stability of the differentiation of Tregs (Takahashi et al., 2012).

Although the dysregulation of miR-10a and BCL6 plays an important role in immunoregulation, no correlation between BCL6 and miR-10a in DLBCL has been reported. In this study, we predicted that BCL6 is a target of miR-10a. After measuring the expression levels of miR-10a and BCL6 in human DLBCL tumor tissues and paired non-neoplastic lymphatic tissues, we confirmed an inverse correlation between miR-10a and the BCL6 protein levels. Furthermore, we experimentally validated the direct inhibition of BCL6 translation by miR-10a through overexpressing or knocking down miR-10a in DLBCL cell lines. Finally, we showed the direct regulation of BCL6 by miR-10a and the biological role of miR-10a targeting BCL6 in human DLBCL.

RESULTS

Upregulation of BCL6 protein, but not mRNA, in DLBCL tissues

The diffuse large B-cell lymphomas (DLBCL) and reactive lymph node hyperplasia (RLH) tissues were embedded in paraffin and then stained with H&E or immunohistochemical staining of Bcl6 for histology examination (Fig. 1A). After measuring the levels of BCL6 protein in DLBCL and RLH tissues via Western blotting, we found that BCL6 protein levels were significantly higher in the DLBCL tissues (Fig.1B, C). Subsequently, we performed quantitative RT-PCR to measure the levels of BCL6 mRNA in the same DLBCL and RLH tissues (Fig. 1D). We found that BCL6 mRNA and protein levels did not correlate between the DLBCL and RLH tissues (Fig. S1). This disparity between the BCL6 protein and mRNA levels in DLBCL tissues strongly suggests that a post-transcriptional mechanism is involved in the regulation of BCL6.

Identification of conserved miR-10a target sites within the 3′-UTR of BCL6

One important mode of post-transcriptional regulation is the repression of mRNA transcripts by miRNAs. miRNAs are therefore likely to play a biologically relevant role in regulating BCL6 expression in DLBCL. Three computational algorithms, including TargetScan (Lewis et al., 2003), miRanda (John et al., 2004) and PicTar (Krek et al., 2005), were used in combination to identify potential miRNAs that can target BCL6. Using these approaches, miR-10a was identified as a candidate regulator of BCL6. The predicted interactions between miR-10a and the targeting sites within the 3′-UTR of BCL6 are illustrated in Fig. 2A. One predicted hybridization was observed between miR-10a and the 3′-UTR of BCL6. There was perfect complementarity between the seed region (the core sequence that encompasses the first 2–8 bases of the mature miRNA) and the putative target sequence. The minimum free energy value of the hybridization between miR-10a and BCL6 was −23.5 kcal/-mol, which is well within the range of genuine miRNA-target pairs. Furthermore, the miR-10a binding sequences in the BCL6 3′-UTR were highly conserved across species. Thus, miR-10a was selected for further experimental verification of its binding to BCL6.

Identification of an inverse correlation between the miR-10a and BCL6 levels in DLBCL tissues

miRNAs are generally thought to have expression patterns that are opposite to those of their targets (Ambros, 2004; Bartel, 2004; He and Hannon, 2004). We next investigated whether miR-10a was inversely correlated with BCL6 in DLBCL. After determining the levels of miR-10a in the same

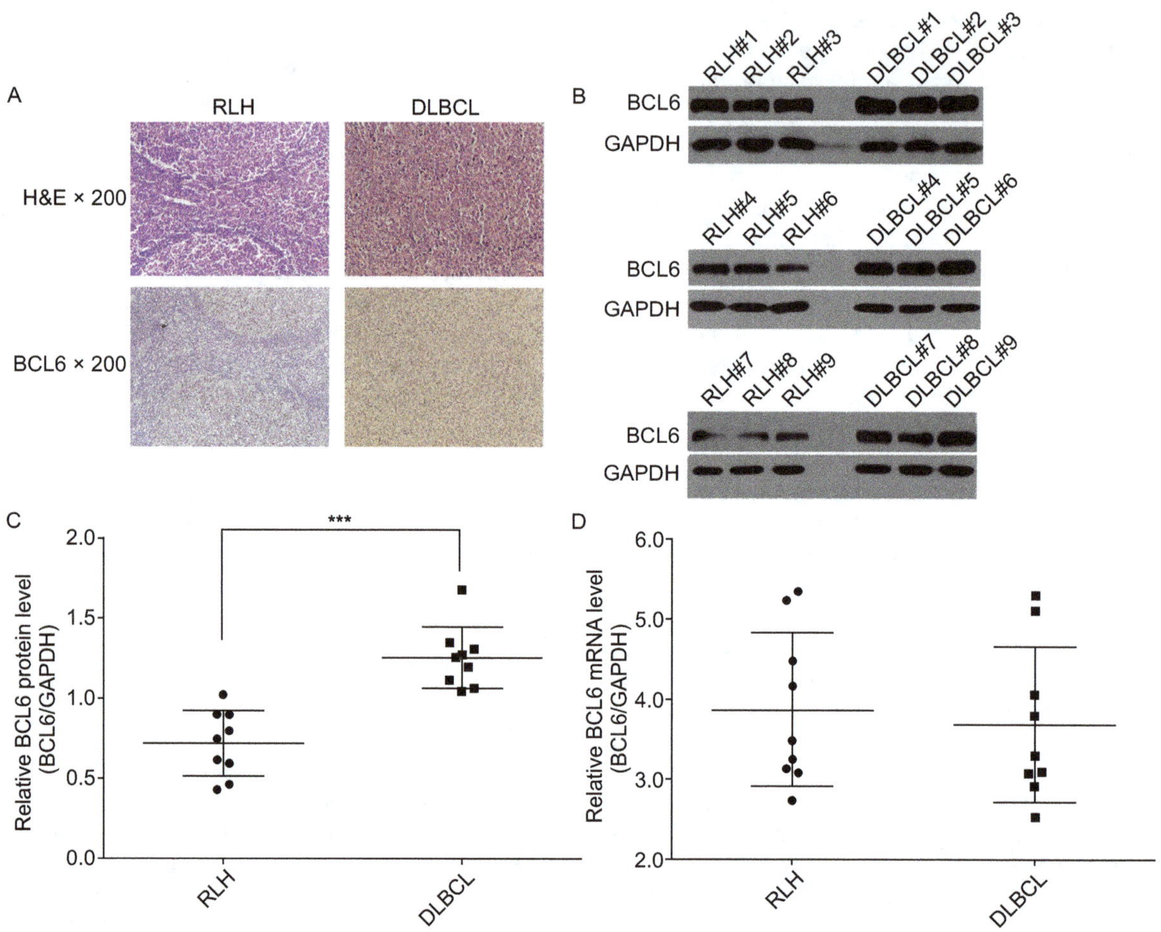

Figure 1. BCL6 protein and mRNA in human tissues. (A) Representative H&E-stained and BCL6-stained sections of the DLBCL&RLH tissues; Western blotting analysis of the expression levels of BCL6 protein in 9 cases of DLBCL and 9 cases of RLH. (B) Representative image. (C) Quantitative analysis; (D) Quantitative RT-PCR analysis of BCL6 mRNA levels in the same DLBCL and RLH tissues, the relative expression was assessed using ΔCt values ($\Delta Ct = Ct_{BCL6} - Ct_{GAPDH}$). The *GAPDH* gene served as the endogenous control. Data (mean ± SEM) are representative of 3 technique replicates. *** $P < 0.001$.

DLBCL and RLH tissues, we found that the miR-10a levels were indeed downregulated in DLBCL tissues (Fig. 2B). Combining the computational prediction with the detection of an inverse correlation between miR-10a and BCL6 *in vivo*, it is quite likely that miR-10a is involved in the post-transcriptional regulation of BCL6.

Validation of BCL6 as a direct target of miR-10a

The correlation between miR-10a and BCL6 expression was further examined by evaluating BCL6 expression in the human DLBCL cell lines OCI-LY7 and OCI-LY3 after the knockdown and overexpression of miR-10a. In these experiments, miR-10a overexpression was achieved by transfecting OCI-LY7 and OCI-LY3 cells with pre-miR-10a (a synthetic RNA oligonucleotide that mimics the miR-10a precursor); and miR-10a knockdown was achieved by transfecting cells with anti-miR-10a, (a chemically modified antisense oligonucleotide designed to specifically target mature miR-10a). As anticipated, cellular miR-10a levels were significantly increased when OCI-LY7 and OCI-LY3 cells were transfected with pre-miR-10a; and were decreased when OCI-LY7 and OCI-LY3 cells were treated with anti-miR-10a (Fig. 3A). The expression of the BCL6 protein was reduced by the overexpression of miR-10a and increased by the knockdown of miR-10a in OCI-LY7 and OCI-LY3 cells (Fig. 3B and 3C). The expression of BCL6 protein was significantly inhibited by the introduction of pre-miR-10a into OCI-LY7 and OCI-LY3 cells, while anti-miR-10a significantly increased the BCL6 protein level in OCI-LY7 and OCI-LY3 cells. To determine the level at which miR-10a regulates BCL6 expression, we repeated the above experiments and examined the expression of BCL6 mRNA after transfection. The overexpression or knockdown of miR-10a did not affect the mRNA stability of BCL6 (Fig. 3D). These results demonstrated that miR-10a specifically regulated

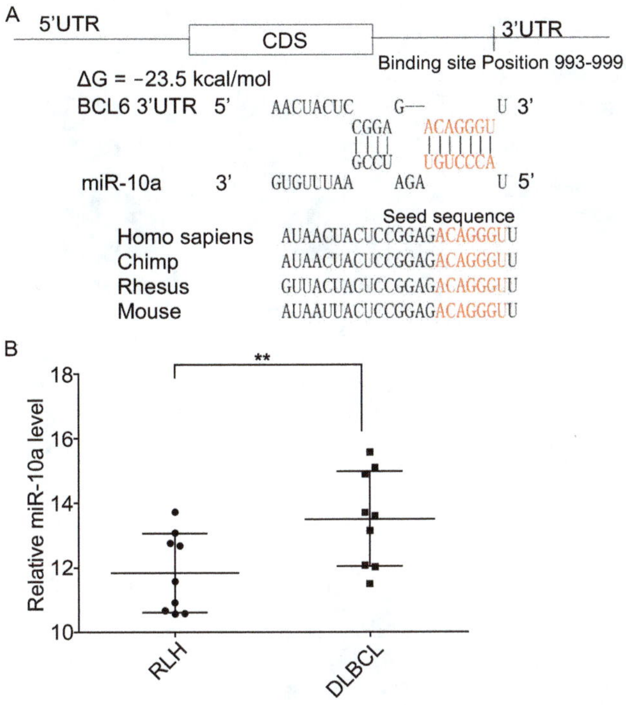

Figure 2. Schematic description of the hypothesized and miR-10a in human tissues. (A) Schematic description of the hypothesized duplex formed by interaction between the BCL6 3′-UTR (top) binding site and miR-10a (bottom). The predicted free energy value of the hybrid is indicated. The seed recognition site is denoted, and all nucleotides in this region are highly conserved across species, including Homo sapiens, chimp, rhesus, and mouse. (B) Quantitative RT-PCR analysis of the expression levels of miR-10a, the relative expression levels were analyzed using ΔCt values ($\Delta Ct = Ct_{miR\text{-}10a} - Ct_{U6}$). The *U6* gene served as the endogenous control in diffuse large B-cell lymphoma(DLBCL) and reactive lymph node hyperplasia (RLH) tissues. Data (mean ± SEM) are representative of 3 technique replicates. ** $P < 0.01$.

BCL6 protein expression at the post-transcriptional level, which is the most common mechanism of animal miRNA action.

To determine whether the negative regulatory effects of miR-10a on BCL6 expression were mediated by the binding of miR-10a to the predicted target sites in the 3′-UTR of the BCL6 mRNA, the full-length 3′-UTR of BCL6 containing the predicted miR-10a binding site was inserted downstream of the firefly luciferase gene in a reporter plasmid. After we proved miR-10a is expressed in HEK293T cells (Fig. S2), the resulting plasmid was co-transfected into HEK293T cells with a transfection control plasmid (β-gal) and either pre-miR-10a or anti-miR-10a. As expected, the luciferase activity was markedly reduced in the cells transfected with pre-miR-10a, whereas the inhibition of miR-10a resulted in an increase in reporter activity compared with transfection with anti-miR control (Fig. 3E). Furthermore, we introduced point mutations into the corresponding complementary sites in the 3′-UTR of BCL6 to disrupt the predicted miR-10a binding sites. This mutated luciferase reporter was unaffected by the overexpression of miR-10a (Fig. 3E). This finding suggested that the binding sites of BCL6 strongly contributed to this miRNA-mRNA interaction, which mediates the post-transcriptional repression of BCL6 expression. In conclusion, our results demonstrate that miR-10a directly binds to the 3′-UTR of the BCL6 mRNA transcript to suppress BCL6 expression.

miR-10a inhibits proliferation and promotes the apoptosis of DLBCL cells by targeting BCL6

We next analyzed the biological consequences of the miR-10a-driven repression of BCL6 expression in DLBCL cells. Because BCL6 is known to be required for DLBCLs to maintain their proliferate and survival (Parekh et al., 2008), we investigated whether the knockdown or overexpression of miR-10a or BCL6 would impact the cell cycle, proliferation and apoptosis of OCI-LY7 cells.

To knock down BCL6 expression, the siRNA sequence targeting different sites of human BCL6 cDNA was designed and transfected into OCI-LY7 cells. To overexpress BCL6, a plasmid expressing the BCL6 ORF was transfected into OCI-LY7 cells. The efficient knockdown or overexpression of BCL6 is demonstrated in Supplementary Figure 3. And the BCL6 protein levels in the samples treated with pre- miR10a or rescued with BCL6 cDNA is demonstrated in Supplementary Figure 4.

The proliferation of OCI-LY7 cells was examined using the cell cycle and the cck8 assay. OCI-LY7 cells transfected with the BCL6 siRNA resulted in the initial accumulation of cells in G0/G1-phase of the cell cycle, in contrast, the OCI-LY7 cells transfected with the BCL6 overexpression plasmid showed a reduction of cells in the G0/G1-phase, whereas the numbers of cells in the S phases increased (Fig. 4A). And OCI-LY7 cells with BCL6 knocked down using siRNA exhibited the promotion of apoptosis, the cells transfected with the BCL6 overexpression plasmid exhibited significantly reduced apoptosis (Fig. 5A). Meanwhile, We evaluated the collective effects of BCL6 on the growth and survival of OCI-LY7 cells using the CCK8 assay. The knockdown of BCL6 inhibited the growth and survival of OCI-LY7 cells; by contrast, to overexpress BCL6 had the opposite effect on the growth and survival of OCI-LY7 cells (Fig. 6A and 6B).

Subsequently, we analyzed the biological consequences of the miR-10a-mediated suppression of BCL6 expression in DLBCL cells. OCI-LY7 cells transfected with pre-miR-10a exhibited G0/G1-phase arrest; by contrast, knocking down of miR-10a the numbers of cells in the S phases increased (Fig. 4B). Moreover, compared with the cells transfected with pre-miR-10a, the cells co-transfected with pre-miR-10a and

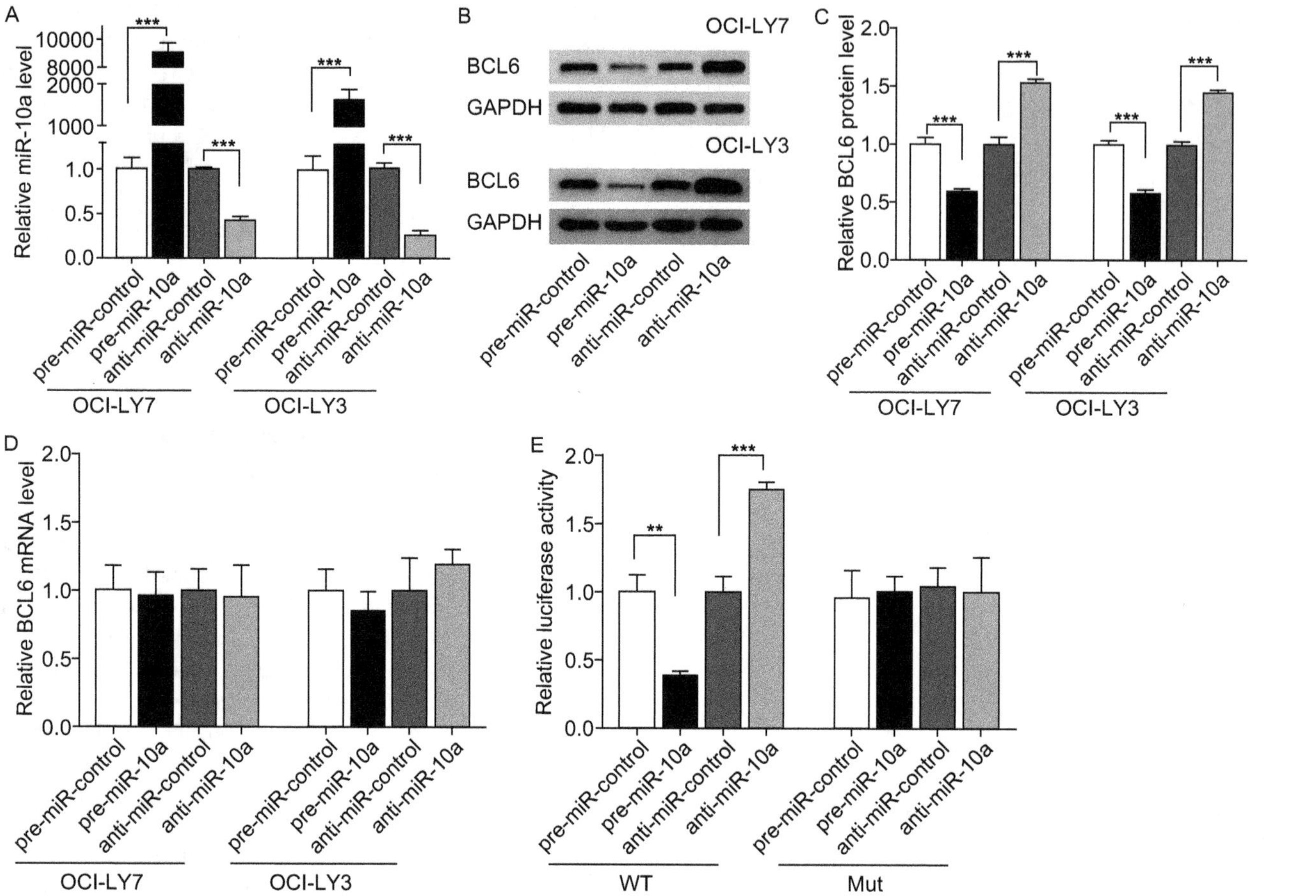

Figure 3. Direct regulation of BCL6 expression by miR-10a at the posttranscriptional level. (A) Quantitative RT-PCR analysis of miR-10a levels in OCI-LY7 and OCI-LY3 cells treated with pre-miR-control, pre-miR-10a, anti-miR-control or anti-miR-10a. U6 snRNA was used as an internal control, and the relative amount of miRNA normalized to the U6 snRNA levels was calculated using the $2^{-\Delta\Delta CT}$ formula, in which $\Delta\Delta C_T = (C_{T\ miRNA} - C_{T\ U6})_{target} - (C_{T\ miRNA} - C_{T\ U6})_{control}$. (B and C) Western blot analysis of BCL6 protein levels in OCI-LY7 and OCI-LY3 cells treated with pre-miR-control, pre-miR-10a, anti-miR-control or anti-miR-10a. B: representative image. C: quantitative analysis. (D) Quantitative RT-PCR analysis of BCL6 mRNA levels in OCI-LY7 and OCI-LY3 cells treated with pre-miR-control, pre-miR-10a, anti-miR-control or anti-miR-10a. (E) Direct recognition of the BCL6 3′-UTR by miR-10a. HEK293T cells were co-transfected with firefly luciferase reporters containing either wild-type (WT) or mutant (MUT) miR-10a binding sites in the BCL6 3′-UTR and pre-miR-control, pre-miR-10a, anti-miR-control or anti-miR-10a, 24 h after transfection, the cells were assayed using a luciferase assay kit. Data are the mean±SEM of 3 independent experiments performed in triplicate, ** $P < 0.01$; *** $P < 0.001$.

the BCL6 overexpression plasmid exhibited reduction of cells in the G0/G1-phase, and the numbers of cells in the S phases increased (Fig. 4C), suggesting that miR-10a-resistant BCL6 expression rescued the suppression of BCL6 expression by miR-10a and attenuated the effect on cell cycle of miR-10a. We investigated the effects of miR-10a on DLBCL cell apoptosis via flow cytometer. The percentage of apoptotic cells was higher among the OCI-LY7 cells transfected with pre-miR-10a but was lower among the OCI-LY7 cells transfected with anti-miR-10a (Fig. 5B). Furthermore, when OCI-LY7 cells were simultaneously transfected with pre-miR-10a and the BCL6 overexpression plasmid, the pro-apoptotic effect of miR-10a was dramatically attenuated (Fig. 5C). Finally, we analyzed the collective effects of miR-10a-mediated suppression of BCL6 expression on the growth and survival of DLBCL cells using the CCK8 assay. OCI-LY7 cells transfected with pre-miR-10a suppressed the growth and survival of OCI-LY7 cells; by contrast, knockdown of miR-10a had the opposite effect on the growth and survival of OCI-LY7 cells (Fig. 6C). Moreover, compared with the cells transfected with pre-miR-10a, the cells co-transfected with pre-miR-10a and the BCL6 overexpression plasmid attenuated the inhibition effect of the growth and survival induced by miR-10a (Fig. 6D), suggesting that miR-10a-resistant BCL6 expression rescued the suppression of BCL6 expression by miR-10a. These results indicate that

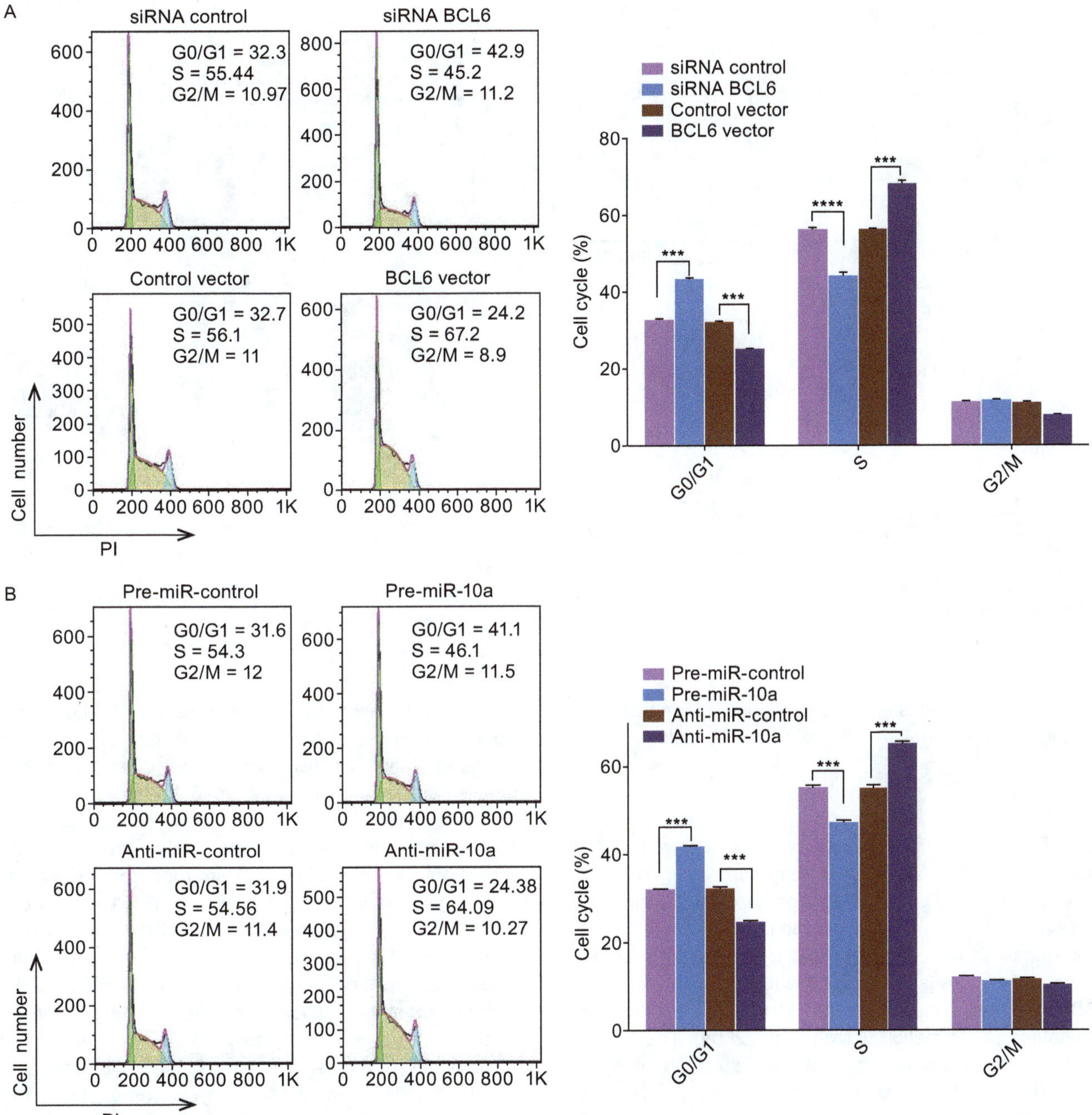

Figure 4. The role of miR-10a targeting BCL6 in the regulation of cell cycle in DLBCL cells. (A) Cell cycle profiles were analyzed using flow cytometry afer the transfection of OCI-LY7 cells with equal dose of control siRNA, BCL6 siRNA, control vector or BCL6 overexpression vector. (B) Cell cycle profiles were analyzed using flow cytometry afer the transfection of OCI-LY7 cells with equal dose of pre-miR-control, pre-miR-10a, anti-miR-control or anti-miR-10a. (C) Cell cycle profiles were analyzed using flow cytometry afer the transfection of OCI-LY7 cells with pre-miR-control plus control vector, a pre-miR-control plus BCL6 overexpression vector, pre-miR-10a plus control vector, or pre-miR-10a plus BCL6 overexpression vector. The panel shows hisograms of cell numbers (y axis) against DNA content (x axis) determined by measuring fluorescence intensity. Numbers denote the percentages of cells in the G0/G1, S and G2/M phases. Left panel: representative image; right panel: quantitative analysis. Data are the mean ± SEM of 3 independent experiments performed in triplicate, ** $P < 0.01$; *** $P < 0.001$.

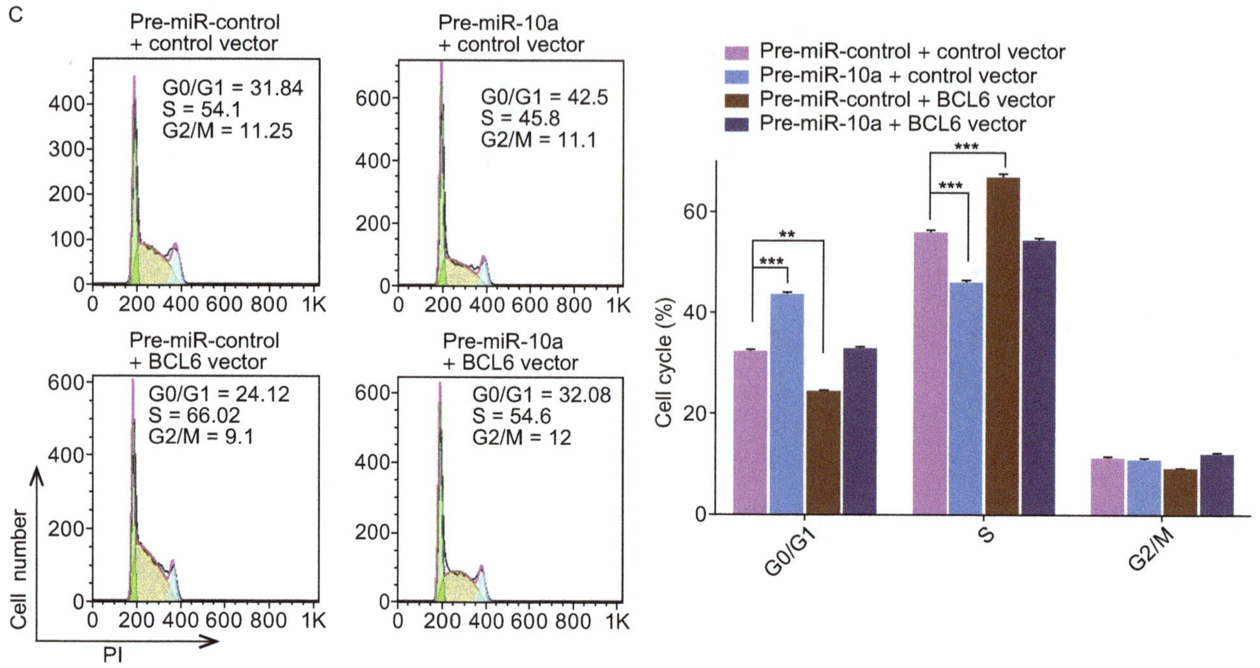

Figure 4. continued.

miR-10a might inhibits proliferation and promotes the apoptosis by downregulating BCL6 in DLBCL cells.

DISCUSSION

BCL6 as a transcriptional repressor, often involved in the development of DLBCL. It is an important gene in B-cell differentiation, which can mediate apoptosis, cell cycle control, survival and inflammatory reaction. BCL6 can influence the prognosis of patients with DLBCL at gene and protein levels and therefore is an independent prognostic factor for DLBCL (Gao et al., 2014; Winter et al., 2006). It is also essential for GC formation (Baron et al., 2004). Thus, most DLBCLs need BCL6 to maintain their survival, and the expression of BCL6 plays a key role in DLBCL. Multiple mechanisms act coordinately to timely modulate BCL6 expression at the transcriptional and post-transcriptional levels (Basso and Dalla-Favera, 2012). BCL6 loss of function can kill DLBCL cells, demonstrating that BCL6 is required for the survival of DLBCL cells and could be an excellent therapeutic target. Recent studies have intended to identify highly specific and non-toxic BCL6 inhibitors (Cerchietti and Melnick, 2013; Duan et al., 2012). Targeting RNA by RNA interference (RNAi) or antisense oligonucleotides (ASOs) has been attempted to eradicate BCL6 from lymphoma cells (Saito et al., 2006). In this study, we found an alternative mechanism—regulating BCL6 expression at the posttranscriptional level in DLBCL.

In this study, we found that silencing BCL6 expression by siRNA could inhibit proliferation and promote apoptosis in DLBCL cells, whereas overexpressing BCL6 induced opposing effects, validating its central role as an essential oncogene during DLBCL tumorigenesis. Interestingly, we identified discordance between the BCL6 protein and mRNA levels in human DLBCL tissues. These results suggested a post-transcriptional regulation mechanism involved in BCL6 repression. One centrally important mode of post-transcriptional regulation is the repression of mRNA transcripts by miRNAs. Therefore, we searched for miRNAs that could target BCL6 and identified miR-10a as a candidate. In addition, by overexpressing or knocking down miR-10a in DLBCL cells, we experimentally validated the direct inhibition of BCL6 translation by miR-10a. Finally, we showed that miR-10a inhibited BCL6 expression resulted in the initial accumulation of cells in G0/G1-phase of cell cycle companied with the inhibited proliferation and promoted apoptosis in cultured DLBCL cells. The results delineate a novel regulatory network employing miR-10a and BCL6 to fine tune cell proliferation and apoptosis. Additionally, a recent study has shown that miR-10a directly targets Bcl-6 to downregulate protein expression level in a murine B cell lymphoma cell line(Takahashi et al., 2012). The latter study, combined with ours, reveals the importance of miR-10a targeting BCL6 as a novel regulatory pathway in DLBCL progression.

miRNAs are aberrantly expressed in cancer and can function as oncogenes or tumor suppressor genes (Calin and Croce, 2006; Esquela-Kerscher and Slack, 2006). In the present study, we found that the levels of miR-10a were lower in DLBCL tumor tissues than in normal adjacent tissues. These results suggest that miR-10a may be involved in the

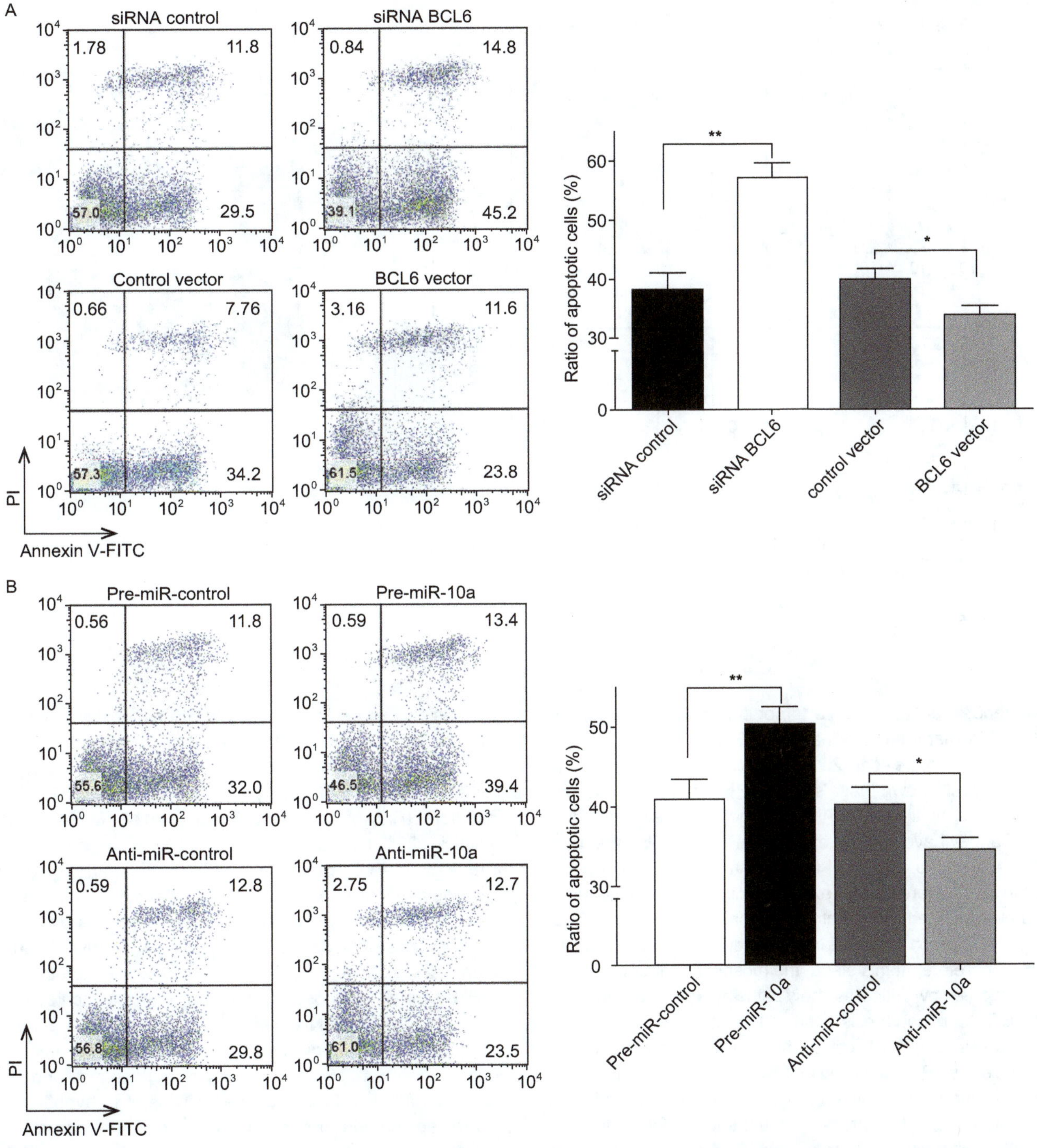

Figure 5. The role of miR-10a targeting BCL6 in the regulation of apoptosis in DLBCL cells. (A)The apoptosis assay was performed 24 h after the transfection of OCI-LY7 cells with equal dose of control siRNA, BCL6 siRNA, control vector or BCL6 overexpression vector; (B) The apoptosis assay was performed 24h after the transfection of OCI-LY7 cells with equal dose of pre-miR-control, pre-miR-10a, anti-miR-control or anti-miR-10a; (C)The apoptosis assay was performed 24h after the transfection of OCI-LY7 cells with pre-miR-control plus control vector, a pre-miR-control plus BCL6 overexpression vector, pre-miR-10a plus control vector, or pre-miR-10a plus BCL6 overexpression vector. Cell apoptosis profiles were analyzed by flow cytometry. The biparametric histogram shows cells in early (bottom right quadrant) and late apoptotic states (upper right quadrant). Viable cells are double negative (bottom left quadrant). Left panel: representative image; right panel: ratio of apoptotic OCI-LY7 cells. Data are the mean ± SEM of 3 independent experiments performed in triplicate, * $P < 0.05$; ** $P < 0.01$.

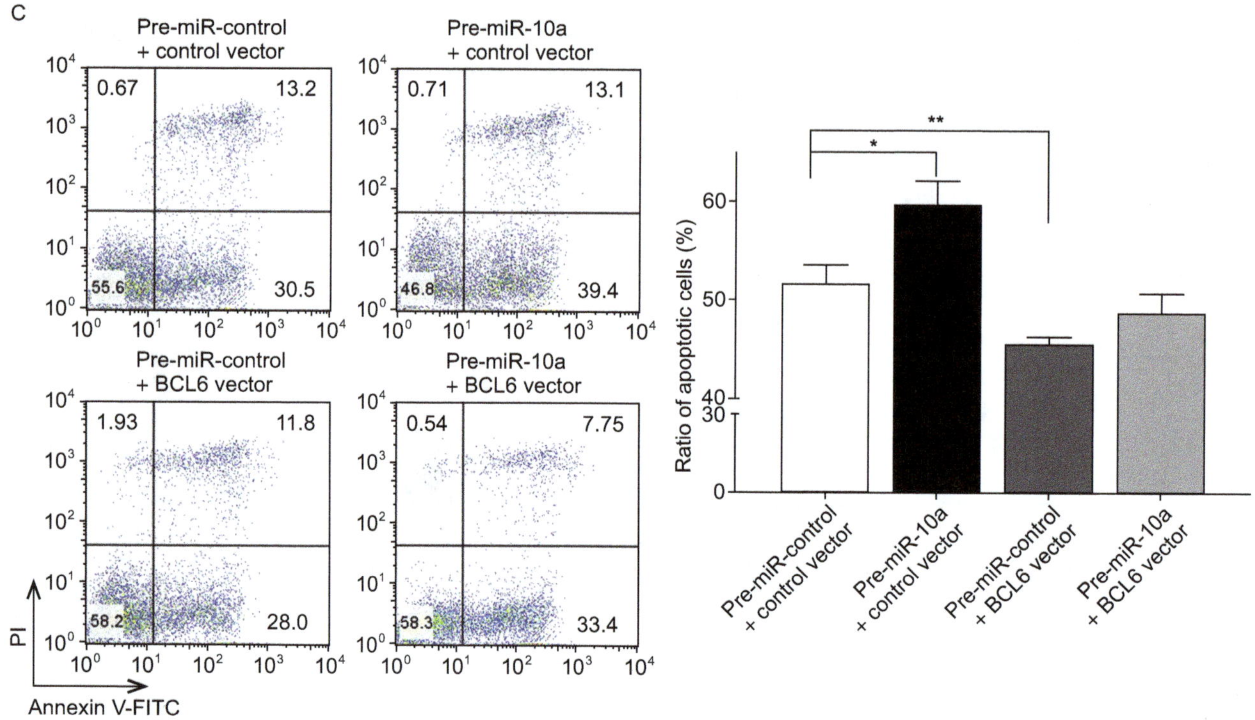

Figure 5. continued.

pathogenesis of DLBCL as a tumor suppressor. Indeed, it is reported in many studies that miR-10a is downregulated in cancers (Havelange et al., 2014; Yan et al., 2013). On the other hand, it is well known that a single miRNA can target multiple genes, while multiple miRNAs can target a single gene. Thus, miR-10a may have multiple different mRNA targets other than BCL6, and these additional targets may also play important roles in carcinogenesis. For example, miR-10a was reported to be downregulated in hematological cancer cell lines (Gaur et al., 2007) such as acute myeloid leukemia (Zhang et al., 2006). Therefore, at this stage, the most important question is to investigate how critical this new pathway would be in the field of DLBCL carcinogenesis. During this study, we found that overexpressing miR-10a could inhibit proliferation and promote apoptosis in DLBCL cells and that BCL6 reduction could mimic miR-10a induction. Interestingly, we observed that the restoration of BCL6 expression could successfully attenuate the anti-proliferative and pro-apoptotic effects of miR-10a on DLBCL cells, although miR-10a has many other targets. These results suggest that the targeting of BCL6 is a major mechanism by which miR-10a exerts its tumor-suppressive function. Therefore, the modulation of BCL6 by miR-10a might explain, at least in part, why the downregulation of miR-10a during DLBCL carcinogenesis can promote cell growth and accelerate DLBCL formation.

BCL6 as a pro-oncogenic has emerged as a critical therapeutic target in DLBCL. The biochemical study of BCL6-mediated gene repression has provided the basis for the design of agents that inhibit BCL6 and kill lymphoma cells (Parekh et al., 2008). Recently, functional and biochemical studies have provided the basis and rational for the development of highly specific and nontoxic BCL6 inhibitors (Basso and Dalla-Favera, 2012; Cerchietti and Melnick, 2013). Thus, BCL6 is likely to become a new target for DLBCL treatment. On the other hand, given the dysregulation of miRNAs in cancer development, the correction of cellular miRNA levels has emerged as a potential therapeutic strategy. The overexpressed miRNAs can be silenced using miRNA ASOs, and the re-expression of miRNAs that are lost in cancers can be achieved by overexpressing miRNA mimics. In this study, the results suggest that targeting BCL6 is a major mechanism by which miR-10a exerts its tumor-suppressive and pro-apoptotic function in DLBCL cells. Thus, it is hypothesized that a replacement treatment with miR-10a mimics may be a promising strategy for cancers characterized by miR-10a downregulation. In summary, as important emerging modulators in cellular pathways, miR-10a and BCL6 may provide attractive, novel therapeutic targets for DLBCL treatment. In future studies, treatments with both miR-10a mimics and BCL6-targeted drugs may offer a viable strategy for DLBCL therapy.

Together, the results of this study delineate a novel regulatory network employing miR-10a and BCL6 to fine tune proliferation and apoptosis in DLBCL cells. This study may provide a potential novel target for future DLBCL therapy.

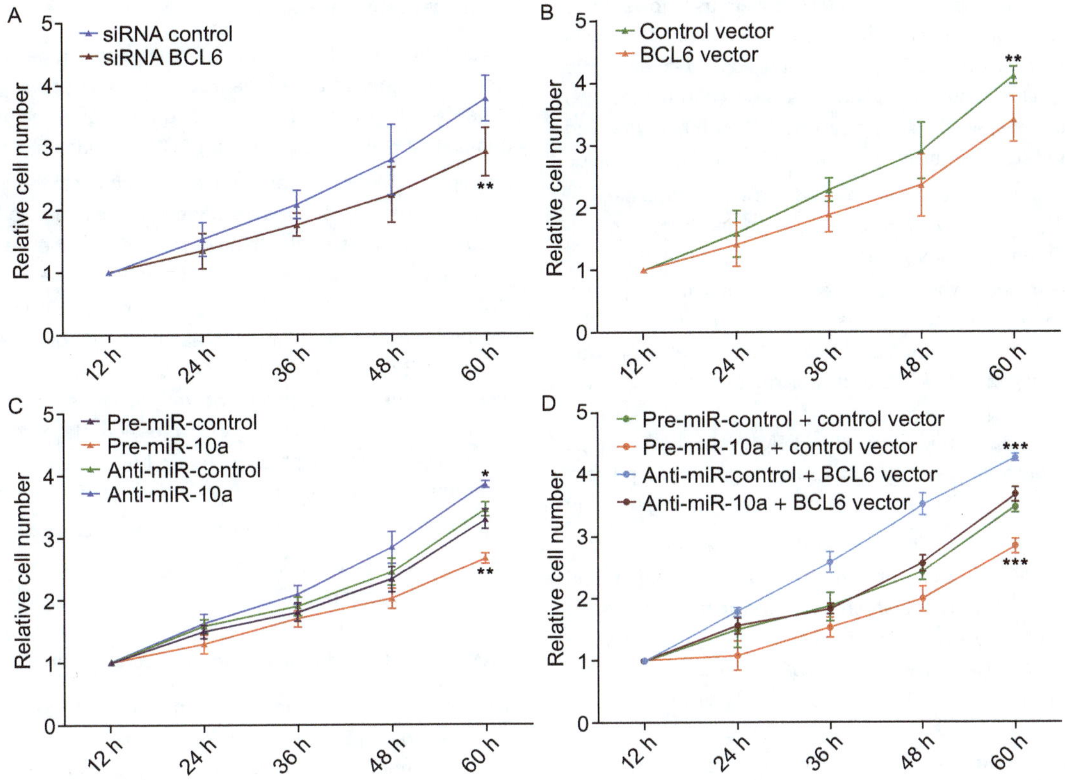

Figure 6. The role of miR-10a targeting BCL6 in the regulation of proliferation in DLBCL cells. (A) CCK8 viability assays were performed 12 h, 24 h, 36 h, 48 h and 60 h after the transfection of OCI-LY7 cells with either a control siRNA or a BCL6 siRNA. (B) CCK8 viability assays were performed 12 h, 24 h, 36 h, 48 h and 60 h after the transfection of OCI-LY7 cells with either a control vector or a BCL6 overexpression vector. (C) CCK-8 viability assays were performed 12 h, 24 h, 36 h, 48 h and 60 h after the transfection of OCI-LY7 cells with pre-miR-control, pre-miR-10a, anti-miR-control or anti-miR-10a. (D) CCK-8 viability assays were performed 12 h, 24 h, 36 h, 48 h and 60 h after the transfection of OCI-LY7 cells with a pre-miR-control plus control vector, pre-miR-10a plus control vector, a pre-miR-control plus BCL6 overexpression vector, or pre-miR-10a plus BCL6 overexpression vector. Data are the mean ± SEM of 3 independent experiments performed in triplicate, * $P < 0.05$; ** $P < 0.01$; *** $P < 0.001$.

MATERIALS AND METHODS

Clinical samples

The patients were eligible if they had previously untreated, biopsy-confirmed diffuse large B-cell lymphoma according to the World Health Organization criteria. In addition, 9 anonymized samples of histopathological verified reactive lymph node hyperplasia (RLH) were included as normal control. The diffuse large B-cell lymphoma tumor tissues and reactive lymph node hyperplasia tissues were derived from patients undergoing a surgical procedure at the Tianjin Medical University Cancer Institute and Hospital (Tianjin, China). Tissue fragments were immediately frozen in liquid nitrogen at the time of surgery and stored at −80°C. The clinical features of the DLBCL patients are listed in Supplementary Table 1.

Cell culture and reagents

The cell lines OCI-LY7, OCI-LY3 and HEK293T were obtained from American Type Culture Collection (ATCC). OCI-LY7 is germinal center B-cell (GCB)-subtype cell line; OCI-LY3 is an activated B-cell (ABC)-subtype cell line; and HEK293T is an embryonic kidney cell line. OCI-LY7 was maintained in complete Iscove's modified essential medium (IMDM; GIBCO, Carlsbad, CA, USA) with 2-mercaptoethanol (1:10000) and 10% fetal bovine serum (GIBCO). OCI-LY3 was cultured in IMDM with 20% fetal bovine serum. HEK293T cells were grown in Dulbecco's Modified Eagle's Medium (DMEM; GIBCO) supplemented with 10% fetal bovine serum. All of the cell lines were supplemented with 1% penicillin/streptomycin (Invitrogen, Carlsbad, CA, USA). All of the cell cultures were maintained at 37°C under 5% CO_2 and 95% air.

RNA isolation and quantitative RT-PCR

Total RNA was extracted from the cultured cells or tissues using Trizol reagent (Invitrogen) according to the manufacturer's instructions. Assays to quantify mature miRNAs were performed using TaqMan miRNA probes (Applied Biosystems, Foster City, CA, USA) according to the manufacturer's instructions. Briefly, 1 μg of total RNA was reverse transcribed to cDNA using AMV reverse transcriptase (TaKaRa, Dalian, China) and a stem-loop RT primer

(Applied Biosystems). The reaction conditions were as follows: 16°C for 30 min, 42°C for 30 min and 85°C for 5 min. Real-time PCR was performed using a TaqMan PCR kit and an Applied Biosystems 7500 Sequence Detection System (Applied Biosystems). The reactions were incubated in a 96-well optical plate at 95°C for 5 min, followed by 40 cycles of 95°C for 15 s and 60°C for 1 min. All of the reactions were performed in triplicate. After the reactions were complete, the cycle threshold (C_T) data were collected using fixed threshold settings, and the mean C_T was determined from triplicate PCRs. A comparative C_T method was used to compare each transcript with the controls. U6 snRNA was used as an internal control, and the relative amount of miRNA normalized to the U6 snRNA levels was calculated using the $2^{-\Delta\Delta CT}$ formula, in which $\Delta\Delta C_T = (C_{T\ miRNA} - C_{T\ U6})_{target} - (C_{T\ miRNA} - C_{T\ U6})_{control}$.

To quantify the BCL6 and GAPDH mRNA levels, 1 μg of total RNA was reversely transcribed to cDNA using oligod (T) 18 primers (TaKaRa) and ThermoScript reverse transcriptase (Invitrogen). The reaction conditions were as follows: 42°C for 60 min and 70°C for 10 min. Real-time PCR was then performed with the RT product, and these reactions included SYBR Green dye (Invitrogen) and specific primers for BCL6 and GAPDH. The sequences of the primers were as follows: 5′-ACTCCCATGTGATAGTGCCA-3′ (BCL6 sense) and 5′-GTGCCTCTTCTGGGATTGTT-3′ (BCL6 antisense); 5′-GATATTGTTGCCATCAATGAC-3′ (GAPDH sense) and 5′-TTGATTTTGGAGGGATCTCG-3′ (GAPDH antisense). The reactions were incubated at 95°C for 5 min, followed by 40 cycles of 95°C for 30 s, 60°C for 30 s and 72°C for 30 s. After the reactions were completed, the C_T values were determined by setting a fixed threshold. The relative amount of BCL6 mRNA was normalized to that of GAPDH.

Overexpression and knockdown of miR-10a

miRNA overexpression was achieved by transfecting cells with a pre-miR-10a (a synthetic RNA oligonucleotide duplex mimicking the miRNA precursor), whereas knockdown was achieved by transfecting cells with an anti-miR-10a (a chemically modified single-stranded antisense oligonucleotide designed to specifically target mature miRNA). Synthetic pre-miR-10a, anti-miR-10a, pre-miR-control and anti-miR-control RNAs were purchased from Genepharma (Shanghai, China). Next, 3×10^6 OCI-LY7 or OCI-LY3 cells were seeded per well in 6-well plates and were transfected with Lipofectamine 2000 (Invitrogen) following the manufacturer's protocols. The cells were harvested 24 h or 48 h after transfection for quantitative RT-PCR or Western blotting. The transfection efficiency was determined by quantitative RT-PCR for hsa-miR-10a. Transfection experiments were repeated three times independently and in each case were done in triplicate.

miRNA target prediction

The miRNAs that may target BCL6 were determined using algorithms from TargetScan (http://genes.mit.edu/targetscan/), PicTar (http://pictar.bio.nyu.edu/), and miRanda (http://cbio.mskcc.org/cgi-bin/mirnaviewer/mirnaviewer.pl).

Luciferase reporter assay

To test the direct binding of miR-10a to the target gene BCL6, a luciferase reporter assay was performed as previously described (Chen et al., 2009). The entire 3′-untranslated region (3′-UTR) of human BCL6 was amplified using PCR with human genomic DNA as a template. The PCR products were inserted into the p-MIR-reporter plasmid (Ambion), and the insertion was confirmed to be correct via sequencing. To test the binding specificity, the sequences that interacted with the miR-10a seed sequence were mutated (from CAGGGTT to GTCCCAA), and the mutant BCL6 3′-UTR was inserted into an equivalent luciferase reporter. For luciferase reporter assays, HEK293T cells were cultured in 24-well plates, and each well was transfected with 1 μg of firefly luciferase reporter plasmid, 1 μg of a β-galactosidase (β-gal) expression plasmid (Ambion), and equal amounts (100 pmol) of pre-miR-10a, anti-miR-10a, or the scrambled negative control RNA using Lipofectamine 2000 (Invitrogen). The β-gal plasmid was used as a transfection control. 24h post-transfection, the cells were assayed using a luciferase assay kit (Promega, Madison, WI, USA).

siRNA and plasmid construction interference assay

The siRNA sequence targeting human BCL6 was designed and synthesized by Genepharma (Shanghai, China). The scrambled siRNA (Genepharma, Shanghai, China) was included as a negative control. A mammalian expression plasmid (pReceiver-M02-BCL6) designed to specifically express the full-length open reading frame (ORF) of human BCL6 without the miR-10a-responsive 3′-UTR was purchased from GeneCopoeia (Germantown, MD, USA). An empty plasmid (pReceiver-M02) served as a negative control. The siRNA or overexpression plasmid was transfected into OCI-Ly7 cells using Lipofectamine2000 (Invitrogen) according to the manufacturer's instructions. Total RNA or protein was isolated 24 h or 48 h after transfection. Quantitative RT-PCR and Western blotting assessed the BCL6 mRNA and protein expression levels.

Protein isolation and Western blotting

All of the cells were rinsed with PBS (pH 7.4) and lysed in RIPA Lysis buffer (Beyotime, China) supplemented with a protease and phosphatase inhibitor cocktail (Thermo Scientific 78440) on ice for 45 min. The DLBCL and RLH tissues were frozen solid with liquid nitrogen, ground into a powder and lysed in RIPA Lysis buffer containing the protease and phosphatase inhibitor cocktail on ice for 30 min. When necessary, sonication was used to facilitate cell lysis. Cell lysates or tissue homogenates were centrifuged for 15 min (12,000 g, 4°C). The supernatant was collected, and the protein concentration was calculated using the Pierce BCA protein assay kit (Thermo Scientific, Rockford, IL, USA). The BCL6 protein levels were analyzed using Western blotting with the corresponding antibodies. The protein levels were normalized by probing the same blots with GAPDH antibody. Experiments were conducted at least in independent triplicates. Densitometry analysis was performed using ImageJ, and normalized by the GAPDH intensity. The antibodies were purchased from the following sources: anti-BCL6 (sc-7388;

Santa Cruz Biotechnology, CA, USA) and anti-GAPDH (sc-365062, Santa Cruz Biotechnology, CA, USA).

Cell cycle assay

To analyze the cell cycle, OCI-LY7 cells were seeded in 6-well plates and transfected with pre-miR-10a, anti-miR-10a, BCL6 siRNA, or the BCL6 overexpression plasmid. Pre-miR-control, anti-miR-control, control siRNA, and control plasmid served as negative controls. Twenty-four hours after transfection, the cells were washed twice with PBS and fixed in 70% ethanol at 4°C overnight. Then, cells were resuspended in 1×PBS with 0.1% Triton X-100, 250 μg/mL RNase A for 1 h at 37°C. Staining for DNA content was performed using 50 mg/mL propidium iodide (BD Biosciences, San Jose, CA). The flow cytometry data was acquired on at least 10,000 cells from each sample on the Becton Dickinson FACS Calibur machine by the Cell Quest Pro software. For the cell cycle analysis, the Dean/Jett/Fox method of Flow Jo software was used.

Apoptosis assays

The apoptosis of OCI-LY7 cells was tested using an Annexin V-FITC/propidium iodide (PI) staining assay. OCI-LY7 cells were cultured in 12-well plates and transfected with pre-miR-10a, anti-miR-10a, BCL6 siRNA, or the BCL6 overexpression plasmid to induce apoptosis. Pre-miR-control, anti-miR-control, control siRNA, and control plasmid served as negative controls. Cells were cultured overnight with serum-depleted medium, and then the cells were harvested. We detect the OCI-LY7 cells apoptosis of under normal or serum deprivation over night (Fig. S5). Flow cytometry analysis of apoptotic cells was performed using an Annexin V-FITC/PI staining kit (BD Biosciences, CA, USA). After washing with cold PBS, the cells were resuspended in binding buffer (100 mM HEPES at pH 7.4, 100 mmol/L NaCl, and 25 mmol/L $CaCl_2$), followed by staining with Annexin V-FITC/PI at room temperature in the dark for 15 min. Apoptotic cells were then evaluated by gating PI- and Annexin V-positive cells using a fluorescence-activated cell-sorting (FACS) flow cytometer (BD Biosciences, San Jose, CA).

Cell proliferation assay

OCI-LY7 cells were cultured in 6-well plates and transfected as mentioned above. Six hours after transfection, the cells were seeded into 96-well plates at a density of 2×10^4 cells per well and then incubated overnight in IMDM supplemented with 10% FBS. Proliferation rates were determined at 12, 24, 36, 48 and 60 h after transfection. Cells were incubated with 10% CCK-8 (CK04-500, Dojindo) at 37°C for 3 h, and then the absorbance of each well was measured at a wave length of 450 nm. The day of transfection was set as Day 0, and the relative cell number was calculated based on the ratio of the absorbance at Day n to that at Day 0.

Statistical analysis

All of the Western blotting images are representative of at least three independent experiments. Quantitative RT-PCR, the luciferase reporter assay, the cell cycle, cell proliferation assay and apoptosis assays were performed in triplicate, and each experiment were repeated three times. Data (mean ± SEM) are representative of at least three independent experiments. The numerical data were statistically analyzed by 2-tailed Student's *t*-test. Bivariate correlation between two independent variables was calculated by Spearman's rank correlation coefficient. Statistically significance was defined as $P < 0.05$.

ACKNOWLEDGEMENTS

This work was supported by grants from the Natural Science Foundation of Jiangsu Province (No. BK2012014).

ABBREVIATIONS

BCL6 B-Cell Lymphoma 6; DLBCL, diffuse large B-cell lymphomas; NHL, non-Hodgkin lymphoma; GC, germinal center; miRNAs, MicroRNAs; RNAi, RNA interference.

COMPLIANCE WITH ETHICS GUIDELINES

Qian Fan, Xiangrui Meng, Hongwei Liang, Huilaizhang, Xianming Liu, Lanfang, Li, Wei Li, Wu Sun, Haiyang Zhang, Ke Zen, Chen-Yu Zhang, Zhen Zhou, Xi Chen and Yi Ba declare that they have no conflict of interest.

All procedures followed were in accordance with the ethical standards of the responsible committee on human experimentation (The ethics committee of Tianjin Medical University Cancer Institute and Hospital, China) and with the Helsinki Declaration of 1975, as revised in 2000 (5). Informed consent was obtained from all patients for being included in the study. Additional informed consent was obtained from all patients for which identifying information is included in this article.

REFERENCES

Agirre X, Jimenez-Velasco A, San Jose-Eneriz E, Garate L, Bandres E, Cordeu L, Aparicio O, Saez B, Navarro G, Vilas-Zornoza A, Perez-Roger I, Garcia-Foncillas J, Torres A, Heiniger A, Calasanz MJ, Fortes P, Roman-Gomez J, Prosper F (2008) Down-regulation of hsa-miR-10a in chronic myeloid leukemia CD34+ cells increases USF2-mediated cell growth. Mol Cancer Res 6 (12):1830–1840

Ambros V (2004) The functions of animal microRNAs. Nature 431 (7006):350–355

Baron BW, Anastasi J, Montag A, Huo D, Baron RM, Karrison T, Thirman MJ, Subudhi SK, Chin RK, Felsher DW, Fu YX, McKeithan TW, Baron JM (2004) The human BCL6 transgene promotes the development of lymphomas in the mouse. Proc Natl Acad Sci USA 101(39):14198–14203

Bartel DP (2004) MicroRNAs: genomics, biogenesis, mechanism, and function. Cell 116(2):281–297

Basso K, Dalla-Favera R (2012) Roles of BCL6 in normal and transformed germinal center B cells. Immunol Rev 247(1):172–183

Bertolo C, Roa S, Sagardoy A, Mena-Varas M, Robles EF, Martinez-Ferrandis JI, Sagaert X, Tousseyn T, Orta A, Lossos IS, Amar S, Natkunam Y, Briones J, Melnick A, Malumbres R, Martinez-Climent JA (2013) LITAF, a BCL6 target gene, regulates autophagy in mature B-cell lymphomas. Br J Haematol 162 (5):621–630

Calin GA, Croce CM (2006) MicroRNA signatures in human cancers. Nat Rev Cancer 6(11):857–866

Cerchietti L, Melnick A (2013) Targeting BCL6 in diffuse large B-cell lymphoma: what does this mean for the future treatment? Expert Rev Hematol 6(4):343–345

Chen X, Guo X, Zhang H, Xiang Y, Chen J, Yin Y, Cai X, Wang K, Wang G, Ba Y, Zhu L, Wang J, Yang R, Zhang Y, Ren Z, Zen K, Zhang J, Zhang CY (2009) Role of miR-143 targeting KRAS in colorectal tumorigenesis. Oncogene 28(10):1385–1392

Duan S, Cermak L, Pagan JK, Rossi M, Martinengo C, di Celle PF, Chapuy B, Shipp M, Chiarle R, Pagano M (2012) FBXO11 targets BCL6 for degradation and is inactivated in diffuse large B-cell lymphomas. Nature 481(7379):90–93

Ding J, Dirks WG, Ehrentraut S, Geffers R, MacLeod RA, Nagel S, Pommerenke C, Romani J, Scherr M, Vaas LA, Zaborski M, Drexler HG, Quentmeier H (2015) BCL6—regulated by AhR/ARNT and wild-type MEF2B—drives expression of germinal center markers MYBL1 and LMO2. Haematologica 100(6):801–809

Esquela-Kerscher A, Slack FJ (2006) Oncomirs—microRNAs with a role in cancer. Nat Rev Cancer 6(4):259–269

Gao P, Li Q, Wang Z, Yan F, Lu C, Cao X (2014) Significance of BCL6, MYC, P53 genes abnormalities for the prognosis of diffuse large B-cell lymphoma. Zhonghua Yi Xue Yi Chuan Xue Za Zhi 31 (5):628–631

Gaur A, Jewell DA, Liang Y, Ridzon D, Moore JH, Chen C, Ambros VR, Israel MA (2007) Characterization of microRNA expression levels and their biological correlates in human cancer cell lines. Cancer Res 67(6):2456–2468

Gerrard M, Waxman IM, Sposto R, Auperin A, Perkins SL, Goldman S, Harrison L, Pinkerton R, McCarthy K, Raphael M, Patte C, Cairo MS (2013) Outcome and pathologic classification of children and adolescents with mediastinal large B-cell lymphoma treated with FAB/LMB96 mature B-NHL therapy. Blood 121 (2):278–285

Hatzi K, Jiang Y, Huang C, Garrett-Bakelman F, Gearhart MD, Giannopoulou EG, Zumbo P, Kirouac K, Bhaskara S, Polo JM, Kormaksson M, MacKerell AD Jr, Xue F, Mason CE, Hiebert SW, Prive GG, Cerchietti L, Bardwell VJ, Elemento O, Melnick A (2013) A hybrid mechanism of action for BCL6 in B cells defined by formation of functionally distinct complexes at enhancers and promoters. Cell Rep 4(3):578–588

Havelange V, Ranganathan P, Geyer S, Nicolet D, Huang X, Yu X, Volinia S, Kornblau SM, Andreeff M, Croce CM, Marcucci G, Bloomfield CD, Garzon R (2014) Implications of the miR-10 family in chemotherapy response of NPM1-mutated AML. Blood 123(15):2412–2415

He L, Hannon GJ (2004) MicroRNAs: small RNAs with a big role in gene regulation. Nat Rev Genet 5(7):522–531

Hollister K, Kusam S, Wu H, Clegg N, Mondal A, Sawant DV, Dent AL (2013) Insights into the role of Bcl6 in follicular Th cells using a new conditional mutant mouse model. J Immunol 191(7):3705–3711

Ivey KN, Srivastava D (2015) microRNAs as developmental regulators. Cold Spring Harb Perspect Biol 7(7):a008144

John B, Enright AJ, Aravin A, Tuschl T, Sander C, Marks DS (2004) Human microRNA targets. PLoS Biol 2(11):e363

Jung I, Aguiar RC (2009) microRNA-155 expression and outcome in diffuse large B-cell lymphoma. Br J Haematol 144(1):138–140

Khan S, Wall D, Curran C, Newell J, Kerin MJ, Dwyer RM (2015) microRNA-10a is reduced in breast cancer and regulated in part through retinoic acid. BMC Cancer 15(1):345

Koscianska E, Krzyzosiak WJ (2014) Current understanding of the role of microRNAs in spinocerebellar ataxias. Cerebellum Ataxias 1:7

Krek A, Grun D, Poy MN, Wolf R, Rosenberg L, Epstein EJ, MacMenamin P, da Piedade I, Gunsalus KC, Stoffel M, Rajewsky N (2005) Combinatorial microRNA target predictions. Nat Genet 37(5):495–500

Krol J, Loedige I, Filipowicz W (2010) The widespread regulation of microRNA biogenesis, function and decay. Nat Rev Genet 11 (9):597–610

Lewis BP, Shih IH, Jones-Rhoades MW, Bartel DP, Burge CB (2003) Prediction of mammalian microRNA targets. Cell 115(7):787–798

Parekh S, Polo JM, Shaknovich R, Juszczynski P, Lev P, Ranuncolo SM, Yin Y, Klein U, Cattoretti G, Dalla Favera R, Shipp MA, Melnick A (2007) BCL6 programs lymphoma cells for survival and differentiation through distinct biochemical mechanisms. Blood 110(6):2067–2074

Parekh S, Prive G, Melnick A (2008) Therapeutic targeting of the BCL6 oncogene for diffuse large B-cell lymphomas. Leuk Lymphoma 49(5):874–882

Pillai RS, Bhattacharyya SN, Filipowicz W (2007) Repression of protein synthesis by miRNAs: how many mechanisms? Trends Cell Biol 17(3):118–126

Polo JM, Dell'Oso T, Ranuncolo SM, Cerchietti L, Beck D, Da Silva GF, Prive GG, Licht JD, Melnick A (2004) Specific peptide interference reveals BCL6 transcriptional and oncogenic mechanisms in B-cell lymphoma cells. Nat Med 10(12):1329–1335

Roehle A, Hoefig KP, Repsilber D, Thorns C, Ziepert M, Wesche KO, Thiere M, Loeffler M, Klapper W, Pfreundschuh M, Matolcsy A, Bernd HW, Reiniger L, Merz H, Feller AC (2008) microRNA signatures characterize diffuse large B-cell lymphomas and follicular lymphomas. Br J Haematol 142(5):732–744

Saito Y, Liang G, Egger G, Friedman JM, Chuang JC, Coetzee GA, Jones PA (2006) Specific activation of microRNA-127 with downregulation of the proto-oncogene BCL6 by chromatin-modifying drugs in human cancer cells. Cancer Cell 9(6):435–443

Shaffer AL, Yu X, He Y, Boldrick J, Chan EP, Staudt LM (2000) BCL-6 represses genes that function in lymphocyte differentiation, inflammation, and cell cycle control. Immunity 13(2):199–212

Takahashi H, Kanno T, Nakayamada S, Hirahara K, Sciume G, Muljo SA, Kuchen S, Casellas R, Wei L, Kanno Y, O'Shea JJ (2012) TGF-beta and retinoic acid induce the microRNA miR-10a, which targets Bcl-6 and constrains the plasticity of helper T cells. Nat Immunol 13(6):587–595

Van Den Neste E, Schmitz N, Mounier N, Gill D, Linch D, Trneny M, Milpied N, Radford J, Ketterer N, Shpilberg O, Duhrsen U, Ma D, Briere J, Thieblemont C, Salles G, Moskowitz CH, Glass B, Gisselbrecht C (2015) Outcome of patients with relapsed diffuse large B-cell lymphoma who fail second-line salvage regimens in the International CORAL study. Bone Marrow Transplant. 51 (1):51–57

Wang K, Xu Z, Wang N, Xu T, Zhu M (2014) microRNA and gene networks in human diffuse large B-cell lymphoma. Oncol Lett 8 (5):2225–2232

Westin JR, Fayad LE (2009) Beyond R-CHOP and the IPI in large-cell lymphoma: molecular markers as an opportunity for stratification. Curr Hematol Malig Rep 4(4):218–224

Winter JN, Weller EA, Horning SJ, Krajewska M, Variakojis D, Habermann TM, Fisher RI, Kurtin PJ, Macon WR, Chhanabhai M, Felgar RE, Hsi ED, Medeiros LJ, Weick JK, Reed JC, Gascoyne RD (2006) Prognostic significance of Bcl-6 protein expression in DLBCL treated with CHOP or R-CHOP: a prospective correlative study. Blood 107(11):4207–4213

Yan Y, Luo YC, Wan HY, Wang J, Zhang PP, Liu M, Li X, Li S, Tang H (2013) microRNA-10a is involved in the metastatic process by regulating Eph tyrosine kinase receptor A4-mediated epithelial-mesenchymal transition and adhesion in hepatoma cells. Hepatology 57(2):667–677

Yatomi Y (2012) From FAB classification to WHO classification of tumors of hematopoietic and lymphoid tissue. Rinsho Byori 60 (6):550–552

Yu D, Rao S, Tsai LM, Lee SK, He Y, Sutcliffe EL, Srivastava M, Linterman M, Zheng L, Simpson N, Ellyard JI, Parish IA, Ma CS, Li QJ, Parish CR, Mackay CR, Vinuesa CG (2009) The transcriptional repressor Bcl-6 directs T follicular helper cell lineage commitment. Immunity 31(3):457–468

Zeng T, Li G (2014) microRNA10a enhances the metastatic potential of cervical cancer cells by targeting phosphatase and tensin homologue. Mol Med Rep 10(3):1377–1382

Zhang L, Huang J, Yang N, Greshock J, Megraw MS, Giannakakis A, Liang S, Naylor TL, Barchetti A, Ward MR, Yao G, Medina A, O'Brien-Jenkins A, Katsaros D, Hatzigeorgiou A, Gimotty PA, Weber BL, Coukos G (2006) microRNAs exhibit high frequency genomic alterations in human cancer. Proc Natl Acad Sci USA 103(24):9136–9141

18

Phosphorylation of Atg31 is required for autophagy

Wenzhi Feng[1], Tong Wu[2], Xiaoyu Dan[2], Yuling Chen[3], Lin Li[4], She Chen[4], Di Miao[3], Haiteng Deng[3], Xinqi Gong[5✉], Li Yu[2✉]

[1] PTN Program, College of Life Science, Peking University, Beijing 100871, China
[2] State Key Laboratory of Biomembrane and Membrane Biotechnology, Tsinghua University-Peking University Joint Center for Life Sciences, School of Life Sciences, Tsinghua University, Beijing 100084, China
[3] Center for Biomedical Analysis, Tsinghua University, Beijing 100084, China
[4] National Institute of Biological Sciences, Beijing 102206, China
[5] Institute for Mathematical Sciences, Renmin University of China, Beijing 100872, China
✉ Correspondence: xinqigong@ruc.edu.cn (X. Gong), liyulab@mail.tsinghua.edu.cn (L. Yu)

ABSTRACT

Autophagy is an evolutionarily conserved cellular process which degrades intracellular contents. The Atg17-Atg31-Atg29 complex plays a key role in autophagy induction by various stimuli. In yeast, autophagy occurs with autophagosome formation at a special site near the vacuole named the pre-autophagosomal structure (PAS). The Atg17-Atg31-Atg29 complex forms a scaffold for PAS organization, and recruits other autophagy-related (Atg) proteins to the PAS. Here, we show that Atg31 is a phosphorylated protein. The phosphorylation sites on Atg31 were identified by mass spectrometry. Analysis of mutants in which the phosphorylated amino acids were replaced by alanine, either individually or in various combinations, identified S174 as the functional phosphorylation site. An S174A mutant showed a similar degree of autophagy impairment as an Atg31 deletion mutant. S174 phosphorylation is required for autophagy induced by various autophagy stimuli such as nitrogen starvation and rapamycin treatment. Mass spectrometry analysis showed that S174 is phosphorylated constitutively, and expression of a phosphorylation-mimic mutant (S174D) in the Atg31 deletion strain restores autophagy. In the S174A mutant, Atg9-positive vesicles accumulate at the PAS. Thus, S174 phosphorylation is required for formation of autophagosomes, possibly by facilitating the recycling of Atg9 from the PAS. Our data demonstrate the role of phosphorylation of Atg31 in autophagy.

KEYWORDS autophagy, Atg31, phosphorylation, autophagosome, pre-autophagosomal structure (PAS)

INTRODUCTION

Autophagy is an evolutionarily conserved, lysosome-based degradation pathway. During autophagy, double-membrane vesicles are formed which engulf cytosol or damaged organelles in a selective or non-selective manner. Autophagy plays important roles in various physiological settings, and disruption of autophagy has been shown to lead to many pathological conditions (Winslow and Rubinsztein, 2008; Hussey et al., 2009; Mizushima and Komatsu, 2011; Jiang and Mizushima, 2014; Martin et al., 2014).

In yeast, autophagy is initiated at a specific site based on a multi-protein complex named the pre-autophagosomal structure (PAS). Formation of a double-membrane structure, named the isolation membrane, is initiated at the PAS. The isolation membrane extends and surrounds cytosolic cargoes before sealing to form the completed autophagosome.

More than 30 Atg proteins involved in autophagy have been identified using *Saccharomyces cerevisiae* as a model organism since the 1990s (Tsukada and Ohsumi, 1993; Thumm et al., 1994; Harding et al., 1995). Most of those Atg proteins can be recruited to the PAS (Suzuki and Ohsumi, 2010). At the core of the PAS is a stable ternary complex of

Atg17, Atg29 and Atg31 (Kabeya et al., 2009). Atg17 interacts with Atg31 and Atg29 independent of nutrient conditions. Under nutrient starvation conditions, the Tor complex is inactivited, which causes dephosphorylation of Atg13, followed by binding of dephosphorylated Atg13 to Atg1. The Atg1-Atg13 complex is then recruited to the Atg17 complex, thus activating the autophagy pathway. Atg31 was originally found as a partner of Atg17 from yeast two-hybrid assays and global mass spectrometry analysis (Kabeya et al., 2007). Atg31 has been reported to be a phosphorylated protein, but the phosphorylation site has not been identified and the function of this phosphorylation remains to be elucidated.

In this study, we demonstrate that Atg31 is constitutively phosphorylated. Mass spectrometry identified 11 phosphorylation sites in Atg31, and analysis of mutants created by alanine swapping confirmed that S174 is the functional phosphorylation site. Autophagy is impaired to a similar degree in the S174A mutant as in the Atg31 deletion mutant. S174 phosphorylation is required for autophagy induced by nitrogen starvation, amino acid starvation and rapamycin treatment. Expression of a phosphorylation-mimic mutant (S174D) in the Atg31 deletion strain restores autophagy. Finally, we show that S174 phosphorylation is required for recycling of Atg9 from the PAS. Our data demonstrate the role of phosphorylation of Atg31 in autophagy.

RESULTS

Atg31 is a phosphorylated protein

We noticed that when cells were grown in both nutrient-rich and starvation conditions, the Atg31 protein displayed multiple bands of higher molecular weight when analyzed by SDS-PAGE (Fig. 1A). Thus, Atg31 appears to undergo some sort of post-translational modification in a nutrient-independent manner. Treating the cell lysate with λ phosphatase elimnated the multiple upper bands, suggesting that Atg31 is modified by phosphorylation (Fig. 1B). To better monitor the phosphorylation level of Atg31 during starvation, we used a phos-tag detection assay which enhances the mobility shifts of phosphorylated proteins on SDS-PAGE (Kinoshita et al., 2004). We found the phosphorylation level of Atg13 is similar in starved and un-starved cells (Fig. 1C).

Identification of Atg31 phosphorylation sites

To identify the Atg31 phosphorylation sites, we tagged Atg31 with an N-terminal GST tag and purified it from yeast under nutrient-rich conditions and starvation conditions. When we analyzed the protein by mass spectrometry (MS), we identified 11 phosphorylation sites (Fig. 1D).

Screening of functional phosphorylation sites in Atg31

Next, we screened the phosphorylation sites for their effect on autophagosome formation using GFP-Atg8 as a marker. We mutated each amino acid individually to alanine, and we also generated mutants in which various combinations of phosphorylation sites were changed to alanine. Then we assessed autophagy activity by monitoring the ability of each mutant to transport GFP-Atg8 into the vacuole. In yeast carrying the S174A single mutant, a lower percentage of cells had vacuolar Atg8. Therefore, our screening method identified the Serine at 174 mutant as a potential functional phosphorylation site (Fig. 2A and 2B).

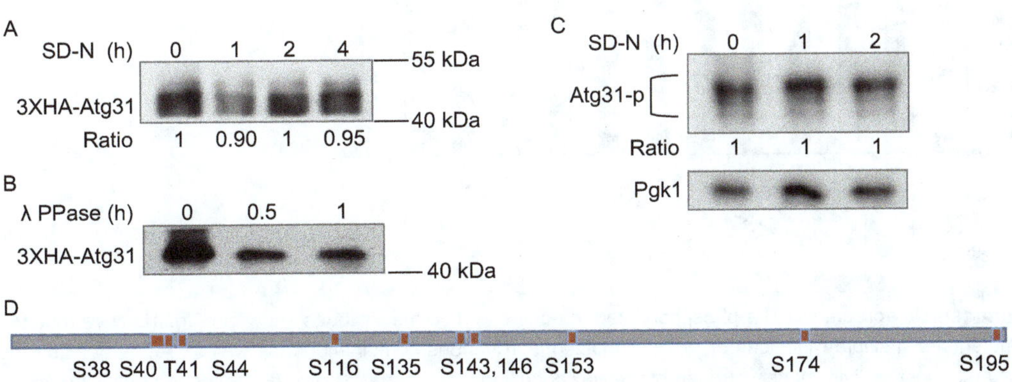

Figure 1. Identification and analysis of phosphorylation sites on Atg31. (A) Yeast cells expressing 3XHA-Atg31 were transferred to SD-N medium for 1 h, 2 h or 4 h. Cell lysates were assessed by Western blotting with HA antibody. the grey value ratio of sample 1 h, 2 h, 4 h compared to 0 h is shown. (B) Cell lysates under full (SD) medium were treated with lambda PPase at 30°C for 0 h, 0.5 h and 1 h, then samples were assessed by Western blotting with HA antibody. (C) Yeast cells expressing 3XHA-Atg31 were transferred to SD-N medium for 2 h. Cells lysates were assessed by Phos-tag Western blotting to detect phosphorylation. The ratio is the same as (A). (D) Atg31 protein was isolated by GST-tag purification from yeast grown in full or SD-N medium and analyzed by mass spectrometry. Atg31 phosphorylation sites are shown as red rectangles.

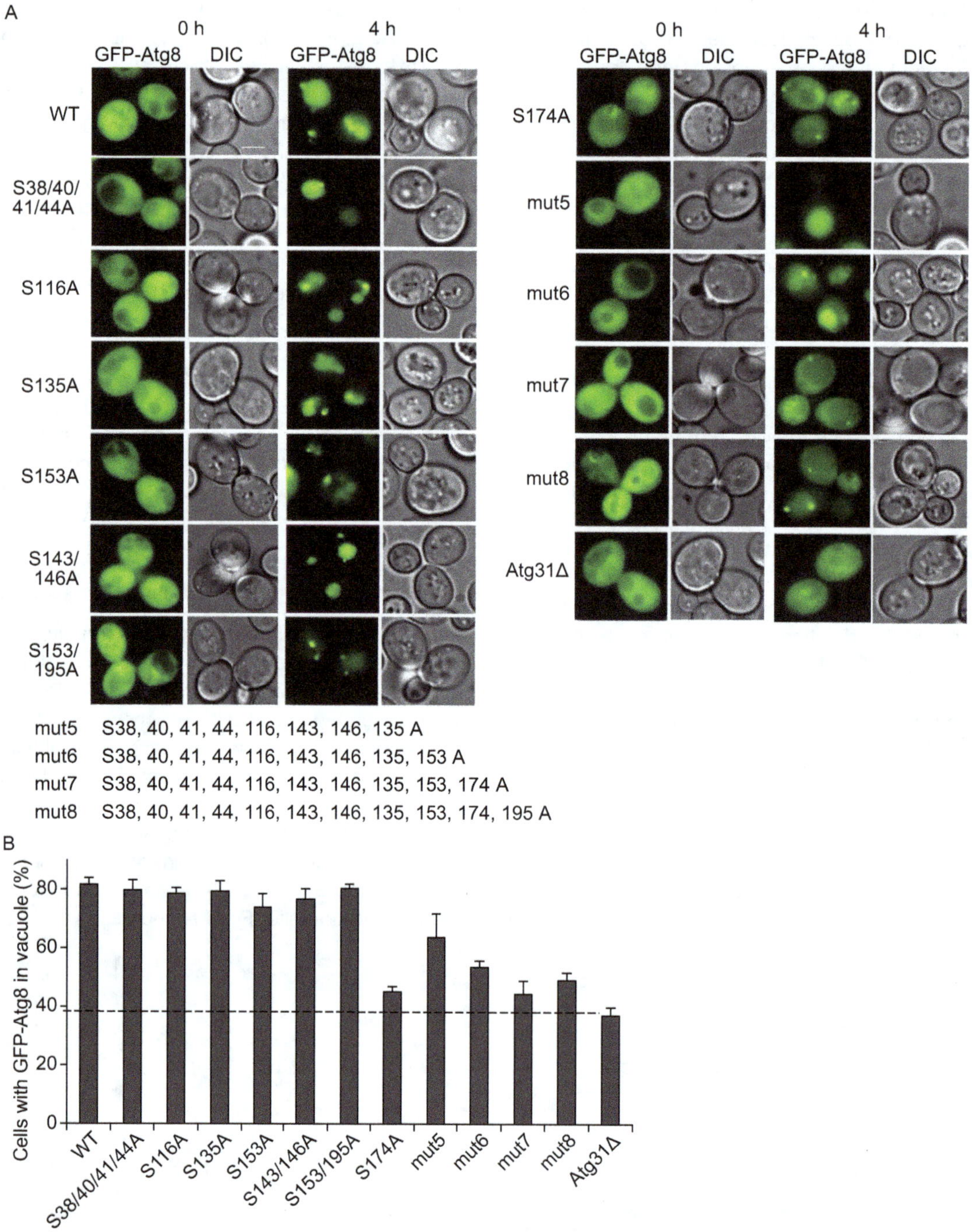

Figure 2. Mutagenisis screen. (A) The phosphorylated threonine and serine residues shown in Fig. 1D were mutated to alanine, and Atg31 mutants with single mutations or multiple mutations in various combinations, as well as Atg31 deletion mutants, were assessed for autophagy activity under nitrogen starvation conditions. Scale bar, 2 μm. (B) Autophagy activity was assessed by monitoring the translocation of GFP-Atg8 into vacuoles. 100 cells were assessed in a blind fashion and quantified. Error bars indicate standard deviation (s.d.) (n = 3).

Phosphorylation at S174 is required for autophagy induced by various cues

To confirm the role of S174 in nitrogen starvation-induced autophagy, we compared the autophagy activity in wild-type cells and cells carrying the Atg31 S174A mutant using various stimuli including nitrogen starvation, amino acid starvation and rapamycin treatment. Autophagy activity was monitored microscopically by translocation of GFP-Atg8 into vacuoles (Fig. 3A), and biochemically by cleavage of GFP-Atg8 as detected by Western blotting (Fig. 3C). Both assays

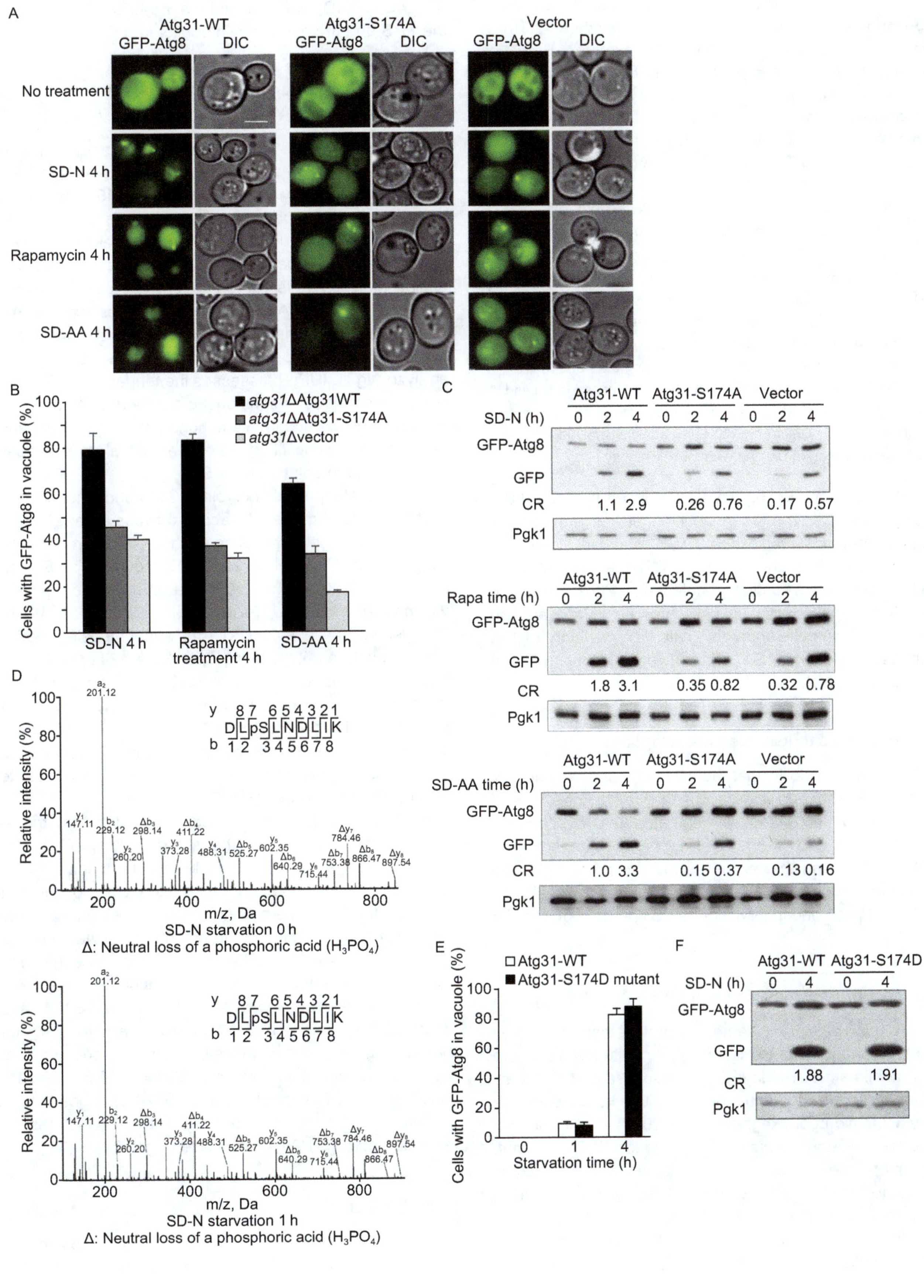
A
Atg31-WT
Atg31-S174A
Vector
GFP-Atg8
DIC
No treatment
SD-N 4 h
Rapamycin 4 h
SD-AA 4 h
B
Cells with GFP-Atg8 in vacuole (%)
atg31ΔAtg31WT
atg31ΔAtg31-S174A
atg31Δvector
SD-N 4 h
Rapamycin treatment 4 h
SD-AA 4 h
C
SD-N (h)
Rapa time (h)
SD-AA time (h)
GFP-Atg8
GFP
CR
Pgk1
1.1 2.9 0.26 0.76 0.17 0.57
1.8 3.1 0.35 0.82 0.32 0.78
1.0 3.3 0.15 0.37 0.13 0.16
D
Relative intensity (%)
D L pS L N D L I K
m/z, Da
SD-N starvation 0 h
Δ: Neutral loss of a phosphoric acid (H_3PO_4)
SD-N starvation 1 h
Δ: Neutral loss of a phosphoric acid (H_3PO_4)
E
Atg31-WT
Atg31-S174D mutant
Cells with GFP-Atg8 in vacuole (%)
Starvation time (h)
F
Atg31-WT Atg31-S174D
SD-N (h)
GFP-Atg8
GFP
CR
1.88 1.91
Pgk1

◀ **Figure 3. Ser174 phosphorylation is required for autophagy under various conditions.** (A) Cells of the Atg31Δ strain expressing wild-type (WT) Atg31, the Atg31-Ser174 mutant (S174A), or a control vector were treated by nitrogen starvation (SD-N), rapamycin or amino acid starvation (SD-AA) and assessed for autophagy activity. Scale bar, 2 μm. (B) Cells from (A) were assessed for autophagy activity by monitoring the translocation of GFP-Atg8 into vacuoles. 100 cells were assessed in a blind fashion and quantified. Error bars indicate standard deviation (s.d.) (n = 3). (C) Cells from (A) were analyzed for GFP-Atg8 cleavage by Western blotting with GFP antibody. CR means the cleavage ratio between GFP-Atg8 and GFP, using grey value statistic analysis of three independent experiments. (D) Mass spectrometry analysis of Ser174 phosphorylation on Atg31 from cells grown in full medium (0 h) or SD-N medium (1 h). (E) The phosphomimic mutant S174D was assessed for autophagy activity by monitoring the translocation of GFP-Atg8 into vacuoles. 100 cells were assessed in a blind fashion and quantified. Error bars indicate standard deviation (s.d.) (n = 3). (F) Cells from (E) were analyzed for GFP-Atg8 cleavage by Western blotting with GFP antibody.

showed that autophagy activity was reduced by about 60% (Fig. 3B and 3C). It is worth noting that this reduction is similar to that observed when ATG31 deleted. Thus, we concluded that phosphorylation at S174 is essential for Atg31 to carry out its function in autophagy. Phosphorylation of S174 can be detected in cells growing in nutrient-rich conditions and in cells undergoing starvation, which implies that phosphorylation of S174 is not regulated by nitrogen starvation (Fig. 3D).

A phosphomimic mutant rescues autophagy

To rule out the possibility that reduction of autophagy activity is due to the serine-to-alanine change, rather than loss of phosphorylation at S174, we generated a phosphomimic mutant strain in which serine 174 is replaced by aspartic acid (S174D). We found that expression of the S174D plasmid can rescue autophagy in the Atg31 deletion mutant (Fig. 3E and 3F). Thus, phosphorylation at S174 is required for autophagy.

Impairment of Atg9 recycling in the S174A mutant

Atg9 is a multi-pass transmembrane protein that plays a key role in autophagosome formation. Atg9-positive vesicles are highly mobile structures in the cytoplasm (Yamamoto et al., 2012) that recycle Atg9 and other molecular from the PAS to the cytoplasmic pool (Reggiori et al., 2004). We found that the number of Atg9 puncta is reduced in the S174A mutant. Furthermore, the fluorescence intensity of the Atg9 puncta is dramatically enhanced, indicating that the dynamic recycling of Atg9 between different pools is impaired and Atg9 accumulates in these puncta (Fig. 4A and 4B). Since Atg9 puncta in S174A mutants are co-localized with the PAS marker Atg8 (Fig. 4C and 4D), we conclude the recycling of Atg9 between the PAS and the cytoplasmic pool is impaired.

S174 phosphorylation affects the interface between Atg31and Atg17

The migration of Atg31 on SDS-PAGE is abnormal and much slower than typical globular proteins. 3XHA-tagged Atg31 shows a molecular weight of about 40–45 kDa by SDS-PAGE, and the dephosphorylated form is about 40 kDa. However, the molecular weight of Atg31 calculated from its amino acid sequence is 22 kDa. Since many proteins containing intrinsically disordered regions have similar abnormal migration on SDS-PAGE gels, we hypothesized that Atg31 may have characteristics of an intrinsically disordered protein (IDP). We used a disorder prediction tool, IUPred, to analyze Atg31. IUPred assesses the tendency of a protein to contain disordered regions based on whether the amino acid sequence allows stable interactions (Dosztanyi et al., 2005). The prediction shows that almost half of the Atg31 sequence is disordered in solution (Fig. 5A).

To elucidate the possible role of S174 phosphorylation, we built a structural model of *Saccharomyces cerevisiae* Atg31 using a homologous modeling method and a threading modeling method (Pronk et al., 2013). As shown in Fig. 5B, the *S. cerevisiae* Atg31 has a very similar structure to *Lachancea thermotolerans* Atg31 (Protein Data Bank code 4PHQ:B) (Ragusa Michael et al., 2012). After superimposing the Atg31 model structure onto the Atg17-Atg31-Atg29 complex structure (4HPQ), it is very clear that the S174 phosphorylation site is located at the interface between Atg17 and Atg31. No other phosphorylation site locates to the interface. Phosphorylation of S174 increases the number of atoms in the side chain and enlarges its volume, which will change the interaction of Atg31 with its binding partners (Fig. 5C).

Furthermore, from this model, we found that the secondary structure (SS) of the six residues 167–172 was changed from a loop to a helix by phosphorylation of S174 (Fig. 5D). The increased number of atoms in phospho-serine changes the local interactions between residue side chains and makes the flexible loop transform into a stable helix, as shown in Fig. 4D. This will also change the local motions of the C-terminal helix. S174 phosphorylation also changes the interaction between Atg31 and Atg17. Near S174, amino acids 171 (arginine) and 175(leucine) are the two most buried residues in the interface between non-phosphorylated Atg31 and Atg17. Their buried areas are 150.5 Å and 113.3 Å respectively. Phosphorylation of S174 enlarges their buried areas by 12.1 Å and 8.5 Å respectively, which results in an enlarged Atg31/Atg17 interface (Fig. 5E).

DISCUSSION

In this study, we identified 11 phosphorylation sites on Atg31. Mutagenesis analysis showed that phosphorylation at serine

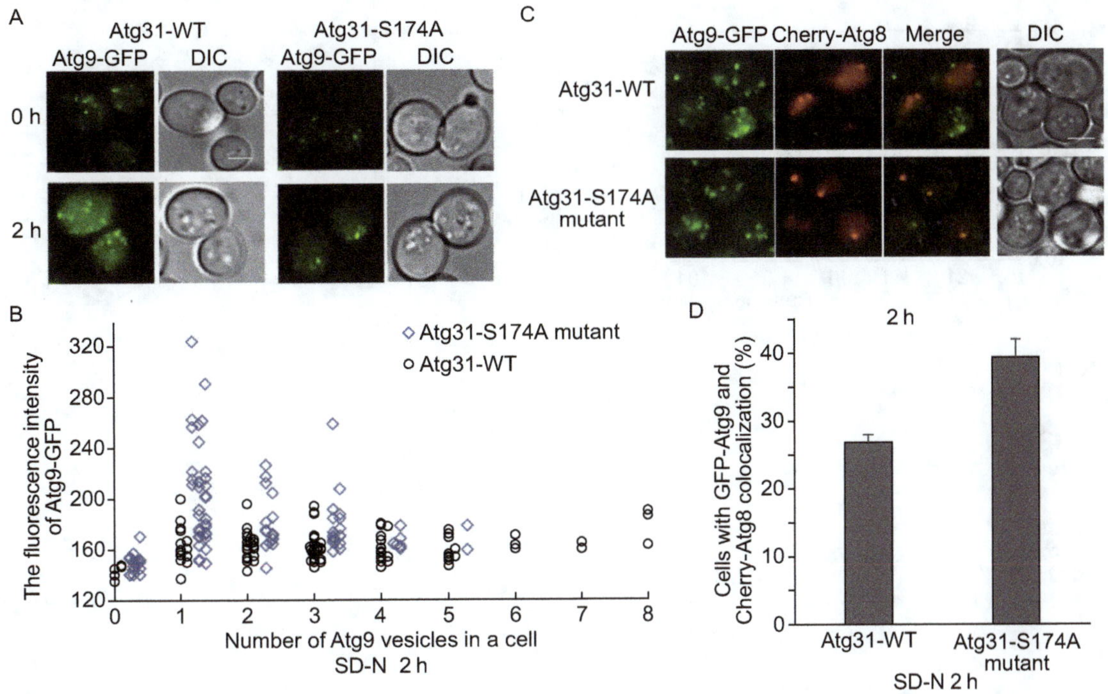

Figure 4. Impairment of Atg9 recycling in the S174A mutant. (A) Wild-type or S174A mutant cells expressing Atg9-GFP were transferred to nitrogen starvation for 0 h or 2 h and observed by confocal microscopy. Scale bar, 2 μm. (B) Cells from (A) were assessed for the number and intensity of Atg9 puncta. (C) Wild-type or S174 mutant cells expressing Atg9-GFP and Cherry red-Atg8 were transferred to SD-N medium for 2 h and observed by confocal microscopy. Scale bar, 2 μm. (D) Cells from (C) were assessed for co-localization between Atg9 and Atg8. 100 cells were assessed in a blind fashion and quantified. Error bars indicate standard deviation (s.d.) (n = 3).

174 is required for Atg31 to carry out its function, while the other phosphorylation sites have no function in the regulation of autophagy. We further demonstrated that phosphorylation at S174 is required for efficient Atg9 recycling, and the impairment of this phosphorylation in the S174A mutant causes accumulation of Atg9 puncta and impaired autophagy.

One obvious question remains unsolved: what is the kinase responsible for phosphorylation of Atg31? So far, our efforts to identify the kinase have been hampered by the lack of a specific antibody against phospho-S174. We have failed to generate such an antibody despite repeated attempts.

Atg31 has intrinsically disordered regions which means that it can easily be phosphorylated (Tompa, 2002; Iakoucheva et al., 2004). Our structure simulations illustrate how phosphorylation at S174 changes the C-terminal loop into a helix. This helix makes additional contacts with the crescent-shaped Atg17, thus enlarging the interface between Atg31 and Atg17. In the S174A mutant, the lack of phosphorylation at S174 causes the C-terminal helix to become a flexible loop. This may result in part of the Atg17 protein becoming exposed, thus leading to abnormal PAS assembly, which eventially causes impaired Atg9 recycling.

MATERIALS AND METHODS

Strains and plasmids

Standard protocols were used for yeast manipulations (Kaiser, 1994). Cells were cultured at 30°C in SD medium (0.17% yeast nitrogen base without amino acids and ammonium sulfate, 0.5% ammonium sulfate, 0.5% casamino acids and 2% glucose) supplemented with appropriate nutrients. Autophagy was induced by transferring the cells to SD-N medium (0.17% yeast nitrogen base, without amino acids and ammonium sulfate, and 2% glucose). Otherwise, to induce autophagy, cells were treated with 0.2 and 0.5 μg/mL rapamycin (Sigma-Aldrich), or transferred to SD-AA medium respectively.

Yeast strains and media

BY4741 (*MATa his3D leu2D met15Dura3D*), ScLY4 (*BY4741 atg31::kanMX*), ScLY5 (*BY4741, ATG31-HA::HIS3*) and ScLY6 (*ScLY4, YEPlac181* [*Gal1-GST-ATG31*]) were used in this study. The other Atg disruptants are listed in Table S1. BY4741 was purchased from Invitrogen. Media and methods for gene disruption have been described previously (Longtine et al., 1998).

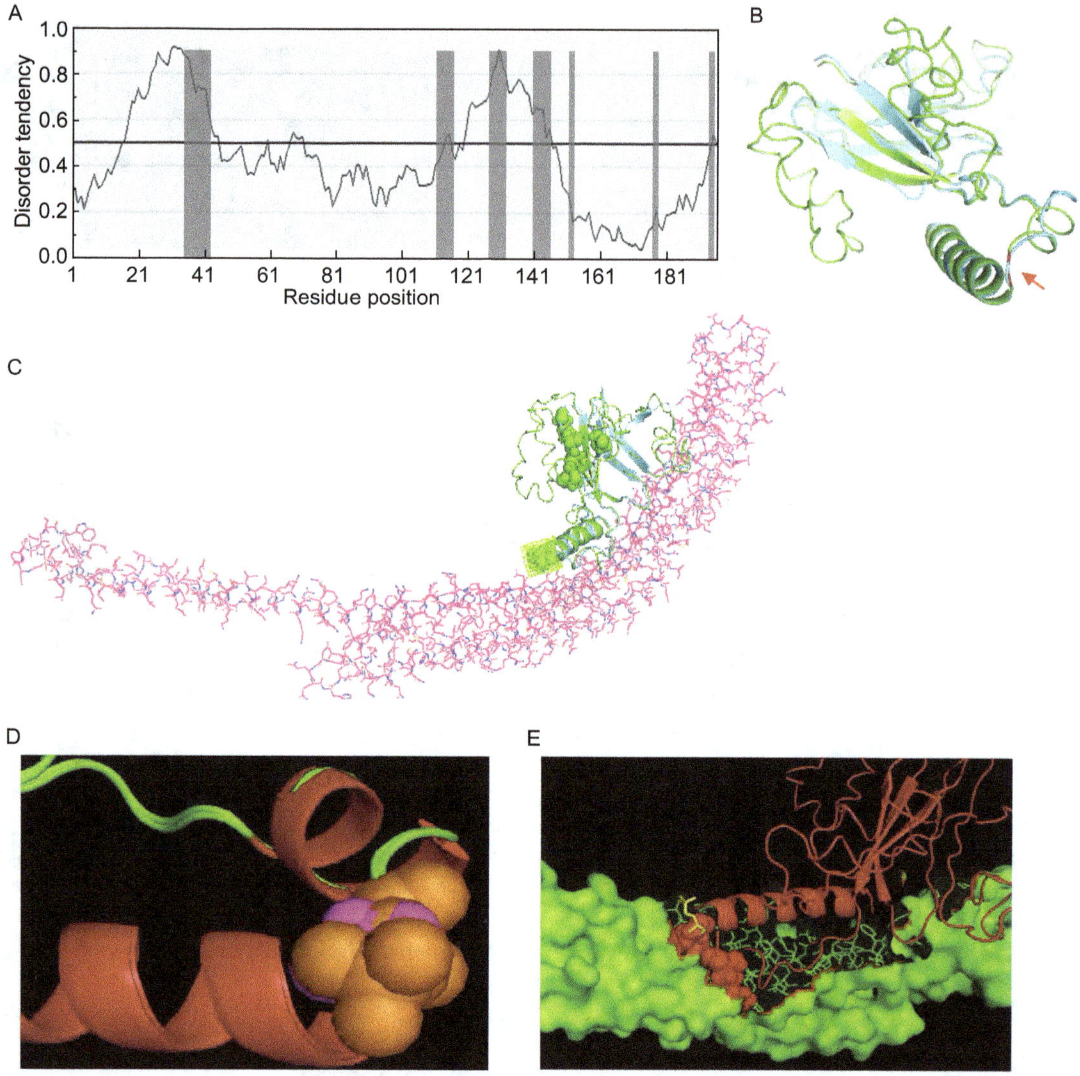

Figure 5. Structure simulation of Atg31. (A) Tendency for intrinsic disorder within the amino acid sequence of Atg31, as predicted by IUPed. The phosphorylation sites are marked by grey columns. (B) Structural modelling of Atg31. A model of Atg31 from *Saccharomyces cerevisiae* was constructed with Modeller 9.11 and superimposed on the crystallographic structure of Atg31 from the *Lachancea thermotolerans* Atg17-Atg31-Atg29 complex (Protein Data Bank code 4HPQ). The major secondary structure motifs (alpha-helixes and beta-strands) are closely aligned in the two structures. The model of *S. cerevisiae* Atg31 is in green and the X-ray-solved structure of *L. thermotolerans* Atg31 is in blue. The red arrow indicates S174. (C) Superimposition of the built model of *S. cerevisiae* Atg31 (green) on the crystallographic structure of *L. thermotolerans* Atg31 (blue) complexed with Atg17 (pink). The *L. thermotolerans* proteins are from the Atg17-Atg31-Atg29 complex (4HPQ). The phosphorylated serine residues in Atg31 are shown as spheres, and S174 is additionally indicated by a mesh. S174 is located at the interface between Atg31 and Atg17. (D) Superimposition of the local structure around S174 with or without phosphorylation. Non-phosphorylated S174 is shown as pink spheres, and phosphorylated S174 is shown as orange spheres. The loop (green) near to S174 in the unmodified structure is changed to a helix (red) by phosphorylation. (E) The arginine 171 and leucine 175 residues near the phosphorylation site are deeply buried after S174 phosphorylation. Red spheres show the two buried residues, and the yellow stick shows phospho-S174.

Plasmids and other materials

The full length *atg31* gene with its endogenous promoter and terminator was amplified by PCR and ligated into pRS316, Yeplac181 and Ycplac111 plasmids with appropriate restriction endonucleases. The *atg31* ORF region and the downstream 600 bp was amplified by PCR and inserted into YEplac181 plasmid after the GAL4 promoter and the N-GST tag sequence. Site-specific mutagenesis was performed with a simple PCR method. Plasmids containing *atg31* were amplified using primers containing sequences 15 bp upstream and downstream of the mutation site. The products were cut by Dpn I

(New England Biolabs) and transformed into competent *E. coli*. Mutations were confirmed by sequencing.

Phos-tag Western blotting assay

The phos-tag assay was performed as described before with modifications (Kosako et al., 2009). For phosphate-affinity polyacrylamide gel electrophoresis, an 8%–10% separating gel containing 25 mmol/L phos-tag acrylamide (AAL-107, Wako) and 50 mmol/L $MnCl_2$ was prepared with a normal stacking gel. After samples were loaded, the gel was run with a current of 15–20 mA for about 2 h. The gel was washed with transfer buffer containing 1 mmol/L EDTA to move the Mn^{2+}, then proteins were transferred to PVDF membranes. The membranes were blocked, incubated with antibodies and processed according to standard procedures.

Fluorescence microscopy

For fluorescence microscopy, cells were grown to OD_{600} = 0.8 − 1.0 in appropriate selective medium and shifted to SD-N medium for various lengths of time as described (Cheong et al., 2005). The cells were observed at room temperature using FV-1000 (Olympus) confocal microscopes. The percentages of cells with vacuolar GFP-Atg8 fluorescent signals were determined by counting 100 cells in three separate experiments (Yi et al., 2012).

GST-tag protein purification

Plasmid Yeplac181, containing the GAL4 promoter and the ORF of the *atg31* gene, was transformed into BY4741 to purify Atg31 protein. Cells were incubated in up to 2 liters of SD-Leu medium containing 2% raffinose instead of glucose from OD_{600} = 0.2 until OD_{600} = 0.8 − 1.2 with rotation in a 30°C incubator. Galactose was added at a final concentration of 2% to induce GST-Atg31 expression for 2–4 h. One liter of the culture was harvested as non-starved cells. The remaining cells were washed three times with SD-N medium containing 2% raffinose and glactose. The cells were then starved for 1 h in this SD-N medium and then harvested and lysed together with the non-starved cells. GST-Atg31 protein was purified with Glutathione Sepharose TM 4B (GE Healthcare) as previously described (Lu et al., 2011). Cells were vortexed with glass beads to break them open, then centrifuged at high speed. The supernatant was collected and incubated with Glutathione Sepharose for 2 h. The column containing the lysate was washed slowly by wash buffer with a high NaCl concentration, and then eluted by reduced GSH (Sigma).

Disorder prediction

The Atg31 protein sequence was used for protein disorder prediction, which was performed using online disorder prediction software, including IUPed, Pondr-FIT and ANCHOR.

Mass spectrum analysis

For LC–MS/MS analysis, peptides were separated by a 90 min gradient elution at a flow rate of 0.250 µL/min with a Thermo-Dionex Ultimate 3000 HPLC system, which was directly interfaced with a Thermo LTQ-Orbitrap Velos pro mass spectrometer. The analytical column was a homemade fused silica capillary column (75 µm ID, 150 mm length; Upchurch, Oak Harbor, WA) packed with C-18 resin (300 A, 5 µm; Varian, Lexington, MA). Mobile phase A consisted of 0.1% formic acid, and mobile phase B consisted of 80% acetonitrile and 0.1% formic acid. An LTQ-Orbitrap mass spectrometer was operated in the data-dependent acquisition mode using Xcalibur 2.2 software and there was a single full-scan mass spectrum in the Orbitrap (400–1800 m/z, 30,000 resolution) followed by 20 data-dependent MS/MS scans in an ion trap at 35% normalized collision energy (CID).

MS/MS spectra from each LC–MS/MS run were searched against Atg31 in the *Saccharomyces cerevisiae* database using the Proteome Discoverer (Version 1.4) searching algorithm. The search criteria were as follows: full tryptic specificity was required; two missed cleavages were allowed; carbamidomethylation was set as fixed modification; oxidation (M) was set as a variable modification; precursor ion mass tolerance was 10 ppm for all MS acquired in the Orbitrap mass analyzer; and fragment ion mass tolerance was 0.8 Da for all MS2 spectra acquired in the LTQ. A high confidence score filter (FDR < 1%) was used to select the "hit" peptides and their corresponding MS/MS spectra were manually inspected.

Computational modeling of Atg31 structure

The comparative modeling tool Modeller 9.11 was first implemented to build a model based on the homologous *Lachancea thermotolerans* Atg31 structure from the Protein Data Bank (code 4HPQ:B) (Joosten et al., 2011). In order to obtain a good model, the I-TASSER (Roy et al., 2010) server was also used, which takes advantage of multiple-threading alignments and iterative template fragment assembly simulations. The best model from Modeller and the best model from I-TASSER were picked out and minimized using the molecular dynamics simulation package Gromacs 4.5 (Gong et al., 2010) using OPLS force field[3]. Then the model with the better energy score was selected as the built model. We used Pymol[4] to superimpose the built Atg31 model structure onto the structure of the Atg17-Atg31-Atg29 complex (Protein Data Bank code 4HPQ) (DeLano, 2002).

ACKNOWLEDGMENTS

This research was supported by the National Natural Science Foundation of China (Grant Nos. 31030043, 30971484 and 31321003), the National Basic Research Program (973 Program) (Nos. 2010CB833704 and 2011CB910100), and Tsinghua University Grants 2010THZ0 and 2009THZ03071 to Li Yu.

COMPLIANCE WITH ETHICS GUIDELINES

Wenzhi Feng, Tong Wu, Xiaoyu Dan, Yuling Chen, Lin Li, She Chen, Di Miao, Haiteng Deng, Xinqi Gong and Li Yu declare that they have no conflict of interest. This article does not contain any studies with human or animal subjects performed by the any of the authors.

ABBREVIATIONS

Atg, autophagy-related (Klionsky et al., 2003); IDP, intrinsically disordered protein; MS, mass spectrometry; PAS, pre-autophagosomal structure; SS, secondary structure.

REFERENCES

Cheong H, Yorimitsu T, Reggiori F, Legakis JE, Wang CW, Klionsky DJ (2005) Atg17 regulates the magnitude of the autophagic response. Mol Biol Cell 16:3438–3453

DeLano WLT (2002) The PyMOL molecular graphics system. CCP4 Newslett Protein Crystallogr 40:82–92

Dosztanyi Z, Csizmok V, Tompa P, Simon I (2005) IUPred: web server for the prediction of intrinsically unstructured regions of proteins based on estimated energy content. Bioinformatics 21:3433–3434

Gong X, Wang P, Yang F, Chang S, Liu B, He H, Cao L, Xu X, Li C, Chen W, et al. (2010) Protein-protein docking with binding site patch prediction and network-based terms enhanced combinatorial scoring. Proteins 78:3150–3155

Harding TM, Morano KA, Scott SV, Klionsky DJ (1995) Isolation and characterization of yeast mutants in the cytoplasm to vacuole protein targeting pathway. J Cell Biol 131:591–602

Hussey S, Travassos LH, Jones NL (2009) Autophagy as an emerging dimension to adaptive and innate immunity. Semin Immunol 21:233–241

Iakoucheva LM, Radivojac P, Brown CJ, O'Connor TR, Sikes JG, Obradovic Z, Dunker AK (2004) The importance of intrinsic disorder for protein phosphorylation. Nucleic Acids Res 32:1037–1049

Jiang P, Mizushima N (2014) Autophagy and human diseases. Cell Res 24:69–79

Joosten RP, te Beek TAH, Krieger E et al (2011) A series of PDB related databases for everyday needs. Nucleic Acids Res 39: D411–D419

Kabeya Y, Kawamata T, Suzuki K, Ohsumi Y (2007) Cis1/Atg31 is required for autophagosome formation in Saccharomyces cerevisiae. Biochem Biophys Res Commun 356:405–410

Kabeya Y, Noda NN, Fujioka Y, Suzuki K, Inagaki F, Ohsumi Y (2009) Characterization of the Atg17–Atg29–Atg31 complex specifically required for starvation-induced autophagy in Saccharomyces cerevisiae. Biochem Biophys Research Commun 389:612–615

Kaiser C, Michaelis S, Mitchell A (1994) Methods in yeast genetics: a Cold Spring Harbor Laboratory course manual. Cold Spring Harbor Laboratory Press, Cold Spring Harbor, NY

Kinoshita E, Takahashi M, Takeda H, Shiro M, Koike T (2004) Recognition of phosphate monoester dianion by an alkoxide-bridged dinuclear zinc(ii) complex. Dalton Trans 8:1189–1193

Klionsky DJ, Cregg JM, Dunn WA Jr, Emr SD, Sakai Y, Sandoval IV, Sibirny A, Subramani S, Thumm M, Veenhuis M, Ohsumi Y (2003) A unified nomenclature for autophagy-related genes. Dev Cell 5(4):539–545

Kosako H, Yamaguchi N, Aranami C, Ushiyama M, Kose S, Imamoto N, Taniguchi H, Nishida E, Hattori, S (2009) Phosphoproteomics reveals new ERK MAP kinase targets and links ERK to nucleoporin-mediated nuclear transport. Nat Struct Mol Biol 16:1026–1035

Longtine MS, McKenzie A III, Demarini DJ, Shah NG, Wach A, Peter Philippsen AB, Pringle JR (1998) Additional modules for versatile and economical PCR-based gene deletion and modification in Saccharomyces cerevisiae. Yeast 14(8):953–961

Lu J-Y, Lin Y-Y, Sheu J-C, Wu J-T, Lee F-J, Chen Y, Lin M-I, Chiang F-T, Tai T-Y, Berger SL et al (2011) Acetylation of yeast AMPK controls intrinsic aging independently of caloric restriction. Cell 146:969–979

Martin DDO, Ladha S, Ehrnhoefer DE, Hayden MR (2014) Autophagy in huntington disease and huntingtin in autophagy. Trends Neurosci 38(1):26–35

Mizushima N, Komatsu M (2011) Autophagy: renovation of cells and tissues. Cell 147:728–741

Pronk Sander, Pail Szilard, Schulz Roland et al (2013) GROMACS 4.5: a high-throughput and highly parallel open source molecular simulation toolkit. Bioinformatics 29(7):845

Ragusa Michael J, Stanley Robin E, Hurley James H (2012) Architecture of the Atg17 complex as a scaffold for autophagosome biogenesis. Cell 151:1501–1512

Reggiori F, Tucker KA, Stromhaug PE, Klionsky DJ (2004) The Atg1–Atg13 complex regulates Atg9 and Atg23 retrieval transport from the pre-autophagosomal structure. Dev Cell 6:79–90

Roy A, Kucukural A, Zhang Y (2010) I-TASSER: a unified platform for automated protein structure and function prediction. Nat Protoc 5:725–738

Suzuki K, Ohsumi Y (2010) Current knowledge of the pre-autophagosomal structure (PAS). FEBS Lett 584:1280–1286

Thumm M, Egner R, Koch B, Schlumpberger M, Straub M, Veenhuis M, Wolf DH (1994) Isolation of autophagocytosis mutants of Saccharomyces cerevisiae. FEBS Lett 349:275–280

Tompa P (2002) Intrinsically unstructured proteins. Trends Biochem Sci 27:527–533

Tsukada M, Ohsumi Y (1993) Isolation and characterization of autophagy-defective mutants of Saccharomyces cerevisiae. FEBS Lett 333:169–174

Winslow AR, Rubinsztein DC (2008) Autophagy in neurodegeneration and development. Biochim Biophys Acta 1782:723–729

Yamamoto H, Kakuta S, Watanabe TM, Kitamura A, Sekito T, Kondo-Kakuta C, Ichikawa R, Kinjo M, Ohsumi Y (2012) Atg9 vesicles are an important membrane source during early steps of autophagosome formation. J Cell Biol 198:219–233

Yi C, Ma M, Ran L, Zheng J, Tong J, Zhu J, Ma C, Sun Y, Zhang S, Feng W et al (2012) Function and molecular mechanism of acetylation in autophagy regulation. Science 336:474–477

19

Deacetylation of TFEB promotes fibrillar Aβ degradation by upregulating lysosomal biogenesis in microglia

Jintao Bao[1], Liangjun Zheng[1], Qi Zhang[1], Xinya Li[1], Xuefei Zhang[1], Zeyang Li[1], Xue Bai[1], Zhong Zhang[1], Wei Huo[1], Xuyang Zhao[2], Shujiang Shang[3], Qingsong Wang[1✉], Chen Zhang[3✉], Jianguo Ji[1✉]

[1] State Key Laboratory of Protein and Plant Gene Research, College of Life Sciences, Peking University, Beijing 100871, China
[2] Institute of Systems Biomedicine, Department of Pathology, School of Basic Medical Sciences, Center for Age-Related Diseases, Peking University Health Science Center, Beijing 100191, China
[3] State Key Laboratory of Biomembrane and Membrane Biotechnology, Peking University, Beijing 100871, China
✉ Correspondence: wangqingsong@pku.edu.cn (Q. Wang), ch.zhang@pku.edu.cn (C. Zhang), jijg@pku.edu.cn (J. Ji)

ABSTRACT

Microglia play a pivotal role in clearance of Aβ by degrading them in lysosomes, countering amyloid plaque pathogenesis in Alzheimer's disease (AD). Recent evidence suggests that lysosomal dysfunction leads to insufficient elimination of toxic protein aggregates. We tested whether enhancing lysosomal function with transcription factor EB (TFEB), an essential regulator modulating lysosomal pathways, would promote Aβ clearance in microglia. Here we show that microglial expression of TFEB facilitates fibrillar Aβ (fAβ) degradation and reduces deposited amyloid plaques, which are further enhanced by deacetylation of TFEB. Using mass spectrometry analysis, we firstly confirmed acetylation as a previously unreported modification of TFEB and found that SIRT1 directly interacted with and deacetylated TFEB at lysine residue 116. Subsequently, SIRT1 overexpression enhanced lysosomal function and fAβ degradation by upregulating transcriptional levels of TFEB downstream targets, which could be inhibited when TFEB was knocked down. Furthermore, overexpression of deacetylated TFEB at K116R mutant in microglia accelerated intracellular fAβ degradation by stimulating lysosomal biogenesis and greatly reduced the deposited amyloid plaques in the brain slices of APP/PS1 transgenic mice. Our findings reveal that deacetylation of TFEB could regulate lysosomal biogenesis and fAβ degradation, making microglial activation of TFEB a possible strategy for attenuating amyloid plaque deposition in AD.

KEYWORDS Alzheimer's disease, microglia, lysosomes, TFEB, SIRT1, deacetylation

INTRODUCTION

The most common form of Alzheimer's disease (AD) appears sporadically during the aging process and is characterized by deposition of β-amyloid (Aβ) peptides within the brain. An imbalance between Aβ clearance and production results in its accumulation as amyloid plaques that usually lead to perturbations of neural network (Palop and Mucke, 2010; Querfurth and LaFerla, 2010). Recent studies indicated that the inefficient Aβ removal might be the significant underlying mechanism in sporadic AD (Mawuenyega et al., 2010). Therefore, how to stimulate Aβ degradation within the brain needs to be thoroughly investigated.

Microglia, the resident macrophages in the central nervous system (CNS), play a key role in surveying the brain for abnormalities and are quickly activated in response to certain stimuli such as cellular debris or misfolded proteins (Lucin and Wyss-Coray, 2009; Nimmerjahn et al., 2005). Previous studies indicated that microglia were recruited to and activated by amyloid plaque mainly containing fibrillar Aβ, which is the clinical hallmark of AD in advanced stage

(Meyer-Luehmann et al., 2008; Rogers et al., 2002). Moreover, inhibition of microglia accumulation accelerates the burden of plaques deposition within the brain (El Khoury et al., 2007). Several lines of evidence showed that microglia take up and degrade both soluble Aβ (sAβ) and fibrillar Aβ (fAβ), which contributes to the clearance of Aβ in addition to extracellular proteolysis pathway (Lucin and Wyss-Coray, 2009). Recent studies revealed that lipidated apoliprotein E (ApoE) contributes to microglial degradation of sAβ through neprilysin (NEP), a process that is regulated by PPARγ/LXRs signalling pathway (Jiang et al., 2008; Mandrekar-Colucci et al., 2012). Despite these findings, little is known about the mechanisms underlying fAβ degradation within microglia.

Lysosomes are regarded as the critical cellular digestive compartments that are responsible for degradation and recycling of a variety of intracellular and extracellular metabolites, depending on their various soluble acidic hydrolases. Piles of work on this cellular organelle demonstrated that numerous lysosomal storage disorders were triggered by the inheritated deficits of specific lysosomal function (Settembre et al., 2013b). Apart from these early-onset genetic diseases, it is notable that lysosomal dysfunction always leads to various age-related neurodegenerative diseases such as Alzheimer's (AD), Parkinson's (PD) and Huntington's diseases (HD) (Schultz et al., 2011). Therefore, upregulation of lysosomal biogenesis, especially in microglia, may exert beneficial effects on attenuating Aβ pathogenesis in AD.

Transcription factor EB (TFEB) acts as a master regulator of lysosomal function by inducing over 500 target genes related with lysosomal biogenesis and autophagy, which is induced by its translocation from cytoplasm to nucleus under certain cellular stresses (Sardiello et al., 2009; Settembre et al., 2011). Further studies revealed that activation of TFEB meets the cellular needs to protect against the toxicities of the abnormal aggregates. Researchers discovered that the neural toxicity from α-synuclein or htt-aggregates could be markedly ameliorated when TFEB is overexpressed or pharmacologically activated in PD or HD animal models, respectively (Decressac et al., 2013; Tsunemi et al., 2012). Recently, TFEB is suggested to be participated in the Aβ-induced pathogenesis of AD by regulating lysosomal pathway (Zhang and Zhao, 2015). Further evidence indicated that TFEB enhances lysosomal biogenesis in both astrocytes and neuron, contributing to increased sAβ clearance and reduced Aβ generation, respectively (Xiao et al., 2014; Xiao et al., 2015). Despite these influential observations, whether TFEB would function in microglia and the mechanisms underlying TFEB functional pathway, especially its upstream regulation within nucleus, remain to be elucidated.

Accumulating evidence suggests that protein acetylation plays a role in regulating lysosome-mediated autophagy process (Banreti et al., 2013). SIRT1 is the most conserved member of sirtuins, a family of NAD^+-dependent protein deacetylases, and contributes to deacetylation of multiple mammalian transcription factors such as p53, FOXOs, and NF-κB in nucleus, making it a key metabolic regulator involved in a vast number of cellular processes associated with age-related diseases including AD (Guarente, 2011; Herskovits and Guarente, 2014). Previous work indicated that SIRT1 protects against Aβ toxicity and cognitive deficits in animal models mainly through inhibiting inflammatory responses in microglia, because overexpression of SIRT1 in microglia suppresses NF-κB-signaling while the reduced level of SIRT1 during the aging process of microglia, upregulates IL-1β expression (Chen et al., 2005; Cho et al., 2015). Apart from the inflammatory perspective, the molecular mechanisms underlying the protective effects of SIRT1 in microglia are still being investigated.

Considering the critical role of TFEB in lysosomal biogenesis, we intended to explore whether TFEB would increase microglial clearance of Aβ and unveil the mechanisms underlying the upstream regulation of TFEB in lysosomal biogenesis in microglia.

In this study, we firstly identify protein acetylation as a novel post-translational modification of TFEB and demonstrate that TFEB is deacetylated by SIRT1 at K116. Moreover, SIRT1-mediated deacetylation of TFEB facilitates microglial degradation of fAβ by enhancing lysosomal biogenesis. Therefore, we identify the upstream molecular mechanism by which deacetylation stimulates TFEB induction of lysosomal biogenesis to enhance microglial degradation of fAβ and reduction of amyloid plaques, suggesting a novel strategy for treatment of AD.

RESULTS

TFEB enhances microglial degradation of fibrillar Aβ in lysosomes

To investigate the contribution of microglia in clearance of fibrillar Aβ (fAβ), the major constituent of amyloid plaque deposited in AD brains, we first incubated fAβ that was generated and identified *in vitro* with BV2 cells and primary microglia (Ma et al., 2013). Our results revealed that fAβ was rapidly taken up and trafficked into lysosomes within 30 min (Fig. 1A–C). As time prolonged, the internalized fAβ level increased to the peak level at 3 h and then gradually disappeared at 18 h (Fig. 1B). By conducting this set of preliminary experiment, 3 h and 18 h were interpreted as the time points representing microglial capabilities of fAβ phagocytosis and degradation, respectively. Indeed, the fAβ originally added into the media was immediately and thoroughly internalized by microglia and little did we observe any resecretion in the media (Fig. S1A). Interestingly, we confirmed fAβ is exclusively degraded within lysosomes, for the reason that inhibitors of lysosomes such as chloroquine or leupeptin remarkably weaken microglial degradation of fAβ while phosphoramidon, inhibitor of NEP that is reported for sAβ degradation (Jiang et al., 2008), exerts little impact on this process (Fig. 1D). TFEB, as a critical transcription factor

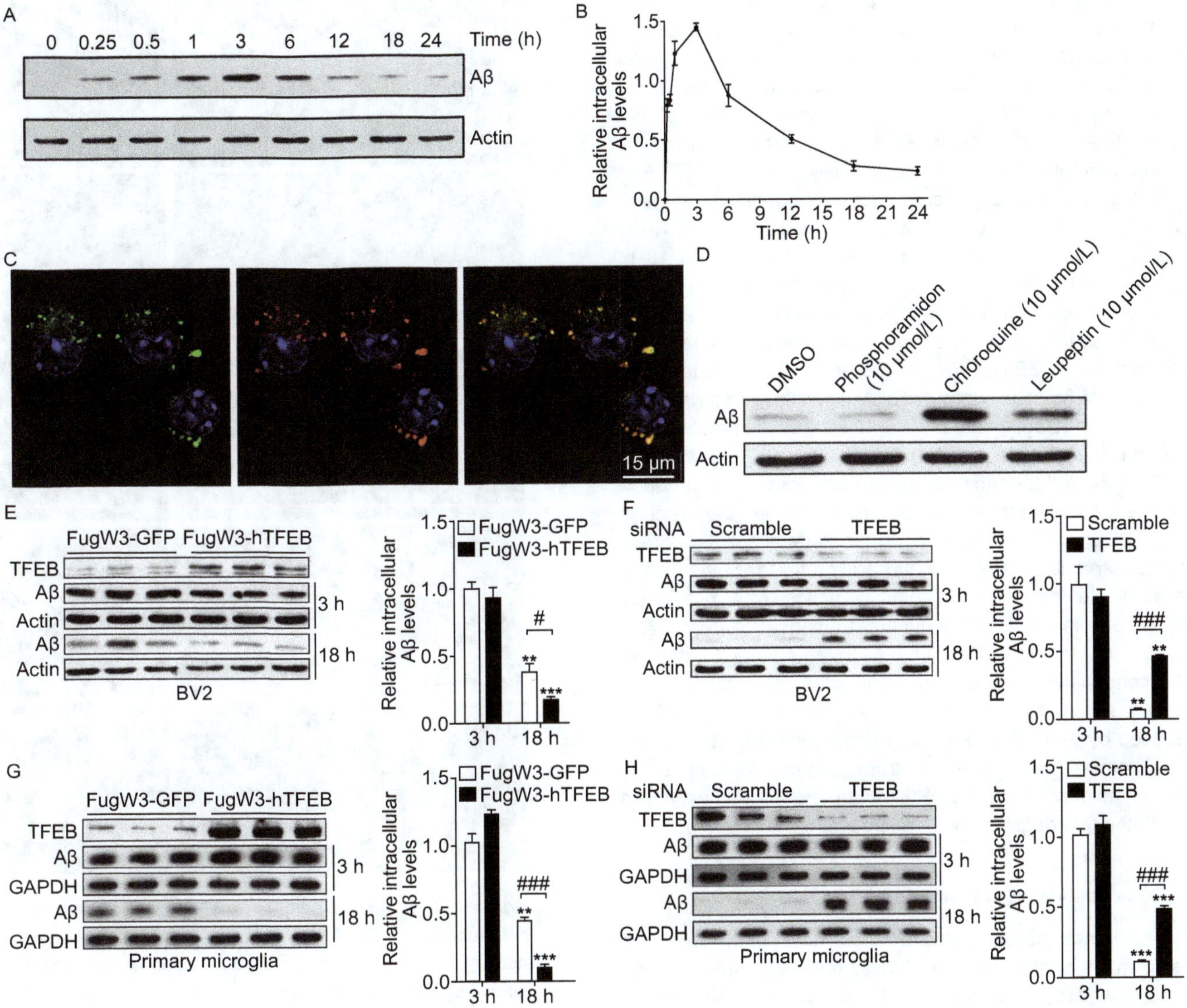

Figure 1. TFEB enhances microglial degradation of fibrillar Aβ in lysosomes. (A and B) Microglia internalize and efficiently degrade fibrillar Aβ. BV2 cells were incubated with fAβ (500 nmol/L) at 37°C and the cells were harvested and lysed at different time points, followed by detection of intracellular Aβ levels by Western blotting analysis (A). The band intensity was measured in three independent experiments indicating relative intracellular Aβ levels and the mean ± SEM are shown in (B). (C) Fibrillar Aβ is rapidly trafficked into lysosomes. Confocal imaging of live BV2 cells 30 min after addition of Hilyte488-labeled fAβ (500 nmol/L) showed localization of fAβ (Green) within lysosomes stained with LysoTracker (Red). Scale bar, 15 μm. (D) Internalized fAβ is degraded in lysosomes. Primary microglia from wild-type mice were pretreated with DMSO, Phosphoramidon (NEP inhibitor, 10 μmol/L), Chloroquine or Leupeptin (Lysosome inhibitor, 10 μmol/L) for 18 h. The cells were then incubated with fAβ (500 nmol/L) in the presence of DMSO or inhibitors for an additional 18 h. The band intensity was measured in three independent experiments indicating relative intracellular Aβ levels. (E and G) TFEB overexpression increases fAβ degradation in microglia. GFP or human TFEB was overexpressed in BV2 cells (E) or primary microglia (G) by lentiviral system. The cells were incubated with fAβ (500 nmol/L) for 3 h or 18 h and were harvested for detection of intracellular Aβ levels by Western blotting analysis. The band intensity was measured in three independent experiments indicating relative intracellular Aβ levels and the mean ± SEM are shown in the right panel. Two-way ANOVA, comparison between different time points, $^{**}P < 0.01$; $^{***}P < 0.001$. Unpaired Student's *t*-test, comparison against the Fugw3-GFP, $\#P < 0.05$; $\#\#\#P < 0.001$. (F and H) TFEB knockdown inhibits fAβ degradation in microglia. Scramble or TFEB siRNA was delivered into BV2 cells (F) or primary microglia (H) for 72 h. The cells were treated as in (E and G). Two-way ANOVA, comparison between different time points, $^{**}P < 0.01$; $^{***}P < 0.001$. Unpaired Student's *t*-test, comparison against the Scramble, $\#\#\#P < 0.001$.

regulating lysosomal biogenesis and lysosomal degradative pathway, is demonstrated to be involved in the pathogenesis of neurodegenerative diseases. Recent studies revealed that TFEB could facilitate oligomeric sAβ clearance by enhancing astrocytic lysosomal biogenesis (Xiao et al., 2014). To examine whether TFEB has an effect on microglial degradation of fAβ, we first exogenously expressed TFEB in BV2 cells and primary microglia by using lentiviral system. We observed less intracellular fAβ remained in the TFEB infected cells than that in the GFP infected cells at 18 h, indicating an enhancement of microglial degradation of fAβ. Meanwhile, microglial phagocytosis remains the same as intracellular fAβ levels at 3 h are comparable between cells infected with TFEB or GFP (Fig. 1E and 1G). Consistent with the gain-of-function data, siRNA specific knockdown of TFEB in microglia greatly reduce their capabilities to degrade fAβ (Fig. 1F and 1H). Intriguingly, we observed that TFEB has a tendency to translocate into nucleus upon stimulation of fAβ which is coincided with previous reports that TFEB will be activated under certain cellular stress (Figs. 2 and S2A). However, we proved that fAβ stimulation failed to inhibit mTORC1 activity which was previously reported to facilitate TFEB nuclear translocation (Fig. S2B), for the reason that fAβ stimulation could not inhibit the phorsphorylation status at specific sites of mTORC1 substrates as compared with the obvious inhibitory effects induced by mTORC1 inhibitor Torin1. Taken together, these data demonstrate that TFEB translocates into nucleus by fAβ stimulation in a mTORC1-independent pathway and facilitates fAβ degradation in microglia.

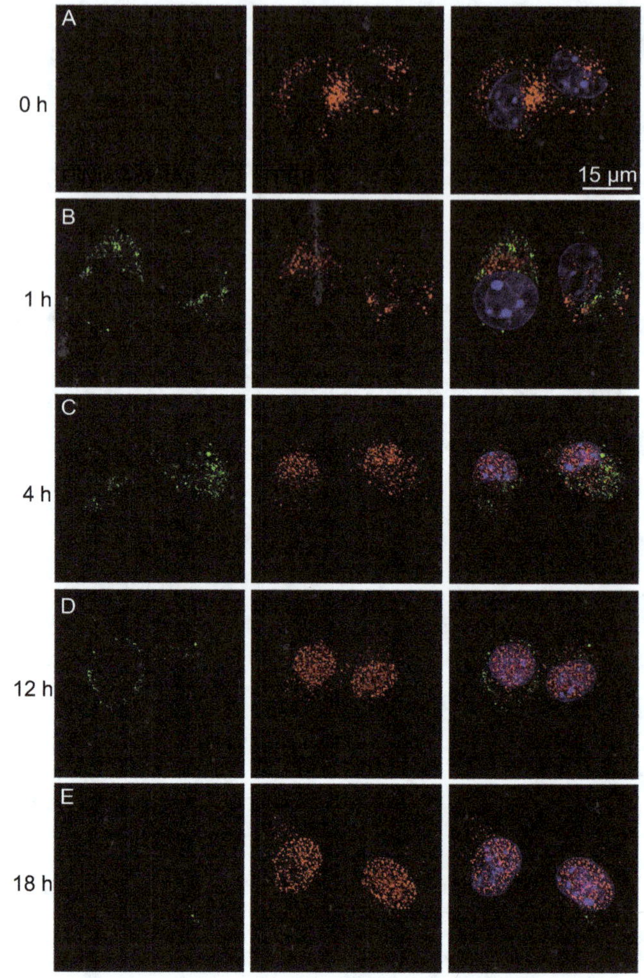

Figure 2. Fibrillar Aβ stimulates TFEB to translocate into nucleus. (A–E) TFEB translocates into nucleus under incubation of fibrillar Aβ. Confocal imaging of BV2 cells incubated with Hilyte488-labeled fAβ (500 nmol/L, Green) at different time points and immunostained for endogenous TFEB (Red). Scale bar, 15 μm.

TFEB is acetylated at lysine 116

A wide range of acetylated proteins have been identified using mass spectrometry, including numerous transcription factors whose activities are proved to be regulated by acetylation. To investigate acetylation status of TFEB, Flag-tagged human TFEB was transiently transfected into human embryonic kidney (HEK) 293T cells and its acetylation level was detected by an anti-acetylated lysine antibody. We observed a presence of acetylation on this exogenously expressed TFEB when the cells were treated with nicotinamide (NAM) that is an inhibitor of sirtuins family as well as trichostatin A (TSA) which inhibits activity of histone deacetylase (HDAC) classes I, II, and IV (Fig. 3A). This result prompted us to explore the potential acetylation sites on TFEB by the means of mass spectrometry analysis of TFEB purified from 293T cells treated with or without NAM and TSA (Fig. 3B). We repeatly identified an acetyl-lysine-containing peptide (FAAHISPAQGSPKPPPAASPGVR) that is matched with a region containing K116 on immunoprecipitated TFEB, in the cells treated with NAM and TSA (Fig. 3C). By performing sequence alignment of TFEB homologues from different species, we found that K116 is highly conserved in evolution (Fig. 3D). To further ascertain whether K116 site is acetylated, we generated single acetylation-deficient TFEB mutant (K-to-R mutation) by site-directed mutagenesis. As we expected, in contrast with wild-type TFEB (TFEB-WT), acetylation level of K116R mutant (TFEB-K116R) can no longer be elevated by NAM treatment (Fig. 3E). These data document that TFEB can be acetylated at K116 site which has not been previously recognized.

TFEB is a deacetylation substrate of SIRT1

Given the evidence that TFEB can be acetylated at K116 site and the knowledge that acetylation status is regarded as a critical mechanism underlying regulation of activities of numerous transcription factors, we intended to identify the upstream regulator that targets TFEB. By performing co-immunoprecipitation (Co-IP) of exogenously expressed TFEB from 293T cells followed by mass spectrometry

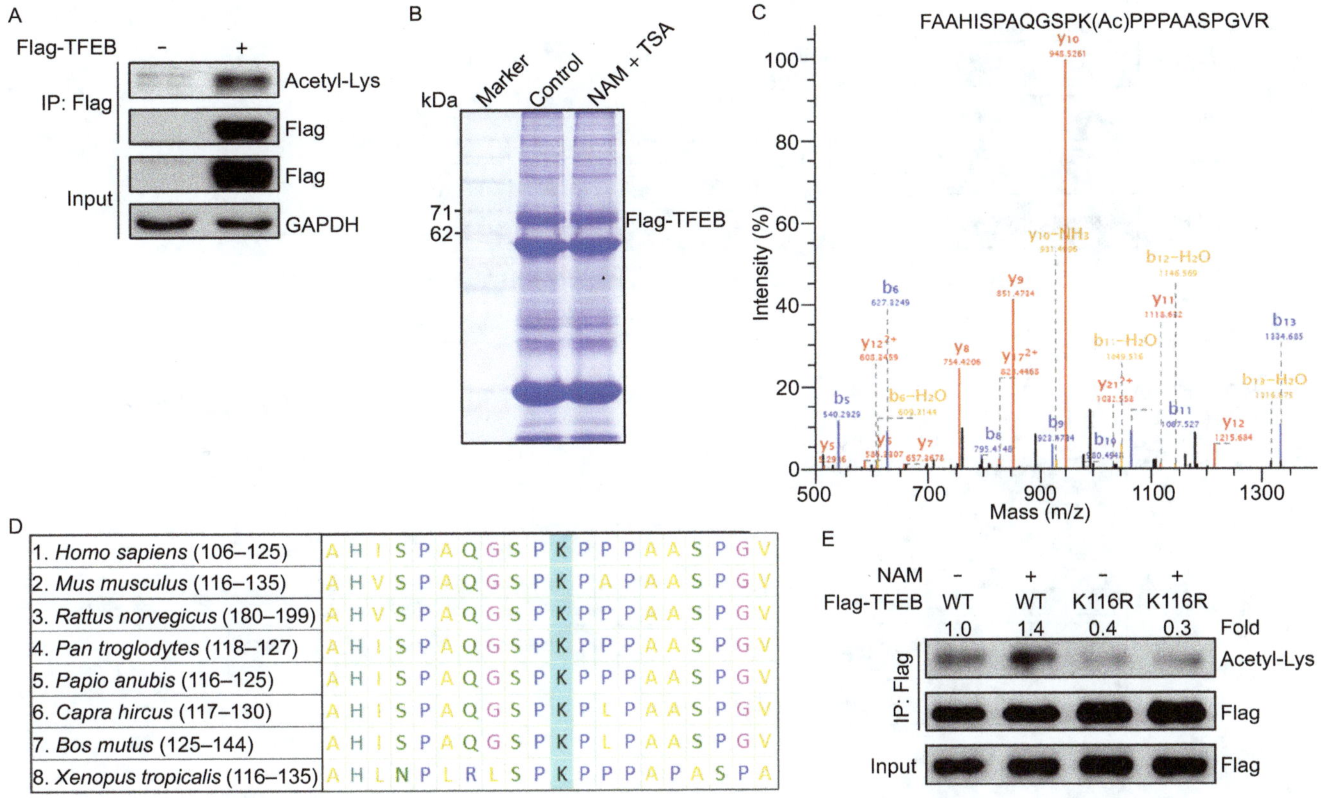

Figure 3. TFEB is acetylated at lysine 116. (A) Exogenous expressed TFEB is acetylated. Empty vector or Flag-tagged TFEB was transfected into HEK293T cells for 24 h treated with sirtuins inhibitor NAM (10 mmol/L) and HDAC I, II, and IV inhibitor TSA (0.5 μmol/L) and was immunoprecipitated for acetylation Western blotting analysis with the antibody against acetylated lysine. (B) Immunoprecipitated Flag-tagged TFEB for mass spectrometry analysis. Flagged-tagged TFEB was transfected into HEK293T cells for 24 h treated with or without NAM (10 mmol/L) and TSA (0.5 μmol/L) and was immunoprecipitated with anti-flag affinity agrose beads, followed by Commassie Blue R250 staining. (C) Identification of TFEB K116 acetylation using mass spectrometry analysis. Flagged-tagged TFEB was purified as depicted in (B) and then analyzed using mass spectrometry. (D) Alignment of the protein sequences of TFEB homologues in different species. The blue mark indicates the identified acetylated lysine residues of TFEB. (E) Deacetylation mimic of TFEB (K116R) counteracts NAM acetylation effects on TFEB. Flag-tagged TFEB (WT or K116R) were transfected into HEK293T cells for 24 h with or without additional 12 h treatment of NAM (10 mmol/L). Values were expressed as fold changes relative to TFEB-WT without NAM treatment and normalized to IP-Flag.

analysis, we detected SIRT1 as a potential candidate. To verify this result, we first overexpressed TFEB or SIRT1 in 293T cells and found endogenous SIRT1 (Fig. S3A) or TFEB (Fig. S3B) in the immunoprecipitates, respectively. Then we performed endogenous Co-IP with SIRT1 antibody and also detected TFEB in the immunoprecipitates (Fig. 4A), indicating that TFEB physically interacts with SIRT1. Next, we wondered whether SIRT1 could change the acetylation levels on TFEB. Treating 293T cells with resveratrol, a chemical activator of SIRT1, resulted in a reduction of acetylation on the immunoprecipitated TFEB, while its acetylation level was augmented by NAM treatment (Fig. S3C). In addition, the acetylation level on Flag-tagged TFEB was much lower in cells coexpressing wild-type SIRT1 (SIRT1-WT) than in cells expressing a catalytically inactive mutant SIRT1 (SIRT1-HY) (Fig. 4B). In parallel with these results, cells deficient of SIRT1 by knocking down with specific siRNA exhibited a higher level of acetylation on exogenously expressed TFEB-WT, but its deacetylation mimic TFEB-K116R counteracted SIRT1 knockdown effects, indicating K116 would be the site deacetylated by SIRT1 (Fig. 4C). To further ascertain whether TFEB is a substrate of SIRT1, we carried out *in vitro* deacetylation assay with purified SIRT1 and acetylated Flag-tagged TFEB (Fig. S3E). The data showed that SIRT1-WT, but not the catalytically inactive mutant SIRT1-HY, deacetylates TFEB in a NAD^+-dependent manner, while treatment with SIRT1 inhibitor NAM reversed the effect of SIRT1 on TFEB (Fig. 4D). Despite these intriguing data, we intended to examine whether other HDAC families are involved in deacetylation of

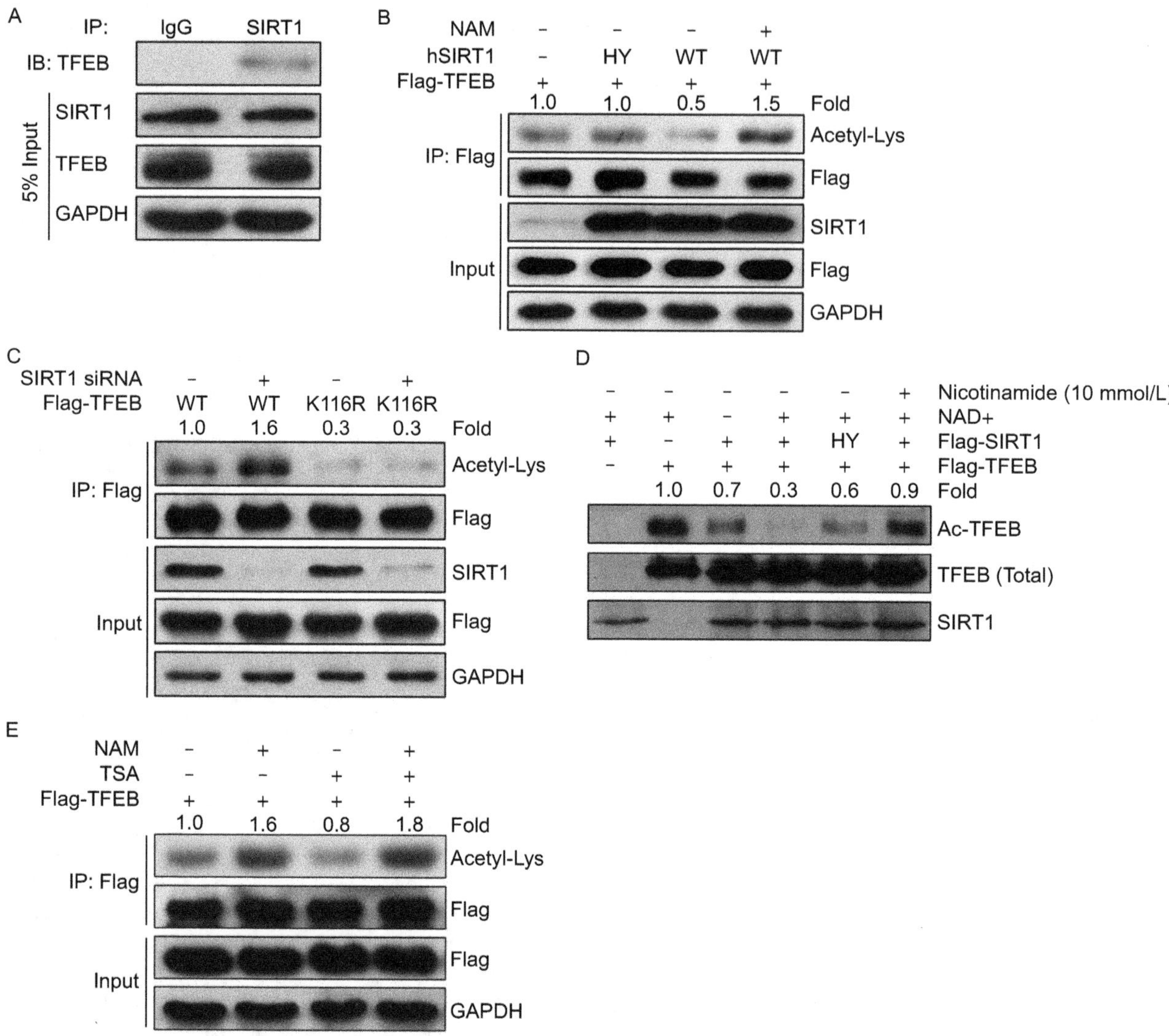

Figure 4. TFEB is a deacetylation substrate of SIRT1. (A) TFEB interacts with SIRT1. Endogenous coimmunoprcipitation assay was conducted with SIRT1 antibody and endogenous TFEB was detected by its specific antibody. (B) Catalytic activity of SIRT1 is required for deacetylation of TFEB. Flag-tagged TFEB was co-transfected into HEK293T cells with wild-type SIRT1 (WT) or its catalytically inactive mutant H363Y (HY), followed by additional treatment with or without NAM before harvest. Acetylation levels of TFEB were measured by IP Western blotting analysis. Values were expressed as fold changes relative to TFEB-WT without SIRT1 overexpression and normalized to IP-Flag. (C) Deacetylation mimic of TFEB (K116R) counteracts SIRT1 knockdown effects on acetylation of TFEB. Flag-tagged TFEB (WT or K116R) were co-transfected into HEK293T cells for 24 h with Scramble or SIRT1 siRNA. Acetylation levels of TFEB were measured by IP Western blotting analysis. Values were expressed as fold changes relative to TFEB-WT with scramble siRNA transfection and normalized to IP-Flag. (D) *In vitro* TFEB deacetylation by SIRT1. Purified Flag-TFEB and Flag SIRT1 (WT or HY) were co-incubated at 30°C for 6 h treated with or without NAD (5 mmol/L) or NAM (10 mmol/L). Acetylation levels of purified Flag-TFEB were measured by Western blotting analysis. Values were expressed as fold changes relative to TFEB-WT with NAD treatment and without SIRT overexpression and normalized to purified Flag-TFEB. (E) NAM, not TSA, increases TFEB acetylation. Flag-tagged TFEB was transfected into HEK293T cells treated with or without NAM (10 mmol/L) or TSA (0.5 μmol/L). Acetylation levels of TFEB were measured by IP Western blotting analysis. Values were expressed as fold changes relative to TFEB-WT without NAM or TSA treatment and normalized to IP-Flag.

TFEB by using deacetylase inhibitors NAM and TSA. SIRT1 inhibitor NAM exerted a greater impact on increases of acetylation status of TFEB than HDACI/II/IV inhibitors TSA, indicating that TFEB is specifically acetylated by SIRT1 rather than other HDACs (Fig. 4E). Furthermore, SIRT6 and SIRT7 could not reduce acetylation levels on TFEB as compared with SIRT1 (Fig. S3D), excluding the possibility that SIRT6 or SIRT7 might involve in deacetylation of TFEB in the nucleus. Collectively, these findings demonstrate that SIRT1 is the primary deacetylase of TFEB *in vivo* and *in vitro*.

SIRT1 accelerates fAβ degradation in microglia by enhancing lysosomal biogenesis

Based on our evidence that TFEB can be deacetylated by SIRT1, we hypothesized that SIRT1 may increase lysosomal biogenesis by activating TFEB, which probably facilitates fAβ degradation in microglia. For this purpose, we first infected BV2 cells and primary microglia with lentivirally expressed SIRT1-WT as well as its catalytically inactive mutant SIRT1-HY. The intracellular Aβ levels reduced more dramatically after 18-h-incubation with microglia overexpressing SIRT1-WT than those overexpressing GFP as a vector control or SIRT1-HY (Figs. 5A, 5C, and S4A), indicating that only the catalytically active SIRT1 is able to facilitate microglial degradation of fAβ efficiently. Consistent with this result, siRNA-mediated SIRT1 deficiency in microglia attenuated the degradation of fAβ in both BV2 cells and primary microglia (Figs. 5B and S4B). In addition, pretreating microglia with resveratrol, the agonist of SIRT1, increased the clearance of intracellular fAβ in a dose-dependent manner (Fig. S4C), while SIRT1 inhibitor NAM abolished its capability to degrade fAβ (Fig. S4D). Meanwhile, we found that TSA could not affect microglial degradative capability on fAβ (Fig. S4E), indicating that it is SIRT1, but not other HDACs, that is responsible for fAβ degradation in microglia. These results prompted us to further examine whether SIRT1 could enhance lysosomal biogenesis. In BV2 microglia, we observed an expansion in abundance of lysosomal compartment stained with LysoTracker Red or lysosome membrane marker LAMP1 when exogenously transfecting with GFP-tagged SIRT1-WT, as compared with vector control or GFP-tagged SIRT1-HY (Fig. 5D). This result is in accordance with that in HeLa cells (Fig. S6A). The number of lysosomes, however, decreased in microglia with SIRT1 knockdown (Fig. S4G). Further, we observed microglia pretreated with NAM exhibited a diminished lysosomal biogenesis compared with the control, while TSA-treated microglia changed little (Fig. S4F). In addition, qPCR test revealed that exogenously overexpressed SIRT1-WT is able to stimulate expression of TFEB downstream targets which are involved in the process of lysosomal biogenesis as well as lysosome-mediated degradative pathways (Fig. 5E), while SIRT1 knockdown decreases expression of these genes (Fig. S4H). Given these evidence, we determined to investigate whether the function of SIRT1 in fAβ clearance in microglia depends on its deacetylation on TFEB. Simultaneously overexpressing SIRT1 by lentivirus and knocking down endogenous TFEB neither promote fAβ degradation nor enrich the abundance of lysosomes in BV2 microglia, indicating TFEB is required for SIRT1's effects on facilitating fAβ degradation and lysosomal biogenesis in microglia (Fig. 5F and 5G). Collectively, these evidence suggest that SIRT1 accelerates fAβ degradation by enhancing lysosomal biogenesis in microglia.

Deacetylation status at K116 of TFEB stimulates fAβ degradation by facilitating lysosomal biogenesis

Given our previous observations that SIRT1 is the primary deacetylase of TFEB and its enhancement of fAβ degradation in microglia counts on TFEB, we determined to explore the possibility that deacetylated TFEB might increase lysosomal function and facilitate fAβ degradation. To address this question, we infected BV2 microglia with lentivirally overexpressed TFEB-WT or TFEB-K116R. The result noted that K116R mutant displayed a stronger capability in degradation of fAβ than its wild-type (Fig. 6A), which is accompanied by observations that more lysosomal compartments were induced in microglia and HeLa cells transfected with TFEB-K116R than TFEB-WT (Figs. 6B and S6B). Accordingly, analysis of these acidic organelles by fluorescence activated cell sorting (FACS) stained with LysoTracker Red revealed that there was an expansion of the quantity of lysosomes in the microglia infected with lentivirally-expressed TFEB-K116R (Fig. 6C). Furthermore, transcription of TFEB downstream targets was significantly upregulated in the presence of TFEB-K116R mutant (Fig. 6D). In order to confirm that the deacetylated K116R mutant promotes TFEB binding to its target genes, we performed chromatin immunoprecipitation (ChIP) assay and found that TFEB-K116R had a much stronger binding in the promoter of TFEB target gene *CLCN7* than the TFEB-WT (Fig. 6E). To further confirm it is the deacetylation status of TFEB that facilitates fAβ degradation by upregulating lysosomal function in microglia, we also evaluated the effects of TFEB-K116Q mutant (acetylation mimic). The results indicated that microglia with TFEB-K116Q overexpression could neither facilitate fAβ degradation (Fig. S5B) nor upregulate the lysosomal biogenesis and function (Fig. S5C and S5D). Collectively, these data indicate that deacetylation of TFEB intensifies its contribution to microglial degradation of fAβ.

Microglial expression of deacetylated TFEB reduces amyloid plaques in brain slices from APP/PS1 mice

To evaluate whether deacetylated TFEB could enhance microglial degradation of amyloid plaques, we performed *ex vivo* brain slice assay as described previously (Bard et al., 2000), where microglial cells were co-cultured with brain slices from aged APP/PS1 transgenic mice (Fig. 7A). Quantitative analysis of thioflavine-S staining indicated that BV2 microglia lentivirally overexpressed with TFEB-K116R were significantly more efficient at clearing the aggregates deposited in the cortex (Fig. 7B and 7C) and hippocampus

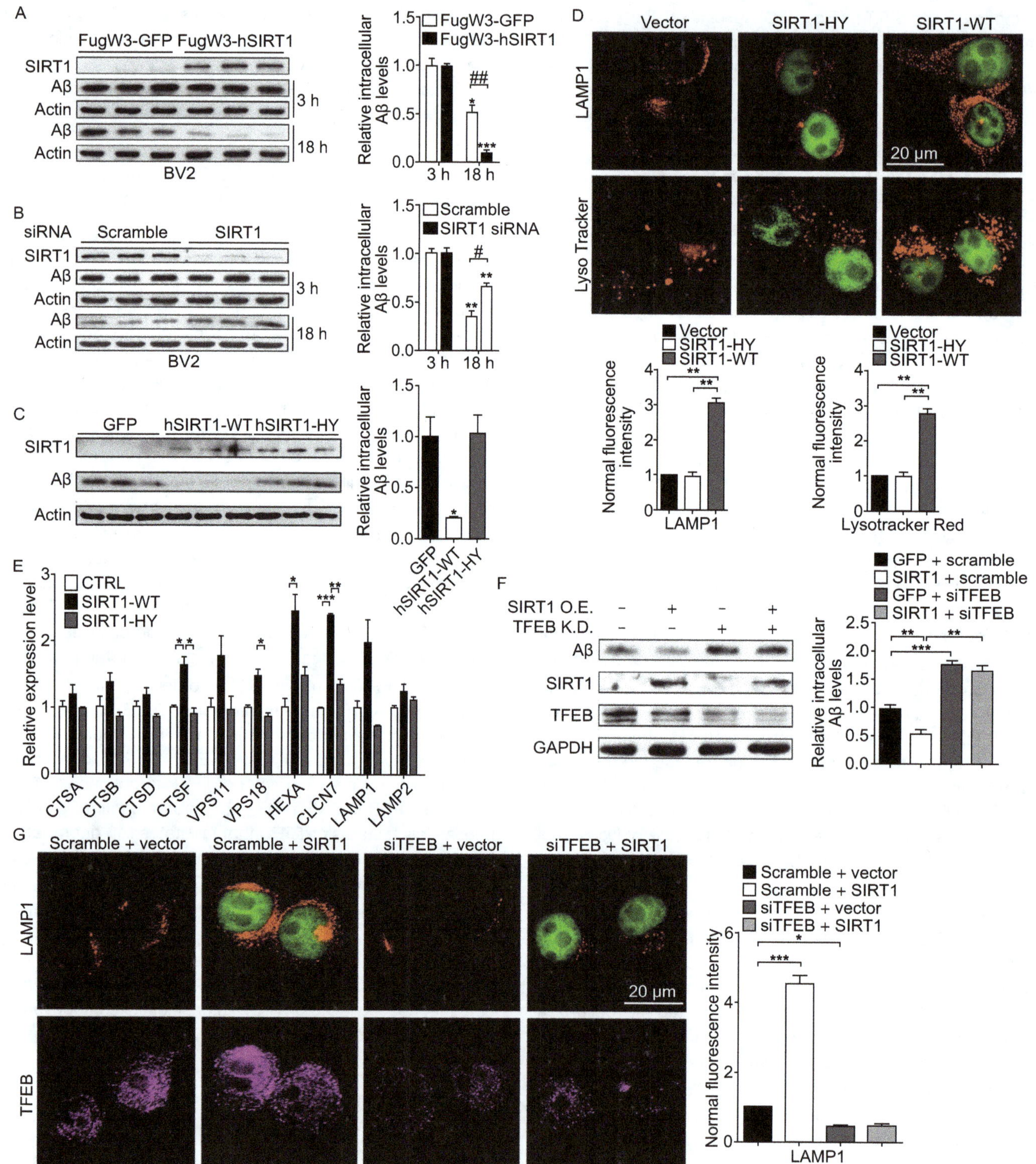

(Fig. 7D) when compared with its wild-type. In addition, independent Western blotting measurements of Aβ levels in the brain slices achieved a similar result (Fig. 7E and 7F). These *ex vivo* findings were consistent with the enhancement of microglial degradation of fAβ *in vitro*, suggesting a strengthened capability of microglial removal of Aβ aggregates which is triggered by deacetylation of TFEB. According to our results, deacetylation of TFEB, carried out by its deacetylase SIRT1, promotes microglial degradation of fAβ and amyloid plaques by upregulating lysosomal function and

Figure 5. SIRT1 accelerates fAβ degradation in microglia by enhancing lysosomal biogenesis. (A and B) SIRT1 accelerates fAβ degradation in microglia. GFP or human SIRT1 (SIRT1) was overexpressed in BV2 cells by lentiviral system (A). Scramble or SIRT1 siRNA was delivered into BV2 cells for 72 h (B). Then the cells were incubated with fAβ (500 nmol/L) for 3 h or 18 h and were harvested for detection of intracellular Aβ levels by Western blotting analysis. Two-way ANOVA, comparison between different time points, $^{*}P < 0.05$; $^{**}P < 0.01$; $^{***}P < 0.001$. Unpaired Student's *t*-test, comparison against the Fugw3-GFP or the Scramble, #$P < 0.05$, ##$P < 0.01$. (C) Catalytic activity of SIRT1 is required for degradation of fAβ. GFP or SIRT1 (WT or H363Y) was overexpressed in BV2 cells by lentiviral system. The cells were incubated with fAβ (500 nmol/L) for 18 h and were harvested for detection of intracellular Aβ levels by Western blotting analysis. One-way ANOVA with Turkey's test, comparison against GFP, $^{*}P < 0.05$. (D) SIRT1 stimulates lysosome biogenesis in microglia. Empty vector or GFP-tagged SIRT1 (WT or HY) was overexpressed in BV2 cells, followed by staining with an antibody against LAMP1 (Red) or LysoTracker Red, respectively. The fluorescence signal of each cell was estimated by examining more than 50 cells. One-way ANOVA with Turkey's test, $^{**}P < 0.01$; (E) SIRT1 upregulates TFEB induction of its target genes. Quantitative PCR (qPCR) analysis of TFEB target genes in BV2 cells overexpressed with empty vector as control or SIRT1 (WT or HY). Values represent mean ± SEM of three independent experiments. One-way ANOVA with Turkey's test, comparison against CTRL, $^{*}P < 0.05$; $^{**}P < 0.01$; $^{***}P < 0.001$. (F) TFEB is required for SIRT1's effects on accelerating fAβ degradation. BV2 cells were infected with lentiviral FugW3-GFP or FugW3-SIRT1 and were simultaneously transfected with scramble or TFEB siRNA for 72 h. The cells were incubated with fAβ (500 nmol/L) for an additional 18 h and were harvested for detection of intracellular Aβ levels by Western blotting analysis. One-way ANOVA with Turkey's test, comparison against GFP with scramble siRNA, $^{**}P < 0.01$; $^{***}P < 0.001$. (G) TFEB is required for SIRT1's stimulation of lysosomal biogenesis. BV2 cells were transfected with scramble or TFEB siRNA for 72 h and were transfected with empty vector or GFP-tagged SIRT1-WT, followed by staining with an antibody against LAMP1 (Red). The fluorescence signal of each cell was estimated by examining more than 50 cells. One-way ANOVA with Turkey's test, $^{*}P < 0.05$; $^{***}P < 0.001$.

biogenesis, which highlights acetylation status of TFEB as a critical role in the degradative pathway in microglia (Fig. 8).

DISCUSSION

Microglia are the principle immune phagocytic cells in the CNS which constantly survey their adjacent environment for maintainance of the homeostasis within the brain. Previous work revealed that functional deficiency in microglia is associated with much more severe deposition of amyloid plaques in AD mouse models, indicating a protective role of microglia in AD (Zhang et al., 2007). Indeed, we and others have observed that microglia internalize Aβ through membrane receptors and transport them into lysosomes for degradation (Doens and Fernandez, 2014; Ma et al., 2013). In this paper, we demonstrated that the ability of microglia to degrade fAβ could be facilitated by TFEB-mediated upregulation of intracellular degradative systems.

Several lines of evidence pointed out that the amyloid deposits cannot be fully eliminated or even remain stable albeit by surrounding coverage of activated microglia (Bolmont et al., 2008; Wegiel et al., 2001), which may be explained in part by the dysfunction of lysosomes in the presence of Aβ stimulation (Hickman et al., 2008; Majumdar et al., 2007). Subsequent studies further clarified that attempts to upregulating intracellular lysosomal hydrolytic enzymes or stabilizing a fitting lysosomal acidic environment eventually enhance or restore microglial capability to degrade Aβ (Majumdar et al., 2011; Majumdar et al., 2008), which may be suggestive of strengthening lysosomal function as a novel approach to increase Aβ clearance. In this regard, our results validated that TFEB, the major regulator of lysosomal biogenesis, facilitates or rejuvenates microglial capability in degradation of fAβ and amyloid plaques by upregulating lysosomal quantity and enhancing lysosomal function.

Emerging evidence pointed out that defective activation of TFEB probably underlies aberrant lysosomal function failing to degrade aggregated toxic proteins in prevalent neurodegenerative diseases (Decressac et al., 2013; Tsunemi et al., 2012; Xiao et al., 2014). Therefore, appropriate regulation of TFEB activity has attracted more and more attention in countering the pathogenesis of these diseases (Settembre et al., 2013b). Despite the fact that TFEB can be activated by its own stimulation in a transcriptional autoregulatory pathway (Settembre et al., 2013a), post-translational modification play an essential role in modulating TFEB activity, among which phosphorylation is the first to be studied. Under basal conditions, TFEB is mainly sequestered in cytosol where it remains to be phosphorylated by rapamycin complex mTORC1 (Settembre et al., 2011). However, once upon specific conditions such as starvation or mTOR suppression, TFEB is rapidly dephosphorylated and translocates into nucleus for transcriptional activation of lysosomal system, a process proved to be dependent on autophagosome-lysosome fusion (Zhou et al., 2013). Nevertheless, it still remains a mystery that whether TFEB transcriptional activity needs proper regulation within the nucleus and could other modifications be involved in such regulatory mechanisms. In this regard, for the first time, we identified acetylation as a prominent post-translational modification in regulating TFEB transcriptional activity within the nucleus. Our results indicated that TFEB translocates into nucleus in a mTORC1-independent pathway once microglia are under exposure to fAβ. Although further investigation is needed to unveil the mechnisms underlying this observation, our findings suggest that such translocation may enable TFEB to be deacetylated by nuclear-located SIRT1, which enhances its transcriptional

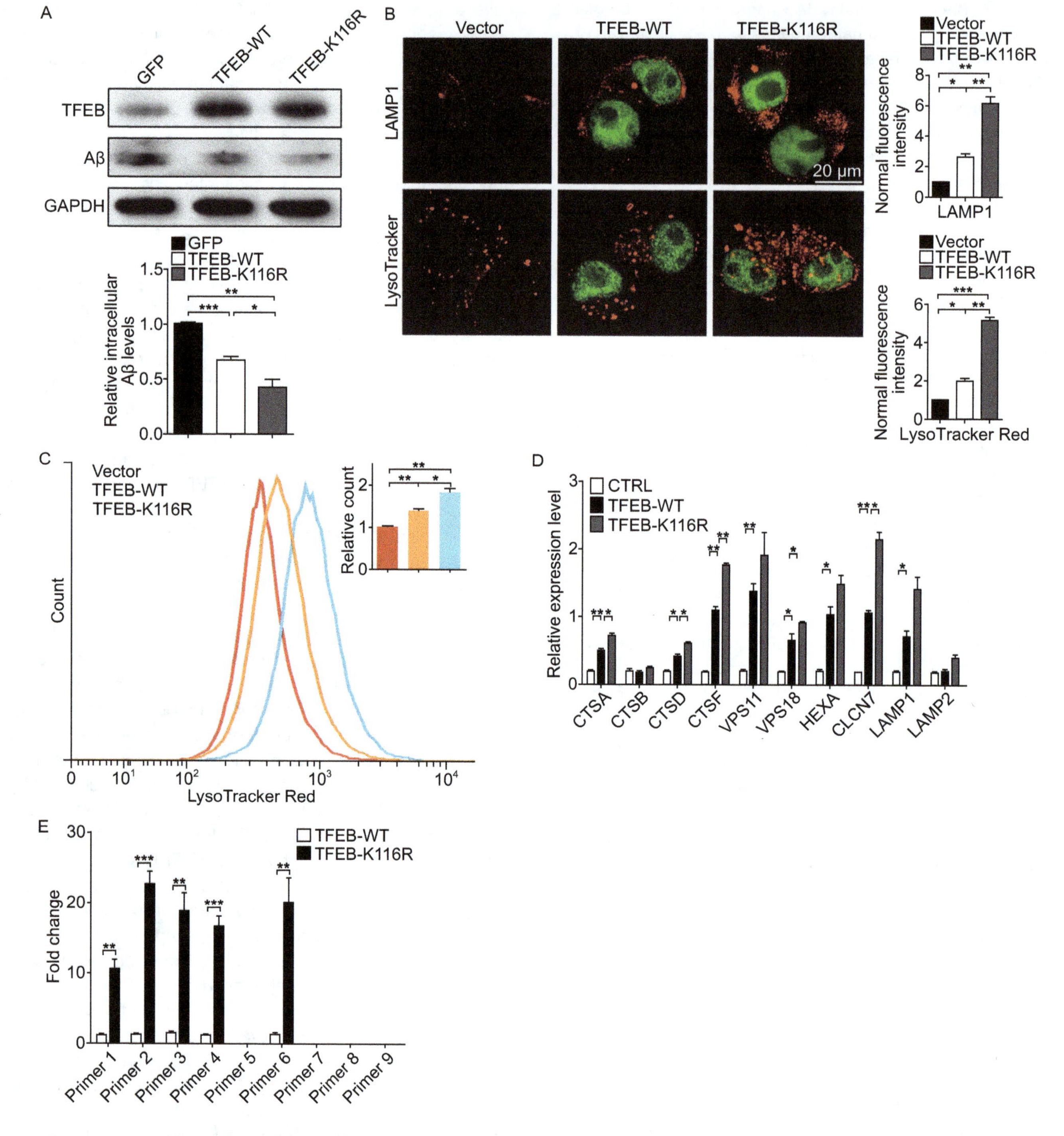

induction of target genes closely associated with lysosomal biogenesis, ultimately resulting in a robust degradation of intracellular fAβ.

Protein acetylation has been recently recognized as a significant approach in controlling autophagic processes such as elimination of damaged organelles or toxic protein aggregates (Banreti et al., 2013). Previous studies revealed that acetylation of ATG proteins, the autophagy core components, by acetyltransferase p300 results in autophagy inhibition (Lee and Finkel, 2009), whereas this effect can be reversed by NAD^+-dependent deacetylase SIRT1 under particular cellular stresses (Lee et al., 2008). In addition, SIRT1-mediated deacetylation of FOXO transcription factors is able to activate autophagy by upregulation of core autophagy genes in the process of autophagosome formation (Chakrabarti et al., 2011; Zhao et al., 2007). In the present study, we demonstrated that deacetylation of TFEB by SIRT1 facilitates lysosomal biogenesis, which suggests a

Figure 6. Deacetylated TFEB facilitates fAβ degradation by enhancing lysosomal biogenesis. (A) Deacetylation of TFEB at K116 futher facilitates fAβ degradation. BV2 cells were infected with lentivirally overexpressed FugW3-GFP or FugW3-TFEB (WT or K116R). The cells were incubated with fAβ (500 nmol/L) for an additional 18 h and were harvested for detection of intracellular Aβ levels by Western blotting analysis. The band intensity was measured in three independent experiments indicating relative intracellular Aβ levels and the mean ± SEM. are shown in the right panel. One-way ANOVA with Turkey's test, $*P < 0.05$; $**P < 0.01$; $***P < 0.001$. (B) Deacetylation of TFEB at K116 further stimulates lysosomal biogenesis in microglia. Empty vector or GFP-tagged TFEB (WT or K116R) was overexpressed in BV2 cells, followed by staining with an antibody against LAMP1 (Red) or LysoTracker Red, respectively. The fluorescence signal of each cell was estimated by examining more than 50 cells. One-way ANOVA with Turkey's test, $*P < 0.05$; $**P < 0.01$; $***P < 0.001$. (C) Flow cytometric analysis of lyososomes stained with LysoTracker Red in BV2 cells treated as in (B). Values of mean fluorescence were expressed as fold changes. One-way ANOVA with Turkey's test, $*P < 0.05$; $**P < 0.01$. (D) Deacetylation status at K116 further upregulates expression of TFEB target genes. qPCR analysis of TFEB target genes in BV2 cells overexpressed with empty vector as control or TFEB (WT or K116R). Values represent mean ± SEM of three independent experiments. One-way ANOVA with Turkey's test, comparison against CTRL, $*P < 0.05$; $**P < 0.01$. (E) ChIP-qPCR analysis for TFEB binding to its target gene *CLCN7*. The histogram shows the amount of immunoprecipitated DNA as detected by qPCR assay. Values were normalized to the input and displayed as relative enrichment over a mock control. Values represent mean ± SEM of three independent experiments. One-way ANOVA with Turkey's test, $**P < 0.01$; $***P < 0.001$.

novel mechanism regulating intracellular degradative pathways and validating the importance of protein acetylation in these processes. However, the exact regulatory role of protein acetylation in autophagy-lysosome system still needs to be defined.

SIRT1, the well-studied NAD^+-dependent deacetylase, has been realized as a master regulator of multiple cellular metabolic processes, including various common neurodegenerative disorders. In particular, AD transgenic mouse models displayed a strikingly reduced Aβ burden in the hippocampus and cortex when SIRT1 was chemical activated or overexpressed by lentiviral infection (Karuppagounder et al., 2009). The underlying mechanism was further explained by observations that neuronal SIRT1 promotes the nonamyloidogenic APP-processing pathway either by directly enhancing α-secretase activity or by upregulating expression of α-secretase, respectively (Donmez et al., 2010; Qin et al., 2006). In contrast with the knowledge that SIRT1 prevents neuronal Aβ production, little is known about SIRT1's function in clearance of Aβ. Previous studies suggested that SIRT1 in microglia contributes to amelioration of Aβ-induced inflammatory response by suppressing NF-κB and IL-1β signaling pathway (Chen et al., 2005; Cho et al., 2015). In this study, we found that SIRT1 is able to regulate microglial degradation of fAβ. Moreover, we identified TFEB as a novel substrate of SIRT1 and provided evidence that SIRT1 upregulates lysosomal biogenesis by stimulating TFEB induction of its downstream targets, ultimately resulting in increased fAβ clearance in microglia. However, considering about the fact that several substrates of SIRT1 were reported to positively regulate autophagy such as Foxo3 and Atg proteins (Lee and Finkel, 2009; Zhao et al., 2007), our present data cannot exclude the possibility that SIRT1 may promote fAβ degradation by regulating other molecules in lysosomal-autophagy degradative pathway. Further investigation is needed to achieve a better understanding of the underlying mechanisms of SIRT1-mediated fAβ degradation. Overall, our findings uncover a novel mechanism that SIRT1 protects against Aβ deposits by facilitating microglial clearance of Aβ other than preventing neuronal production of Aβ.

It is noteworthy that our results are in accordance with a recent work indicating that TFEB accelerates astrocytic capability in Aβ removal (Xiao et al., 2014). In fact, these two glia cells may play distinct roles in Aβ clearance depending on amyloid aggregation states and pathogenesis stages. Astrocytes possibly play a constitutive role in the very early AD stages by taking up oligomeric sAβ, while microglia mainly function in the late stages as fAβ accumulates to form fibrillar deposits (Condello et al., 2015; Mulder et al., 2014; Weldon et al., 1998). In fact, we did conduct the experiment by culturing sAβ with microglia in our degradative model and found that deacetylated TFEB could degrade sAβ more efficiently as well (Fig. S5A), which collectively verify the vital role of deacetylation status of TFEB in Aβ degradation. Moreover, TFEB overexpression in neuron is recently reported to reduce Aβ generation by enhancing APP degradation in lysosomes (Xiao et al., 2015). In this scenario, our results favour a plausible novel therapy that upregulation of TFEB activities in both glia cells and neuron may help to coordinately ameliorate AD pathogenesis, providing a better healing of AD.

In summary, we identify protein acetylation as a novel mechanism underlying regulation of TFEB transcriptional activity for the first time, and demonstrate that deacetylation of TFEB in microglia enhances lysosomal biogenesis by stimulating expression of its target genes, which ultimately facilitates fAβ degradation and decreased amyloid plaque deposition. Our findings highlight that TFEB is a central regulator in lysosome-mediated clearance of toxic protein aggregates. Also, distinct post-translational modifications of TFEB occurring at different cellular localization may regulate TFEB activity coordinately in the autophagy-lysosome pathway. These findings may provide an appealing avenue in countering pathogenesis of AD and other neurodegenerative diseases.

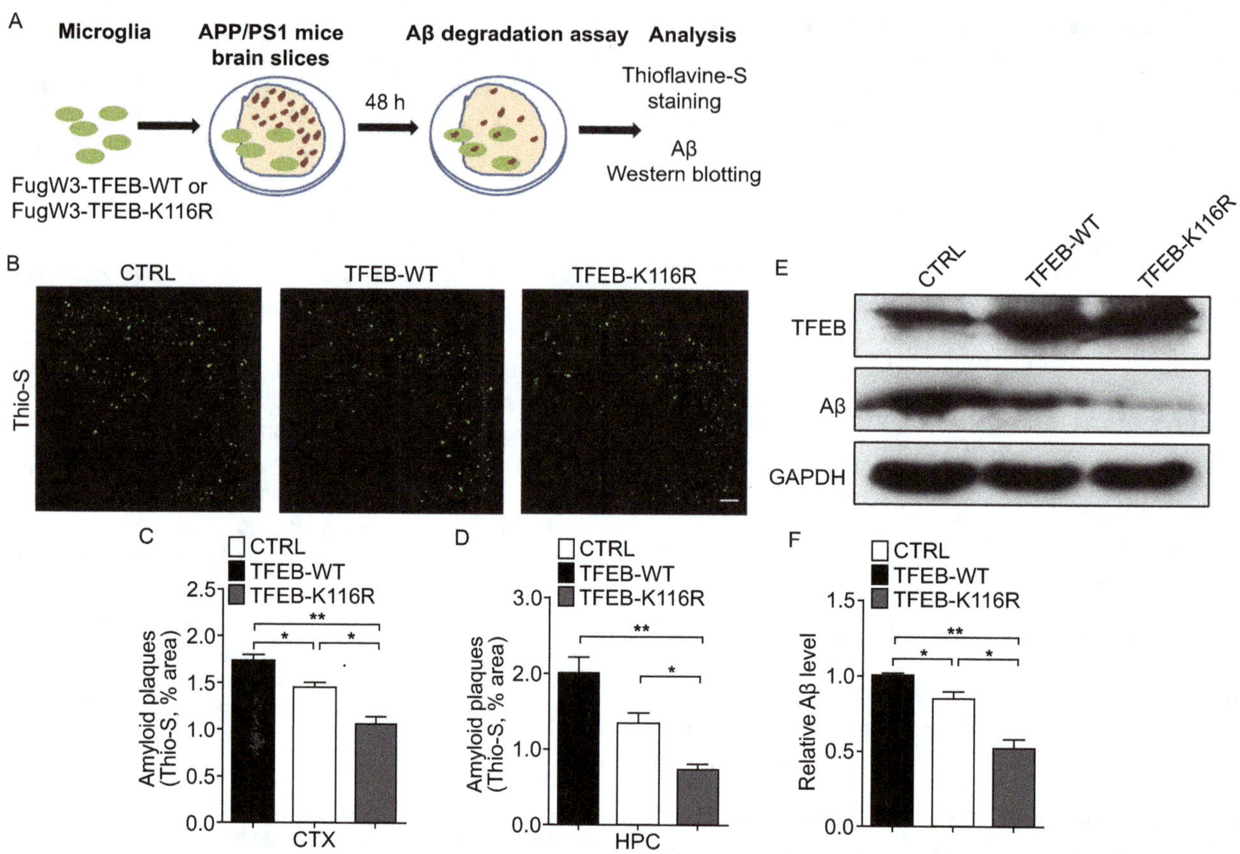

Figure 7. Microglial expression of deacetylated TFEB reduces amyloid plaques in the brain slices from APP/PS1 mice. (A) *Ex vivo* degradation assay. BV2 microglia were infected with lentivirally overexpressed FugW3-TFEB (WT or K116R) and co-cultured with 10 μm thick freshly cut, unfixed brain slices from aged APP/PS1 transgenic mice. After 48 h the slices were used for subsequent analysis. (B–E) Deacetylation of TFEB at K116 further reduces amyloid plaques. BV2 microglia overexpressed with vector control (CTRL) or TFEB (WT or K116R) were co-cultured with brain sections prepared as depicted in (A), followed by staining amyloid plaques with thioflavine-S (Green) (B–D) or by Western blotting analysis of fAβ levels (E and F). CTX, cortex; HPC, hippocampi. Results are mean area fraction of amyloid plaques ± SEM for 3 mice per group. The band intensity was measured in three independent experiments indicating relative fAβ levels and the mean ± SEM. One-way ANOVA with Turkey's test, **P* < 0.05; ***P* < 0.01. Scale bar, 200 μm.

MATERIALS AND METHODS

Cell culture

BV2, HEK293T, and HeLa cell lines were obtained from Cell Resource Center of Peking Union Medical College Hospital and cultured in Dulbecco's Modified Eagle Medium (DMEM) supplemented with 10% fetal bovine serum (FBS) (Hyclone, USA) and antibotics (100 U/mL penicillin, 100 μg/mL streptomycin). Cells were incubated at 37°C in a 95% humidified atmosphere with 5% CO_2.

Primary microglia were prepared from the whole brains of 24-h-old Sprague-Dawley (SD) rats as described previously (Ma et al., 2013). Briefly, pups were decapitated and then the meninges and blood vessels were removed from the cortices which were minced and digested with Dnase I (0.01%) and trypsin (0.25%) for 30 min at 37°C. The digestion was stopped by DMEM containing 10% FBS and 1% penicillin-streptomycin. The cell mixture were then triturated and plated on T75 flask (Corning, USA). The media were changed to fresh DMEM with 10% FBS 24 h later and were cultured for another 14–21 days at 37°C. Primary microglia were isolated by shaking the plates for 2 h at 260 rpm. Cells were then counted and plated for experiments at the appropriate density in DMEM/F12 containing 2% FBS. The media were changed to serum-free DMEM/F12 overnight before conducting the experiment. The enriched microglia were >98% pure as determined by immunoflurescence shown with Iba1-immunoreactive (IR) and GFAP-IR cells (Fig. S1C).

Preparation of fibrillar Aβ

Human $A\beta_{1-42}$ (20276; AnaSpec, USA) or Hilyte-Fluor-488 fibrillar $A\beta_{1-42}$ (60479; AnaSpec, USA) were dissolved in 1 mmol/L phosphate-buffered saline buffer (PBS) and incubated at 37°C for 7 days

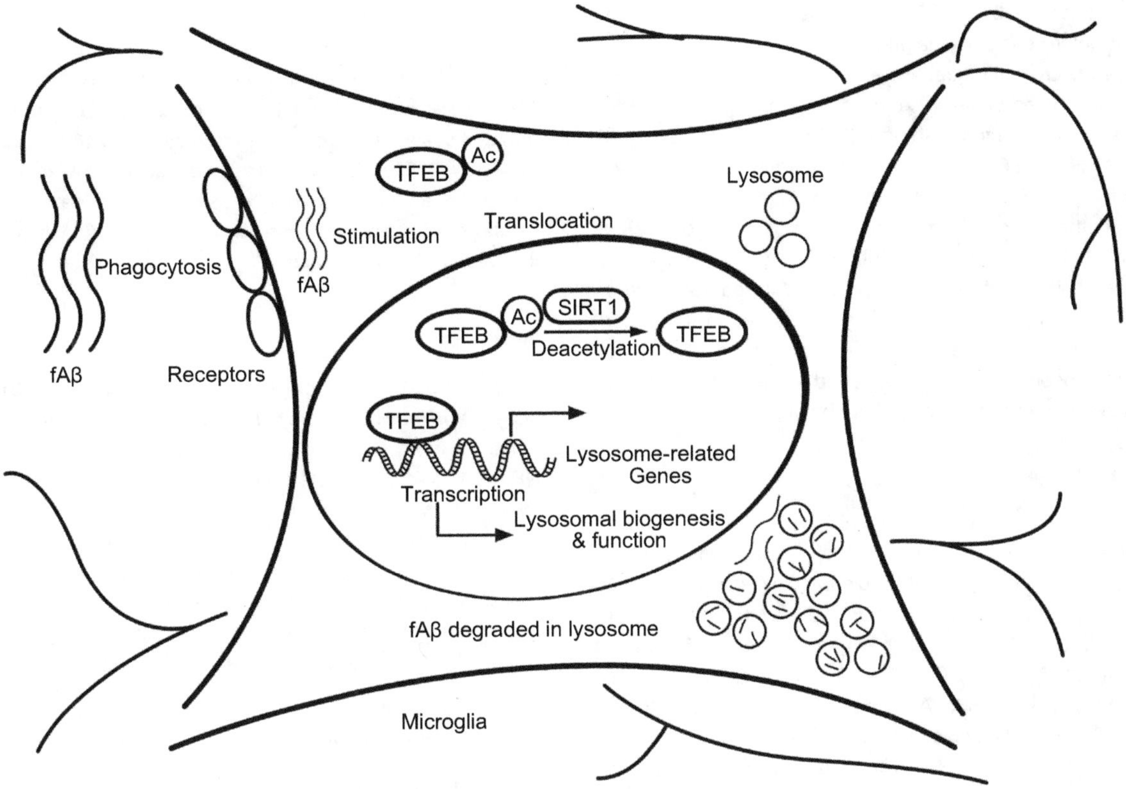

Figure 8. A schematic model of TFEB deacetylation on facilitating fAβ degradation in microglia by enhancing lysosomal biogenesis. In the presence of fAβ stimulation, microglia take up fAβ and transport them into lysosomes. Meanwhile, TFEB translocates into nucleus where it interacts with SIRT1 and is deacetylated at K116. Deacetylation of TFEB enhances its transcriptional induction of its downstream target genes closely associated with lysomomal biogenesis, which eventually upregulates microglial capability in degradation of fAβ.

and stored at −20°C before use. The ultra structure of the peptide aggregates was examined by transmission electron microscopy.

Plasmids and transfection

Human TFEB was subcloned into pCMV-Tag2B and pEGFP-N3, respectively. Human SIRT1-WT and SIRT1-HY was subcloned into pCMV-3×HA and pEGFP-N3, respectively. TFEB-K116R was mutated from TFEB- pCMV-Tag2B and TFEB-pEGFP-N3, respectively. Transient transfection of plasmids were conducted using Lipofectamine LTX according to the instruction (Invitrogen), followed by changing the medium 4 h later. The cells were harvested 24 h later for the subsequent experiments.

Lentivirus infection

The FugW3 four-plasmid-lentivirus system were utilized to overexpress SIRT1 (WT or HY) and TFEB (WT or K116R) in BV2 and primary microglia. FugW3-GFP as a vector control and FugW3-SIRT1-WT, -SIRT1-HY, -TFEB-WT, and -TFEB-K116R constructs were transfected into 293T-packaging cells with three other helper plasmids (RRE, REV, and VsVg) respectively, using Lipofectamine LTX. Medium was changed 12 h later for another 60 h incubation before collecting the supernatant which was filtered through a 0.45 μm filter, aliquated and stored at −80°C. BV2 or primary microglia was infected with the supernatant in the presence of 2 μg/mL Polybrene for 24 h before changing the medium. Cells were harvested 48–72 h later for the subsequent experiments.

RNA interference

BV2 or primary microglia in 12-well plates were transfected with siRNAs against target genes (SIRT1 siRNA: sc-40987; TFEB siRNA: sc-38510; Santa Cruz Biotechnology, CA) or scramble siRNA as control with Lipofectamine RNAiMAX (Invitrogen, Carlsbad, CA). Cells were harvested 48–72 h later for the subsequent experiments.

fAβ uptake and degradation

Measurements of fAβ uptake and degradation were performed as depicted previously (Jiang et al., 2008). Briefly, to measure localization of the internalized fAβ, BV2 cells or primary microglia were plated in 12-well plates, and 500 nmol/L Hilyte-Fluor-488 fibrillar

$A\beta_{1-42}$ was added for 30 min. The cells were stained with LysoTracker Red before confocal imaging.

To analyse microglial degradation, 500 nmol/L *in vitro* synthetic $fA\beta_{1-42}$ was incubated with BV2 and primary microglia for varying times. Cells were infected with lentivirally overexpressed GFP, SIRT1 (WT or HY), TFEB (WT or K116R) or were transfected with siRNA for SIRT1 or TFEB for 72 h before incubation with 500 nmol/L $fA\beta_{1-42}$ for another 3 h or 18 h, indicating microglial phagocytosis or degradation of fAβ, respectively. Subsequently, cells were thoroughly washed with DMEM three times and lysed for Western blotting Analysis. The band intensity was measured in three independent experiments indicating relative intracellular fAβ levels. Inhibitors for lysosome or NEP were added to the cells 2 h before adding fAβ and were kept in the medium before harvest.

Antibodies and reagents

Mouse monoclonal anti-β-actin (ab3280; Abcam, MA); mouse monoclonal anti-GAPDH (cw0266; CWBio; China); mouse monoclonal anti-HA (H9658; Sigma, St. Louis, MO); mouse monoclonal anti-Flag M2 antibody (F3165; Sigma, St. Louis, MO); mouse monoclonal anti-SIRT1 (8469; Cell Signaling Technology, MA); rabbit polyclonal anti-SIRT1 (2493; Cell Signaling Technology, MA); goat polyclonal anti-TFEB (ab2636; Abcam, MA); rabbit polyclonal anti-TFEB (sc48784; Santa Cruz Biotechnology, CA); mouse monoclonal anti-$A\beta_{1-16}$ (6E10) (SIG39320, Covance, USA); rabbit monoclonal acetylated-lysine antibody (9814; Cell Signaling Technology, MA); rabbit polyclonal anti-Lamp1 (ab 24170; Abcam, MA); LysoTracker Red (C1046; Beyotime, China); Leupeptin (L2884; Sigma, St. Louis, MO); Chloroquine (C6628; Sigma, St. Louis, MO); Phosphoramidon (R7385; Sigma, St. Louis, MO); Thioflavine-S (T1892; Sigma, St. Louis, MO).

Western blotting analysis

Cells were harvested and lysed with lysis buffer consisting of 50 mmol/L Tris-HCl (pH 7.5), 150 mmol/L NaCl, 1% NP-40, 1% SDS, protease inhibitors cocktail (Sigma, St. Louis, MA) and 10% the mixture of deoxyribonuclease I (1 mg/mL) and ribonuclease A (0.25 mg/mL) at 4°C. The lysates were sonicated, followed by centrifugation at 14,800 ×*g* at 4°C for 10 min to remove the insoluble substances. The protein concentration was determined by 2-D Quantitative Kit (GE Healthcare, USA). Equal amount the lysates were resolved by 10% SDS-PAGE and transferred onto PVDF membranes using a semi-dry blotting apparatus (Bio-Rad, USA). After blocking with 5% non-fat milk in phosphate buffered saline (PBS) with 0.05% Tween 20, the membranes were incubated overnight at 4°C with primary antibodies, followed by incubation with HRP-conjugated secondary antibodies (1031-05 or 4050-05; Southern Biotech, MA). Protein bands were detected by chemiluminescence (Millipore, Billerica, MA).

Ex vivo degradation assay

Ex vivo degradation assays were performed as previous described (Bard et al., 2000). Briefly, 12-month-old APP/PS1 transgenic mice were perfused with saline and cryostat sections (10 μm in thickness) were cut from snap-frozened brain and mounted onto glass coverslips pre-coated with poly-L-lysine. Brain sections were dried for at least 2 h at room temperature and washed with hybridoma serum-free media (H-SFM; Thermo Fisher) containing 1% FBS and 1% penicillin/streptomycin. BV2 microglia were seeded (8×10^5 cell/mL) and co-cultured with the brain sections in 24-well plates for 48 h at 37°C with 5% CO_2. After incubation, sections were washed twice with PBS and fixed with 4% paraformaldehyde for subsequent thioflavine-S staining or lysed in 6.25 mol/L guanidinhydrochloride for Western blotting analysis.

Immunoprecipitation

Cells were pretreated with TSA (0.5 μmol/L) and NAM (10 mmol/L), or seperately. Then cells were harvested and lysed in five volume of lysis buffer (25 mmol/L Tris-HCl, 150 mmol/L NaCl, 3 mmol/L $MgCl_2$, 0.5% NP-40, 1 mmol/L DTT, 5% Glycerol, 1% PI, pH 7.5) for rotating at 4°C for 2 h, followed by centrifugation at 21,000 ×*g* for 30 min. Then the supernatant were mesured by 2-D Quantitative Kit. Equal amount of protein was incubated with the ANTI-FLAG M2 Affinity Gel (Sigma, St. Louis, MO) at 4°C overnight. Immunoprecipitates were washed with lysis buffer for four times and eluted with SDS loading buffer, followed by SDS-PAGE separation and Western blotting Analysis.

Mass spectrometry analysis

HEK293T cells were transfected with Flag-tagged TFEB for 24 h, then the cells were pretreated with 0.5 μmol/L TSA and 10 mmol/L NAM for 12 h before harvest. Cells were lysed and immunoprecipitated as depicted above. The eluted mixture was resolved by 10% SDS-PAGE and stained with Commassie Blue R250 and the bands were excised from the gel, followed by tryptic digestion and mass spectrometry. Modification identification was measured with the database search accordingly and peptide identifications were validated with Proteome Discoverer 1.4.

In vitro deacetylation assay

Flag-tagged SIRT1 (WT or HY) and TFEB were purified from 293T cells by immunoprecipitation. The deacetylation reaction was performed in 100 μL of reaction buffer containing 5 μg of puridied Flag-tagged TFEB proteins and 15 μg purified Flag-SIRT1-WT or SIRT1-HY proteins in Tris-HCl buffer (Tris-HCl, PH 8.5, 50 mmol/L; NaCl 50 mmol/L; $MgCl_2$ 4 mmol/L; DTT 1 mmol/L; Glycerol 5%; NAD^+ 5 mmol/L; NAM 10 mmol/L) at 30°C for 6 h. During the incubation, the buffer was mixed every 5 min and the reaction was stopped by SDS loading buffer, followed by SDS-PAGE separation and Western blotting Analysis.

Immunofluorescence

Cells were 4% PFA fixed for 20 min, permeabilised in 0.1% PBST for 20 min, and then blocked in 10% BSA/0.05% PBST for 1 h at room temperature. The cells were then incubated with primary antibody in 1% BSA/PBST overnight at 4°C, followed by a further incubation at room temperature for 1 h with AlexaFluor secondary antibodies. Images were acquired with LSM 710 NLO & DuoScan System confocal laser-scanning microscope (Zeiss, Jena, Germany).

Flow cytometric analysis

Cells were incubated with LysoTracker Red (0.1 μmol/L) for 10 min and washed by PBS thoroughly, followed by flow cytometric analysis on BD LSRFortessa instrument, 30,000 events per run. FlowJo software was used to perform data analysis.

RNA isolation and quantitative real-time PCR

Total RNA was isolated from cells with different treatments by Trizol reagent (Invitrogen, USA), and 1 μg of total RNA was reverse transcribed to cDNA using HiFi-MMLV cDNA first strand synthesis Kit (CW Bio, China). Quantitative PCR (qPCR) was carried out by mixing cDNA with GoTaq qPCR Master Mix (Promega, USA) and the reaction was performed on the CFX96 Real Time PCR Detection system (Bio-Rad, USA), under the following conditions: 95°C, 2 min; 95°C, 15 s; 60°C, 30 s; 72°C, 25 s; followed by 40 cycles of 95°C, 15 s; 60°C, 2 min.

The primers for qPCR were designed according to gene sequences published in theGenBank (SIRT1, forward, 5′-TGTGTC ATAGGTTAGGTGGTGA-3′, reverse, 5′-AGCCAATTCTTTTTGTGT TCGTG-3′; TFEB, forward, 5′-AAGGTTCGGGAGTATCTGTCTG-3′, reverse, 5′-GGGTTGGAGCTGATATGTAGCA-3′; CTSA, forward, 5′-TCCCAGCATGAACCTTCAGG-3′, reverse, 5′-AGTAGGCAAAGTA-GACCAGGG-3′; CTSB, forward, 5′-ACAACGTGGACATGAGCTA CT-3′, reverse, 5′-TCGGTAAACATAACTCTCTGGGG-3′; CTSD, forward, 5′-TGCTCAAGAACTACATGGACGC-3′, reverse, 5′-CGAA GACGACTGTGAAGCACT-3′; CTSF, forward, 5′-AGAGAGGCCCA ATCTCCGT-3′, reverse, 5′-GCATGGTCAATGAGCCAAGG-3′; VPS11, forward, 5′-CAAGCCTACAAACTACGGGTG-3′, reverse, 5′-GAGTGCAGAGTGGATTGCCA-3′; VPS18, forward, 5′-CACTCGG GGTATGTGAATGCC-3′, reverse, 5′-TCGGAAGGGGTGAAGTCAA TG-3′; HEXA, forward, 5′-ACGTCCTTTACCCGAACAACT-3′, reverse, 5′-CGAAAAGCAGGTCACGATAGC-3′; CLCN7, forward, 5′-CCACGTTCACCCTGAATTTTGT-3′, reverse, 5′-AAACCTTCCGAA GTTGATGAGG-3′; LAMP1, forward, 5′-CAGATGTGTTAGTGG-CACCCA-3′, reverse, 5′-TTGGAAAGGTACGCCTGGATG-3′; LA MP2, forward, 5′-GCACAGTGAGCACAAATGAGT-3′, reverse, 5′-CAGTGGTGTGTATGGTGGGT-3′).

Chromatin immunoprecipitation assay

Chromatin immunoprecipitation assay (ChIP) was performed as depicted previously (Hug et al., 2004). In brief, cells were transfected with Flag-tagged TFEB-WT or TFEB-K116R for 24 h. The cells were harvested for immunoprecipitation with ANTI-FLAG M2 Affinity Gel (Sigma, St. Louis, MO) and were then subjected to the ChIP assay following the manufacturer's instruction of Chromatin immunoprecipitation kit (9002, Cell Signaling Technology, MA). qPCR was performed to figure out the TFEB binding in promoter. The data were analysed and displayed as described previously (Settembre et al., 2013a). Primer sequences of *CLCN7* for amplifying purified DNA were listed in the Table S1.

Statistical analysis

All functional experiments were repeated at least in three independent experiments. The band intensity in Western blot analysis was quantified by ImageJ. Data were analyzed by GraphPad Prism 5.0 software and shown by the mean ± SEM. Multiple sets of data were analyzed by one-way or two-way ANOVA, and unpaired Student's *t* test was used to analyze two sets of data. Significance level was set at $P < 0.05$.

ACKNOWLEDGEMENTS

This work was supported by grants from the National Natural Science Foundation of China (Grant Nos. 31470807, 31270872, and 31200610), the National Basic Research Program (973 Program) (Nos. 2010CB912203 and 2011CB915504) and Founds from State Key Laboratory of Protein and Plant Gene Research, College of Life Sciences, Peking University.

AUTHORS' CONTRIBUTIONS

BJT conceived and performed experiments on fibrillar Aβ degrading modeling, biochemical assay, gene cloning, data analysis, and manuscript preparation. ZLJ contributed to the confocal imaging. ZQ and LXY performed qPCR/CHIP-qPCR design and experiment. LZY and ZXY contributed to analysis of mass spectrometry data. ZXF and BX helped with gene cloning. HW helped with CoIP experiments. ZZ performed data analysis. SSJ and ZC contributed to the intrastriatal administration of adenoviruses. WQS analyzed data and prepared the manuscript. JJG designed all experiments, analyzed data, and wrote the manuscript. All authors read and approved the final manuscript.

ABBREVIATIONS

Aβ, β-amyloid; AD, Alzheimer's disease; ApoE, Apoliprotein E; fAβ, fibrillar Aβ; HDAC, histone deacetylase; CNS, central nervous system; NAM, nicotinamide; NEP, neprilysin; sAβ, soluble Aβ; TFEB, transcription factor EB; TSA, trichostatin A.

COMPLIANCE WITH ETHICS GUIDENLINES

Jintao Bao, Liangjun Zheng, Qi Zhang, Xinya Li, Xuefei Zhang, Zeyang Li, Xue Bai, Zhong Zhang, Wei Huo, Xuyang Zhao, Shujiang Shang, Qingsong Wang, Chen Zhang, and Jianguo Ji declare that they have no conflict of interest.

All institutional and national guidelines for the care and use of laboratory animals were followed. All participants gave informed consent to participate in research.

REFERENCES

Banreti A, Sass M, Graba Y (2013) The emerging role of acetylation in the regulation of autophagy. Autophagy 9:819–829

Bard F, Cannon C, Barbour R, Burke RL, Games D, Grajeda H, Guido T, Hu K, Huang J, Johnson-Wood K et al (2000) Peripherally administered antibodies against amyloid beta-peptide enter the central nervous system and reduce pathology in a mouse model of Alzheimer disease. Nat Med 6:916–919

Bolmont T, Haiss F, Eicke D, Radde R, Mathis CA, Klunk WE, Kohsaka S, Jucker M, Calhoun ME (2008) Dynamics of the microglial/amyloid interaction indicate a role in plaque maintenance. J Neurosci 28:4283–4292

Chakrabarti P, English T, Karki S, Qiang L, Tao R, Kim J, Luo Z, Farmer SR, Kandror KV (2011) SIRT1 controls lipolysis in adipocytes via FOXO1-mediated expression of ATGL. J Lipid Res 52:1693–1701

Chen J, Zhou Y, Mueller-Steiner S, Chen LF, Kwon H, Yi S, Mucke L, Gan L (2005) SIRT1 protects against microglia-dependent amyloid-beta toxicity through inhibiting NF-kappaB signaling. J Biol Chem 280:40364–40374

Cho SH, Chen JA, Sayed F, Ward ME, Gao F, Nguyen TA, Krabbe G, Sohn PD, Lo I, Minami S et al (2015) SIRT1 deficiency in microglia contributes to cognitive decline in aging and neurodegeneration via epigenetic regulation of IL-1beta. J Neurosci 35:807–818

Condello C, Yuan P, Schain A, Grutzendler J (2015) Microglia constitute a barrier that prevents neurotoxic protofibrillar Abeta42 hotspots around plaques. Nat Commun 6:6176

Decressac M, Mattsson B, Weikop P, Lundblad M, Jakobsson J, Bjorklund A (2013) TFEB-mediated autophagy rescues midbrain dopamine neurons from alpha-synuclein toxicity. Proc Natl Acad Sci USA 110:E1817–E1826

Doens D, Fernandez PL (2014) Microglia receptors and their implications in the response to amyloid beta for Alzheimer's disease pathogenesis. J Neuroinflammation 11:48

Donmez G, Wang D, Cohen DE, Guarente L (2010) SIRT1 suppresses beta-amyloid production by activating the alpha-secretase gene ADAM10. Cell 142:320–332

El Khoury J, Toft M, Hickman SE, Means TK, Terada K, Geula C, Luster AD (2007) Ccr2 deficiency impairs microglial accumulation and accelerates progression of Alzheimer-like disease. Nat Med 13:432–438

Guarente L (2011) Sirtuins, aging, and metabolism. Cold Spring Harbor Symp Quan Biol 76:81–90

Herskovits AZ, Guarente L (2014) SIRT1 in neurodevelopment and brain senescence. Neuron 81:471–483

Hickman SE, Allison EK, El Khoury J (2008) Microglial dysfunction and defective beta-amyloid clearance pathways in aging Alzheimer's disease mice. J Neurosci 28:8354–8360

Hug BA, Ahmed N, Robbins JA, Lazar MA (2004) A chromatin immunoprecipitation screen reveals protein kinase Cbeta as a direct RUNX1 target gene. J Biol Chem 279:825–830

Jiang Q, Lee CY, Mandrekar S, Wilkinson B, Cramer P, Zelcer N, Mann K, Lamb B, Willson TM, Collins JL et al (2008) ApoE promotes the proteolytic degradation of Abeta. Neuron 58:681–693

Karuppagounder SS, Pinto JT, Xu H, Chen HL, Beal MF, Gibson GE (2009) Dietary supplementation with resveratrol reduces plaque pathology in a transgenic model of Alzheimer's disease. Neurochem Int 54:111–118

Lee IH, Finkel T (2009) Regulation of autophagy by the p300 acetyltransferase. J Biol Chem 284:6322–6328

Lee IH, Cao L, Mostoslavsky R, Lombard DB, Liu J, Bruns NE, Tsokos M, Alt FW, Finkel T (2008) A role for the NAD-dependent deacetylase Sirt1 in the regulation of autophagy. Proc Natl Acad Sci USA 105:3374–3379

Lucin KM, Wyss-Coray T (2009) Immune activation in brain aging and neurodegeneration: too much or too little? Neuron 64:110–122

Ma Y, Bao J, Zhao X, Shen H, Lv J, Ma S, Zhang X, Li Z, Wang S, Wang Q et al (2013) Activated cyclin-dependent kinase 5 promotes microglial phagocytosis of fibrillar beta-amyloid by up-regulating lipoprotein lipase expression. Mol Cellular Proteomics: MCP 12:2833–2844

Majumdar A, Cruz D, Asamoah N, Buxbaum A, Sohar I, Lobel P, Maxfield FR (2007) Activation of microglia acidifies lysosomes and leads to degradation of Alzheimer amyloid fibrils. Mol Biol Cell 18:1490–1496

Majumdar A, Chung H, Dolios G, Wang R, Asamoah N, Lobel P, Maxfield FR (2008) Degradation of fibrillar forms of Alzheimer's amyloid beta-peptide by macrophages. Neurobiol Aging 29:707–715

Majumdar A, Capetillo-Zarate E, Cruz D, Gouras GK, Maxfield FR (2011) Degradation of Alzheimer's amyloid fibrils by microglia requires delivery of ClC-7 to lysosomes. Mol Biol Cell 22:1664–1676

Mandrekar-Colucci S, Karlo JC, Landreth GE (2012) Mechanisms underlying the rapid peroxisome proliferator-activated receptor-gamma-mediated amyloid clearance and reversal of cognitive deficits in a murine model of Alzheimer's disease. J Neurosci 32:10117–10128

Mawuenyega KG, Sigurdson W, Ovod V, Munsell L, Kasten T, Morris JC, Yarasheski KE, Bateman RJ (2010) Decreased clearance of CNS beta-amyloid in Alzheimer's disease. Science 330:1774

Meyer-Luehmann M, Spires-Jones TL, Prada C, Garcia-Alloza M, de Calignon A, Rozkalne A, Koenigsknecht-Talboo J, Holtzman DM, Bacskai BJ, Hyman BT (2008) Rapid appearance and local toxicity of amyloid-beta plaques in a mouse model of Alzheimer's disease. Nature 451:720–724

Mulder SD, Nielsen HM, Blankenstein MA, Eikelenboom P, Veerhuis R (2014) Apolipoproteins E and J interfere with amyloid-beta uptake by primary human astrocytes and microglia in vitro. Glia 62:493–503

Nimmerjahn A, Kirchhoff F, Helmchen F (2005) Resting microglial cells are highly dynamic surveillants of brain parenchyma in vivo. Science 308:1314–1318

Palop JJ, Mucke L (2010) Amyloid-beta-induced neuronal dysfunction in Alzheimer's disease: from synapses toward neural networks. Nat Neurosci 13:812–818

Qin W, Yang T, Ho L, Zhao Z, Wang J, Chen L, Zhao W, Thiyagarajan M, MacGrogan D, Rodgers JT et al (2006) Neuronal SIRT1 activation as a novel mechanism underlying the prevention of Alzheimer disease amyloid neuropathology by calorie restriction. J Biol Chem 281:21745–21754

Querfurth HW, LaFerla FM (2010) Alzheimer's disease. New Engl J Med 362:329–344

Rogers J, Strohmeyer R, Kovelowski CJ, Li R (2002) Microglia and inflammatory mechanisms in the clearance of amyloid beta peptide. Glia 40:260–269

Sardiello M, Palmieri M, di Ronza A, Medina DL, Valenza M, Gennarino VA, Di Malta C, Donaudy F, Embrione V, Polishchuk RS et al (2009) A gene network regulating lysosomal biogenesis and function. Science 325:473–477

Schultz ML, Tecedor L, Chang M, Davidson BL (2011) Clarifying lysosomal storage diseases. Trends Neurosci 34:401–410

Settembre C, Di Malta C, Polito VA, Garcia Arencibia M, Vetrini F, Erdin S, Erdin SU, Huynh T, Medina D, Colella P et al (2011)

TFEB links autophagy to lysosomal biogenesis. Science 332:1429–1433

Settembre C, De Cegli R, Mansueto G, Saha PK, Vetrini F, Visvikis O, Huynh T, Carissimo A, Palmer D, Klisch TJ et al (2013a) TFEB controls cellular lipid metabolism through a starvation-induced autoregulatory loop. Nat Cell Biol 15:647–658

Settembre C, Fraldi A, Medina DL, Ballabio A (2013b) Signals from the lysosome: a control centre for cellular clearance and energy metabolism. Nat Rev Mol Cell Biol 14:283–296

Tsunemi T, Ashe TD, Morrison BE, Soriano KR, Au J, Roque RA, Lazarowski ER, Damian VA, Masliah E, La Spada AR (2012) PGC-1alpha rescues Huntington's disease proteotoxicity by preventing oxidative stress and promoting TFEB function. Science translational medicine 4:142ra197.

Wegiel J, Wang KC, Imaki H, Rubenstein R, Wronska A, Osuchowski M, Lipinski WJ, Walker LC, LeVine H (2001) The role of microglial cells and astrocytes in fibrillar plaque evolution in transgenic APP(SW) mice. Neurobiol Aging 22:49–61

Weldon DT, Rogers SD, Ghilardi JR, Finke MP, Cleary JP, O'Hare E, Esler WP, Maggio JE, Mantyh PW (1998) Fibrillar beta-amyloid induces microglial phagocytosis, expression of inducible nitric oxide synthase, and loss of a select population of neurons in the rat CNS in vivo. J Neurosci 18:2161–2173

Xiao Q, Yan P, Ma X, Liu H, Perez R, Zhu A, Gonzales E, Burchett JM, Schuler DR, Cirrito JR et al (2014) Enhancing astrocytic lysosome biogenesis facilitates Abeta clearance and attenuates amyloid plaque pathogenesis. J Neurosci 34:9607–9620

Xiao Q, Yan P, Ma X, Liu H, Perez R, Zhu A, Gonzales E, Tripoli DL, Czerniewski L, Ballabio A et al (2015) Neuronal-targeted TFEB accelerates lysosomal degradation of APP, reducing abeta generation and amyloid plaque pathogenesis. J Neurosci 35:12137–12151

Zhang YD, Zhao JJ (2015) TFEB participates in the abeta-induced pathogenesis of Alzheimer's disease by regulating the autophagy-lysosome pathway. DNA Cell Biol 34:661–668

Zhang J, Shi XQ, Echeverry S, Mogil JS, De Koninck Y, Rivest S (2007) Expression of CCR2 in both resident and bone marrow-derived microglia plays a critical role in neuropathic pain. J Neurosci 27:12396–12406

Zhao J, Brault JJ, Schild A, Cao P, Sandri M, Schiaffino S, Lecker SH, Goldberg AL (2007) FoxO3 coordinately activates protein degradation by the autophagic/lysosomal and proteasomal pathways in atrophying muscle cells. Cell Metab 6:472–483

Zhou J, Tan SH, Nicolas V, Bauvy C, Yang ND, Zhang J, Xue Y, Codogno P, Shen HM (2013) Activation of lysosomal function in the course of autophagy via mTORC1 suppression and autophagosome-lysosome fusion. Cell Res 23:508–523

20

Structure of the catalytic domain of a state transition kinase homolog from *Micromonas* algae

Jiangtao Guo[1,2], Xuepeng Wei[1,2], Mei Li[1], Xiaowei Pan[1], Wenrui Chang[1]✉, Zhenfeng Liu[1]✉

[1] National Laboratory of Biomacromolecules, Institute of Biophysics, Chinese Academy of Sciences, Beijing 100101, China
[2] University of Chinese Academy of Sciences, Beijing 100039, China
✉ Correspondence: liuzf@sun5.ibp.ac.cn (Z. Liu), wrchang@sun5.ibp.ac.cn (W. Chang)

ABSTRACT

Under natural environments, plants and algae have evolved various photosynthetic acclimation mechanisms in response to the constantly changing light conditions. The state transition and long-term response processes in photosynthetic acclimation involve remodeling and composition alteration of thylakoid membrane. A chloroplast protein kinase named Stt7/STN7 has been found to have pivotal roles in both state transition and long-term response. Here we report the crystal structures of the kinase domain of a putative Stt7/STN7 homolog from *Micromonas sp.* RCC299 (MsStt7d) in the apo form and in complex with various nucleotide substrates. MsStt7d adopts a canonical protein kinase fold and contains all the essential residues at the active site. A novel hairpin motif, found to be a conserved feature of the Stt7/STN7 family and indispensable for the kinase stability, interacts with the activation loop and fixes it in an active conformation. We have also demonstrated that MsStt7d is a dual-specificity kinase that phosphorylates both Thr and Tyr residues. Moreover, preliminary *in vitro* data suggest that it might be capable of phosphorylating a consensus N-terminal pentapeptide of light-harvesting proteins *Micromonas* Lhcp4 and *Arabidopsis* Lhcb1 directly. The potential peptide/protein substrate binding site is predicted based on the location of a pseudo-substrate contributed by the adjacent molecule within the crystallographic dimer. The structural and biochemical data presented here provide a framework for an improved understanding on the role of Stt7/STN7 in photosynthetic acclimation.

KEYWORDS Stt7/STN7 kinase, state transition, phosphorylation

INTRODUCTION

In order to adapt to constantly changing light conditions under natural environments, photosynthetic organisms have developed the ability to regulate distribution of excitation energy between photosystem II (PSII) and photosystem I (PSI), through a process known as state transition (Bonaventura and Myers, 1969; Murata, 1969). The efficiency of photosynthesis under fluctuating light conditions is thereby maintained at optimal level. During state transition, the major light-harvesting complex (LHCII) is reversibly phosphorylated and migrates between PSII and PSI to balance the excitation level of two photosystems (Bennett, 1991; Allen, 1992). The activity of the kinase responsible for LHCII phosphorylation was found to be regulated by the redox state of plastoquinone (PQ) pool (Allen et al., 1981) and cytochrome b_6f (cyt b_6f) (Vener et al., 1997; Zito et al., 1999). Based on the present model of state transition (Wollman, 2001; Rochaix, 2007; Kargul and Barber, 2008), when the PSII is over-excited, the PQ pool is highly reduced and plastoquinol PQH_2 binds to the Q_o site of cyt b_6f, leading to activation of LHCII kinase which then phosphorylates LHCII. The phospho-LHCII migrates from PSII to PSI and redistributes energy to PSI (state II). Over-excitation of PSI leads to oxidation of the PQ pool and deactivation of LHCII kinase. The LHCII phosphatase dephosphorylates phospho-LHCII, followed by migration of LHCII from PSI back to PSII. Thus, the energy absorbed by LHCII is delivered back to PSII (state I).

Searching for the LHCII kinase is a central and long-standing goal in the field of state transition research (Lin et al., 1982; Coughlan and Hind, 1987; Snyders and Kohorn, 1999, 2001; Allen and Race, 2002). Recently, Stt7 from *Chlamydomonas reinhardtii* (CrStt7) and its ortholog STN7 from *Arabidopsis thaliana* (AtSTN7) have been reported as protein kinases located on the thylakoid membrane with crucial roles

Table 1. Multiple Stt7/STN7 homologs from *Micromonas sp.* RCC299[a]

Accession	Name	Max score	Total score	Query coverage	E-value	Max identity
XP_002499717	MsStt7a	291	291	98%	2×10^{-93}	51%
XP_002504036	MsStt7b	231	231	94%	3×10^{-69}	39%
XP_002500340	MsStt7c	221	221	98%	1×10^{-67}	40%
XP_002504185	MsStt7d	149	149	98%	8×10^{-41}	33%
XP_002505498	MsStt7e	145	145	97%	4×10^{-40}	32%

[a] The top five results obtained by searching the Non-redundant protein sequence (nr) database of *Micromonas sp.* RCC299 in NCBI with the BLAST program using the kinase domain of AtSTN7 (Phe 134–Phe 452) as the query sequence (Altschul et al., 1997).

in state transition and phosphorylation of LHCII (Depège et al., 2003; Bellafiore et al., 2005). In the Stt7/STN7 deletion mutants, the state transition from state I to II is blocked and the phosphorylation of LHCII stays at low levels under changing light conditions (Depège et al., 2003; Bellafiore et al., 2005). The Stt7/STN7 protein contains a catalytic domain resembling Ser/Thr kinases, a predicted transmembrane helix, and a lumen-exposed N-terminal region with a pair of conserved Cys residues which are important for the regulation of Stt7 activity (Depège et al., 2003; Lemeille et al., 2009). Although co-immunoprecipitation data showed that Stt7 may interact with LHCII (Lemeille et al., 2009), so far it remains unclear whether Stt7/STN7 phosphorylates LHCII directly or indirectly by activating an unidentified LHCII kinase through an unknown signaling cascade. The structural basis for the phosphorylation of LHCII, as well as the regulatory mechanism of LHCII kinase activity, is also largely unknown.

In addition to its role in the state transition (acting on a timescale of seconds to minutes), STN7 might also be involved in the long-term response (LTR, on a timescale of hours, days and seasons) of plants toward changes in light quality which leads to modification of thylakoid composition (Bonardi et al., 2005; Wagner et al., 2008; Pesaresi et al., 2009). It was proposed that STN7 phosphorylated an unidentified thylakoid protein that triggered LTR signaling processes (Pesaresi et al., 2009). Nevertheless, the detailed mechanism of LTR and the molecular basis for the function of STN7 in LTR await further investigation.

An in-depth understanding of the mechanistic roles of Stt7/STN7 in state transition and LTR has been impeded by the lack of three-dimensional structure of Stt7/STN7. Here we report for the first time the crystal structures of the kinase domain of a putative Stt7/STN7 homolog from *Micromonas sp.* RCC299 (MsStt7d) in the apo form and in complex with nucleotide substrates. The phosphorylation behavior of MsStt7d, as well as the peptide substrate specificity and its potential binding site on MsStt7d, has been investigated in this work.

RESULTS

The kinase domain of MsStt7d adopts the canonical protein kinase fold

In an effort to characterize the structure and function of Stt7/STN7 proteins, numerous homologs of Stt7/STN7 including those from *Arabidopsis thaliana*, *Oryza sativa*, *Chlamydomonas reinhardtii* and *Micromonas sp.* RCC299 have been cloned and overexpressed in N-terminal truncation form covering the entire kinase domain. Among more than 50 constructs made, MsStt7d exhibited the best behavior with high expression level in soluble and stable form, while other homologs either expressed as inclusion body, or precipitated readily during purification. As one of the multiple putative Stt7/STN7 homologs found in *Micromonas sp.* RCC299 (Table 1), the MsStt7d contains a kinase domain (MsStt7d-KD, Arg 172–Phe 488) sharing 47% sequence similarity with the corresponding region of CrStt7 or AtSTN7 (Fig. 1). To reveal the molecular architecture of Stt7/STN7 kinase domain and to gain insights into its catalytic and regulatory mechanisms, we have determined the high-resolution crystal structures of MsStt7d-KD in the apo form, and in complex with ADP, ATP or AMP-PNP. Data collection and refinement statistics are listed in Table 2.

In the crystallographic asymmetric unit, there are two MsStt7d-KD molecules related by pseudo-C_2 symmetry (Fig. 2A). Each monomer contains the kinase domain of MsStt7d (Arg 172–Phe 488), and 21 additional amino acid residues (Leu 151–Ile 171) at the N-terminal region (Fig. 2B). The kinase domain adopts a canonical protein kinase fold which can be divided into two lobes. The small N-terminal lobe consists of a sheet of six antiparallel β-strands (β1–β6) and a single α-helix (α1), while the large C-terminal lobe contains nine α helices (α2–α10) and a pair of short β-strands (β7 and β8) (Fig. 2B). The nucleotide substrate binding site is located in the cleft sandwiched between the two lobes. Within the N-lobe region, residues 178–186 linking β2 and β3 form a highly conserved glycine-rich loop (with LGXGXXGXV motif, also known as the P-loop) (Fig. 2B) that is usually involved in the binding of ATP, substrate recognition, catalysis and kinase activity regulation (Bossemeyer, 1994; Hanks and Hunter, 1995). In the C-lobe region, there is another functionally important loop between residue 338–370 (Fig. 2B) corresponding to the activation loop in protein kinases. The activation loop may adopt variable conformations under inactive and active states, exerting an important role in the control of kinase activity (Nolen et al., 2004).

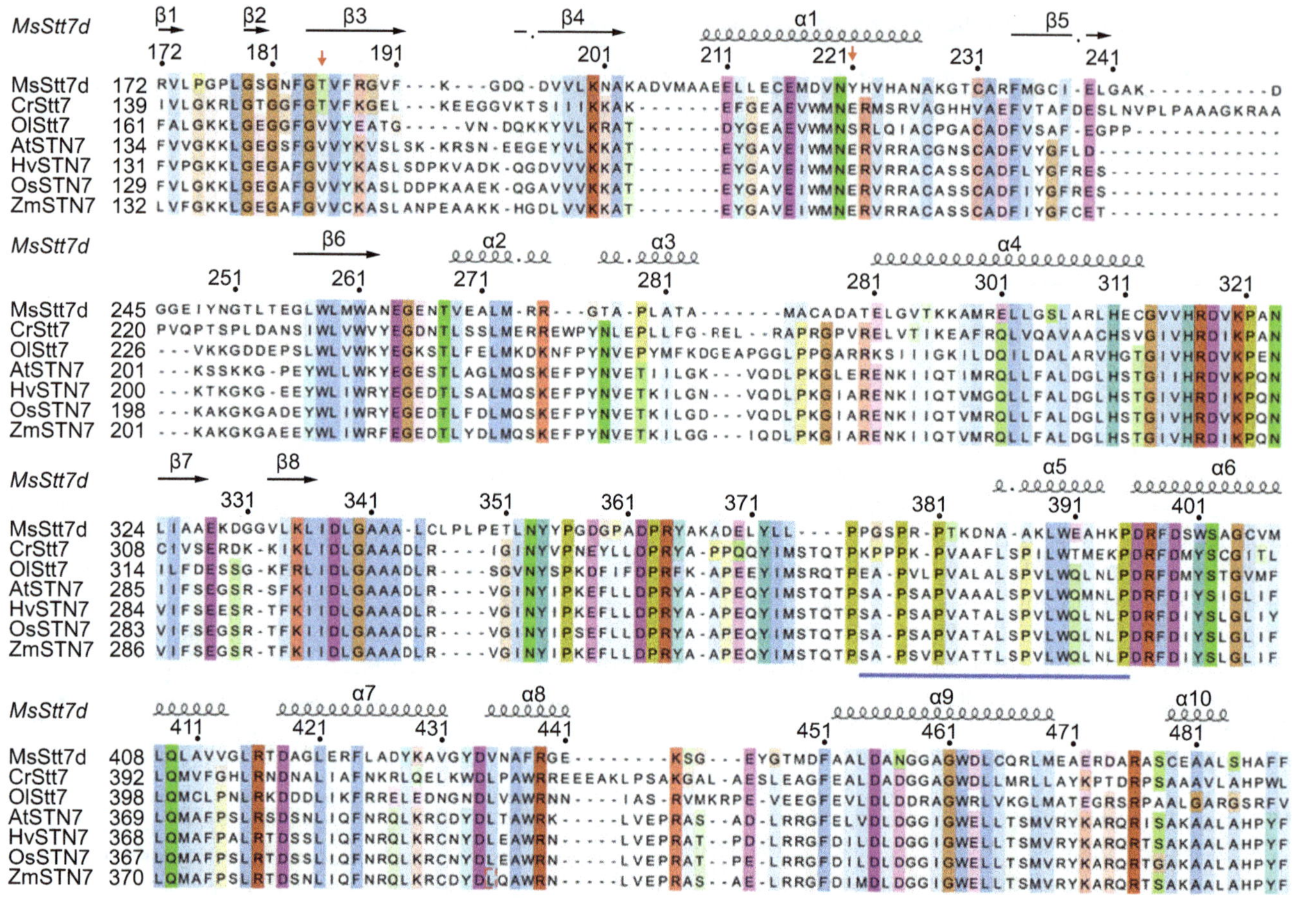

Figure 1. Sequence alignment of the protein kinase domains of various Stt7/STN7 family members. Multiple sequence alignment is performed with ClustalW2 followed by manual adjustment and displayed with Jalview using the coloring mode by conservation. The autophosphorylation sites on MsStt7d-KD are indicated with red arrows; the novel α5 hairpin region is highlighted by thick blue underline. Abbreviations: MsStt7d, a Stt7 homolog from *Micromonas sp.* RCC299; CrStt7, Stt7 from *Chlamydomonas reinhardtii*; OlStt7, Stt7 from *Ostreococcus lucimarinus*; AtSTN7, STN7 from *Arabidopsis thaliana*; HvSTN7, STN7 from *Hordeum vulgare*; OsSTN7, STN7 from *Oryza sativa*; ZmSTN7, STN7 from *Zea mays*.

As shown in Fig. 2C, MsStt7d-KD contains all the conserved residues essential for the binding of ATP and catalysis of phosphoryl-transfer reaction at the active site located within the inter-lobe cleft. Previous work on the cyclin-dependent protein kinase 2 (CDK2, a Ser/Thr kinase) has demonstrated that residues Lys 33, Asn 132 and Asp 145 are involved in the binding of ATP phosphate groups directly or through Mg^{2+} ion (De Bondt et al., 1993). The corresponding conserved residues in MsStt7d-KD are residues Lys 200, Asn 323 and Asp 338 according to the superposition of MsStt7d-KD structure with that of CDK2 (Fig. 2C). The role of these residues in binding nucleotide molecule is further confirmed in the structure of MsStt7d-KD-ADP complex. In this structure, one ADP and one Mg^{2+} ion per monomer are found inside the active site (Fig. 2D). The Mg^{2+} ion is coordinated by the carbonyl of Asn 323 side chain, the carboxyl of Asp 338, two oxygen atoms from α- and β-phosphate groups, respectively, and two water molecules (Fig. 2D). The α-phosphate group of ADP molecule forms strong ionic interaction with the side chain amine group of Lys 200. A nearby conserved acidic residue, Glu 216 from the α1 helix stabilizes the orientation of Lys 200 side chain by forming ionic interaction with it (Fig. 2D). The Asp 318 in MsStt7d-KD is deduced to be the catalytic residue with a similar role as the Asp 127 in CDK2, which coordinates the hydroxyl of the substrate for accepting the γ-phosphate from ATP (Fig. 2C) (De Bondt et al., 1993).

The structures of MsStt7d-KD in complex with ATP and AMP-PNP are nearly identical to the one in complex with ADP. In the MsStt7d-KD-ATP complex, the γ-phosphate group of ATP could not be detected in the electron density map, presumably due to hydrolysis of ATP by the kinase during the process of crystal soaking (Fig. 2E). Previously, the intrinsic ATPase activity was also observed in other protein kinases and interpreted as the activity of catalyzing phosphoryl-transfer reaction toward water molecules (Ward and O'Brian, 1992; Mendelow et al., 1993). As for the MsStt7d-KD-AMP-PNP complex,

Table 2. Data collection and refinement statistics

Parameters	MsStt7d-KD-KI	Native MsStt7d-KD	MsStt7d-KD-ADP	MsStt7d-KD-ATP	MsStt7d-KD-AMP-PNP
Data collection					
Wavelength (Å)	1.5418	0.97939	0.97853	0.97853	0.97853
Space group	$P2_1$	$P2_1$	$P2_1$	$P2_1$	$P2_1$
a, *b*, *c* (Å)	54.10, 133.31, 55.30	53.71, 133.14, 55.05	53.93, 133.03, 55.16	53.67, 132.64, 55.01	53.86, 132.61, 55.18
β (°)	116.65	116.30	116.49	116.26	116.64
Resolution (Å)	33.3–2.26	50–1.35	50–1.50	33.16–1.60	33.15–1.70
Unique reflections	32604	150593	110828	90626	75812
Completeness (%)	99.4 (97.7)[a]	99.9 (100.0)	99.9 (100.0)	100.0 (100.0)	100.0 (100.0)
Redundancy	12.9 (12.4)	4.0 (3.7)	7.5 (7.4)	6.9 (7.1)	7.2 (7.2)
$I/\sigma(I)$	20.9 (7.7)	29.0 (2.8)	32.9 (4.5)	14.4 (3.5)	14.8 (4.1)
R_{sym}[b]	0.094 (0.315)	0.043 (0.475)	0.063 (0.542)	0.080 (0.518)	0.081 (0.474)
Refinement					
Resolution (Å)	-	50–1.35	50–1.50	33.16–1.60	33.15–1.70
R_{work}/R_{free}[c]	-	0.133/0.174	0.157/0.199	0.168/0.193	0.172/0.201
Protein atoms	-	5295	5281	5293	5280
Ligand/ion atoms	-	4	57	56	65
Water atoms	-	876	654	572	625
B-factors ($Å^2$)					
all	-	20.3	23.5	21.6	26.6
protein	-	18.8	22.4	21.0	25.9
ligand/ion	-	17.4	23.4	36.1	36.5
water	-	29.5	32.2	26.1	31.2
rmsd bond length (Å)		0.012	0.005	0.006	0.006
rmsd bond angle (°)	-	1.738	1.019	1.076	0.999
Ramachandran plot (%)[d]	-	92.5, 7.0, 0.5, 0	91.6, 8.4, 0, 0	92.0, 7.9, 0.2, 0	91.8, 8.0, 0.2, 0

[a] Values in parentheses refer to data in the highest resolution shell.

[b] $R_{sym} = \Sigma_{hkl} \Sigma_i |I_i(hkl) - <I(hkl)>| / \Sigma_{hkl} \Sigma_i I_i(hkl)$.

[c] $R_{work} = \Sigma_{hkl}|F_o(hkl) - F_c(hkl)| / \Sigma_{hkl}F_o(hkl)$. R_{free} was calculated with 5% of the data that were not used for refinement.

[d] Percentage of residues in most favored, additional allowed, generously allowed and disallowed regions in Ramachandran plot.

the γ-phosphate group of AMP-PNP molecule is found to be in a highly bent conformation (Fig. 2E). The adenine groups of all nucleotide molecules are bound deep in a hydrophobic pocket within the cleft between the N- and C-lobes, while the tail groups are partially exposed to the solvent region. Superposition of all six nucleotide molecules (two molecules per dimeric MsStt7d-KD-nucleotide complex) indicates a fairly rigid adenine group and highly flexible polyphosphate groups with multiple conformations (Fig. 2E). When the phosphorylatable peptide substrate binds to the kinase, the terminal phosphate group will be fixed in an optimal orientation for the phosphoryl-transfer reaction, as observed in the ternary complex structure of cAMP-dependent protein kinase, nucleotide and peptide substrates (Madhusudan et al., 1994).

A novel hairpin motif stabilizes the activation loop in an active conformation

Is the structure of MsStt7d-KD in an active or inactive state? To predict the conformational state of MsStt7d-KD in the crystal, the structure of MsStt7d-KD was compared with those of CDK2 in either active or inactive state (Schulze-Gahmen et al., 1996; Brown et al., 1999) (Fig. 3). The structure of MsStt7d-KD closely resembles that of active-state CDK2 in two major aspects that are related to the activation of protein kinases (Endicott et al., 2012). Firstly, the α1 helix of MsStt7d-KD and the equivalent helix of active-state CDK2 (also named C helix previously) are aligned in the same orientation, allowing the Glu residue (Glu 216 in MsStt7d-KD) to stabilize the Lys resi-

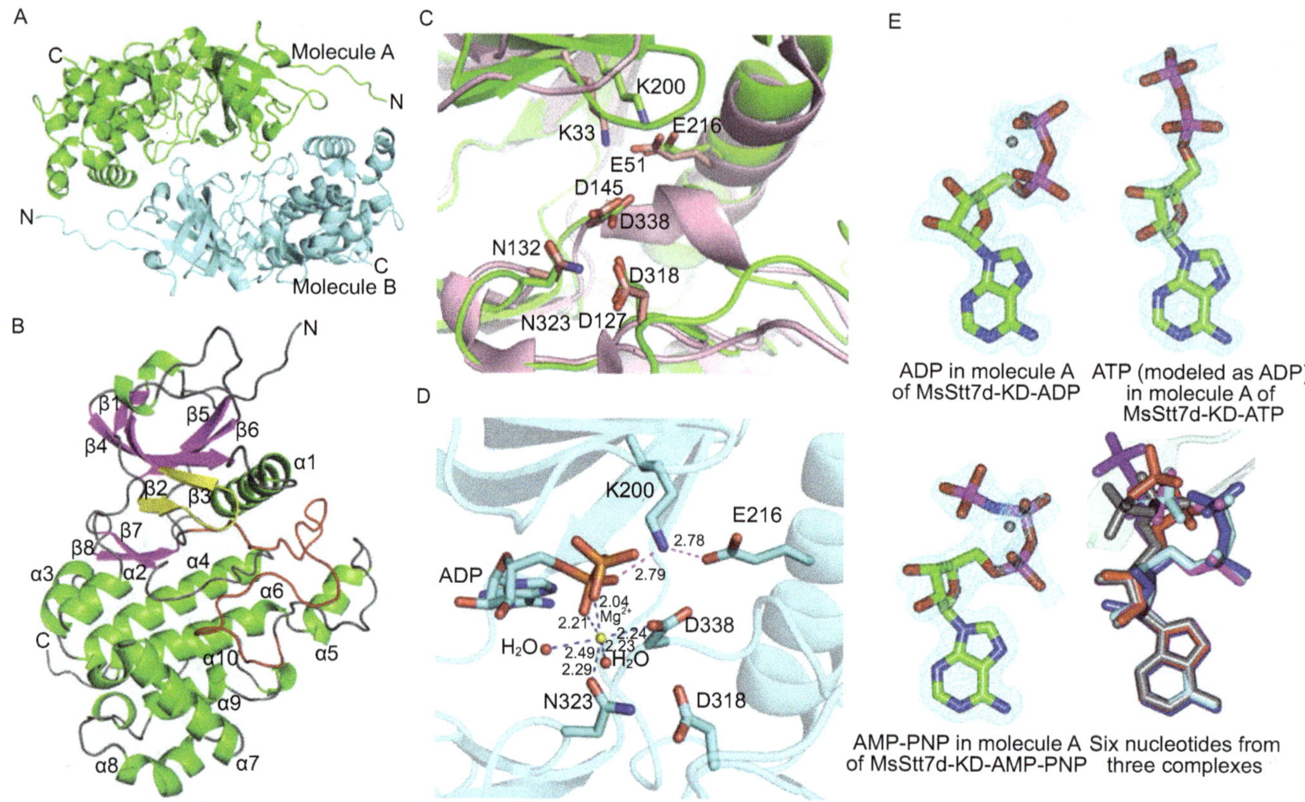

Figure 2. Structures of MsStt7d-KD apoprotein and in complex with nucleotide substrates. (A) A dimer of the MsStt7d-KD apoprotein within the asymmetric unit. (B) The structure of monomeric MsStt7d-KD. The α-helices are colored in green, the β-strands are in magenta and loops are in grey. The functionally important glycine-rich loop and activation loop regions are highlighted by yellow and red colors, respectively. (C) Conserved residues in the active site important for ATP binding and phosphoryl-transfer reaction. Green, MsStt7d-KD; pink, CDK2 (PDB code: 1QMZ). The labels for residues from MsStt7d-KD and CDK2 are in green and pink colors, respectively. (D) The ADP binding site of MsStt7d-KD-ADP complex structure. One ADP and one Mg^{2+} per monomer are assigned in the complex structure. The blue dashes indicate the coordination bonds around Mg^{2+}; the magenta dashes indicate the ionic interactions between K200 and E216 or ADP α-phosphate group. The numbers near the dash lines show the distances between two atoms (Å). (E) $2F_o$-F_c (1.0 σ level) electron densities of ADP, ATP and AMP-PNP molecules bound to MsStt7d-KD and superposition of all six nucleotide molecules observed in the complex structures. The ATP molecules in the MsStt7d-KD-ATP complex structures have no detectable features for the γ-phosphate groups in the electron density map (presumably due to hydrolysis by the kinase) and thereby modeled as ADP molecules instead.

due (Lys 200 in MsStt7d-KD) in a position optimal for the binding of ATP α-phosphate group (Figs. 2D and 3A). Secondly, the activation loops of MsStt7d-KD and active-state CDK2 are both positioned beside the catalytic cleft to form a polypeptide substrate-binding platform on one hand and open up the active site for the access of ATP molecule on the other hand (Fig. 3A). In comparison, the inactive-state CDK2 has a tilted C helix and its conserved Glu and Lys residues drift away from the ATP binding site. Moreover, the catalytic cleft in the inactive-state CDK2 is obstructed by the activation loop (Fig. 3B). Therefore, the structure of MsStt7d-KD is most likely in an active conformation ready for accepting substrates and performing catalysis.

In the case of CDK2, when the activator, cyclin A, binds to the C helix and the activation loop region, the C helix is reoriented and the activation loop moves away from the catalytic cleft, leading to the activation of CDK2 (Jeffrey et al., 1995). Why is the MsStt7d-KD trapped in an active state? A novel hairpin motif was found adjacent to the activation loop region of MsStt7d-KD (Figs. 3 and 4A). This motif consists of the α5 helix (Ala 387–Pro 395) and a short loop (Pro 376–Ala 386) antiparallel to the α5 helix, and is named α5 hairpin hereby. The α5 hairpin forms extensive hydrophobic interactions and hydrogen bonds with residues from the activation loop region. On the hairpin surface which contacts the activation loop, two hydrophobic pockets accommodate the side chains of Tyr 354 and Pro 356 from the activation loop region, respectively (Fig. 4B). Furthermore, the residues Ala 386, Ala 387, Trp 390 and Pro 395 from α5 hairpin form extensive hydrophobic interactions with residues Cys 345, Leu 346 and Pro 347 from the activation loop, enhancing the interactions between the two regions (Fig. 4B). Besides, water molecules located at the

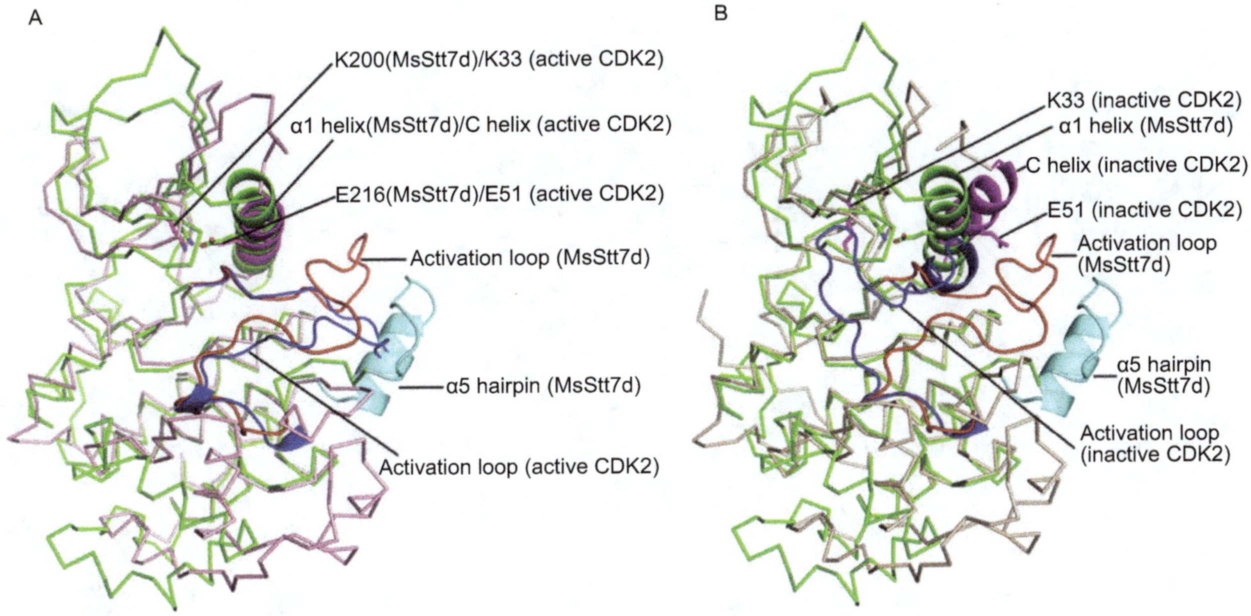

Figure 3. Superposition of MsStt7d-KD structure with that of CDK2 in the active or inactive state. (A) Superposition of MsStt7d-KD (green) with the active-state CDK2 (PDB code: 1QMZ, pink). The α1 helix (green), activation loop (red) and α5 hairpin (cyan) of MsStt7d-KD, and the C helix (magenta) and the activation loop (blue) of active CDK2 are shown as cartoon models. The side chains of the key residues Lys and Glu from two molecules are shown as stick models. (B) Superposition of MsStt7d-KD (green) with the inactive-state CDK2 (PDB code: 1HCL, wheat). The C helix, the activation loop, the key residues Lys and Glu of inactive CDK2 are shown in the same colors as those of active CDK2 in (A).

interface between the activation loop and the hairpin contribute to the formation of hydrogen bonds which further strengthen the interactions between the two regions (Fig. 4C). Thus, the activation loop is fixed in a conformation which closely resembles that of the active-state CDK2 (Brown et al., 1999) (Fig. 3A), but differs dramatically from that of the inactive-state CDK2 (Schulze-Gahmen et al., 1996) (Fig. 3B). When the α5 helix was deleted from the MsStt7d-KD protein (the Δ386–393 mutant), the majority of target protein ended in the insoluble fraction when expressed in *E. coli*, suggesting that the deletion mutant was highly unstable or did not fold properly. According to the sequence alignment result, the α5 hairpin is a conserved feature also present in other Stt7/STN7 family members, but absent in CDK2 or other protein kinases (Fig. 4D). Although it appears that the seven residues between Pro 381 and Leu 389 in MsStt7d are not well conserved when compared to the other Stt7/STN7 members, a close examination on the structure reveals that the side chains of these residues point away from the interface between the α5 hairpin and the activation loop. They are exposed on the protein surface, thus explaining why the sequence of this region is relatively variable. Furthermore, sequence-based prediction on the secondary structures of CrStt7/AtSTN7 indicates that their corresponding regions aligning to the α5 hairpin of MsStt7d form similar coil-helix structures. Therefore, the hairpin motif containing a "$PX_{7/8}LW$" consensus sequence (Fig. 4D) is a characteristic feature for the Stt7/STN7 family which not only has an important role in stabilizing the activation loop in the active state, but also is essential for protein stability.

MsStt7d-KD exhibits basal autophosphorylation activity toward both Thr and Tyr residues

To further analyze whether the MsStt7d-KD protein is active in converting ATP to ADP in solution, a luminescent ADP-detection assay was performed on the wild-type, K200R, D318A and D318N mutant proteins. The three active-site mutants had nearly abolished activity in comparison with the wild-type protein (Fig. 5A). Evidently, residues Lys 200 and Asp 318 are essential for its ADP-producing activity. As no exogenous peptide substrates were added in the luminescent ADP-detecting assay, the activity of wild-type MsStt7d-KD is therefore attributed to the intrinsic ATPase activity and/or autophosphorylation activity. When the kinase was incubated with radioactive substrate [γ-^{32}P] ATP, we found that the purified MsStt7d-KD protein displayed basal autophosphorylation activity (Fig. 5B). The active-site mutants including K200R, D318A and D318N all lost their autophosphorylation activity (Fig. 5B). Previous *in vivo* study showed that the K167R and K167Q mutants of AtSTN7 had abolished function in state transition (Bellafiore et al., 2005). Lys 200 in MsStt7d is the equivalent residue of Lys 167 in AtSTN7. Therefore, the abolished kinase activity of the MsStt7d-KD K200R mutant observed through our *in vitro* assays is consistent with the loss-of-function phenotype of

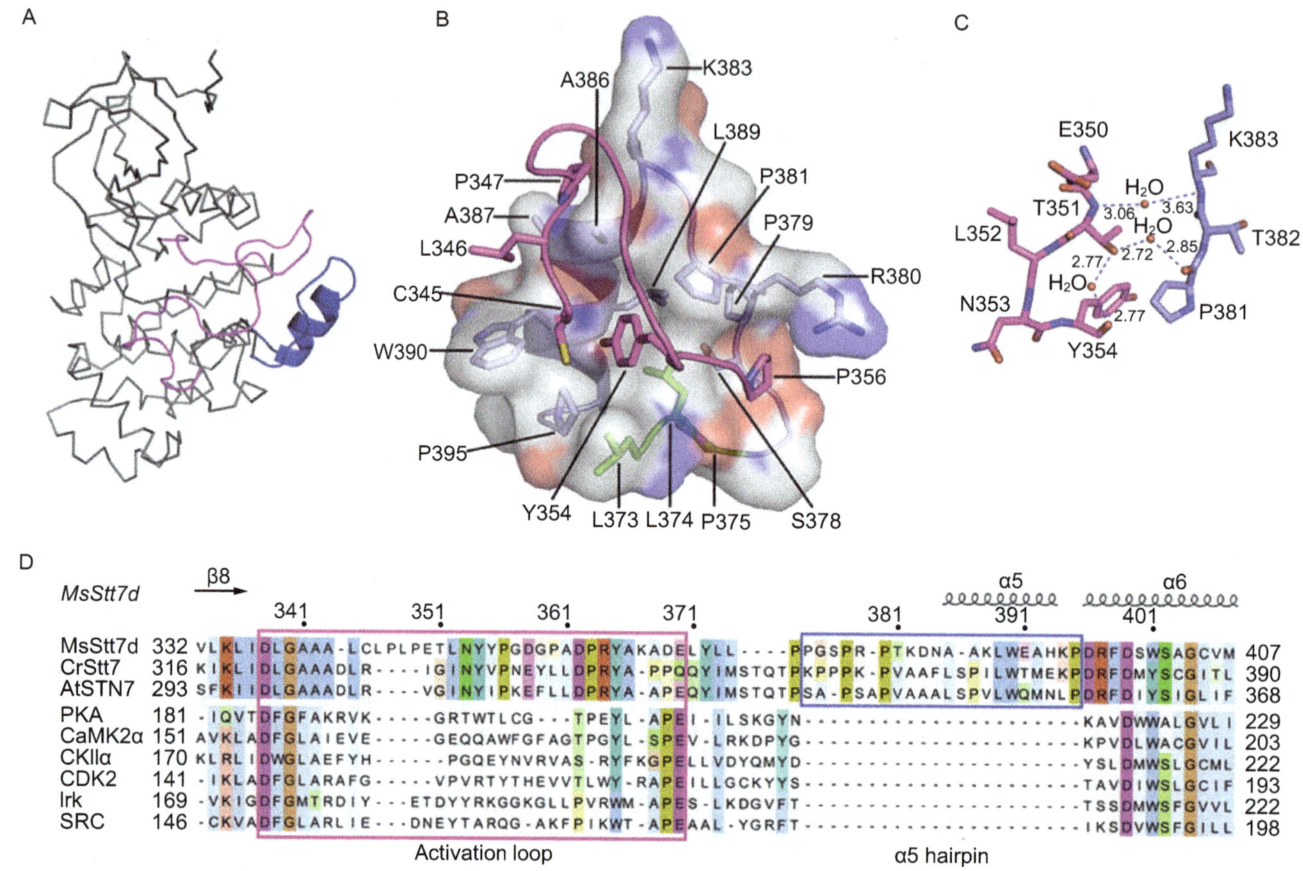

Figure 4. A novel hairpin motif discovered in the structure of MsStt7d-KD. (A) The α5 hairpin motif is positioned next to the activation loop. The activation loop (magenta) and the α5 hairpin (blue) are shown as cartoon models. (B) The interactions between the α5 hairpin and the activation loop. The α5 hairpin is shown as a cartoon model superposed with a semi-transparent surface presentation. Amino acid residues involved in the hydrophobic interactions at the α5 hairpin-activation loop interface are highlighted as stick models. Color codes: blue, residues from the α5 hairpin; magenta, residues from the activation loop; green, residues from other regions. (C) Hydrogen bond interactions bridged by water molecules at the α5 hairpin-activation loop interface. The numbers near the dash lines show the lengths of the hydrogen bonds (Å). (D) Sequence alignment showing that other members of Stt7/STN7 family (CrStt7 and AtSTN7) also contain stretches of amino acid sequences similar to that of the α5 hairpin motif in MsStt7d. Other protein kinases are included as negative controls showing the absence of such a feature in their sequences. Abbreviations: PKA, cAMP-dependent protein kinase catalytic subunit α isoform 1 from *Homo sapiens*; CaMK2α, calcium/calmodulin-dependent protein kinase type II subunit α isoform 1 from *Homo sapiens*; CKIIα, casein kinase II subunit α from *Mus musculus*; CDK2, cyclin-dependent kinase 2 from *Homo sapiens*; Irk, Tyr kinase domain of insulin receptor from *Homo sapiens*; SRC, Tyr protein kinase C-Src from *Homo sapiens*.

K167R/K167Q mutant of AtSTN7.

The autophosphorylation reaction products of MsStt7d-KD were analyzed through western blots with anti-phosphoserine, anti-phosphothreonine or anti-phosphotyrosine antibody to identify which type/types of amino acid residues on the protein was/were phosphorylated. The results indicate that the dominating autophosphorylation sites in MsStt7d-KD are on Thr and Tyr residues, but not Ser residues (Fig. 5C). Subsequently, identification of the phosphorylation sites using mass spectrometry has led to the finding that MsStt7d-KD is primarily autophosphorylated on residue Thr 185 in the glycine-rich loop and Tyr 221 in the α1 helix (Fig. 5D and 5E). Further analysis on the autophosphorylation level of T185A and Y221F mutants (Fig. 5F) indicates that Thr 185 is the major autophosphorylation site, while Tyr 221 is a minor site. The results demonstrate that MsStt7d-KD exhibits basal autophosphorylation activity toward both Thr and Tyr residues. Thus, MsStt7d-KD belongs to the dual-specificity protein kinase family which is capable of phosphorylating both Thr and Tyr residues (Lindberg et al., 1992).

The peptide substrate binding site of MsStt7d-KD

In order to test whether MsStt7d-KD is capable of phosphorylating the N-terminal peptides of the light-harvesting complexes directly, we have constructed, expressed and purified various fusion proteins with the RKXVA pentapeptide (X being T/Y for phosphorylation) fused to MsStt7d-KD at the N-terminus. The RKTVA is a consensus N-terminal sequence shared between

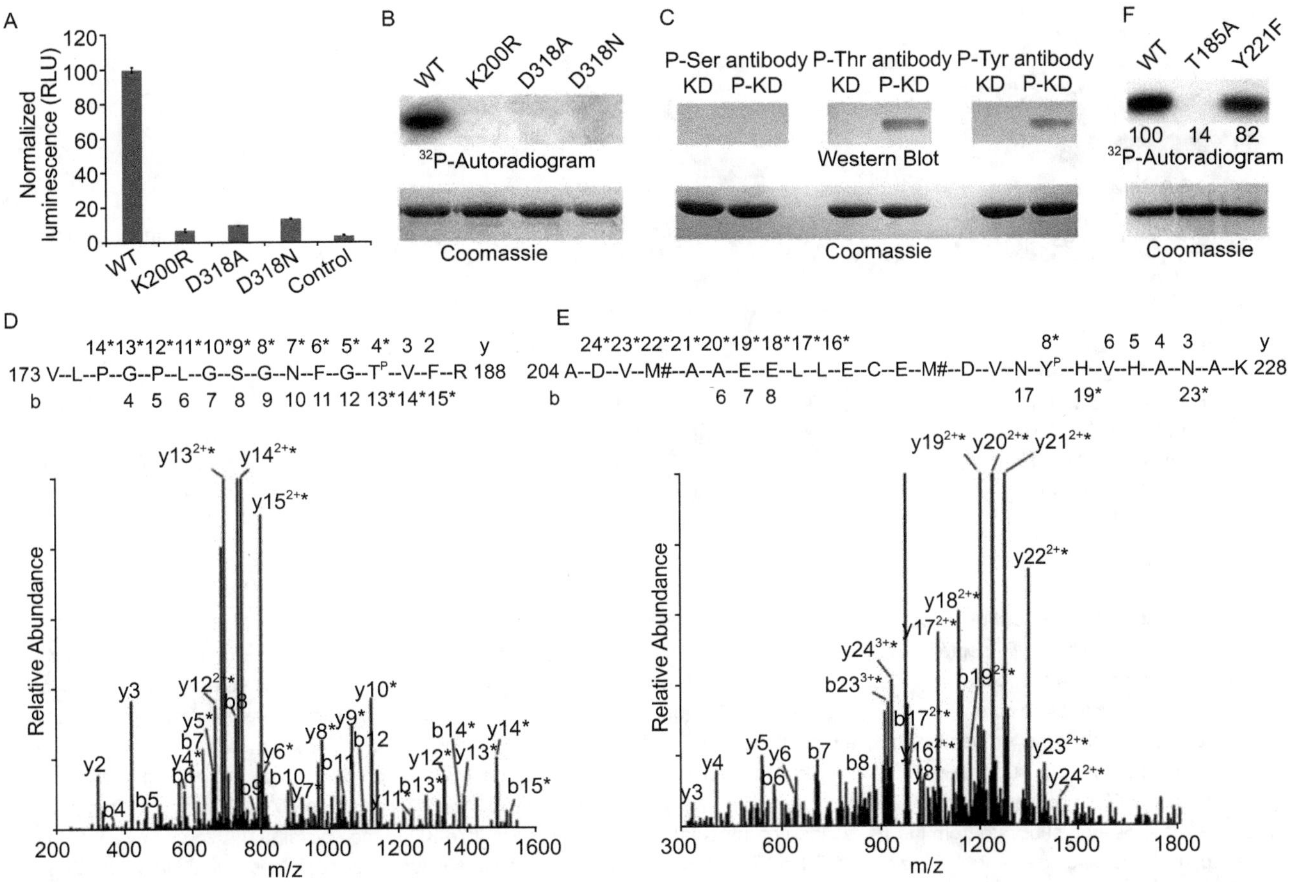

Figure 5. Activity assays on MsStt7d-KD and its mutants. (A) The kinase activity of wild-type (WT) and active-site mutants of MsStt7d-KD measured through luminescent ADP-Glo assay (Promega). The activity data is present as the normalized relative light unit (RLU) of luminescence signal correlated to the amount of ADP produced by the kinase. The data are the mean values of three independent experiments ± standard deviations (SDs). (B) Autophosphorylation of the wild-type MsStt7d-KD and active-site mutants as detected by autoradiography using [γ-^{32}P] ATP as the kinase substrate. The gels stained with Coomassie Brilliant Blue are shown as loading controls in all the figures with "Coomassie" labels throughout the paper. (C) Western blots analysis of the autophosphorylation products using anti-phosphoserine, anti-phosphothreonine and anti-phosphotyrosine antibodies, respectively. The "KD" and "P-KD" indicate the MsStt7d-KD samples before and after autophosphorylation reaction, respectively. (D) The MS/MS spectrum of a doubly charged ion at m/z = 849.53 for MH_2^{2+} corresponding to the mass of the phosphorylated peptide VLPGPLGSGNFGTPVFR (173–188). The * labeled peaks correspond to masses of b and y ions of phosphorylated peptide fragments. (E) The MS/MS spectrum of a triply charged ion at m/z = 992.05 for MH_3^{3+} corresponding to the mass of the phosphorylated and oxidized peptide ADVM#AAEELLECEM#DVNYPHVHANAK (204–228). The * labeled peaks correspond to masses of b and y ions of phosphorylated peptide fragments; the # indicates the oxidized methionine residues. (F) Verification of the MS-detected autophosphorylation sites through mutagenesis and autoradiography. The relative integrated intensity over the area of each band is shown below autoradiogram.

the light harvesting proteins Lhcb1 from *Arabidopsis thaliana* (AtLhcb1) and the Lhcp4 from *Micromonas sp.* RCC299 (MsLhcp4). Compared with the wild-type MsStt7d-KD, the phosphorylation level of RKTVA-MsStt7d-KD increased dramatically (Fig. 6A). The RKYVA-MsStt7d-KD protein also displayed enhanced phosphorylation at a level even higher than that of RKTVA-MsStt7d-KD. In contrast, when the residue Thr/Tyr was mutated to Ser/Ala/Phe, the phosphorylation of these three fusion proteins (RKSVA-MsStt7d-KD, RKAVA-MsStt7d-KD and RKFVA-MsStt7d-KD) decreased to a level close to that of the wild-type protein (Fig. 6A). Therefore, the observed enhanced phosphorylation of RKTVA-MsStt7d-KD and RKYVA-MsStt7d-KD is most likely due to direct phosphorylation on the exogenous peptides fused to the N-terminus of MsStt7d-KD. Consistent with the observed basal autophosphorylation on endogenous Thr and Tyr residues (Fig. 5), MsStt7d-KD may also phosphorylate exogenous Thr and Tyr residues covalently linked to the N-terminal region of the kinase domain (Fig. 6A).

To gain insights into the peptide/protein substrate binding site on MsStt7d-KD, the structure of CDK2-peptide complex (Brown et al., 1999) or insulin receptor kinase (Irk, a tyrosine kinase)-peptide complex (Hubbard, 1997) is superposed onto

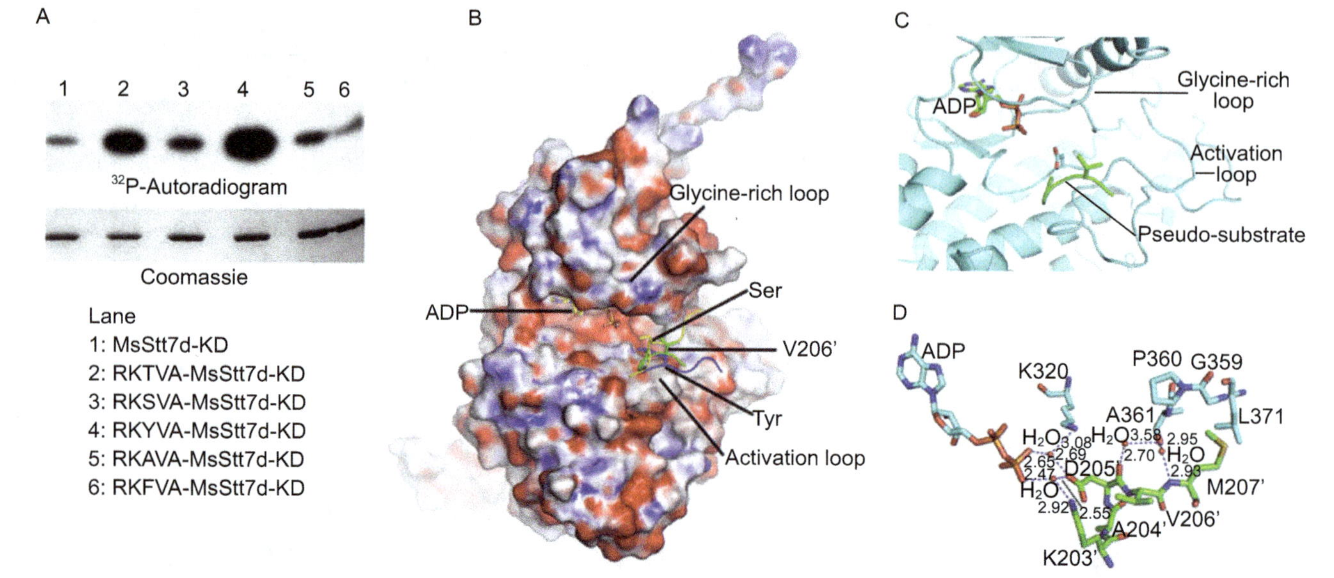

Figure 6. The potential peptide substrate binding site of MsStt7d-KD. (A) The phosphorylation activity of MsStt7d-KD toward the "RKXVA" (X being T/S/Y/A/F) pentapeptides covalently fused to its N-terminus. The "RKTVA" pentapeptide is a consensus N-terminal sequence shared by *Micromonas* Lhcp4 and *Arabidopsis* Lhcb1. (B) The MsStt7d-KD is shown as electrostatic potential surface model with calculated electrostatic potential in the range from -15 kT/e (deep red, electronegative) to 15 kT/e (deep blue, electropositive). k, the Boltzmann constant; T, the temperature; e, the magnitude of the electron charge. ADP molecule is shown as a stick model. The yellow and blue peptides are the substrates of CDK2 (PDB code: 1QMZ) and Irk (PDB code: 1IR3), respectively. They are docked on the surface of MsStt7d-KD by superposing CDK2 and Irk structures (omitted from the figure for clarity) with the MsStt7d-KD structure. The green peptide is the pseudo-substrate from the β4–α1 loop of the adjacent molecule of MsStt7d-KD in the crystal. (C) A zoom-in view of the pseudo-substrate binding site of MsStt7d-KD. MsStt7d-KD is shown in cyan; the pseudo-substrate polypeptide is colored in green. D318 and V206′ indicate key residues from the MsStt7d-KD enzyme and pseudo-substrate, respectively. (D) Interactions between the pseudo-substrate (green) polypeptide and MsStt7d-KD (cyan). The hydrogen bonds are indicated by the blue dashes. The numbers near the dash lines show the distances between two adjacent atoms forming hydrogen bonds (Å). The residues from pseudo-substrate are labeled with ′ symbols.

that of MsStt7d-KD-ADP complex. The result shows that the potential binding site for the peptide substrate on MsStt7d-KD is around the flat surface area between the glycine-rich loop and activation loop region (Fig. 6B). The α-carbon atom of Tyr residue on the Irk substrate peptide moves outwards by 4.9 Å as compared with the α-carbon atom of Ser residue on the CDK2 substrate peptide. The terminal hydroxyl groups of both Ser and Tyr residues are located merely 3.6 Å and 3.3 Å from the side chain of the catalytic residue Asp 318 in MsStt7d-KD, suggesting that they are in a position ready to accept the γ-phosphate group from the ATP molecule.

A straightforward way of observing the interactions between the peptide substrates and MsStt7d-KD is to solve the structures of MsStt7d-KD in complex with peptide substrates. Extensive attempts have been made either by soaking the MsStt7d-KD crystals in a solution with excess peptide substrates or by co-crystallizing the MsStt7d-KD with peptide substrates. Nevertheless, no peptide substrates were found in the structures solved, indicating the binding site might not be accessible to the peptide substrates within the crystal. Further analysis on the structure of MsStt7d-KD-ADP complex reveals the presence of a pseudo-substrate occupying the potential peptide binding site. In the dimer of MsStt7d-KD-ADP complex within the asymmetric unit, residue 203–207 (with the sequence of KADVM) on the β4–α1 loop from one monomer binds to the potential peptide substrate binding site of the other monomer (Fig. 6C), thereby mimicking the substrate-bound state (Fig. 6B). In this pseudo-substrate, the phosphorylatable residue has been replaced by a non-phosphorylatable Val residue (Val 206′ with the ′ symbol indicating that the residue is from the adjacent molecule) whose side chain points away from the active site. The interactions between MsStt7d-KD and its pseudo-substrate peptide are mainly hydrogen bonds between Asp 205′/Lys 203′ and the water molecules surrounding the β-phosphate group of ADP, hydrogen bonds between Asp 205′/Met 207′ and Ala 361 bridged by water molecules, and hydrophobic interactions between Met 207′ and Gly 359/Ala 361/Leu 371 (Fig. 6D). Coincidently, the pseudo-substrate binding site lies between the predicted binding sites for Ser/Thr-containing peptide and Tyr-containing peptide (Fig. 6B), which further supports that this area is most likely for the binding of polypeptide/protein substrates.

As shown in Fig. 6B, it is predictable that the position of peptide backbone docking site might be adjustable on the flat surface of MsStt7d depending on the size of phosphorylatable residue, so that the terminal hydroxyl group can be positioned close enough to both the catalytic residue (Asp 318) and the ATP γ-phosphate for phosphorylation reaction to occur (Fig. 6B). The open and flat surface feature of the peptide-binding site might account for the capability of MsStt7d-KD in phosphorylating both Thr and Tyr residues.

DISCUSSION

When the MsStt7d protein sequence is used to search the Ref-Seq protein database of *Arabidopsis thaliana* for its potential homologs through the BLASTP program (Altsch et al., 1997), the result shows top two hits being STN7 and STN8 with E-values of 2×10^{-38} and 3×10^{-43}, respectively. STN8 is a close homolog of STN7 which is essential for the phosphorylation of photosystem II core proteins (Bonardi et al., 2005). Since MsStt7d does not contain a long STN8-specific loop inserted between the β3 and β4 strands (Vainonen et al., 2005), it is evident that MsStt7d resemble STN7 instead of STN8. Unlike CrStt7 or AtSTN7 which has a predicted N-terminal transmembrane helix and a pair of lumen-exposed Cys residues, the full-length MsStt7d has a predicted PDZ domain fused to the kinase domain at the N-terminal region instead. The presence of a PDZ domain at the N-terminal region of MsStt7d might provide an alternative mechanism for kinase activity regulation. Nevertheless, the BLAST result strongly suggests that MsStt7d is a homolog of the station transition kinase family, but with a different N-terminal region.

Through structural analysis of MsStt7d-KD, we have discovered that the canonical protein kinase domain and the newly identified α5 hairpin motif are both the conserved features of Stt7/STN7 family (Fig. 1). The MsStt7d-KD structure reported here serves as the homologous model of the core domain of Stt7/STN7. Based on the following observation, we believe the structure of MsStt7d-KD represents the kinase domain of Stt7/STN7 in the active but autoinhibited state. Firstly, the structure of MsStt7d-KD closely resembles that of CDK2 in the active state (Brown et al., 1999) (Fig. 3) whose activation loop swings away from the active site, favoring the binding of both nucleotide and peptide substrates. Secondly, the MsStt7d-KD dimer observed in the crystal has its peptide binding sites occupied by the pseudo-substrates from the adjacent molecules (Fig. 6C), suggesting that the kinase is in an active but autoinhibited state when it forms a dimer.

Sequence alignment of various Stt7/STN7 homologs shows that CrStt7 contains a phosphorylatable Thr residue in the glycine-rich loop region equivalent to Thr 185 (confirmed autophosphorylation site) in MsStt7d (Fig. 1). The relative conservation of phosphorylatable site in the glycine-rich loop may imply a potential role of this site for the regulation of kinase activity. In CDK2, residues Tyr 15 and Thr 14 in the glycine-rich loop are phosphorylated by other kinases, and their phosphorylations negatively regulate the catalytic activity of CDK2 through reducing the affinity of kinase toward the peptide/protein substrate (Welburn et al., 2006). Recently, autophosphorylation within the glycine-rich loop of Brassinosteroid insensitive 1 (BRI1) receptor kinase from *Arabidopsis thaliana* has been reported to reduce the kinase activity of BRI1 too (Oh et al., 2012). Therefore, it is likely that autophosphorylation at the glycine-rich loop in Stt7 may also have a similar regulatory effect on the kinase activity.

It was reported earlier that the C-terminal region of STN7 contained four phosphorylation sites (Ser 526, Thr 537, Thr 539, and Thr 541) identified through a large-scale *Arabidopsis* phosphoproteome profiling study (Reiland et al., 2009). A follow-up work suggested that phosphorylation of STN7 on these sites might have a role in regulating its turnover during state transition, while loss of the phosphorylation sites did not affect its function in state transition (Willig et al., 2011). For both studies, the phosphorylation of STN7 was not assigned as a consequence of autophosphorylation activity due to the complexity of *in vivo* system when many other protein kinases are present (Reiland et al., 2009; Willig et al., 2011). By using the *in vitro* assay system with highly purified MsStt7d-KD kinase sample, we are now able to demonstrate that the *Micromonas* Stt7d homolog exhibits autophosphorylation behavior. Our data suggest a possibility that autophosphorylation might be responsible, at least partly, for the phosphorylation of STN7 under *in vivo* conditions.

Previous studies suggested that STN7 may have important roles in LTR in addition to its well-established role in state transition (Bonardi et al., 2005; Wagner et al., 2008; Pesaresi et al., 2009). It is proposed that STN7 senses the redox signal of PQ pool and transduces it to the level of gene expression in chloroplast and nucleus (Bonardi et al., 2005; Rochaix, 2007; Dietzel et al., 2008; Wagner et al., 2008; Pesaresi et al., 2009; Lemeille and Rochaix, 2010). The discovery of dual-substrate specificity of MsStt7d opens a new avenue for searching downstream Thr/Ser and/or Tyr-containing substrate proteins of STN7 kinases involved in the signaling process during LTR. Although it remains to be verified if the other STN7 members are also dual-specificity kinases, such a property might account for the versatile functions of STN7 in photosynthetic acclimation.

MATERIALS AND METHODS

Protein expression, purification and crystallization

The gene encoding MsStt7d kinase domain (Asp 140–Asp 489) was ligated into pET28a vector (Novagen) between *Nco* I and *Not* I sites. Mutants were constructed using the QuikChange Site-Directed Mutagenesis method (Agilent). For protein fused with pentapeptide substrate, the DNA sequence encoding the pentapeptide RKXVA (X being T/S/Y/A/F) was introduced to the 5'-end of MsStt7d-KD gene. The final constructs were confirmed by DNA sequencing. The vectors were transformed into *E. coli* BL21 (DE3) and protein expression was induced at 20°C with 0.4 mmol/L IPTG when the cell density reached an OD_{600nm} of 0.6–0.8. Protein was purified through immobilized metal ion affinity chromatography (IMAC) using Nickel-Chelating Sepharose

Fast Flow (GE Healthcare), followed by size exclusion chromatography using Superdex 200 10/300 GL (GE Healthcare) with a buffer containing 20 mmol/L Tris pH 8.0 and 200 mmol/L NaCl. The eluted major peak fraction was concentrated to 10 mg/mL protein and used for crystallization trials. The best crystals of MsStt7d-KD were grown at 18°C using hanging-drop vapor diffusion method by mixing 1 μL of protein sample and 1 μL of well solution containing 0.1 mol/L HEPES pH 7.5, 200 mmol/L $MgCl_2$ and 25% PEG3350. The MsStt7d-KD-nucleotide complex crystals were prepared by soaking the MsStt7d-KD crystals overnight in an artificial mother liquor (0.1 mol/L HEPES pH 7.5, 200 mmol/L $MgCl_2$, 27% PEG3350) supplemented with 10 mmol/L nucleotide substrate (ADP/ATP/AMP-PNP).

Data collection and structure determination

For phasing, a high-quality MsStt7d-KD crystal was soaked briefly in an iodide-containing derivatizing solution (0.1 mol/L HEPES pH 7.5, 200 mmol/L $MgCl_2$, 27% PEG3350 and 1 mol/L KI) for 45 s at room temperature and then flash frozen in a 100 K cold nitrogen gas stream. A high-redundancy set of diffraction data of the MsStt7d-KD-KI derivative crystal were collected on a MicroMax 007 rotation-anode X-ray source with Rigaku Raxis IV++ image plate area detector. The data for the native MsStt7d-KD and complexes with various nucleotide substrates were collected at BL17U of Shanghai Synchrotron Radiation Facility with ADSC Q315r CCD detector. Data were processed and scaled with iMOSFLM (Battye et al., 2011) and Scala (Evans, 2006) or with the HKL2000 package (Otwinowski and Minor, 1997). The initial phases were solved by the single wavelength anomalous dispersion (SAD) method using the MsStt7d-KD-KI derivative data with the AutoSol program in the PHENIX software suite (Adams et al., 2010). In the asymmetric unit, a total of twenty three iodide atoms with occupancy above 0.20 were located. The figure of merit (FOM) of the initial phases was 0.39 which was further improved to 0.68 after density modification. Around 90% of the structural model was automatically built by the AutoBuild program in PHENIX suite (Adams et al., 2010). The output model was corrected and readjusted manually in the COOT program (Emsley and Cowtan, 2004). The structures were refined using CNS (Brunger et al., 1998), Phenix.refine (Adams et al., 2010) and Refmac (Murshudov et al., 1997) in the CCP4 package (CCP4, 1994). The structures of native MsStt7d-KD and in complex with nucleotide substrates were solved by the difference Fourier method using the structure of MsStt7d-KD-KI derivative as the initial model. The final refined structures were checked with program PROCHECK (Laskowski et al., 1993). Structural figures were prepared using PyMOL (DeLano, 2002) or CCP4MG (McNicholas et al., 2011). Data collection and structure refinement statistics is summarized in Table 2.

ADP-Glo kinase assay

ADP-Glo kinase assay was performed as described in the protocol (Promega). Each reaction mixture contained 2 μmol/L kinase and 100 μmol/L ultrapure ATP in 25 μL of reaction buffer (40 mmol/L Tris pH 7.5, 20 mmol/L $MgCl_2$ and 0.1 mg/mL BSA). The reaction was initiated by adding ultrapure ATP, and after incubation at 20°C for 1 h it was terminated by adding equal volume of ADP-Glo Reagent. After incubation at room temperature for 40 min to deplete the remaining original ATP, 50 μL of Kinase Detection Reagent was added to convert ADP into ATP and the newly synthesized ATP was detected through the luciferase/luciferin reaction which produced luminescence signal. After incubation at room temperature for 60 min, luminescence was recorded using Varioskan Flash Multimode Reader (Thermo). The control experiment was carried out under the same reaction condition without adding the kinase. For each reaction, three repeats were performed for the calculation of mean values and standard deviations (SDs).

Kinase autophosphorylation assay and autoradiography

The phosphorylation reaction mixture (25 μL volume) contained 20 mmol/L HEPES, pH 7.5, 50 mmol/L NaCl, 10 mmol/L $MgCl_2$, 10 mmol/L NaF, 0.05% n-Dodecyl-β-D-Maltoside, 4 μmol/L kinase, 0.4 mmol/L ATP and 10 μCi [γ-^{32}P] ATP. The phosphorylation reaction was initiated by the addition of ATP, and after incubation at room temperature for 2 h the reaction was terminated by adding SDS-PAGE sample loading buffer. The sample was heated at 100°C for 10 min and then loaded onto the gel for SDS-PAGE. Afterwards, the gels were subjected to autoradiography using X-OMAT BT film (Kodak) and then stained with Coomassie Brilliant Blue.

Mass spectrometry

To identify the autophosphorylation sites of kinases, samples were separated with SDS-PAGE. The bands corresponding to kinases were excised from the gel, reduced with 10 mmol/L DTT and alkylated with 55 mmol/L iodoacetamide. The in-gel digestion was carried out by adding the sequence-grade modified trypsin (Promega) in 50 mmol/L ammonium bicarbonate at 37°C overnight. The peptides were extracted twice with 1% formic acid in 50% acetonitrile aqueous solution for 30 min. The extraction was then centrifuged in a speedvac to reduce the sample volume. The sample was loaded onto a homemade C18 column (100 mm × 100 μm) packed with Sunchrom packing material (SP-120-3-ODS-A, 3 μm) for the separation of peptide products, and the nano-LC-ESI-MS/MS analysis was carried out subsequently. The peptides were eluted from the HPLC column with a gradient of 0%–90% of Buffer B (acetonitrile: water: acetic acid = 80:19.9:0.1, volume ratio) mixed with Buffer A (acetonitrile: water: acetic acid = 5:94.9:0.1, volume ratio) at a flow rate of approximately 500 nL/min. The eluted peptides were sprayed directly from the tip of the capillary column to the LTQ mass spectrometer (Thermo Finnigan) for mass spectrometry analysis. The LTQ mass spectrometer was operated in the data-dependent mode in which the initial MS scan recorded the mass to charge (m/z) ratios of ions over the mass range from 350 Da to1800 Da. The five most abundant ions were automatically selected for subsequent collision-activated dissociation with the multistage activation enabled. All MS/MS data were searched against a database with MsStt7d-KD protein sequence using the SEQUEST (v.28) program (Thermo). The SEQUEST outputs were then analyzed using the commercial software BioWorks (v.3.3.1) (Thermo).

Western blot

The MsStt7d-KD protein samples before or after phosphorylation reactions were separated by SDS-PAGE and then transferred to the PVDF membranes. The membranes were blocked with nonfat milk (Blotting-Grade Blocker, BIO-RAD) and immunoblotted with anti-phosphothreonine antibody (Cell Signaling Technology), anti-phosphoserine antibody (Millipore) and anti-phosphotyrosine antibody (Cell Signaling

Technology), respectively. The horseradish peroxidase (HRP)-conjugated secondary antibody and 3, 3′-Diaminobenzidine (DAB) substrate were used for the detection of signals on the blot.

ACCESSION NUMBERS

Sequence data from this article can be found in the NCBI databank under the following accession numbers: XP_002504185 (MsStt7d), Q84V18 (CrStt7), NP_564946 (AtSTN7), XP_002499717 (MsStt7a), XP_002504036 (MsStt7b), XP_002500340 (MsStt7c), XP_002505498 (MsStt7e), XP_001420129 (OlStt7), BAJ86343 (HvSTN7), NP_001056233 (OsSTN7), NP_001148075 (ZmSTN7), NP_002721 (PKA), NP_057065.2 (CaMK2α), AAA96795 (CKIIα), AAP36159 (CDK2), 1IRK_A (Irk) and 1FMK_A (SRC). Coordinates for protein structures can be found in the Protein Data Bank under the following accession numbers: 1QMZ (active CDK2), 1HCL (inactive CDK2), 1IR3 (Irk). Coordinates and structure factors for MsStt7d-KD (PDB: 4IX3), MsStt7d-KD-ADP (PDB: 4IX4), MsStt7d-KD-ATP (PDB: 4IX6), MsStt7d-KD-AMP-PNP (PDB: 4IX5) have been deposited in the Protein Data Bank.

ACKNOWLEDGEMENTS

We thank the staffs at Shanghai Synchrotron Radiation Facility (SSRF) as well as Yi Han and Shengquan Liu at the X-ray core facility at the Institute of Biophysics (IBP), Chinese Academy of Sciences (CAS) for their technical support during crystal screening and data collection. We also thank Hongmei Zhang for the assistance with molecular biology and biochemistry, Zhensheng Xie and Fuquan Yang at the Proteomic Core Facility Center of IBP, CAS and Jieyuan Liu at the Center of Biomedical Analysis of Tsinghua University for their technical assistance at the Mass Spectrometry analysis. This work was supported by the National Basic Research Program (973 Program) (Nos. 2011CBA00902 and 2011CBA00903) awarded to W. R. C. and Z. F. L., respectively, and the National Natural Science Foundation (Grant No. 31021062).

ABBREVIATIONS

AMP-PNP, adenosine 5′-(β, γ-imido) triphosphate; AtLhcb1, light-harvesting protein Lhcb1 from *Arabidopsis thaliana*; AtSTN7, STN7 from *Arabidopsis thaliana*; CaMK2α, calcium/calmodulin-dependent protein kinase type II subunit α isoform 1 from *Homo sapiens*; CDK2, cyclin-dependent kinase 2 from *Homo sapiens*; CKIIα, casein kinase II subunit α from *Mus musculus*; CrStt7, Stt7 from *Chlamydomonas reinhardtii*; HvSTN7, STN7 from *Hordeum vulgare*; Irk, Tyr kinase domain of insulin receptor from *Homo sapiens*; KD, kinase domain; LHCII, major light-harvesting complex II; LTR, long-term response; MsLhcp4, light-harvesting protein Lhcp4 from *Micromonas sp.* RCC299; MsStt7a-d, Stt7 homolog from *Micromonas sp.* RCC299; OlStt7, Stt7 from *Ostreococcus lucimarinus*; OsSTN7, STN7 from *Oryza sativa*; PKA, cAMP-dependent protein kinase catalytic subunit α isoform 1 from *Homo sapiens*; PSI, photosystem I; PSII, photosystem II; SRC, Tyr protein kinase C-Src from *Homo sapiens*; ZmSTN7, STN7 from *Zea mays*

COMPLIANCE WITH ETHICS GUIDELINES

Jiangtao Guo, Xuepeng Wei, Mei Li, Xiaowei Pan, Wenrui Chang and Zhenfeng Liu declare that they have no conflict of interest.

This article does not contain any studies with human or animal subjects performed by any of the authors.

REFERENCES

Adams, P.D., Afonine, P.V., Bunkoczi, G., Chen, V.B., Davis, I.W., Echols, N., Headd, J.J., Hung, L.W., Kapral, G.J., Grosse-Kunstleve, R.W., et al. (2010). PHENIX: a comprehensive Python-based system for macromolecular structure solution. Acta Crystallogr D 66, 213–221.

Allen, J.F. (1992). Protein phosphorylation in regulation of photosynthesis. Biochim Biophys Acta 1098, 275–335.

Allen, J.F., Bennett, J., Steinback, K.E., and Arntzen, C.J. (1981). Chloroplast protein phosphorylation couples plastoquinone redox state to distribution of excitation energy between photosystems. Nature 291, 25–29.

Allen, J.F., and Race, H.L. (2002). Will the real LHC II kinase please step forward? Sci STKE 2002, pe43.

Altschul, S.F., Madden, T.L., Schaffer, A.A., Zhang, J., Zhang, Z., Miller, W., and Lipman, D.J. (1997). Gapped BLAST and PSI-BLAST: a new generation of protein database search programs. Nucleic Acids Res 25, 3389–3402.

Battye, T.G., Kontogiannis, L., Johnson, O., Powell, H.R., and Leslie, A.G. (2011). iMOSFLM: A new graphical interface for diffraction-image processing with MOSFLM. Acta Crystallogr D 67, 271–281.

Bellafiore, S., Barneche, F., Peltier, G., and Rochaix, J.D. (2005). State transitions and light adaptation require chloroplast thylakoid protein kinase STN7. Nature 433, 892–895.

Bennett, J. (1991). Protein phosphorylation in green plant chloroplasts. Annu Rev Plant Physiol Plant Mol Biol 42, 281–311.

Bonardi, V., Pesaresi, P., Becker, T., Schleiff, E., Wagner, R., Pfannschmidt, T., Jahns, P., and Leister, D. (2005). Photosystem II core phosphorylation and photosynthetic acclimation require two different protein kinases. Nature 437, 1179–1182.

Bonaventura, C., and Myers, J. (1969). Fluorescence and oxygen evolution from *Chlorella pyrenoidosa*. Biochim Biophys Acta 189, 366–383.

Bossemeyer, D. (1994). The glycine-rich sequence of protein kinases: a multifunctional element. Trends Biochem Sci 19, 201–205.

Brown, N.R., Noble, M.E., Endicott, J.A., and Johnson, L.N. (1999). The structural basis for specificity of substrate and recruitment peptides for cyclin-dependent kinases. Nat Cell Biol 1, 438–443.

Brunger, A.T., Adams, P.D., Clore, G.M., DeLano, W.L., Gros, P., Grosse-Kunstleve, R.W., Jiang, J.S., Kuszewski, J., Nilges, M., Pannu, N.S., et al. (1998). Crystallography and NMR system: a new software suite for macromolecular structure determination. Acta Crystallogr D 54, 905–921.

CCP4 (1994). The CCP4 suite: programs for protein crystallography. Acta Crystallogr D 50, 760–763.

Coughlan, S., and Hind, G. (1987). Phosphorylation of thylakoid proteins by a purified kinase. J Biol Chem 262, 8402–8408.

De Bondt, H.L., Rosenblatt, J., Jancarik, J., Jones, H.D., Morgan, D.O., and Kim, S.H. (1993). Crystal structure of cyclin-dependent kinase 2. Nature 363, 595–602.

DeLano, W.L. (2002). The PyMOL Molecular Graphics System. (San Carlos, CA: DeLano Scientific). http://www.pymol.org.

Depège, N., Bellafiore, S., and Rochaix, J.D. (2003). Role of chloroplast protein kinase Stt7 in LHCII phosphorylation and state transition in *Chlamydomonas*. Science 299, 1572–1575.

Dietzel, L., Brautigam, K., and Pfannschmidt, T. (2008). Photosynthetic acclimation: state transitions and adjustment of photosystem stoichiometryfunctional relationships between short-term and long-term light quality acclimation in plants. FEBS J 275, 1080–1088.

Emsley, P., and Cowtan, K. (2004). Coot: model-building tools for molecular graphics. Acta Crystallogr D 58, 1948–1954.

Endicott, J.A., Noble, M.E., and Johnson, L.N. (2012). The structural basis for control of eukaryotic protein kinases. Annu Rev Biochem 81, 587–613.

Evans, P. (2006). Scaling and assessment of data quality. Acta Crystallogr D 62, 72–82.

Hanks, S.K., and Hunter, T. (1995). Protein kinases 6. The eukaryotic protein kinase superfamily: kinase (catalytic) domain structure and classification. FASEB J 9, 576–596.

Hubbard, S.R. (1997). Crystal structure of the activated insulin receptor tyrosine kinase in complex with peptide substrate and ATP analog. EMBO J 16, 5572–5581.

Jeffrey, P.D., Russo, A.A., Polyak, K., Gibbs, E., Hurwitz, J., Massague, J., and Pavletich, N.P. (1995). Mechanism of CDK activation revealed by the structure of a cyclinA-CDK2 complex. Nature 376, 313–320.

Kargul, J., and Barber, J. (2008). Photosynthetic acclimation: structural reorganisation of light harvesting antenna--role of redox-dependent phosphorylation of major and minor chlorophyll a/b binding proteins. FEBS J 275, 1056–1068.

Laskowski, R.A., MacArthur, M.W., Moss, D.S., and Thornton, J.M. (1993). PROCHECK: a program to check the stereochemical quality of protein structures. J Appl Cryst 26, 283–291.

Lemeille, S., Willig, A., Depege-Fargeix, N., Delessert, C., Bassi, R., and Rochaix, J.D. (2009). Analysis of the chloroplast protein kinase Stt7 during state transitions. PLoS Biol 7, e45.

Lin, Z.F., Lucero, H.A., and Racker, E. (1982). Protein kinases from spinach chloroplasts. I. Purification and identification of two distinct protein kinases. J Biol Chem 257, 12153–12156.

Lindberg, R.A., Quinn, A.M., and Hunter, T. (1992). Dual-specificity protein kinases: will any hydroxyl do? Trends Biochem Sci 17, 114–119.

Madhusudan, Trafny, E.A., Xuong, N.H., Adams, J.A., Ten Eyck, L.F., Taylor, S.S., and Sowadski, J.M. (1994). cAMP-dependent protein kinase: crystallographic insights into substrate recognition and phosphotransfer. Protein Sci 3, 176–187.

McNicholas, S., Potterton, E., Wilson, K.S., and Noble, M.E.M. (2011). Presenting your structures: the CCP4mg molecular-graphics software. Acta Crystallogr D 67, 386–394.

Mendelow, M., Prorok, M., Salerno, A., and Lawrence, D.S. (1993). ATPase-promoting Dead End Inhibitors of the CAMP-dependent Protein Kinase. J Biol Chem 268, 12289–12296.

Murata, N. (1969). Control of excitation transfer in photosynthesis. I. Light-induced change of chlorophyll a fluorescence in Porphyridium cruentum. Biochim Biophys Acta 172, 242–251.

Murshudov, G.N., Vagin, A.A., and Dodson, E.J. (1997). Refinement of Macromolecular structures by maximum-likelihood method. Acta Crystallogr D 53, 240–255.

Nolen, B., Taylor, S.S., and Ghosh, G. (2004). Regulation of Protein Kinases Controlling Activity through Activation Segment Conformation. Mol Cell 15, 661–675.

Oh, M.H., Wang, X., Clouse, S.D., and Huber, S.C. (2012). Deactivation of the Arabidopsis BRASSINOSTEROID INSENSITIVE 1 (BRI1) receptor kinase by autophosphorylation within the glycinerich loop. Proc Natl Acad Sci U S A 109, 327–332.

Otwinowski, Z., and Minor, W. (1997). Processing of X-ray diffraction data collected in oscillation mode. Methods Enzymol 276, 307–326.

Pesaresi, P., Hertle, A., Pribil, M., Kleine, T., Wagner, R., Strissel, H., Ihnatowicz, A., Bonardi, V., Scharfenberg, M., Schneider, A., et al. (2009). *Arabidopsis* STN7 kinase provides a link between short- and long-term photosynthetic acclimation. Plant Cell 21, 2402–2423.

Reiland, S., Messerli, G., Baerenfaller, K., Gerrits, B., Endler, A., Grossmann, J., Gruissem, W., and Baginsky, S. (2009). Large-scale *Arabidopsis* phosphoproteome profiling reveals novel chloroplast kinase substrates and phosphorylation networks. Plant Physiol 150, 889–903.

Rochaix, J.D. (2007). Role of thylakoid protein kinases in photosynthetic acclimation. FEBS Lett 581, 2768–2775.

Schulze-Gahmen, U., De Bondt, H.L., and Kim, S.H. (1996). High-resolution crystal structures of human cyclin-dependent kinase 2 with and without ATP: bound waters and natural ligand as guides for inhibitor design. J Med Chem 39, 4540–4546.

Snyders, S., and Kohorn, B.D. (1999). TAKs, thylakoid membrane protein kinases associated with energy transduction. J Biol Chem 274, 9137–9140.

Snyders, S., and Kohorn, B.D. (2001). Disruption of thylakoid-associated kinase 1 leads to alteration of light harvesting in *Arabidopsis*. J Biol Chem 276, 32169–32176.

Vainonen, J.P., Hansson, M., and Vener, A.V. (2005). STN8 protein kinase in *Arabidopsis thaliana* is specific in phosphorylation of photosystem II core proteins. J Biol Chem 280, 33679–33686.

Vener, A.V., van Kan, P.J.M., Rich, P.R., Ohad, I., and Andersson, B. (1997). Plastoquinol at the quinol oxidation site of reduced cytochrome bf mediates signal transduction between light and protein phosphorylation: Thylakoid protein kinase deactivation by a single-turnover flash. Proc Natl Acad Sci U S A 94, 1585–1590.

Wagner, R., Dietzel, L., Brautigam, K., Fischer, W., and Pfannschmidt, T. (2008). The long-term response to fluctuating light quality is an important and distinct light acclimation mechanism that supports survival of *Arabidopsis thaliana* under low light conditions. Planta 228, 573–587.

Ward, N.E., and O'Brian, C.A. (1992). The intrinsic ATPase activity of protein kinase C is catalyzed at the active site of the enzyme. Biochemistry 31, 5905–5911.

Welburn, J.P.I., Tucker, J.A., Johnson, T., Lindert, L., Morgan, M., Willis, A., Noble, M.E.M., and Endicott, J.A. (2006). How tyrosine 15 phosphorylation inhibits the activity of cyclin-dependent kinase 2-cyclin A. J Biol Chem 282, 3173–3181.

Willig, A., Shapiguzov, A., Goldschmidt-Clermont, M., and Rochaix, J.D. (2011). The phosphorylation status of the chloroplast protein kinase STN7 of *Arabidopsis* affects its turnover. Plant Physiol 157, 2102–2107.

Wollman, F.A. (2001). State transitions reveal the dynamics and flexibility of the photosynthetic apparatus. EMBO J 20, 3623–3630.

Zito, F., Finazzi, G., Delosme, R., Nitschke, W., Picot, D., and Wollman, F.A. (1999). The Qo site of cytochrome b6f complexes controls the activation of the LHCII kinase. EMBO J 18, 2961–2969.

Permissions

All chapters in this book were first published in P&C, by Springer; hereby published with permission under the Creative Commons Attribution License or equivalent. Every chapter published in this book has been scrutinized by our experts. Their significance has been extensively debated. The topics covered herein carry significant findings which will fuel the growth of the discipline. They may even be implemented as practical applications or may be referred to as a beginning point for another development.

The contributors of this book come from diverse backgrounds, making this book a truly international effort. This book will bring forth new frontiers with its revolutionizing research information and detailed analysis of the nascent developments around the world.

We would like to thank all the contributing authors for lending their expertise to make the book truly unique. They have played a crucial role in the development of this book. Without their invaluable contributions this book wouldn't have been possible. They have made vital efforts to compile up to date information on the varied aspects of this subject to make this book a valuable addition to the collection of many professionals and students.

This book was conceptualized with the vision of imparting up-to-date information and advanced data in this field. To ensure the same, a matchless editorial board was set up. Every individual on the board went through rigorous rounds of assessment to prove their worth. After which they invested a large part of their time researching and compiling the most relevant data for our readers.

The editorial board has been involved in producing this book since its inception. They have spent rigorous hours researching and exploring the diverse topics which have resulted in the successful publishing of this book. They have passed on their knowledge of decades through this book. To expedite this challenging task, the publisher supported the team at every step. A small team of assistant editors was also appointed to further simplify the editing procedure and attain best results for the readers.

Apart from the editorial board, the designing team has also invested a significant amount of their time in understanding the subject and creating the most relevant covers. They scrutinized every image to scout for the most suitable representation of the subject and create an appropriate cover for the book.

The publishing team has been an ardent support to the editorial, designing and production team. Their endless efforts to recruit the best for this project, has resulted in the accomplishment of this book. They are a veteran in the field of academics and their pool of knowledge is as vast as their experience in printing. Their expertise and guidance has proved useful at every step. Their uncompromising quality standards have made this book an exceptional effort. Their encouragement from time to time has been an inspiration for everyone.

The publisher and the editorial board hope that this book will prove to be a valuable piece of knowledge for researchers, students, practitioners and scholars across the globe.

List of Contributors

Cécile Apert, Paola Romagnoli and Joost P. M. van Meerwijk
CPTP, Université de Toulouse, CNRS, Inserm, UPS, Toulouse, France

Yi Feng, Hai-yan Ying, Ying Qu, Xiao-bo Cai, Ming-yi Xu and Lun-gen Lu
Department of Gastroenterology, Shanghai General Hospital, Nanjing Medical University, Shanghai 200080, China

Anna Gortat, Mónica Sancho, Laura Mondragón and Mar Orzáez
Laboratory of Peptide and Protein Chemistry, Centro de Investigación Príncipe Felipe, 46012 Valencia, Spain

Enrique Pérez-Payá
Laboratory of Peptide and Protein Chemistry, Centro de Investigación Príncipe Felipe, 46012 Valencia, Spain
Instituto de Biomedicina de Valencia, IBV-CSIC, 46010 Valencia, Spain

Àngel Messeguer
Department of Chemical and Biomolecular Nanotechnology, Instituto Química Avanzada de Cataluña (CSIC), 08034 Barcelona, Spain

Xiang Li, Duanqing Pei and Hui Zheng
CAS Key Laboratory of Regenerative Biology, Guangdong Provincial Key Laboratory of Stem Cell and Regenerative Medicine, Guangzhou Institutes of Biomedicine and Health, Chinese Academy of Sciences, Guangzhou 510530, China

Wenzhi Li, Peizhe Wang, Jing Zhang, Jia Ming and Jie Na
Center for Stem Cell Biology and Regenerative Medicine, School of Medicine, Tsinghua University, Beijing 100084, China

Bingjie Zhang and Wei Xie
Center for Stem Cell Biology and Regenerative Medicine, MOE Key Laboratory of Bioinformatics, THU-PKU Center for Life Sciences, School of Life Sciences, Tsinghua University, Beijing 100084, China

Qianqian Li, Zewen Gao, Ye Chen and Min-Xin Guan
Division of Clinical Genetics and Genomics, The Children's Hospital, Zhejiang University School of Medicine, Hangzhou 310058, China
Institute of Genetics, Zhejiang University and Department of Genetics, Zhejiang University School of Medicine, Hangzhou 310058, China

Yang Li
Department of Pharmacology, Key Laboratory of Metabolism and Molecular Medicine (The Ministry of Education), School of Basic Medical Science, Fudan University, Shanghai 200032, China
State Key Laboratory of Molecular Developmental Biology, Institute of Genetics and Developmental Biology, Chinese Academy of Sciences, Beijing 100101, China

Yu Zhang, Xiao Ding and Lingli Guo
State Key Laboratory of Phytochemistry and Plant Resources in Western China, Kunming Institute of Botany, Chinese Academy of Sciences, Kunming 650021, China

Xiaojiang Hao
State Key Laboratory of Phytochemistry and Plant Resources in Western China, Kunming Institute of Botany, Chinese Academy of Sciences, Kunming 650021, China
The Key Laboratory of Chemistry for Natural Product of Guizhou Province, Chinese Academy of Science, Guiyang 550002, China

Qiwen Gan, Meng Xu, Jingjing Liang, Kai Liu, Xuezhao Liu and Zhiyang Gao
State Key Laboratory of Molecular Developmental Biology, Institute of Genetics and Developmental Biology, Chinese Academy of Sciences, Beijing 100101, China

Chonglin Yang
State Key Laboratory of Molecular Developmental Biology, Institute of Genetics and Developmental Biology, Chinese Academy of Sciences, Beijing 100101, China
State Key Laboratory for Conservation and Utilization of Bio-Resources in Yunnan, Center for Life Sciences, School of Life Sciences, Yunnan University, Kunming 650091, China

Xin Wang
State Key Laboratory for Conservation and Utilization of Bio-Resources in Yunnan, Center for Life Sciences, School of Life Sciences, Yunnan University, Kunming 650091, China

Yan Chen, Xue Xiao, Fei Liu, Jing Zou and Jun Wu
CAS Key Laboratory of Pathogenic Microbiology and Immunology, Institute of Microbiology, Chinese Academy of Sciences, Beijing 100101, China
University of Chinese Academy of Sciences, Beijing 100049, China

Peipei Liu, Xi Yang, Jing Gao, Xuyu Zhou and Baoli Zhu
CAS Key Laboratory of Pathogenic Microbiology and Immunology, Institute of Microbiology, Chinese Academy of Sciences, Beijing 100101, China
University of Chinese Academy of Sciences, Beijing 100049, China

George F. Gao
CAS Key Laboratory of Pathogenic Microbiology and Immunology, Institute of Microbiology, Chinese Academy of Sciences, Beijing 100101, China
University of Chinese Academy of Sciences, Beijing 100049, China
Chinese Center for Disease Control and Prevention (China CDC), Beijing 102206, China
Research Network of Immunity and Health (RNIH), Beijing Institutes of Life Science, Chinese Academy of Sciences, Beijing 100101, China

Di Liu and Juncai Ma
Network Information Center, Institute of Microbiology, Chinese Academy of Sciences, Beijing 100101, China

Fangqing Zhao
Research Network of Immunity and Health (RNIH), Beijing Institutes of Life Science, Chinese Academy of Sciences, Beijing 100101, China

Mengqi Lv, Chongyuan Wang, Fudong Li, Junhui Peng, Bin Wen, Qingguo Gong,Yunyu Shi and Yajun Tang
Hefei National Laboratory for Physical Sciences at Microscale and School of Life Sciences, University of Science and Technology of China, Hefei 230027, China

Juan Ma, Changzhen Liu, Pan Ma, Xiaomei Li and Hua Tao
CAS Key Laboratory of Pathogenic Microbiology and Immunology, Institute of Microbiology, Chinese Academy of Sciences, Beijing 100101, China
Key Laboratory of Infection and Immunity, Institute of Biophysics, Chinese Academy of Sciences, Beijing 100101, China

Huamin Han
CAS Key Laboratory of Pathogenic Microbiology and Immunology, Institute of Microbiology, Chinese Academy of Sciences, Beijing 100101, China
Key Laboratory of Infection and Immunity, Institute of Biophysics, Chinese Academy of Sciences, Beijing 100101, China

Li Ma
Department of Obstetrics and Gynecology, China-Japan Friendship Hospital, Beijing 100029, China

Xin Xue
Department of Immunology, Basic Medical Theory of Chinese Medicine, China Academy of Chinese Medical Sciences, Beijing 100700, China

Jiangtao Ren and Yangbing Zhao
Center for Cellular Immunotherapies, Perelman School of Medicine, University of Pennsylvania, Philadelphia, PA 19104-5156, USA

Xiaomin Song, Sheng Wang and Lin Li
State Key Laboratory of Molecular Biology, Institute of Biochemistry and Cell Biology, Shanghai Institutes for Biological Sciences, Chinese Academy of Sciences, Shanghai 200031, China

Luming Zhao, Ming Xu and Na Xiong
Centre for Molecular Immunology and Infectious Diseases and Department of Veterinary and Biomedical Sciences, The Pennsylvania State University, University Park, PA 16802, USA

Jie Yang
Centre for Molecular Immunology and Infectious Diseases and Department of Veterinary and Biomedical Sciences, The Pennsylvania State University, University Park, PA 16802, USA
Department of Immunology, The University of Texas MD Anderson Cancer Center, 7455 Fannin Street, Houston, TX 77054, USA

Liguo Zhang
Key Laboratory of Immunity and Infection, Institute of Biophysics, Chinese Academy of Sciences, Beijing 100101, China

Haisheng Yu and Xiangyun Yin
Key Laboratory of Immunity and Infection, Institute of Biophysics, Chinese Academy of Sciences, Beijing 100101, China
Graduate School of the Chinese Academy of Sciences, Beijing 100080, China

Ya Cui and Taijiao Jiang
Key Laboratory of Protein and Peptide Pharmaceuticals, Institute of Biophysics, Chinese Academy of Sciences, Beijing 100101, China

Peng Zhang
Key Laboratory of Protein and Peptide Pharmaceuticals, Institute of Biophysics, Chinese Academy of Sciences, Beijing 100101, China
Graduate School of the Chinese Academy of Sciences, Beijing 100080, China

Zhao Yin, Quanxing Shi and Shouli Wang
Department of Cardiology, 306th Hospital of PLA, Beijing 100101, China

Guanyuan Liu
Department of Gynecology and Obstetrics, Beijing Chaoyang Hospital, Capital Medical University, Beijing 100020, China

Pier Paolo Piccaluga
Department of Experimental, Diagnostic, and Specialty Medicine, Hematopathology & Hematology Sections, Molecular Pathology Laboratory, S. Orsola-Malpighi Hospital, Bologna University, Bologna 40126, Italy

Kejing Zhang, Tomoaki Hishida and Concepcion Rodriguez Esteban
Gene Expression Laboratory, Salk Institute for Biological Studies, 10010 North Torrey Pines Road, La Jolla, CA 92037, USA

Juan Carlos Izpisua Belmonte
Gene Expression Laboratory, Salk Institute for Biological Studies, 10010 North Torrey Pines Road, La Jolla, CA 92037, USA
Center for Regenerative Medicine in Barcelona, Dr. Aiguader 88, 08003 Barcelona, Spain

Guang-Hui Liu
National Laboratory of Biomacromolecules, Institute of Biophysics, Chinese Academy of Sciences, Beijing 100101, China

Fei Yi
Department of Molecular and Cellular Physiology, Stanford University School of Medicine, 265 Campus Drive, Stanford, CA 94305, USA

Nuria Montserrat
Center for Regenerative Medicine in Barcelona, Dr. Aiguader 88, 08003 Barcelona, Spain
Biomedical Research Networking Center in Bioengineering, Biomaterials and Nanomedicine (CIBER-BBN), Barcelona, Spain

Wenzhi Feng
PTN Program, College of Life Science, Peking University, Beijing 100871, China

Juan Feng, Silin Lü, Yanhong Ding, Ming Zheng and Xian Wang
Department of Physiology and Pathophysiology, School of Basic Medical Science, Peking University, Key Laboratory of Molecular Cardiovascular Science, Ministry of Education, Beijing 100191, China

Qian Fan, Xiangrui Meng, Huilai Zhang, Xianming Liu, Lanfang Li and Wei Li
Department of Lymphoma, Sino-US Center for Lymphoma and Leukemia, Tianjin Medical University Cancer Institute and Hospital, National Clinical Research Center of Cancer, Key Laboratory of Cancer Prevention and Therapy, Tianjin 300060, China

Hongwei Liang, Ke Zen, Chen-Yu Zhang, Zhen Zhou and Xi Chen
State Key Laboratory of Pharmaceutical Biotechnology, NJU Advanced Institute of Life Sciences, Jiangsu Engineering Research Center for MicroRNA Biology and Biotechnology, School of Life Sciences, Nanjing University, Nanjing 210093, China

Wu Sun, Haiyang Zhang and Yi Ba
Department of Digestion, Tianjin Medical University Cancer Institute and Hospital, National Clinical Research Center of Cancer, Key Laboratory of Cancer Prevention and Therapy, Tianjin 300060, China

Tong Wu, Xiaoyu Dan and Li Yu
State Key Laboratory of Biomembrane and Membrane Biotechnology, Tsinghua University-Peking University Joint Center for Life Sciences, School of Life Sciences, Tsinghua University, Beijing 100084, China

Yuling Chen, Di Miao and Haiteng Deng
Center for Biomedical Analysis, Tsinghua University, Beijing 100084, China

Lin Li and She Chen
National Institute of Biological Sciences, Beijing 102206, China

Xinqi Gong
Institute for Mathematical Sciences, Renmin University of China, Beijing 100872, China

Jintao Bao, Liangjun Zheng, Qi Zhang, Xinya Li, Xuefei Zhang, Zeyang Li, Xue Bai, Zhong Zhang, Wei Huo, Qingsong Wang and Jianguo Ji
State Key Laboratory of Protein and Plant Gene Research, College of Life Sciences, Peking University, Beijing 100871, China

Xuyang Zhao
Institute of Systems Biomedicine, Department of Pathology, School of Basic Medical Sciences, Center for Age-Related Diseases, Peking University Health Science Center, Beijing 100191, China

Shujiang Shang and Chen Zhang
State Key Laboratory of Biomembrane and Membrane Biotechnology, Peking University, Beijing 100871, China

Mei Li, Xiaowei Pan, Wenrui Chang and Zhenfeng Liu
National Laboratory of Biomacromolecules, Institute of Biophysics, Chinese Academy of Sciences, Beijing 100101, China

Jiangtao Guo and Xuepeng Wei
National Laboratory of Biomacromolecules, Institute of Biophysics, Chinese Academy of Sciences, Beijing 100101, China
University of Chinese Academy of Sciences, Beijing 100039, China

Qihui Wang, Jun Liu, Jinghua Yan and Jianxun Qi
CAS Key Laboratory of Pathogenic Microbiology and Immunology, Institute of Microbiology, Chinese Academy of Sciences, Beijing 100101, China

Hao Song
CAS Key Laboratory of Pathogenic Microbiology and Immunology, Institute of Microbiology, Chinese Academy of Sciences, Beijing 100101, China
University of Chinese Academy of Sciences, Beijing 100049, China

George F Gao
CAS Key Laboratory of Pathogenic Microbiology and Immunology, Institute of Microbiology, Chinese Academy of Sciences, Beijing 100101, China
University of Chinese Academy of Sciences, Beijing 100049, China
Research Network of Immunity and Health (RNIH), Beijing Institutes of Life Science, Chinese Academy of Sciences, Beijing 100101, China

Gol Nam and Myongchol Ryu
CAS Key Laboratory of Pathogenic Microbiology and Immunology, Institute of Microbiology, Chinese Academy of Sciences, Beijing 100101, China
University of Chinese Academy of Sciences, Beijing 100049, China
Institute of Microbiology, State Academy of Sciences, Pyongyang, Democratic People's Republic of Korea

Yi Shi
CAS Key Laboratory of Pathogenic Microbiology and Immunology, Institute of Microbiology, Chinese Academy of Sciences, Beijing 100101, China
Research Network of Immunity and Health (RNIH), Beijing Institutes of Life Science, Chinese Academy of Sciences, Beijing 100101, China

Jinhua Xu, Xiaochao Wei, Dan Liu, Chune Peng and Wenqing Shui
College of Life Sciences and Tianjin State Laboratory of Protein Science, Nankai University, Tianjin 300071, China
High-throughput Molecular Drug Discovery Center, Tianjin Joint Academy of Biotechnology and Medicine, Tianjin 300457, China

Yuanyuan Ma, Limin Yan and Zhiyong Lou
Laboratory of Structural Biology and MOE Laboratory of Protein Science, School of Medicine and Life Sciences, Tsinghua University, Beijing 100084, China

Yu Guo, Honggang Zhou and Cheng Yang
High-throughput Molecular Drug Discovery Center, Tianjin Joint Academy of Biotechnology and Medicine, Tianjin 300457, China
College of Pharmacy and State Key Laboratory of Medicinal Chemical Biology, Nankai University, Tianjin 300071, China

Index

www.ingramcontent.com/pod-product-compliance
Lightning Source LLC
LaVergne TN
LVHW081032210926
848485LV00004B/1006
* 9 7 8 1 6 3 9 8 9 0 0 2 6 *